Earth Surface Modeling: Tools, Techniques and Applications

Earth Surface Modeling: Tools, Techniques and Applications

Edited by **Russell Sands**

SYRAWOOD
PUBLISHING HOUSE

New York

Published by Syrawood Publishing House,
750 Third Avenue, 9th Floor,
New York, NY 10017, USA
www.syrawoodpublishinghouse.com

Earth Surface Modeling: Tools, Techniques and Applications
Edited by Russell Sands

International Standard Book Number: 978-1-68286-196-7 (Hardback)

Printed in the United States of America.

Contents

Preface

Earth's surface is complex and dynamic. This all inclusive book on earth surface modeling gives comprehensive insights into the movements, changes and interactions of the earth's surface and enables the readers to understand various processes that happen within its crust. It elucidates the concepts and innovative models around prospective developments with respect to earth surface modeling, such as experimental and numerical modelling of earth surface processes, remote sensing, etc. This book consists of contributions made by international experts. It will be an apt resource for students pursuing graduation and post-graduation in earth sciences and allied disciplines.

After months of intensive research and writing, this book is the end result of all who devoted their time and efforts in the initiation and progress of this book. It will surely be a source of reference in enhancing the required knowledge of the new developments in the area. During the course of developing this book, certain measures such as accuracy, authenticity and research focused analytical studies were given preference in order to produce a comprehensive book in the area of study.

This book would not have been possible without the efforts of the authors and the publisher. I extend my sincere thanks to them. Secondly, I express my gratitude to my family and well-wishers. And most importantly, I thank my students for constantly expressing their willingness and curiosity in enhancing their knowledge in the field, which encourages me to take up further research projects for the advancement of the area.

Editor

Short Communication: TopoToolbox 2 – MATLAB-based software for topographic analysis and modeling in Earth surface sciences

W. Schwanghart[1,2] and D. Scherler[3]

[1]Geohazards, University of Potsdam, Institute of Earth and Environmental Science, Karl-Liebknecht-Str. 24–25, 14476 Potsdam-Golm, Germany
[2]Department of Agroecology, Aarhus University, Blichers Allé 20, 8830 Tjele, Denmark
[3]Geological and Planetary Sciences, California Institute of Technology, 1200 E California Blvd, Mailcode 100-23, Pasadena, CA 91125, USA

Correspondence to: W. Schwanghart (w.schwanghart@geo.uni-potsdam.de)

Abstract. TopoToolbox is a MATLAB program for the analysis of digital elevation models (DEMs). With the release of version 2, the software adopts an object-oriented programming (OOP) approach to work with gridded DEMs and derived data such as flow directions and stream networks. The introduction of a novel technique to store flow directions as topologically ordered vectors of indices enables calculation of flow-related attributes such as flow accumulation ~ 20 times faster than conventional algorithms while at the same time reducing memory overhead to 33 % of that required by the previous version. Graphical user interfaces (GUIs) enable visual exploration and interaction with DEMs and derivatives and provide access to tools targeted at fluvial and tectonic geomorphologists. With its new release, TopoToolbox has become a more memory-efficient and faster tool for basic and advanced digital terrain analysis that can be used as a framework for building hydrological and geomorphological models in MATLAB.

1 Introduction

An increasing number of research studies use digital elevation models (DEMs) for automated spatial analysis and advanced process-based modeling. Although geographical information system (GIS) software packages such as ESRI's ArcGIS include various interfaces for implementing user-specific codes and models, many users prefer programming environments such as MATLAB that already include large libraries for different computational tasks. Since its release in 2010, TopoToolbox, a code library for MATLAB, has been used in various studies of the Earth's surface and surficial water and material fluxes. Applications include topics such as supraglacial meltwater dynamics (Clason et al., 2012), contaminant transport in streams and groundwater (Messier et al., 2012), tectonic geomorphology (Shahzad and Gloaguen, 2011; Scherler et al., 2013), soil nutrient dynam- ics (Schwanghart and Jarmer, 2011), karst hydrology (Borghi et al., 2011) and flood modeling (de Moel et al., 2012).

Despite broad attention and positive feedback, the lack of a user-friendly graphical user interface (GUI) may also have discouraged interested users who favor a visual approach towards DEM processing and analysis. Moreover, processing very large DEMs, or, within a modeling framework, processing certain computations a large number of times, is a common challenge in modern geomorphological studies and requires efficient computational techniques. While very large DEMs are most easily handled with software that implements efficient input/output (I/O) communication between internal and external memory, a strategy pursued by ArcGIS, GRASS (Jasiewicz, 2011) and more specialized DEM processing software such as terraflow (Arge et al., 2003), the memory constraint exerted by MATLAB's primary use of the main memory is an obstacle even for medium-sized

DEMs. Although the spread of 64 bit computers has allayed the memory problem somewhat, MATLAB's heavy use of double-precision data imposes limits that are too rigid for larger analyses and modeling tasks.

Here we introduce several important modifications and various extensions that come with the release of version 2 of the TopoToolbox. Specifically, we report the development of an object-oriented design towards the representation of grids, flow routing, and stream networks and give account of a suite of GUIs that provide easy access to a variety of new and existing methods frequently used in topographic studies. Our strategy to reduce memory constraints, which have been particularly pertinent to flow-related algorithms, enables much faster and memory-efficient work flow and thus the processing of much larger DEMs than with previous versions of TopoToolbox. We believe that these advances will promote the development of tools and models in Earth surface sciences.

2   Object-oriented design

Gridded DEMs and terrain attributes incorporate a variety of information other than a gridded set of points. A DEM has physical units and is georeferenced; that is, it is associated with physical space by a set of geographical or map coordinates. In addition, DEM derivatives (or terrain attributes) are tightly linked to algorithms and models and lend themselves to specific types of analysis (Pike et al., 2009). These task- and object-specific properties provide the basis for an object-oriented approach to DEMs. In object-oriented programming (OOP), data, attributes, and procedures are coupled into objects that encapsulate the model of what each object represents (Register, 2007). Typical objects in DEM analysis are surface flow paths and the resulting stream networks. In TopoToolbox the aim of using OOP is to organize data and functions into classes with specific object-related attributes that organize and encapsulate information on spatial referencing, flow topology, and geometry and thereby facilitate data handling and analysis.

TopoToolbox 2 includes three main object classes: GRIDobj, a class to store and analyze gridded data (grid objects); FLOWobj, a class for flow-path networks (flow objects) that is derived from an instance of GRIDobj; and STREAMobj, which represents the channelized fraction (stream objects) of the flow-path network from which it is derived. Additional object classes, such as SWATHobj, which contains swath profiles that are derived along directed linear features (e.g., an instance of STREAMobj), are more tool-specific and not as tightly coupled within TopoToolbox as the three main object classes. These classes are templates for the creation of specific instances of each class. Whereas an instance of GRIDobj simply contains the data grid as well as georeferencing data, instances of FLOWobj and STREAMobj implement a novel storage approach that

represents the directed network of flow paths and stream networks, respectively.

3   A graph-theoretical representation of flow directions

Topography governs the transport mode and the magnitude of lateral material and water fluxes on the Earth's surface and in the near-surface underground. Among topographic attributes, the flow direction probably belongs to the most frequently used, because it forms the basis for many other hydrology-related variables that explain or control many key environmental processes (Wilson and Gallant, 2000). Algorithms to derive flow directions are classified into single-neighbor and multiple-neighbor flow algorithms (Gruber and Peckham, 2009). Single-neighbor algorithms pass the flow from each cell to its lowest neighboring cell (usually referred to as the SFD or D8 algorithm; O'Callaghan and Mark, 1984). Multiple-neighbor algorithms, such as MFD (Quinn et al., 1991), $D\infty$ (Tarboton, 1997), DEMON (Costa-Cabral and Burges, 1994) or the mass-flux method (Gruber and Peckham, 2009), encompass a number of different approaches to model dispersive flow, which is particularly important for simulating water distribution on hillslopes.

TopoToolbox supports the SFD and MFD algorithms. In contrast to other GIS software, in the first version of TopoToolbox, flow directions were stored in a sparse, weighted adjacency matrix to represent the flow network. While this strategy allows for calculation of flow accumulation and other flow-related terrain attributes using MATLAB's built-in sparse matrix routines (Schwanghart and Kuhn, 2010), it also requires large memory space. The MFD and SFD matrix require approximately six and two times the memory space of the DEM, respectively, if all variables are stored in in the same data formats.

In the newly introduced flow direction class (FLOWobj), the storage of flow directions takes advantage of the fact that flow networks can be modeled as directed acyclic graphs (DAGs). Our approach is similar to MATLAB's sparse matrix storage organization but uses the "tuple" format, which is a collection of row, column and value 3-tuples of the nonzero elements of the flow direction matrix (Kepner, 2011). The important difference is that row and column indices are stored in topologically descending order, an approach that was recently adopted in a numerical landscape evolution model (LEM) (Braun and Willett, 2012) and a dynamic overland flow-routing model (Huang and Lee, 2013). Topological sorting is only possible for DAGs and refers to the permutation of nodes that brings the flow direction matrix to a lower triangular form (Chen and Jacquemin, 1988). The storage of flow directions in topological order in a FLOWobj speeds up the computation of flow-related topographic attributes by about 500 % compared to the previous version of TopoToolbox. Furthermore, the FLOWobj requires 89 %

Table 1. Performance comparison between TopoToolbox 2 and similar software. Performance is evaluated as computation time in seconds required to run a specific tool on a computer with an Intel Core i5 CPU (M540) with 2.53 GHz and 4 GB RAM and a solid-state drive.

Tool/grid size (rows × columns)	ESRI ArcGIS[1]	TauDEM[2]	SAGA GIS[3]	Whitebox[4]	TopoToolbox 2
1385 × 1371					
Fill sinks	3	3.1 (2.3)	4	5	0.22
D8 flow direction	6	24 (23)	–	2	2.51
Flow accumulation	44	1.5 (1.3)	5	3	0.04
2769 × 2742					
Fill sinks	12	10.5 (7.7)	17	19	0.90
D8 flow direction	16	157 (153)	–	5	10.1
Flow accumulation	188	4.2 (3.6)	18	8	0.15
5537 × 5484					
Fill sinks	43	30 (26)	66	80	4.03
D8 flow direction	50	1344 (1331)	–	18	40.3
Flow accumulation	363	15 (13)	58	29	0.66

[1] http://www.esri.com/software/arcgis/extensions/spatialanalyst; version 10.1; benchmarking excludes default pyramid and statistics calculation and output data type is set to integer.
[2] http://hydrology.usu.edu/taudem/taudem5/index.html; version 5; we used the command line interface to run the application with 8 parallel processes. Numbers in parentheses refer to computation time without overhead time for reading and writing.
[3] http://www.saga-gis.org/; version 2.1; fill sinks algorithm proposed by Wang and Liu (2006) (XXL) with minimum of slope of 0.001°; recursive catchment area algorithm.
[4] http://www.uoguelph.ca/~hydrogeo/Whitebox/; version 3.0.8; fill sinks algorithm proposed by Wang and Liu (2006) with minimum of slope of 0.001°.

of the memory of the associated multiple-flow-direction matrix and 33 % of the single-flow-direction matrix. Because the nodes can be divided into independent subsets according to drainage basin affiliation, they can be distributed to different processors, so that flow-related terrain attributes can be evaluated in parallel, an important aspect for reducing computation times in LEMs (Braun and Willett, 2012).

In a topologically sorted FLOWobj, each grid cell's index ix appears before its downstream neighbor ixc if there is a directed link from ix to ixc. For routing algorithms that simulate flow through the directed network, the topological order entails that each link or edge is only traversed once, which simplifies flow computations and decreases running time to $O(n)$. For example, flow accumulation is calculated by sequentially updating an initial weight raster A with rows x cols $= n$ cells connected by a number of nrEdges directed links. The links connect upstream cells addressed with the vector of linear indices ix with their respective downstream neighbors indexed by the vector ixc.

$$A = \text{ones}(\text{row, cols}); \qquad (1)$$

for $i = 1 : \text{nrEdges}$,

$$A(ixc(i)) = A(ix(i)) + A(ixc(i)). \qquad (2)$$

end

Note that the calculation is performed by in-place operations (that is, the function input A is directly overwritten) which significantly reduces overhead memory of the algorithm. The flow direction object (FLOWobj) can be created

from an existing flow direction matrix (SFD or MFD) or a DEM. The latter is currently supported only for D8 flow routing, but allows users to simultaneously define if and how flow directions across flat terrain and topographic depressions are derived. Sinks and contiguous, flat areas in a DEM can be traversed using an auxiliary topography calculated by a gray-weighted distance transform algorithm (Soille et al., 2003; Metz et al., 2011; Schwanghart et al., 2013). The resulting flow directions tend to correspond more closely to actual flow paths compared to those that are derived from a DEM that was hydrologically conditioned using "flood filling" (Poggio and Soille, 2011).

4 Performance comparison

To evaluate the performance of TopoToolbox 2, we compared the runtimes of tools implemented in TopoToolbox with those of similar software. We emphasize that times to perform these computations fail to provide a concise, comparable measure since different programs pursue different strategies to store, process and access the data. TAUDEM 5, for example, does not hold the entire DEM in the main memory but partitions it into domains that are processed in parallel (Tesfa et al., 2011) and thus shows its particular strength with data sets that would exceed main memory capacity. Unlike TopoToolbox 2 and most other programs that calculate flow accumulation based on a data set that contains flow directions, SAGA GIS calculates flow accumulation directly from the DEM.

We compared the performance between TopoToolbox 2 and ArcGIS 10.1, TauDEM 5, SAGA GIS 2.1 and Whitebox GAT 3.0.8. We used a 6° × 6° tile of the 3 arcsec DEM made available by Ferranti (2013) covering the Alps and their foreland, which we reprojected into a Universal Transverse Mercator projection. We chose this region since it includes a high diversity of landscapes ranging from high-mountain areas to flat and low-lying topography and encompasses frequent sinks and flat areas that either arise from valley narrowing or lakes or as a result of the integer representation of elevations. The resulting DEM has 5537 rows and 5484 columns and a cell size of 79 m. We measured the times that each program requires (1) to hydrologically condition the DEM by filling sinks, (2) to derive D8 flow directions and (3) to calculate flow accumulation. We repeated this procedure after resampling the DEM to a cell size of 158 and 316 m, corresponding to 2769 × 2742 and 1385 × 1371 rows and columns, respectively.

Table 1 contains the benchmark results and shows that the elapsed times required by the software to perform the respective operations are highly variable. Sink processing is performed the fastest with TopoToolbox, which resorts to MATLAB's image-processing toolbox's built-in morphological reconstruction algorithm (Vincent, 1993). Besides the adoption of different algorithms by the other programs, reasons for not being as fast may include data swapping between the main memory and the hard disk to avoid memory problems (ArcGIS, TauDEM) or due to minimum slope imposition onto flat sections as implemented in SAGA GIS and Whitebox GAT. Whitebox GAT performs better in deriving D8 flow directions, which is likely attributed to TopoToolbox's computationally more demanding approach that enforces flow convergence in flat terrain (Soille et al. 2003). Finally, the new approach in TopoToolbox to represent flow directions leads to approximately 20 times faster calculation of flow accumulation than TauDEM 5, which we found to be second fastest among the programs that we compared.

5 Further changes and enhancements

The main goal of TopoToolbox is to provide access to numerical tools for DEM analysis in the form of functions that can be used as a library in custom scripts or functions for advanced analyses or modeling purposes. Most functions available with TopoToolbox 1 are now methods associated with the new classes GRIDobj and FLOWobj. Changes to calling syntaxes are few and were mainly made since the new classes encapsulate information such as cell size and spatial referencing in their properties, which previously had to be supplied to functions as additional input arguments. Thus, function syntaxes have become simpler and more intuitive. A new suite of functions was developed around the class STREAMobj that allows for creating, modifying and analyzing stream networks. Emphasis is placed on tools that facilitate the study of drainage networks and their relation to the tectonic setting, e.g., the analysis of longitudinal river profiles using slope–area plots (Wobus et al., 2006) or χ plots (Harkins et al., 2007; Perron and Royden, 2013; Scherler et al., 2013).

Additional performance increases were achieved by writing many flow-related algorithms as subroutines in C programming language executed by MATLAB (C-MEX), which provides 5–10 times faster evaluation of algorithms such as flow accumulation or drainage basin delineation. Computation speed is particularly significant when repeatedly calculating flow networks and accumulation areas, such as in numerical LEMs (e.g., Willgoose, 2005; Pelletier, 2008; Tucker and Hancock, 2010). While rivers act as major pathways for sediment or upstream-migrating knickpoints in mountain environments, they encompass only a small portion of the entire landscape. Accordingly, in LEMs the heterogeneous distribution of upslope areas leads to numerical solutions that require very small time steps to be numerically stable. Though implicit methods exist that relax the time step constraints (Fagherazzi et al., 2001; Braun and Willet, 2013), they fail to include the planform changes of river and flow network development, thus making frequent updating of the flow directions necessary. With its new implementation, TopoToolbox 2 can serve as a function library for the development of LEMs in MATLAB by providing fast-executing codes in a programming language that already supports a large library of numerical tools.

Visual exploration and interpretation of DEMs and data derived during their analysis constitute an integral work step in most studies. Furthermore, certain tasks are often easier and more rapidly accomplished through direct interaction with DEMs and their derivatives, as compared to writing and running scripts. Thus, TopoToolbox 2 now comes with several applications that allow usage and exploration of various toolbox functions with a GUI (Fig. 1).

The application topoapp is an explorative tool for visual analysis of DEMs and provides rapid access to many basic and several advanced functions of the TopoToolbox 2. Importantly, this application is defined as an own object class that, when working on an instance of topoapp with the GUI, remains in the workspace and is constantly updated with every action of the user. Additionally, it was designed in a way that allows users to add their own functions or modules to it and therefore provides an expandable platform for displaying the graphical outcomes of DEM analysis with the TopoToolbox and other MATLAB tools. With the currently embedded functions, users can quickly and easily extract and plot simple profiles; entire drainage networks, or subsets of them; watershed boundaries; and individual channel reaches (Fig. 1). The created features can be used to build swath profiles, to subset a DEM, to compute basic catchment statistics and to create χ plots; furthermore, all features can also be exported to common formats, amongst other options.

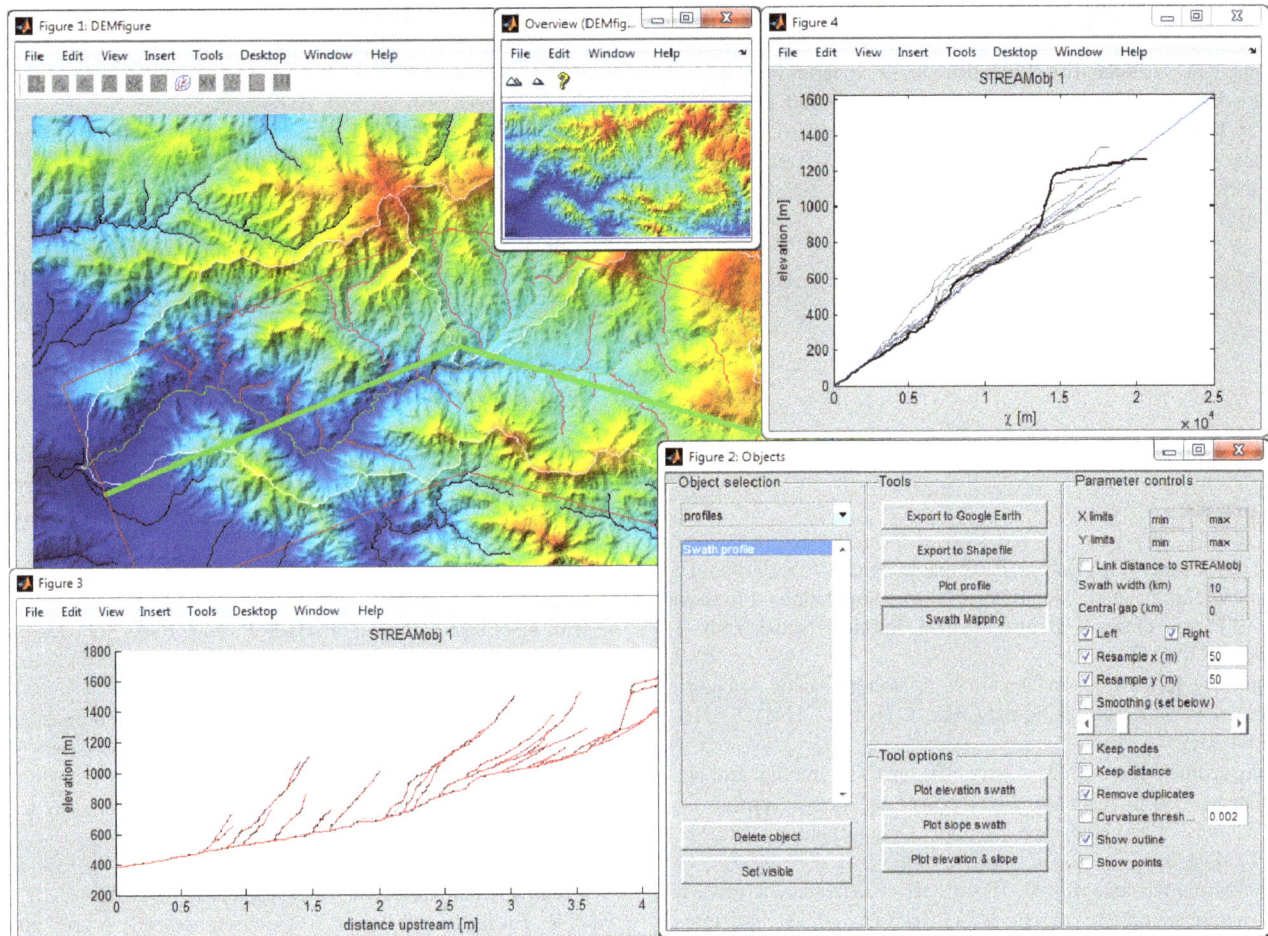

Figure 1. Example layout of topoapp, a graphical user interface that enables access to the majority of TopoToolbox functions.

Manual editing of DEMs with spurious sinks is often required to produce hydrologically correct DEMs. Although automated ways to distinguish between actual and spurious sinks exist (Lindsay and Creed, 2006), choosing which sinks should be filled, carved or retained may be preferred and requires visual interpretation of the data and results. The new application preprocessapp enables a combination of automatic and manual hydrological correction of a DEM by providing interactive tools to fill topographic depressions or carve along least-cost paths to breach features that obstruct digital flow paths. The application flowpathapp is a tool to map individual streams or a stream network by interactively setting channel head locations, and slopeareatool provides an interactive tool to study slope–area relationship for individual streams or stream networks.

6 Conclusions

With the release of version 2, the TopoToolbox for MATLAB has become a more memory-efficient and faster tool for digital terrain analysis. The main difference to the previous version is the adoption of an object-oriented representation of grids, flow direction and stream networks. Topological ordering reduces the memory bottleneck of the previous, sparse matrix approach to flow direction while increasing computation speed and allowing parallel computing. Software quality and particularly reusability are improved so that TopoToolbox can be used as a framework for implementing dynamic, hydrological and geomorphological models. Finally, newly introduced, interactive tools will not only be valuable for researchers to perform rapid analysis but also for conveying and exploring the richness of Earth surface processes in classrooms or labs.

TopoToolbox 2 is platform-independent, free and open software, but requires MATLAB 2011b or later as well as the Image Processing Toolbox. Version 2.0 is included in the supplementary material. The software, future releases and updates are hosted by the Community Surface Dynamics Modeling System (CSDMS) server (http://csdms.colorado.edu/).

Acknowledgements. W. Schwanghart is supported by the Potsdam Research Cluster for Georisk Analysis, Environmental Change and Sustainability (PROGRESS). D. Scherler is supported by a scholarship from the Alexander von Humboldt Foundation and by Caltech's Tectonics Observatory. The support is gratefully acknowledged. We thank P. Steer, T. Hengl and A. J. A. M Temme for their constructive comments on a previous version of the manuscript.

Edited by: A. Temme

References

Arge, L., Chase, J. S., Halpin, P., Toma, L., Vitter, J. S., Urban, D., and Wickremesinghe, R.: Efficient flow computation on massive grid terrain datasets, GeoInformatica, Springer Netherlands, 7, 283–313, 2003.

Borghi, A., Renard, P., and Jenni, S.: A pseudo-genetic stochastic model to generate karstic networks, J. Hydrol., 414/415, 516–529, 2012.

Braun, J. and Willett, S. D.: A very efficient O(n), implicit and parallel method to solve the stream power equation governing fluvial incision and landscape evolution. Geomorphology, 180/181, 170–179, 2013.

Chen, M. and Jacquemin, M.: Footprints of dependency: towards dynamic memory management for massive parallel architectures. Technical report, Yale University, Department of Computer Science, YALEU/DCS/TR-593, 1988.

Clason, C., D. Mair, W. F., Burgess, D. O., and Nienow, P. W.: Modelling the delivery of supraglacial meltwater to the ice/bed interface: application to southwest Devon Ice Cap, Nunavut, Canada. J. Glaciol., 58, 361–374, 2012.

Costa-Cabral, M. and Burges, S.: Digital Elevation Model Networks (DEMON), a model of flow over hillslopes for computation of contributing and dispersal areas. Water Resour. Res., 30, 1681–1692, 1994.

De Moel, H., Asselman, N. E. M., and Aerts, J. C. J. H.: Uncertainty and sensitivity analysis of coastal flood damage estimates in the west of the Netherlands. Nat. Hazard Earth Sys., 12, 1045–1058, 2012.

Dietrich, W. E., Bellugi, D. G., Sklar, L. S., and Stock, J. D.: Geomorphic transport laws for predicting landscape form and dynamics. Prediction in Geomorphology, American Geophysical Union, 135, 1–30, 2003.

Fagherazzi, S., Howard, A. D., and Wiberg, P. L.: An implicit finite difference method for drainage basin evolution. Water Resour. Res., 38, 21.1–21.5, 2002.

de Ferranti, J.: Digital elevation data, Url: http://www.viewfinderpanoramas.org/dem3.html, accessed on 20 Octoer 2013.

Gruber, S. and Peckham, S.: Land-surface parameters and objects in hydrology. In: Hengl, T., Reuter, H. I. (Eds.), Geomorphometry. Concepts, Software, Applications, Elsevier, 33, 171–194, 2009.

Harkins, N., Kirby, E., Heimsath, A., Robinson, R., and Reiser, U.: Transient fluvial incision in the headwater of the Yellow River, northeastern Tibet, China. J. Geophys. Res., 112, F03S04, doi:10.1029/2006JF000570, 2007.

Huang, P.-C. and Lee, K. T.: An efficient method for DEM-based overland flow routing. J. Hydrol., 489, 238–245, 2013.

Jasiewicz, J.: A new GRASS GIS fuzzy inference system for massive data analysis. Comput. Geosci., 37, 1525–1531, 2011.

Kepner, J.: Graphs and matrices, in: Graph algorithms in the language of linear algebra, SIAM series on Software, edited by: Kepner, J. and Gilbert, J., Environments and Tools, Philadelphia, 3–12, 2011.

Lindsay, J. B. and Creed, I. F.: Distinguishing actual and artefact depressions in digital elevation data. Comput. Geosci., 32, 1192–1204, 2006.

Messier, K.P., Akita, Y., and Serre, M. L.: Integrating address geocoding, land use regression, and spatiotemporal geostatistical estimation for groundwater tetrachloroethylene. Envir. Sci. Tech. Lib., 46, 2772–2780, 2012.

Metz, M., Mitasova, H., and Harmon, R. S.: Efficient extraction of drainage networks from massive, radar-based elevation models with least cost path search. Hydrol. Earth Syst. Sci., 15, 667–678, 2011,
http://www.hydrol-earth-syst-sci.net/15/667/2011/.

O'Callaghan, J. F. and Mark, D. M.: The extraction of drainage networks from digital elevation data. Comput. Vision Graph., 28, 323–344, 1984.

Pelletier, J.: Quantitative modeling of Earth surface processes, Cambridge University Press, 295 pp., 2008.

Perron, J. and Royden, L.: An integral approach to bedrock river profile analysis. Earth Surf. Proc. Land., 38, 570–576, 2013.

Pike, R., Evans, I., and Hengl, T.: Geomorphometry: A brief guide. In: Hengl, T., Reuter, H. I. (Eds.), Geomorphometry. Concepts, Software, Applications, Elsevier, 33, 3–30, 2009.

Poggio, L. and Soille, P.: Influence of pit removal methods on river network position. Hydrol. Process., 26, 1984–1990, 2011.

Quinn, P., Beven, K., Chevallier, P., and Planchon, O.: The prediction of hillslope flow paths for distributed hydrological modelling using digital terrain models. Hydrol. Process., 5, 59–79, 1991.

Register, A. H.: A guide to MATLAB object-oriented programming. Chapman & Hall/CRC, SciTech Publishing, Boca Raton, 354 pp., 2007.

Scherler, D., Bookhagen, B., and Strecker, M. R.: Tectonic control on ^{10}Be-derived erosion rates in the Garhwal Himalaya, India. J. Geophys. Res. Earth Surf., in press, 2013.

Schwanghart, W. and Kuhn, N. J.: TopoToolbox: A set of Matlab functions for topographic analysis, Environ. Modell. Softw., 25, 770–781, 2010.

Schwanghart, W. and Jarmer, T.: Linking spatial patterns of soil organic carbon to topography – a case study from south-eastern Spain, Geomorphology, 126, 252–263, 2011.

Schwanghart, W., Groom, G., Kuhn, N. J., and Heckrath, G.: Flow network derivation from a high resolution DEM in a low relief, agrarian landscape. Earth Surf. Proc. Land., 38, 1576–1586, 2013.

Shahzad, F. and Gloaguen, R.: TecDEM: A MATLAB based toolbox for tectonic geomorphology, Part 1: Drainage network pre-processing and stream profile analysis. Comput. Geosci., 37, 250–260, 2011.

Soille, P.: Morphological carving. Pattern Recogn. Lett., 25, 543–550, 2004.

Soille, P., Vogt, J., and Colombo, R.: Carving and adaptive drainage enforcement of grid digital elevation models. Water Resour. Res., 39, SWC 10.1–13, 2003.

Tarboton, D. G.: A new method for the determination of flow directions and upslope areas in grid digital elevation models. Water Resour. Res., 33, 309–319, 1997.

Tesfa, T. K., Tarboton, D. G., Watson, D. W., Schreuders, K. A., Baker, M. E., and Wallace, R. M.: Extraction of hydrological proximity measures from DEMs using parallel processing. Environ. Modell. Softw., 26, 1696–1709, 2011.

Tucker, G. E. and Hancock, G. R.: Modelling landscape evolution. Earth Surf. Proc. Land., 35, 28–50, 2010.

Vincent, L.: Morphological grayscale reconstruction in image analysis: applications and efficient algorithms, IEEE T. Image Process., 2, 176–201, 1993.

Wang, L. and Liu, H.: An efficient method for identifying and ffilli surface depressions in digital elevation models for hydrologic analysis and modelling. Int. J. Geogr. Inf. Sci., 20, 193–213, 2006.

Willgoose, G.: Mathematical modelling of whole landscape evolution. Annu. Rev. Earth Pl. Sc., 33, 443–459, 2005.

Wilson, J. P. and Gallant, J. C.: Digital terrain analysis, in: Terrain Analysis: Principles and Applications, edited by: Wilson, J. P. and Gallant, J. C., Wiley, 1–27, 2000.

Wobus, C., Whipple, K. X., Kirby, E., Snyder, N., Johnson, J., Spyropolou, K., Crosby, B., and Sheehan, D.: Tectonics from topography: procedures, promise, and pitfalls, GSA Special Papers, 398, 55–74, 2006.

Opportunities from low-resolution modelling of river morphology in remote parts of the world

M. Nones[1], M. Guerrero[2], and P. Ronco[3]

[1]Research Center for Constructions – Fluid Dynamics Unit, University of Bologna, Italy
[2]Hydraulic Laboratory, University of Bologna, Italy
[3]University of Padova, Department of Civil, Environmental and Architectural Engineering ICEA, Padova, Italy;
currently at Cà Foscari Venice University, Department of Environmental Sciences, Informatics and Statistics,
Venice, Italy

Correspondence to: M. Nones (michael.nones@unibo.it)

Abstract. River morphodynamics are the result of a variety of processes, ranging from the typical small-scale of fluid mechanics (e.g. flow turbulence dissipation) to the large-scale of landscape evolution (e.g. fan deposition). However, problems inherent in the long-term modelling of large rivers derive from limited computational resources and the high level of process detail (i.e. spatial and temporal resolution). These modelling results depend on processes parameterization and calibrations based on detailed field data (e.g. initial morphology). Thus, for these cases, simplified tools are attractive. In this paper, a simplified 1-D approach is presented that is suited for modelling very large rivers. A synthetic description of the variations of cross-sections shapes is implemented on the basis of satellite images, typically also available for remote parts of the world. The model's flexibility is highlighted here by presenting two applications. In the first case, the model is used for analysing the long-term evolution of the lower Zambezi River (Africa) as it relates to the construction of two reservoirs for hydropower exploitation. In the second case, the same model is applied to study the evolution of the middle and lower Paraná River (Argentina), particularly in the context of climate variability. In both cases, having only basic data for boundary and initial conditions, the 1-D model provides results that are in agreement with past studies and therefore shows potential to be used to assist sediment management at the watershed scale or at boundaries of more detailed models.

1 Introduction

One-dimensional (1-D) models for river hydraulics simulations have been used since the 1970s by many agencies around the world dealing with water resources management and flood control. Aiming to describe the dynamics of free water surface and riverbed (i.e. bed level changes), a variety of open source and commercial 1-D codes has been realized. These models solve the equations of water flow and, in several cases, also the sediment continuity equation in the longitudinal direction. Among others, the HEC-RAS (Hydrologic Engineering Centers River Analysis System) by the US Army Corps of Engineers; the Full Equations (FEQ) model by the US Geological Survey; the Sedimentation and

River Hydraulics (SRH) 1-D by the US Bureau of Reclamation; the Telemac Mascaret 1-D by a European consortium based in France, Germany and the UK; the Mike11 by the Danish Hydraulic Institute; and the SOBEK 1-D by Deltares (NL) are well tested codes for river engineering that apply the 1-D longitudinal simplification. It is worth noting that, while open source codes have been optimized for specific applications (e.g. the FEQ is particularly suited for simulations of unsteady flow in open channel networks with a variety of control structures), commercial software packages have greatly improved to easily integrate detailed data from geographic information systems (GIS) and boundary conditions from climate–hydrology models and for downscaling to more detailed simulations (i.e. 2-D and 3-D models). Albeit in the last decades 1-D and 2-D codes have been provided with modelling of bed level changes to be coupled with hydraulic computations, the simulation of large rivers at the watershed scale remains a difficult task because of the involved space and time lengths, which are within 10^2 and 10^3 kilometres and years, respectively (Di Silvio and Nones, 2013).

Sediment management of large rivers is particularly relevant when evaluating damming or climate change impacts, for instance. This is most challenging in remote parts of the world, where frequent and detailed topographic surveys of river cross sections are uncommon and which may largely limit the application of detailed numerical codes. While conventional 1-D and 2-D morphodynamic models require detailed inputs describing initial morphology and boundary conditions, the model proposed here is able to describe the longitudinal evolution of a river reach at non-detailed spatial and temporal scales by using a basic database. To reduce the computational effort and simulation results dependence on field data (e.g. river channel bathymetry, detailed spatial distribution of alluvial deposition), a series of testable simplifications are introduced in the hydraulic equations, such as the kinematic propagation of the water flow and the local uniform flow (LUF) hypothesis (Fasolato et al., 2011). Under these assumptions, it is possible to simulate riverbed dynamics by implementing a simplified description of the river morphology, which can be derived from maps and satellite images. The use of satellite images and digital terrain models (DTMs) rather than detailed topographic surveys noticeably extends the 1-D model applicability to large rivers in undeveloped countries by drastically reducing modelling efforts and results that depend on detailed field data (Ronco et al., 2010; Nones, 2013).

Two applications of this simplified 1-D code are presented here. These case studies show the 1-D model's performance in providing long-term predictions to support sediment management in large rivers, which is related to natural and anthropogenic pressures (e.g. climate variation, hydropower exploitation). Given the lack of detailed and extensive data for remote parts of the world, the simplified model can substitute and/or support multi-parameter 2-D modelling of river watersheds (Coulthard and Van de Wiel, 2013) or be coupled to detailed 2-D modelling of some river features (e.g. junctions and bifurcations), such as the typical downscaling approach used in Guerrero et al. (2013b). 2-D modelling requires detailed field data for accurate calibrations and validations, which fix a model's parameters and therefore influence the reliability of the results (Guerrero and Lamberti, 2013; Guerrero et al., 2013a; Williams et al., 2013). The model applied here avoids these limitations by coupling a 1-D hydro-morphodynamic model (physically-based on water flow and riverbed profile equations) with a cross-sections sub-model. The 1-D model simulates the riverbed aggradation–degradation rate and the corresponding sediment sorting along the watercourse, although at non-

detailed scale, while the sub-model describes the empirical relationship between river width and hydrological cycle.

This contribution aims to demonstrate the potential of non-detailed models for analysing the long-term impact of anthropogenic and natural pressures, especially in the case of large rivers in remote parts of the world where detailed field data are usually not available for the calibration of high-resolution models.

2 Case studies

The first application concerns a lower branch of the Zambezi River (Africa) that has been regulated by the construction of various dams during the last century. These dams strongly modify the natural pattern of flow variations in time and quantity (Suschka and Napica, 1986; McCartney et al., 2013). Several authors have studied the impact of the two major reservoirs (namely, Kariba and Cahora Bassa dams) on the Zambezi River system. Most of them focused on the biological, ecological and economical effects (biodiversity, fisheries, wetlands, etc.) of these reservoirs (Bowmaker, 1960; Attwell, 1970; Hall et al., 1977; Du Toit, 1984; Dunham, 1989; Beilfuss and Davies, 1999; Beilfuss et al., 2001; Scodanibbio and Mañez, 2005; Tilmant et al., 2012; McCartney et al., 2013). Fewer works addressed morphological changes correlated to the dams' impacts on sediment and water discharges (Guy 1981; Suschka and Napica, 1986; Beilfuss and Davies, 1999; Davies et al., 2000; Basson, 2004; Ronco et al., 2010; Brown and King, 2012; Nones et al., 2013). These two major dams have noticeably changed the hydrological cycle of the lower Zambezi River. Data collected at the gauging station of Tete (Mozambique, about 135 km downstream of Cahora Bassa) show a modified pattern of monthly runoff characterized by the increase of minimum and decrease of maximum values (Ronco, 2008; Ronco et al., 2010). As far as the effects on river morphology are concerned, some authors have evaluated changes of the lower part of the Zambezi after the dams' construction (Suschka and Napica, 1986; Beilfuss and Davies, 1999; Davies et al., 2000; Ronco et al., 2010). The removal of sediments from gorge sections, the stabilization of braids and bars in some sections and of an individual channel in others, the delta erosion and the concurrent increase of salt-water intrusion have all been attributed to these impoundments (Chenje, 2000). Unfortunately, no systematic, detailed surveys of observed morphological changes are reported in the literature, which might have been applied to validate a detailed 2-D model. As an alternative, a simplified 1-D modelling was performed on the basis of the riverbed profile before damming, river stage records, a few sedimentological samples and available satellite images.

The second application addresses climate variability's effectiveness in modifying the morphology of the middle and lower Paraná River reaches in Argentina. Amsler et al. (2005) and Castro et al. (2007) analysed how the streamflow variation has affected the Paraná morphology at its middle and lower reaches, respectively, during the 20th century. These authors related the climate inter-decadal variability to the discharge values most effectively modifying morphology (namely, the effective discharge) and, as a consequence, to river morphological changes observed. In particular, Amsler at al. (2005) observed that the dry midst of the last century (1930–1970) was characterized by low effective discharge, which promoted a decrease in width, braided index, thalweg sinuosity, width-to-depth ratio and channel volume in the middle Paraná, with opposite patterns found in the beginning and end of the same century. Differently, Castro et al. (2007) investigated the lower Paranà morphology and pointed out that it was not as straightforwardly correlated to the hydrological variability as was the middle Paranà. In particular, a continuous and progressive oversimplification of the river channel planimetric morphology toward a smaller width-to-depth ratio was observed regardless of the oscillation in hydrology that had occurred. Reasons for these different responses to the same climate variabilities were found in alluvial plains, which noticeably enlarged and degraded when passing from the middle to the lower Paranà. These morphological constraints, together with the driving increase of precipitations during the latter part of the 20th century, would have produced sediment deposition at secondary reaches and therefore the streamflow gathering in a single straight and deep reach. Although these channel divagations can be simulated with a 2-D morphodynamic model, the large-scale morphology (i.e. river slope and width, sediment sorting) was modelled by means of a simplified 1-D code. Thus, in this study, the combined effects of climate variation and morphological constraints during the 20th and 21st centuries were simulated for the middle and lower Paranà. Future hydrology was computed by applying four climate scenarios spanning the period 1991–2098, which were provided from different combinations of global and regional climate models (Saurral et al., 2013). A significant increase of the effective discharge and flow–discharge variability can be observed passing from the 20th century time series to forecasts, which slightly increases the rate of change of riverbed levels. However, the bed profile appears rather stable, confirming that the present river-alluvial slopes will remain fixed constraints to channel divagation.

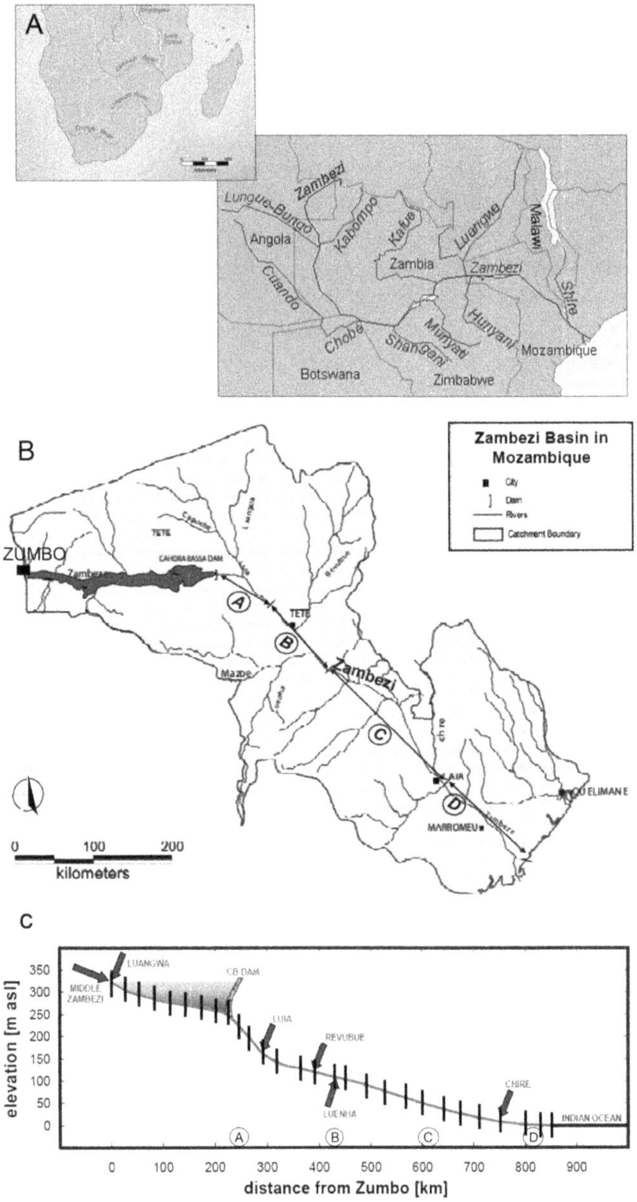

2.1 The Zambezi River

The Zambezi River (Fig. 1) is the fourth-longest river in Africa, with a basin of about 1.4×10^6 km^2. This river, with a total length of about 2574 km, has its source in Zambia and flows through Angola, along the borders of Namibia, Botswana, Zambia and Zimbabwe, to Mozambique. The Zambezi system can be subdivided into three reaches, each differing in its geological template, biodiversity and landscape characteristics: the upper, middle and lower Zambezi (Hughes and Hughes, 1992; Main, 1992; Timberlake, 1998).

Figure 1. (**A**) Main African rivers' basin, focus on the Zambezi River. (**B**) Lower Zambezi downstream the Cahora Bassa Dam. (**C**) Long profile of the main channel and principle tributaries of the lower reach (adapted from Ronco et al., 2010).

The upper Zambezi flows through Angola and Zambia for about 1500 km. The Victoria Falls represent the boundary between the upper and middle reaches. About 240 km downstream of the Victoria Falls, the river flows into the Kariba Lake, created in 1959 by the Kariba Dam, which has a storage capacity of 180.6×10^9 m^3 (Reeve, 1960; Davies et al., 2000). The middle Zambezi flows into the Cahora Bassa Lake (Mozambique), which has a storage capacity (72.5×10^9 m^3, HCB, 2004) resulting from the construction of the Cahora Bassa Dam in 1974. The lower Zambezi

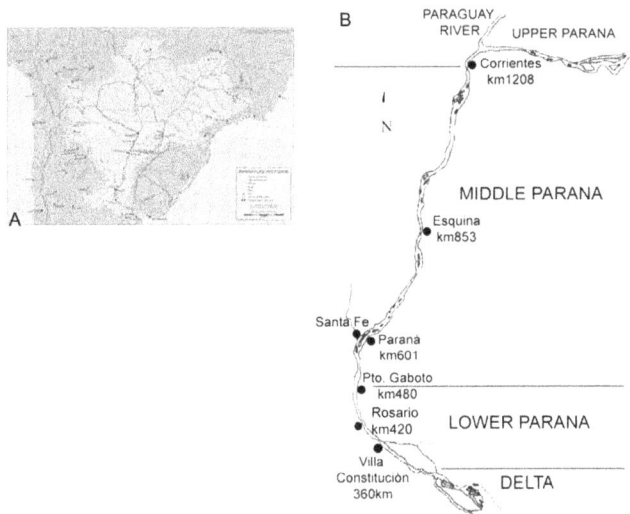

Figure 2. (A) La Plata Basin. **(B)** The middle and lower Paraná with the principal cities and corresponding kilometric labelling of the water way.

is about 650 km long, reaching from the Cahora Bassa Lake to the Indian Ocean. This reach is characterized by a broad floodplain, with parallel channels and shifting sandbanks, while the coastal portion includes extensive grasslands and freshwater swamps, dunes and mangroves (Timberlake, 1998).

As reported in Ronco (2008), the construction of these two dams has considerably changed the hydrological cycles in the lower Zambezi: effective discharge downstream Cahora Bassa has decreased from around 3500 m^3 s^{-1} during the 1950s to around 1800 m^3 s^{-1} during the 1990s. Some authors (Bolton, 1983; Walling, 1984; Walford et al., 2005) have estimated sediment yield of the Zambezi within the range of 10–20 × 10^6 t yr^{-1}, by means of observed delta modifications and sedimentary records.

2.2 The Paraná River

The Paraná River, with a watershed of about 2.6 × 10^6 km^2, is the most important tributary of the La Plata Basin (LPB), which has a basin of 3.2 × 10^6 km^2 within Argentina, Brazil, Paraguay, Uruguay and Bolivia, and outflows into the Atlantic Ocean near Buenos Aires (Fig. 2a). Figure 2b reports the studied river reaches from Corrientes to Villa Constitución, (i.e. the middle and lower Paraná) with main cities and their kilometric labels along the water way.

The middle and lower Paraná reaches flow from north to south through Argentina for about 1000 km; the mean annual discharge is around 12 000–15 000 m^3 s^{-1} and the total sediment transport is around 130–135 × 10^6 t yr^{-1} (Amsler et al., 2005). The Paraguay River joins the Paraná just upstream the city of Corrientes. No other significant tributaries that supply

water and sediments are present downstream of this confluence (Amsler and Drago, 2009).

The middle and lower Paraná play a significant role in the LPB's sediment transport and morphodynamics. Clay and silt materials, coming from the Bermejo River (a tributary that flows from northern Argentina), represent around 80 % of the total solid discharge and are transported as wash load to the Paraná delta and large wetlands during flood periods. The remaining 20 % (around 20–25 × 10^6 t yr^{-1}) is fine sand, which is transported as bed- and suspended-load, modifying the river morphology. The channel bed is composed mostly of fine and medium sand, and its planform pattern is classified as anabranching with meandering thalweg (Latrubesse, 2008). In planform, a succession of wider and narrower nodal sections is observed, with mean channel widths and depths ranging from 600–2500 m and from 5–16 m, respectively.

3 Methods

3.1 1-D mathematical modelling

The 1-D model applied here was firstly developed a few years ago by Di Silvio and Peviani (1989) to analyse the longitudinal evolution (i.e. variations of bed profile, grain size composition and sediment transport) of small rivers at watershed scale. Conventional 1-D morphological models require a detailed description of river geometry. On the contrary, the applied code couples the 1-D equations of water flow with the sediment continuity ones (namely, the De St. Venant Eq. and the Exner and Hirano (1971), reported below) to describe the longitudinal evolution of a river reach at non-detailed scales.

$$\frac{\partial Q}{\partial x} + \frac{\partial A}{\partial t} = 0 \tag{1}$$

$$\frac{\partial}{\partial x}\left[H + Z + \frac{Q^2}{2g \cdot A^2} \right] = -\frac{1}{g} \cdot \frac{\partial U}{\partial t} - J \tag{2}$$

$$\sum_{i=1}^{N} \frac{\partial G s_i}{\partial x} = -B \cdot \frac{\partial Z}{\partial t} \tag{3}$$

In these equations, Q is the river discharge, A represents the wetted area of the channel, having depth H and width B, Z is the bottom elevation, U indicates the current velocity and G_{Si} is the solid discharge of the ith grain size class.

To reduce the computational effort, a series of acceptable simplifications are introduced in the hydraulic equations, such as the local uniform flow (LUF) hypothesis (Fasolato et al., 2011) and the kinematic propagation of water flow. Under the LUF hypothesis, the energy line, water and bed profiles have the same slope, and the variation of the longitudinal bottom elevation $\partial Z/\partial x$, computed by Eq. (2), depends only on the slope of the energy line J (Eq. 5). This equation

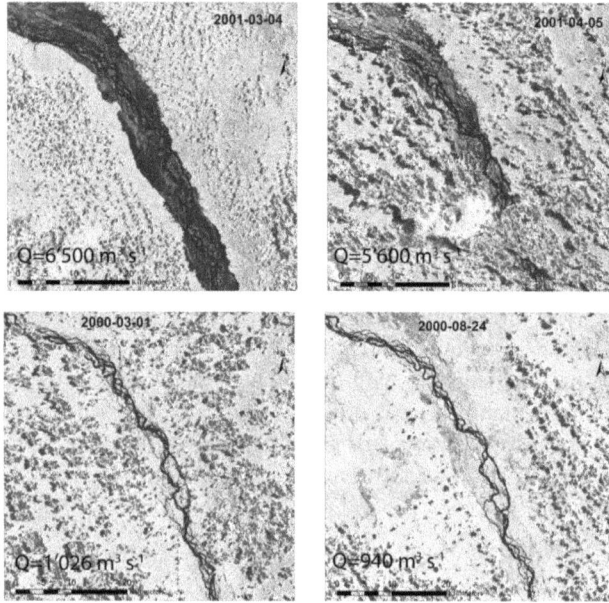

Figure 3. Example of satellite images of the lower Zambezi River upstream Caia (Mozambique): variation of the active river width related to the variation of the river discharge. Discharges vary between $6500 \, \mathrm{m^3 \, s^{-1}}$ and $940 \, \mathrm{m^3 \, s^{-1}}$.

is applied to describe the bed profile evolution by modelling the water profile at appropriate length X and duration scales τ. The spatial and temporal scales, respectively called "morphological box" and "evolution window", are functions of the Froude number of the current (Ronco et al., 2009; Fasolato et al., 2011). It is worth noting that the Local Uniform Flow Morphodynamic (LUFM) model was implemented for long-term analyses that on average fulfil the applied simplifications. In fact, modelled river features are averaged over a relative long period τ (i.e. time window), which was fixed to one year for both performed studies on the basis of resulting Froude numbers. The Eqs. (1–3) respectively become

$$\frac{\mathrm{d}Q}{\mathrm{d}X} = 0, \tag{4}$$

$$\frac{\mathrm{d}Z}{\mathrm{d}X} = -J, \tag{5}$$

$$\sum_{i=1}^{N} \frac{\mathrm{d}Gs_i}{\mathrm{d}X} = -B \cdot \frac{\mathrm{d}Z}{\mathrm{d}\tau}. \tag{6}$$

The sediment transport formula, Eq. (7), applied in the model is derived from the Engelund–Hansen formula for computing the bed- plus suspended-load (i.e. total-load), as described in Ronco (2008) and Nones (2013). In Eq. (7) the rate of bed and suspended sediment transport at yearly

scale, $G_s(\tau)$, depends on energy slope (corresponding to water and bed slopes under the LUF hypothesis), $J(\tau)$, active width, $B(\tau)$, sediment mean diameter, $\mathrm{d}(\tau)$, and effective discharge, $Q_{\mathrm{eff}}(\tau)$, at each morphological box (around 50–80 km for both studies, depending on the Froude number):

$$G_s(\tau) = \alpha' \cdot \left[\frac{Q_{\mathrm{eff}}(\tau)^m \cdot J(\tau)^n}{B(\tau)^p \cdot \mathrm{d}(\tau)^q} \right]. \tag{7}$$

The active cross section is defined as the river channel width that conveys the annual sediment transport, except for wash load (Ronco et al., 2009; Nones, 2013). The effective discharge defines the flow that cumulates most of channel sediments (Biedenharn et al., 1999), i.e. following the Schaffernak approach (Garde and Ranga Raju, 1977), the maximum value of the product of interval frequencies fr with solid discharges G_s occurring in the interval as assessed for corresponding hydrological conditions. The coefficient α' and the exponent m were calibrated by assuming the long-term equilibrium between measured effective discharge and sediment transport rate, while other exponents used in Eq. (7) are functions of m. More details on these parameters and their computation can be found in Di Silvio (1983, 2004).

The initial river widths, $B(\tau = 0)$, were derived from satellite images and the initial slopes, $J(\tau = 0)$, from low-resolution topography (i.e. watershed cartography, DTMs), while sediment distributions were available from some samples taken along the river. Such a basic database is usually also available for remote parts of the world, which noticeably extends the model applicability to large rivers with respect to more advanced 2-D codes.

In order to describe the seasonal widening and narrowing of river sections, a basic calibration method was introduced that combines satellite images with flow discharge records. The 1-D model was coupled to a simplified description of river cross sections (explained in detail in Nones, 2013), which accounts for river width changing at a smaller timescale (i.e. seasonal oscillation) with respect to the bed level evolution simulated. This synthetic cross-sections model (Eq. 8) computes the active river width as a function of the flow discharge. Indeed, the statistical distribution of river width $B(\delta)$ was expressed by means of the statistical distribution of water discharges $Q(\delta)$, where δ is the event occurrence frequency at the analysed section and during the calibration period (i.e. 2000–2010 for both applications). Thus, for discharges lower than the bankfull value (the minimal discharge flowing over alluvial plains), an at-a-site hydraulic geometry relationship was assumed (Eq. 8) for the synthetic description of active cross sections (Singh, 2003):

$$B(\delta) = \alpha \cdot Q(\delta)^{\beta} \tag{8}$$

where α and β are the calibration parameters (see, e.g. Leopold and Maddock, 1953; Yalin, 1992; Parker et al., 2007; Wilkerson and Parker, 2011; Nones, 2013). The river widths for the calibration period were obtained from Landsat 7 satellite images (USGS database), which have a resolution of 15 m. The corresponding discharges were derived from the available river stage records. Images corresponding to flood events were discarded; in those cases, wetted areas did not provide evidence of the active channel geometry but the flooding (Fig. 3).

The LUFM model simulates the evolution of the riverbed profile, which is driven by the fluvial sediment transport. The investigated dynamics were uncoupled from the geological processes (namely, from the LGM, Last Glacial Maximum, to present day) and also from the floodplains transport. The assumption of fixed floodplains is justified by their very slow evolution. Indeed, finest sediment, transported as wash load, has been cumulated in the alluvial plain during flood events recurring over thousands of years, while the active channel morphology was noticeably modified within a few decades.

3.2 Model reliability

Albeit the validation of the proposed model is not within the objectives of this study, a short summary of past works about the application of the LUFM model to gauged watersheds is reported for the convenience of readers.

A first version of the LUFM model was applied to the gauged basins of small rivers located in Italy, such as the Adige River (Nones et al., 2009), the Mallero River (Di Silvio and Peviani, 1989) and the Piave River (Fasolato et al., 2006). This version, assuming that the river width remains constant with time, computed reliable evolutions of the rivers, also by introducing anthropogenic forcing terms (e.g. dredging operations, dams, land use changes).

In the case of the Piave River, the LUFM model was used to analyse the morphology of the river due to flushing operations. A comparison between the measured turbidity values and the computed sediment transport highlighted the reliability of the model to reproduce the periodic flushing peaks. The deposition of sediments downstream of the impoundment and the reduction of the grain size are quite well reproduced, notwithstanding the few available input data.

In the case of the Adige River, the model was applied to analyse the major anthropogenic changes that have occurred during the past century, relating to the reduction of sediment transport due to soil-protection interventions, derivations for industrial and agricultural uses, changes in the duration curve of flow discharge, interception of solid material by hydropower reservoirs, and extraction of fluvial sediments by means of quarries. The results of the LUFM model confirmed (1) the decreasing of sediment transport observed after 1920, chiefly due to the construction of various hydropower plants in the Adige's basin, and (2) the bed erosion in the lower part of the river caused by quarry activities.

The studied rivers have been intensively studied and monitored since the beginning of the past century, which has provided a valuable matching to verify the model reliability with a strict comparison between the computed results and the available measured values. These validations have highlighted a good correlation between model results and the available rich data set from the literature, despite the basic description that is required as input.

4 Results

4.1 Evolution of the Zambezi River during the 20th century

The present paper summarizes the major results of a LUFM model's application to the unsurveyed lower Zambezi River basin published in Ronco et al. (2010) and Nones et al. (2013). The application simulated the morphological behaviour of the lower Zambezi starting from 1907, when first systematic measurements were recorded at the gauging station of Zumbo, Mozambique (Fig. 1b). These measurements were analysed to assess the effective discharge of lower Zambezi and its main tributaries, together with the corresponding sediment input, during the 20th century. The data were applied at model boundaries and for characterizing river width–discharge relationship along the simulated reach. The turbidity measurements performed by Hall et al. (1977) have been utilized for the calibration of the transport formula (Eq. 7), as reported in Ronco et al. (2010).

The mentioned previous simulations performed with the LUFM model on the lower Zambezi River did not considered river width–discharge relationship (Ronco, 2008; Ronco et al., 2010). For that case, the model results show an enduring process of sediment deposition, which slowly propagates downstream. The long-term evolution was apparently dominant with respect to the recent evolution related to the dams' constructions.

The computations reported here also accounted for seasonal oscillations, which simulated river widening and narrowing. These simulations confirmed the previous results: the lower reach presents a widespread tendency to a progressive deposition, especially in the flat zones downstream of the gorges, where some tributaries flow into the main channel (Fig. 1c). This trend was affected by the two reservoirs to some degree (Nones et al., 2013). Dams trap sediments, which slowly modify riverbed morphology and corresponding bed composition. In more detail, the undisturbed river aggradation corresponded to a progressive fining of riverbed sediments. In the cases of dams' implementations, two opposite effects altered the natural process, yielding a more stable bed: (1) the water flow reduction rapidly propagates downstream, with important implications for the local ecosystem; and (2) the sediment sorting is delayed because of trapping (Ronco et al., 2010; Nones et al., 2013).

The simplified modelling of cross-sections changes at seasonal timescale bore out a typical evolution from braided to unicursal morphology (Leopold and Wolman, 1957; Nones, 2013). The simulated flow decreasing because of water

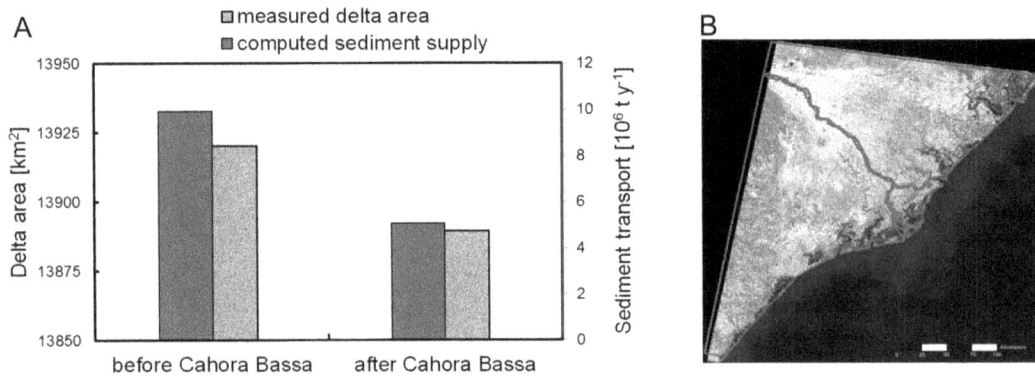

Figure 4. (**A**) Averaged values of the simulated sediment yield and the measured evolution of the Zambezi Delta from satellite images, before and after the construction of the Cahora Bassa Dam. (**B**) Satellite image of the Zambezi Delta (USGS database, 2013).

impounded in the reservoirs corresponded to narrowing of active cross sections. This simplified modification of cross-sections shapes reflects the observed tendency to unicursal morphology and accretion of vegetated floodplains, which limits channel divagations. In other words, lower flows were simulated with sections narrowing, which agree with the observations of vegetation density increase and abandoned secondary channels. This behaviour was also reported by Davies et al. (2000) and Nones et al. (2013).

The performed simulations also gave multiple evidence on the delta shoreline evolution (Fig. 4), which has been recognized as a key indicator of sediment yield in various large rivers of the world. This evolution was evaluated on the basis of seven satellite images (from the USGS's database) spanning the period 1970–2013, which confirmed the general negative trend of the delta surface extension (Ronco, 2008; Ronco et al, 2010; Nones et al., 2013). Figure 4a shows the negative trend after the construction of the Cahora Bassa Dam. In total, erosions prevailed yielding a loss of the delta area of around 50 km^2 in the analysed period. The simulated time series of sediment transport rate in the most downstream morphological box confirmed the observed erosional trend for the Zambezi's delta. In fact, this time series reflects the historical sediment supply to the delta. The simulated rate of sediment transport appeared correlated to the observed variation of the delta area (Fig. 4b), notwithstanding that the satellite images used may be affected by occasional changes in water level and that the shoreline morphology is also affected by the Indian Ocean dynamics (Ronco et al., 2010). More detailed analyses, also with a specific model able to simulate the long-shore drift, are necessary for a thorough characterization of the delta's dynamics.

4.2 Evolution of the Paraná River during the 20th and the 21st centuries

In this case, outputs from four regional climate models (RCMs) were used to produce future hydrological scenar-

ios at the La Plata Basin. These scenarios were prescribed as boundary conditions for the performed modelling, covered much of the 21st century and were provided by the research project "A Europe-South America Network for Climate Change Assessment and Impact Studies in La Plata Basin" (CLARIS-LPB). Specifically, the four RCMs are (1) PROMES (UCLM, Spain) covering the period 1991–2098, (2) RCA (SMHI, Sweden) for 1981–2098, (3) RegCM3 (USP, Brazil) having information for period 1981–2048, and (4) LMDZ (IPSL, France) with information for period 1991–2048. In addition, the 20th century scenario was produced from climatology–hydrology records. The details on these data sets are reported in Saurral et al. (2013). Figure 5 shows a comparison of the different applied scenarios in terms of frequency distribution of monthly flow discharges at Corrientes, which is the upstream boundary of the Paraná reach simulated by means of the LUFM model. It is possible to observe a relatively large dispersion among the different RCMs and a general increase in the frequency of large discharges for predictions with respect to the past century, which is particularly evident for the RegCM3 scenario.

As reported in the literature (Amsler et al., 2007), the alluvial plains morphology changes take place over a longer timescale (geological periods) and are affected by finest sediment (wash load) deposition only during floods recurring in thousands of years time frames. The LUFM model computed the sediment transport in the river channel, i.e. the effective discharges that form the channel morphology over decades, while flooding discharges were disregarded. Actually, the wash load was not computed by the model, but it may also affect the delta growth.

The literature data regarding the middle and lower Paraná River morphology at watershed scale mostly concern the transported sediment volume per year and the corresponding most effective discharge in conveying this volume at some sections (Amsler et al., 2005; Castro et al., 2007). Thus, the LUFM model results were used to assess sediment volumes per year and corresponding effective discharges for future

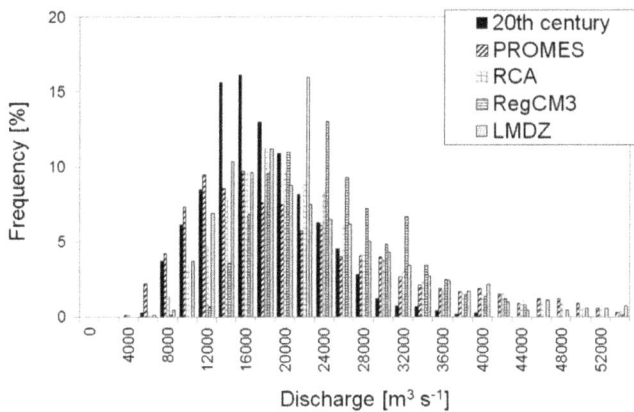

Figure 5. Frequency distributions of monthly flow discharges of applied scenarios for LUFM modelling (20th century and future prediction from listed RCMs) at Corrientes.

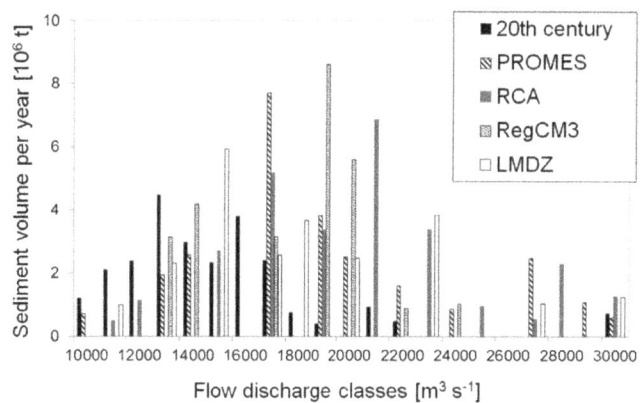

Figure 6. Sediment volume per year from LUFM modelling for corresponding flow discharge classes at Corrientes, characterizing past and future scenarios.

and past scenarios. The average effective discharges for each decade of the 20th century from the literature compared to LUFM corresponding results show a good correlation with maximum deviations in the order of a few per cent, with few exceptions for the largest discharges. As input data, the 1-D model uses river discharge with 1 yr time step, while the literature data are referred to as monthly discharges. Yearly values can be representative of the averaged morphological variations of the channel due to the inter-decadal variability, with a reduction of the computational effort. Flooding events, which have a duration of 1–2 months, were considered unimportant for the morphological evolution of the Paraná River because of their temporal scale (months) with respect to the morphological changes (years).

According to the Schaffernak approach, the sediment volume per year was divided into discharge classes of $2000\,\mathrm{m^3\,s^{-1}}$ width and was compared among the scenarios (Fig. 6). For any analysed scenario, the average volume per year was almost the same (changing within the range of $20–30 \times 10^6\,\mathrm{t\,yr^{-1}}$), but the distributions among the classes resulted rather differently. The distribution for the 20th century did not present a very high, recognizable maximum. The highest sediments volume was conveyed by the flow discharge interval of $13\,000–17\,000\,\mathrm{m^3\,s^{-1}}$ (i.e. the average effective discharge per year for the 20th century), which roughly agrees with the Amsler et al. (2005) investigation. Using records from last century's water levels and monthly flow discharges in Corrientes, Amsler et al. (2005) assessed the effective discharge to vary within the $15\,400–24\,500\,\mathrm{m^3\,s^{-1}}$ range, among five periods having different lengths of time during the years 1904 to 1990, which yields the weighted average of $16\,800\,\mathrm{m^3\,s^{-1}}$. When comparing these values to the LUFM results, the model resolution must be considered. The interval of $13\,000–17\,000\,\mathrm{m^3\,s^{-1}}$ was detected using discharge classes derived from the yearly time series simulated (one year is the LUFM's time win-

dow), while Amsler et al. (2005) applied monthly records that present larger peaks (maximum monthly discharges in the order of $50\,000\,\mathrm{m^3\,s^{-1}}$).

Regarding the resulting morphology, the cross-sections widths appeared with the largest sensitivity to flow regime at the lower Paraná, which exacerbated the deviation between various scenarios in terms of resulting bed profiles and average wetted area. The model gave almost fixed slopes at the lower and middle Paraná for the intervals of around 2–$3\,\mathrm{cm\,km^{-1}}$ and $3–5\,\mathrm{cm\,km^{-1}}$, respectively, and mostly depending on the sub-reach location (morphological box), while the corresponding cross-sections changes were also correlated to the variability of forcing time series (scenarios). The river widths at the lower and middle Paraná varied within the range of 1600–3300 m and 1200–2400 m, respectively.

The resulting bed levels in terms of cumulate variation (i.e. final depositions and erosions at each morphological box) were assessed for the 20th century and the two continuous time series available from future projections (PROMES and RCA scenarios). Despite the noticeable increase in the forcing and effective discharges (Figs. 5 and 6, respectively) when passing from past to future scenarios, the bed level changes along the river maintained almost the same aggradation or degradation tendencies as for the 20th century, with maximal values of a few centimetres within one century (Fig. 7). The bed level result was rather stable along the river, especially at the lower Paraná.

Some morphological data of the Paraná River can be retrieved from the literature: Amsler et al. (2005) and Castro et al. (2007) observed that the same climate variability drove braided and bifurcated patterns of the channel morphology at the middle and lower Paraná, respectively. This difference was ascribed to the alluvial plane morphology change along the river, which noticeably enlarges and degrades downstream. Although these observed channel morphologies may be simulated with detailed 2-D models, the

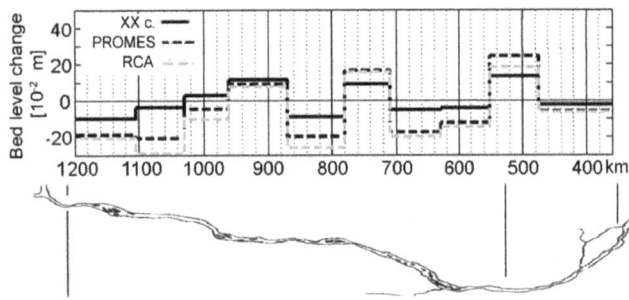

Figure 7. Simulated cumulate variation of the bed level for the continuous time series: past century and 21st century predictions (PROMES and RCA scenarios).

performed simulations corroborated the argument that river slope and width are almost time-fixed constraints.

5   Discussion

The changes in effective discharge, sediment transport rate, riverbed level, bed sediment sorting and cross-section width occurring during the 20th and the 21st centuries along the Zambezi and Paraná rivers were analysed with the LUFM model. This 1-D code introduces some important simplifications in the mathematical description of river hydraulics (among others, kinematic propagation of water flow and local uniform flow hypothesis) and in the schematization of the channel geometry (namely, cross-sections synthetic profile). These simplifications gave some opportunities with respect to more detailed modelling (e.g. 2-D codes). In fact, the LUFM model simulated the long-term morphodynamic processes influenced by natural and anthropogenic pressures at non-detailed spatial and temporal scales by implementing a very basic database. This is particularly relevant for remote parts of the world, such as in the presented case studies.

The performed analyses confirmed that the LUFM model is able to simulate average variations at large temporal and spatial scales (as pointed out by Ronco et al., 2010). As hoped, the schematizations used for the two applications were able to reduce the computational effort and the topographic resolution usually required to implement standard 1-D models and, even more, for detailed 2-D modelling. Notwithstanding the lack of geometrical resolution, the LUFM model responses appeared rather reliable and less sensitive to inaccuracies induced by model parameters calibration on the basis of detailed field data. Indeed, the estimation of induced biases is not trivial, especially for very detailed modelling, which includes a lot of processes parameterization and for long-term simulations that extrapolate to a timeline far from the initial conditions. The operated synthetic description of transversal cross sections, based on satellite images, presents some advantages with respect to occasional surveys. The LUFM model accounted for sediment transport variations due to the active channel modi-

fications driven by hydrological changes at seasonal scale (Nones et al., 2013).

Therefore, the simplified model applied presents a potential method to overcome the lack of detailed field data, providing, at the same time, reliable results for a variety of engineering and environmental practices, ranging from long-term sediment management to environmental impact studies. Basic databases (i.e. historical cartography for initial geometry and river stage records for boundary conditions as well as sporadic geomorphological data) are usually also available for remote parts of the world, which noticeably enlarges the model applicability to large rivers with respect to more advanced 2-D codes.

Furthermore, it is worth noting that the simplified modelling of river cross-sections variations refines the code resolution to seasonal oscillation, although the larger evolution window was used for updating the bed profile and sediment sorting, which fulfils the LUF hypothesis. To this end, the additional river width–discharge database is required, which introduces the model's calibration on statistic distributions from a certain period. Indeed, satellite images, historical cartography and river stage records are largely also available for ungauged river basins and date back from decades (for satellite images) to centuries in the cases of historical records and maps.

6   Conclusions

Two large rivers were analysed with a simplified 1-D model, which highlights the opportunity for using basic databases to simulate river morphology at low-resolution for remote parts of the world. The proposed 1-D code couples the longitudinal modelling of their riverbeds with a synthetic description of cross-sections shapes and was applied for studying past and future river evolution on the basis of topographic basic data, stage records and satellite images. In addition, future projections were produced from available climate–hydrology scenarios.

The results indicate the potential of non-detailed models for analysing the long-term impact of anthropogenic and natural pressures, especially in the case of rivers located in remote parts of the world, for which detailed field data are usually not available. This potential is a particularly relevant option to 2-D modelling for which results depend on processes parameterization and calibrations based on detailed field data.

In our application to the lower Zambezi River only a topographic and granulometric basic survey was applied to describe the initial morphology, while satellite images and stage records provided river cross-sections shapes and boundary conditions. The impacts of hydropower reservoirs on river morphology during the 20th century were simulated. As highlighted by the cited literature concerning the river's evolution, the results reported here agree with the riverbed aggradation and delta erosion observed during the last century.

In the second case study, in addition to basic data, climate–hydrology scenarios were used at model boundaries to study the impacts of future climate variability over the middle and lower Paraná. The results reported here corroborate the hypothesis that climate drivers differently affect the channel divagation because of riverbed slope and width decreasing and increasing, respectively, when passing from the middle to the lower Paraná River.

Acknowledgements. This research has received funding from (1) the European Community's Seventh Framework Programme (FP7/2007–2013) under grant agreement no. 212492 (CLARIS-LPB, a Europe-South America Network for Climate Change Assessment and Impact Studies in La Plata Basin); and (2) the CUIA IT-AR universities consortium under the project "Impatto dei cambiamenti ambientali sull'eco-idromorfodinamica fluviale".

Edited by: J. Turowski

References

Amsler, M. L., Ramonell, C. G., and Toniolo, H. A.: Morphologic changes in the Parana River channel (Argentina) in the light of the climate variability during the 20th century, Geomorphol., 70, 257–278, 2005.

Amsler, M. L., Drago, E. C., and Paira, A. R.: Fluvial sediments: main channel and floodplain interrelationships, in: The Middle Parana River: limnology of a subtropical wetland, edited by: Iriondo, M., Paggi, J. J., Parma, M. J., Springer, Berlin Heidelberg New York, 123–142, 2007.

Amsler, M. L. and Drago, E. C.: A review of the suspended sediment budget at the confluence of the Parana and Paraguay Rivers, Hydrol. Proc., 23, 3230–3235, 2009.

Attwell, R. I. G.: Some effects of Lake Kariba on the ecology of a floodplain of the mid-Zambezi valley of Rhodesia, Biol. Conserv., 2, 189–196, 1970.

Basson, G.: Hydropower Dams and Fluvial Morphological Impacts–An African Perspective, Paper from United Nations Symposium on Hydropower and Sustainable Development, 2004.

Beilfuss, R. D. and Davies, B. R.: Prescribed flooding and wetlands rehabilitation in the Zambezi Delta, Mozambique, An Int. Perspective on Wetland Rehabilitation, 143–158, 1999.

Beilfuss, R. D., Moore, D., Dutton, P., and Bento, C.: Patterns of vegetation change in the Zambezi Delta, Mozambique, Working paper #3 of the Program for the Sustainable Management of Cahora Bassa Dam and the Lower Zambezi Valley, International Crane Foundation, USA, 2001.

Biedenharn, D. S., Thorne, C. R., Soar, P. J., Hey, R. D., and Watson, C. C.: A practical guide to effective discharge calculation (Appendix A), US Army Corps of Eng., Vicksburg, Mississippi, USA, 1999.

Bolton, P.: The regulation of the Zambezi in Mozambique: A study of the origins and impact of the Cabora Bassa project, PhD Thesis. University of Edinburgh, Great Britain, 1983.

Bowmaker, A. P.: A Report on the Kariba Lake Area and Zambezi River Prior to Inundation, and the Initial Effects of Inundation with Particular Reference to the Fisheries Training Centre on Fishery Survey for the Countries of African Region (Report), FAO Library Fiche AN, 59986, Rome, Italy, 1960.

Brown, C. and King, J.: Modifying dam operating rules to deliver environmental flows: experiences from southern Africa, Int. J. River Basin Manage., 10, 13–28, 2012.

Castro, S. L., Cafaro, E. D., Gallego, M. G., Ravelli, A. M., Alarcón, J. J., Ramonell, C. G., and Amsler, M. L.: Evolución morfológica histórica del cauce del río Parana en torno a Rosario (456–406 km), Proceedings of the XXI Congreso Nacional del Agua, CONAGUA 2007, Tucumán, Argentina, 2007 (in Spanish).

Chenje, M.: State of the Environment Zambezi Basin, SADC/IUCN/ZRA/SARDC, Maseru/Lusaka, Harare, 2000.

Coulthard, T. J. and Van de Wiel, M. J.: Climate, tectonics or morphology: what signals can we see in drainage basin sediment yields?, Earth Surf. Dynam. Discuss., 1, 67–91, 2013.

Davies, R. D., Beilfuss, R. D., and Thomas, M. C.: Cahora Bassa Retrospective, 1974-1997, effects of flow regulation on the Lower Zambezi River, Limnol. Develop. World, 27, 1–9, 2000.

Di Silvio, G.: Modelli matematici per lo studio di variazioni morfologiche dei corsi d'acqua a lunga e breve scala temporale, Studi e Ricerche, 356A, 1983 (in Italian).

Di Silvio, G. and Peviani, M.: Modelling Short- and Long-Term evolution of mountain rivers, an application to the torrent Mallero (Italy), Lecture Notes in Earth Sciences n. 37, Fluvial Hydraulics of Mountain Regions, edited by: Armanini, A. and Di Silvio, G., Springer Verlag, 1991, 293–315, 1989.

Di Silvio, G.: Review of state-of-the-art research on erosion and sediment dynamics from catchment to coast (a Northern perspective. Meeting of the Task Force Group of ISI (International Sediment Initiative) of UNESCO-IHP (technical report), Paris, France, 2004.

Di Silvio, G. and Nones, M.: Morphodynamic reaction of a schematic river to sediment input changes: Analytical approaches, Geomorphology, in press, doi:10.1016/j.geomorph.2013.05.021, 2013.

Du Toit, R. F.: Some environmental aspects of proposed hydroelectric schemes on the Zambezi river, Zimbabwe, Biol. Conserv., 28, 73–87, 1984.

Dunham, K. M.: Vegetation-environment relations of a Middle Zambezi floodplain, Vegetatio, 82, 13–24, 1989.

Fasolato, G., Ronco, P., and Tregnaghi, M.: Operazioni di sghiaiamento da un serbatoio alpino ed effetti sulla morfodinamica fluviale. Proceedings of the XXX Convegno di Idraulica e Costruzioni Idrauliche – IDRA 2006, Roma, Italy, 2006 (in Italian).

Fasolato, G., Ronco, P., Langendoen, E. J., and Di Silvio, G.: Validity of Uniform Flow Hypothesis in One-Dimensional Morphodynamic Models, Journal of Hydraulic Engineering, ASCE, 37, 183–195, 2011.

Garde, R.J. and Ranga Raju K.G.: Mechanics of sediment transportation and alluvial stream problems, Wiley Eastern Ltd., New Delhi, India, 1977.

Guerrero, M. and Lamberti, A.: Bed-roughness investigation for a 2-D model calibration: the San Martìn case study at Lower Paranà, Int. J. Sediment Res., 28, in press, 2013.

Guerrero, M., Di Federico, V., and Lamberti, A.: Calibration of a 2-D morphodynamic model using water-sediment flux maps derived from an ADCP recording, J. Hydroinformatics, 15, 813–828, 2013a.

Guerrero, M., Nones, M., Saurral, R., Montroull, N., and Szupiany, R.N.: Parana River morphodynamics in the context of climate change, International J. River Basin Manage., doi:10.1080/15715124.2013.826234, 2013b.

Guy, P. R.: River bank erosion in the mid-Zambezi valley, downstream of Lake Kariba, Biol. Conserv., 19, 199–212, 1981.

Hall, A., Valente, I., and Davies, B. R.: The Zambezi River in Mozambique, the physicochemical status of the Middle and Lower Zambezi prior to the closure of the Cabora Bassa Dam, Freshwater Biol., 7, 187–206, 1977.

Hidroeléctrica de Cahora Bassa (HCB): Technical Reports on Cahora Bassa Project, Songo, Mozambique, 2004.

Hirano, M.: River bed degradation with armouring.Transaction of Japanese Society of Civil Engineering, 3, 194–195, 1971

Hughes, R. H. and Hughes, J. S.: A Directory of African Wetlands, IUUCN/UNEP/WCMC, Gland, Switzerland, ISBN 2-88032-949-3, 1992.

Latrubesse, E.: Patterns of Anabranching channels, the ultimate end-member adjustments of mega-rivers, Geomorphology, 101, 130–145, 2008.

Leopold, L. B. and Maddock Jr., T.: The Hydraulic Geometry of Stream Channels and Some Physiographic Implications, Geol. Survey Paper, 252 pp., 1953.

Leopold L. B. and Wolman M.G.: River Channel Patterns: Braided, Meandering and Straight, US Geol. Survey Prof. Paper, 282 pp., 1957.

Main, M.: Zambezi, Journey of a River. Southern Book Publishers, Halfway House, South Africa, ISBN 1-86812-257-3, 1992.

McCartney, M., Cai, X., and Smakhtin, V.: Evaluating the Flow Regulating Functions of Natural Ecosystems in the Zambezi River Basin, 2013.

Nones, M., Bonaldo, D., Di Silvio, G., and Guarino, L.: Sediment budget of rivers at watershed scale: the case of Adige River. EGU General Assembly Conference Abstracts, 11, p. 1197, 2009.

Nones, M.: Riverine dynamics at watershed scale: hydro-morphobiodynamics in rivers, Eds. LAP Lambert Academic Publishing, p. 140, ISBN-13: 978-3659367854, 2013.

Nones, M., Ronco, P., and Di Silvio, G.: Modelling the impact of large impoundments on the Lower Zambezi River. Int. Journal of River Basin Management 11(??), 221-236, 2013.

Parker, G. P., Wilcock, P. R., Paola, C., Dietrich, W. E., and Pitlick, J.: Physical basis for quasi-universal relations describing bankfull hydraulic geometry of single-thread gravel bed rivers, J. Geophys. Res., 112, F04005, doi:10.1029/2006JF000549, 2007.

Reeve, W. H.: Progress and Geographical Significance of the Kariba Dam, The Geograph. J., 126, 140–146, 1960.

Ronco, P.: Sediment Budget of Unsurveyed Rivers at Watershed Scale: the Case of Lower Zambezi. PhD Thesis. University of Padova, Italy, http://paduaresearch.cab.unipd.it/625, 2008.

Ronco, P., Fasolato, G., and Di Silvio, G.: Modelling evolution of bottom profile and grainsize distribution in unsurveyed rivers, Int. J. Sed. Res., 24, 127–144, 2009.

Ronco, P., Fasolato, G., Nones, M., and Di Silvio, G.: Morphological effects of damming on lower Zambezi River, Geomorphology, 115, 43–55, 2010.

Saurral, R., Montroull, N., and Camilloni, I.: Development of statistically unbiased 21st century hydrology scenarios over La Plata Basin, Accepted by International Journal of River Basin Management, 2013.

Scodanibbio, L. and Mañez, G.: The World Commission on Dams: A fundamental step towards integrated water resources management and poverty reduction? A pilot case in the Lower Zambezi, Mozambique, Phys. Chem. Earth, 30, 976–983, 2005.

Singh, V. P.: On the theories of hydraulic geometry, Int. J. Sed. Res., 18, 196–218, 2003.

Suschka, J. and Napica, P.: Ten years after the conclusion of Cabora Bassa Dam, The impacts of large water projects on the environment, Proceedings of an international symposium, UNEP/UNESCO, Paris, France, 1986.

Tilmant, A., Kinzelbach, W., Juizo, D., Beevers, L., Senn, D., and Casarotto, C.: Economic valuation of benefits and costs associated with the coordinated development and management of the Zambezi river basin, Water Policy, 14, 490–508, 2012.

Timberlake, J.: Biodiversity of the Zambezi basin wetlands, review and preliminary assessment of available information, Final Report IUCN, Harare, Zimbabwe, 1998.

Walford, H. L., White, N. J., and Sydow, J. C.: Solid sediment load history of the Zambezi Delta, Earth Planet. Sci. Lett., 238, 49–63, 2005.

Walling, D. E.: The sediment yields of African rivers. Challenges in African Hydrology and Water Resources, Proceedings of the Harare Symposium, IAHS Publ., 144, 265–283, 1984.

Wilkerson, G. V. and Parker, G.: Physical Basis for Quasi-Universal Relations Describing Bankfull Hydraulic Geometry of Sand-Bed Rivers, J. Hydraul. Engin., 137, 739–753, 2011.

Williams, R. D., Brasington, J., Hicks, M., Measures, R., Rennie, C. D., and Vericat, D.: Hydraulic validation of two-dimensional simulations of braided river flow with spatially continuous aDcp data, Water Resour. Res., 49, 5183–5205, doi:10.1002/wrcr.20391, 2013.

Yalin, M. S.: River mechanics, Eds. Pergamon Press, Oxford, England, 1992.

Effect of self-stratification on sediment diffusivity in channel flows and boundary layers: a study using direct numerical simulations

S. Dutta[1], M. I. Cantero[3], and M. H. Garcia[1,2]

[1]Dept. of Civil and Environmental Engineering, University of Illinois at Urbana-Champaign, Urbana, USA
[2]Dept. of Geology, University of Illinois at Urbana-Champaign, Urbana, USA
[3]Centro Atómico Bariloche and Instituto Balseiro, Consejo Nacional de Investigaciones Científicas y Técnicas
(CONICET) and Comisión Nacional de Energía Atómica (CNEA), San Carlos de Bariloche, Río Negro,
Argentina

Correspondence to: S. Dutta (dutta5@illinois.edu)

Abstract. Sediment transport in nature comprises of bedload and suspended load, and precise modelling of these processes is essential for accurate sediment flux estimation. Traditionally, non-cohesive suspended sediment has been modelled using the advection–diffusion equation (Garcia, 2008), where the success of the model is largely dependent on accurate approximation of the sediment diffusion coefficients. The current study explores the effect of self-stratification on sediment diffusivity using suspended sediment concentration data from direct numerical simulations (DNS) of flows subjected to different levels of stratification, where the level of stratification is dependent on the particle size (parameterized using particle fall velocity $\tilde{V}$) and volume-averaged sediment concentration (parameterized using shear Richardson number Ri_τ). Two distinct configurations were explored, first the channel flow configuration (similar to flow in a pipe or a duct) and second, a boundary-layer configuration (similar to open-channel flow). Self-stratification was found to modulate the turbulence intensity (Cantero et al., 2009b), which in turn was found to reduce vertical sediment diffusivity in portions of the domain exposed to turbulence damping. The effect of particle size on vertical sediment diffusivity has been studied in the past by several authors (Rouse, 1937; Coleman, 1970; Nielsen and Teakle, 2004); so in addition to the effect of particle size, the current study also explores the effect of sediment concentration on vertical sediment diffusivity. The results from the DNS simulations were compared with experiments (Ismail, 1952; Coleman, 1986) and field measurements (Coleman, 1970), and were found to agree qualitatively, especially for the case of channel flows. The aim of the study is to understand the effect of stratification due to suspended sediment on vertical sediment diffusivity for different flow configurations, in order to gain insight of the underlying physics, which will eventually help us to improve the existing models for sediment diffusivity.

1 Introduction

Turbulent mixing and accompanying transport is a prevalent phenomenon in natural and industrial settings. One of the most important transport phenomena in nature is that of sediment, and it can be broadly divided into bedload transport and suspended load transport. In most rivers, the suspended load comprises approximately 80–85 % of the total sediment load, thus playing an important role in morphodynamics of the system. In situ measurement of suspended sediment is still very discontinuous and expensive, so accurate modelling of transport of suspended sediment is essential for correct approximation of the net sediment flux in a river. For the generic case of suspended sediment of constant density and particle size in unsteady turbulent flow, suspended sediment can be modelled using the Reynolds-averaged mass balance equation and the appropriate boundary conditions (Garcia, 2008):

$$\frac{\partial \bar{c}}{\partial t} + \frac{\partial F_i}{\partial x_i} = 0, \quad \text{where} \quad F_i = (u_i - V\delta_{i3})\,\bar{c} + \overline{u_i'c'}. \tag{1}$$

Here $\bar{c}$ is the mean (averaged over turbulence) volumetric concentration of suspended sediment, c' is instantaneous fluctuation of sediment concentration, u_i is the mean fluid velocity, u_i' is turbulent fluctuations, F_i is the Reynolds-averaged suspended sediment flux, V is particle settling velocity in quiescent water and δ_{i3} is the Kronecker delta. With the assumption of the river/stream flowing at a steady state and being confined in a wide channel, Eq. (1) reduces to (Garcia, 2008)

$$\frac{d}{dz}\left(\overline{w'c'} - V\bar{c}\right) = 0. \tag{2}$$

Under typical conditions prevailing in most streams and rivers, the suspended sediment can be safely assumed to be in equilibrium; and combining it with the boundary conditions at free surface, Eq. (2) further reduces to $\overline{w'c'} - V\bar{c} = 0$. The eddy-diffusivity assumption can be used to model $\overline{w'c'}$; the resulting relationship has been widely used for modelling transport of suspended sediment (Rouse, 1937; Vanoni, 1946):

$$K_z \frac{d\bar{c}}{dz} + V\bar{c} = 0, \tag{3}$$

where K_z is the vertical sediment diffusivity due to turbulence mixing. Success of the above model depends on the correct estimation of the sediment diffusivity coefficient. Using Prandtl's analogy, and assuming that the logarithmic velocity profile holds for the full depth of the flow, Rouse (1937) derived a formula for K_z

$$\frac{K_z}{Hu_*} = \kappa \frac{z}{H}\left(1 - \frac{z}{H}\right). \tag{4}$$

In the above equation, H is depth of the flow, κ is the von Karman constant, u_* is the bed shear velocity and z is the normal distance from the bed. Even though Prandtl's analogy might not perfectly hold under all circumstances, the above relation (also known as the Rousian formulation for vertical eddy viscosity) has been used extensively in the field of suspended sediment transport. One of the first studies to question the universal applicability of the Rousian formulation was Coleman (1970); he used suspended sediment measurements from lab experiments and field measurements to calculate K_z/Hu_* for sands with different values of V/u_*. Rearranging Eq. (3) and dividing both sides by Hu_* gives us the formula used for calculating K_z/Hu_*:

$$\frac{K_z}{Hu_*} = -\frac{\bar{\bar{c}}\tilde{V}}{d\bar{\bar{c}}/d\tilde{z}}. \tag{5}$$

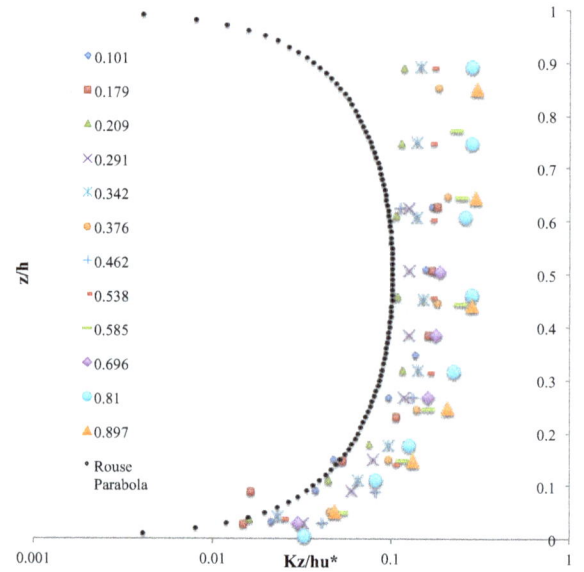

Figure 1. Vertical sediment diffusivity K_z/Hu_* profiles for sediments with different $\tilde{V} = V/u_*$. The data has been reproduced from calculations done by Coleman (1970) on field data of Anderson (1942). The generic Rousian profile of kinematic eddy viscosity has also been plotted. For most cases the Rousian profile underestimates sediment diffusivity. There is a trend that vertical sediment diffusivity increases with increase in $\tilde{V}$; but the trend is not obvious for some of the cases plotted above (e.g. between 0.585 and 0.696).

In the above equation $\tilde{V}$ is V/u_*, $\tilde{z}$ is z/H and $\bar{\bar{c}}$ is the mean volumetric suspended sediment concentration normalized by volume-averaged concentration $c^{(v)}$. In his study Coleman (1970) used field data of Anderson (1942) to calculate K_z/Hu_*, and the same data has been reproduced here (Fig. 1) along with the Rousian profile calculated using Eq. (4). It can be observed in Fig. 1 that K_z/Hu_* is parabolic only in the lower portion of the domain; also for most cases the Rousian profile underestimates vertical sediment diffusivity. Van Rijn (1984) put forward the idea that the ratio of sediment diffusivity and kinematic eddy diffusivity is always greater than 1 and suggested the use of an empirical coefficient to adjust kinematic eddy diffusivity to match the vertical sediment diffusivity. However, Bennett et al. (1998) attributed the disparity to the use of suspended sediment concentration profiles to calculate sediment diffusivity, and instead recommended the use of direct turbulence measurements. In general, the common consensus has been that the Rousian profile is not an appropriate surrogate for vertical sediment diffusivity.

The Rousian profile, though a very good first approximation, does not capture all the ingrained physics present in the interaction of suspended sediment and the ambient fluid. One of the first one to point out the breakdown of Prandtl's analogy for real sediments was Rouse (1938) through his classic jar experiments. Among the different contributing mechanisms that influence the break down of Prandtl's analogy, one of the most prominent one is the inertial effect of relatively large sediment particles (Nielsen, 1992). Further, Nielsen and Teakle (2004) used the finite-mixing-length theory to justify their interpretation of Coleman's (1970) data, in which they point out that vertical sediment diffusivity increases with increase in V/u_* (dubbed as the Rouse number, though exact definition of the Rouse number is $V/\kappa u_*$). In Fig. 1 data from Coleman (1970) has been reproduced; an obvious trend emerges where K_z/Hu_* for cases with higher V/u_* is higher than those with lower V/u_*. But the aforementioned trend is not universal, and there are cases where the sediment with relatively lower V/u_* has higher or almost equal sediment diffusivity when compared with sediment with relatively higher V/u_* (e.g. compare the cases with V/u_* of 0.585 and 0.696). This may be an artefact of an unrecognized competing mechanism that tends to reduce the vertical sediment diffusivity with increase in V/u_*.

The hypothesis is that the aforementioned anomaly can be explained if the effect of self-stratification due to suspended sediment on sediment diffusivity is accounted. The settling sediment particles form a continuous concentration profile, with higher concentration near the bottom and lower at the top. This concentration gradient causes stratification in the fluid, and as the suspended sediment particles themselves cause stratification, the phenomenon is also referred to as self-stratification. This concentration gradient is known to modulate turbulence and affect bulk properties of the flow (Cantero et al., 2009b, 2012; Shringarpure et al., 2012). Wright and Parker (2004) showed the importance of sediment-induced stratification in large low-gradient streams/rivers. Smith and McLean (1977) and McLean (1992), among others, proposed the use of simple algebraic closures based on the gradient Richardson number (Ri_g) to take into account the effect of self-stratification on the Rousian profile. Through laboratory experiments, Cellino and Graf (1999) reported the suppression of turbulence due to presence of suspended sediment. In their experiments, they found the estimated momentum and sediment diffusivity to be smaller than the theoretically predicated value (the Rousian profile). Cellino and Graf (2000) also studied the effect of bed-forms on vertical sediment and momentum diffusivity, and found that presence of bed-forms increases the ratio of vertical sediment and momentum diffusivity from less than 1.0 to greater than 1.0 (Graf and Cellino, 2002).

The aim of the present study is to explore the effect of self-stratification on vertical sediment diffusivity under two different configurations: first for channel flows, which is an analogue for flow in a pipe or a duct; and second for a boundary-layer configuration, which is similar to an open-channel flow. For the first portion of the study, we have used steady-state sediment concentration profiles from direct numerical simulations (DNS) of sediment-laden flows. For the DNS, sediment has been modelled using an Eulerian approach with the assumption that the sediment particles do not have any inertia. Though this is not true for larger sediment in nature, it is a good assumption for fine sediment. It was also done this way to explore the effect of self-stratification without other mechanisms (like inertial effects; see for example Cantero et al., 2008) coming into play. DNS was done for a constant shear Reynolds number (Re_τ) but for different levels of self-stratification, which depend on the sediment particle settling velocity (parameterized using $\tilde{V} = V/u_*$) and volume-averaged suspended sediment concentration (parameterized using shear Richardson number Ri_τ). Traditionally, sediment diffusivity under different circumstances has primarily been studied for the open-channel-like configuration; so the present study also explores it in the channel flow setting. Apart from using data from DNS, data from experiments by Ismail (1952) and Coleman (1986) have been used to study the effect of stratification on sediment diffusivity. The aim of the current study is to extend our understanding of the effects of self-stratification on sediment diffusivity in channel and open-channel-like flows.

2 Mathematical formulation

DNS were conducted for a horizontal channel, where the flow is driven by a constant pressure gradient. The constant pressure gradient here is a surrogate for a constant slope in a stream/river that drives the flow, especially for the open-channel-like configuration. Suspended sediment particles are assumed to be of constant size, negligible inertia, and having a constant settling velocity $\tilde{V}$. Eulerian representation has been used to represent the suspended sediment particles, and this has been found to be valid for sediment particles that are small enough (Ferry and Balachandar, 2001). The flow is assumed to be dilute enough that the Boussinesq approximation holds. The set of dimensionless equations used to model the flow is

$$\frac{\partial \tilde{u}_i}{\partial \tilde{t}} + \tilde{u}_j \frac{\partial \tilde{u}_i}{\partial \tilde{x}_j} = \tilde{G}\delta_{i1} - \frac{\partial \hat{p}}{\partial \tilde{x}_i} + \frac{1}{Re_\tau} \frac{\partial^2 \tilde{u}_i}{\partial \tilde{x}_j \partial \tilde{x}_j} - Ri_\tau \tilde{c}\delta_{i3}, \quad (6a)$$

$$\frac{\partial \tilde{u}_i}{\partial \tilde{x}_i} = 0, \quad (6b)$$

$$\frac{\partial \tilde{c}}{\partial \tilde{t}} + \left(\tilde{u}_j - \tilde{V}\delta_{i3}\right)\frac{\partial \tilde{c}}{\partial \tilde{x}_j} = \frac{1}{Sc\,Re_\tau}\frac{\partial^2 \tilde{c}}{\partial \tilde{x}_j \partial \tilde{x}_j}. \quad (6c)$$

In the above equations, $\tilde{u}_i$ is the velocity of the fluid phase, $\tilde{c}$ is the volumetric concentration of the suspended sediment particles. $\tilde{G}$ is the constant streamwise mean pressure gradient driving the flow and has a magnitude equal to 1 and $\hat{p}$ is the pressure field, which is the combination of the dynamic pressure ($\tilde{p}$) and the hydrostatic component due to the suspended sediment. The mathematical formulation used in the present study is exactly the same as the one used by Cantero et al. (2009b) in their study of turbulence modulation due to self-stratification, and additional details about the model can be found in Cantero et al. (2009b). Sediment has been modelled under the Eulerian framework using the advection–diffusion equation (see Eq. 6c). The diffusion term in Eq. (6c) might look out of place, but it serves multiple purposes. Even though the sediment particles are assumed to be big enough that their Brownian motion can be ignored, it is well established that relatively large particles can also diffuse due to long-range hydrodynamic interactions (Mucha and Brenner, 2003) and the diffusive term takes into account the aforementioned mechanism. The diffusion term also provides a way to resuspend sediment from the bed (Garcia and Parker, 1993), while providing numerical stability (Cantero et al., 2009a) to the simulation.

In the above set of equations, all the variables are dimensionless. Velocity has been made dimensionless using average shear velocity (u_*); the parameter used for scaling length is the channel half-height h (where $2h$ is the height of the channel) and the parameter used for scaling pressure is $\rho_f u_*^2$, where ρ_f is ambient fluid density. Equation (6) has four dimensionless numbers, which together define various properties of the flow; shear Reynolds number (Re_τ), shear Richardson number (Ri_τ), Schmidt number (Sc) and the non-dimensional particle fall velocity ($\tilde{V}$). These non-dimensional numbers are defined as

$$Re_\tau = \tfrac{u_* h}{\nu} \quad Ri_\tau = \tfrac{g R c^{(v)} h}{u_*^2},$$
$$Sc = \tfrac{\nu}{K_s} \quad \tilde{V} = \tfrac{V}{u_*}, \tag{7}$$

where ν is the kinematic viscosity, g is acceleration due to gravity, K_s is diffusivity of the sediment particles (this diffusion of sediments arise from their long-range hydrodynamic interaction; see for example Segre et al., 2001), $c^{(v)}$ is the volume-averaged concentration, $R = \rho_s/\rho_f - 1$ and ρ_s is the density of the sediment particles. In the current study all the DNSs were done for $Re_\tau = 180$. The shear Reynolds number of the flow was kept constant, as the aim of the study was to understand the effect of self-stratification when the flow remains the same. The shear Richardson number (Ri_τ) is used to parameterize the initial volume-averaged suspended sediment concentration ($c^{(v)}$), and Ri_τ has been found to play an important role in influencing the final degree of self-stratification (Dutta, 2012). $\tilde{V}$ has also been found to influence the degree of self-stratification by defining the sediment concentration profile among different cases having constant Ri_τ (initial sediment concentration) and Re_τ (Cantero et al., 2009b). Dutta (2012) showed that both $\tilde{V}$ and Ri_τ have an effect on the final degree of self-stratification, so in the current study both $\tilde{V}$ and Ri_τ have been varied to impose different levels of self-stratification. Based on observations made in previous studies (Cantero et al., 2009b), in the present study the Schmidt number (Sc) has been kept equal to 1.

The above-stated governing equations were solved using a dealiased pseudo-spectral code. The setup is exactly the same as the one used by Cantero et al. (2009b), so further details of the exact numerical methods adopted can be found there. Dimensions of the rectangular domain used for the numerical simulations were $\tilde{L}_x = 4\pi$, $\tilde{L}_y = 4\pi/3$ and $\tilde{L}_z = 2$, and the domain was discretized using a computational grid

having $N_x = 96$, $N_y = 96$ and $N_z = 97$ nodes in the x, y and z (wall-normal) directions respectively. The grid size is uniform in the longitudinal and transverse directions, and in terms of wall units ($\tilde{z}^+$) they are 23.562 and 7.854 respectively. For the wall-normal direction a Chebyshev expansion with Gauss–Lobatto quadrature points has been used. This allows for a very high resolution near the boundaries and relatively lower resolution at the centre of the domain. In terms of wall units, the distance between two nodes is 0.0964 for the nodes near the wall and 5.889 at the centre of the domain. Cantero et al. (2009b) had found the aforementioned computation-grid resolution sufficient for capturing all the relevant flow statistics. Periodic boundary conditions were used in the longitudinal and transverse directions. The top and bottom walls of the domain were assumed to be smooth; and depending on the configuration simulated, a no-slip or slip boundary condition was imposed on the fluid phase at the top wall. At the bottom wall a no-slip condition was employed for all simulations. Sediment particles were assumed to be fine enough to have zero net deposition; thus a boundary condition was imposed, which instantly re-entrains all settled sediment particles. For the channel flow configuration, the imposed boundary conditions are mathematically represented as

$$\tilde{u}_i = 0 \text{ at } \tilde{z} = -1 \text{ and } \tilde{z} = 1, \tag{8a}$$

$$\tilde{c}\tilde{V} + \frac{1}{Re_\tau Sc}\frac{\partial \tilde{c}}{\partial \tilde{z}} = 0 \text{ at } \tilde{z} = -1 \text{ and } \tilde{z} = 1. \tag{8b}$$

For the open-channel-like configuration (boundary-layer configuration), the imposed boundary conditions are

$$\tilde{u}_i = 0 \text{ at } \tilde{z} = -1 \text{ and } \frac{\partial \tilde{u}}{\partial \tilde{z}} = \frac{\partial \tilde{v}}{\partial \tilde{z}} = \tilde{w} = 0 \text{ at } \tilde{z} = 1, \tag{9a}$$

$$\tilde{c}\tilde{V} + \frac{1}{Re_\tau Sc}\frac{\partial \tilde{c}}{\partial \tilde{z}} = 0 \text{ at } \tilde{z} = -1 \text{ and } \tilde{z} = 1. \tag{9b}$$

The boundary condition imposed for suspended sediment allows the net amount of sediment in suspension to remain constant throughout the simulation. When integrated over time, the aforementioned condition allows the flow to reach a statistically steady state (Cantero et al., 2009b).

3 Results

Sixteen DNS simulations were run for the present study. They were all run for the same shear Reynolds number of 180, but different particle fall velocities ($\tilde{V}$) and shear Richardson number (Ri_τ). All the simulated cases have been listed in Table 1. The simulations can be broadly divided into two parts: twelve that were done with the channel flow configurations and four done with the boundary-layer configuration. The sediment concentration profiles obtained from the simulations were used in conjunction with Eq. (5) to obtain sediment diffusivity profiles (K_z). In order to quantify the

Table 1. The table lists all the cases of direct numerical simulations used in the current study. All the simulations have the same Re_τ. Cases 1–12 correspond to the simulations for channel configuration and 13–16 correspond to the simulations for open-channel-like configuration. Case 1 corresponds to the case with no sediment in suspension, and was simulated to compare with the self-stratified cases.

Case	Re_τ	$\tilde{V} = V/u_*$	Ri_τ	Configuration
1	180	0	0	channel
2	180	0.005	18	channel
3	180	0.01	18	channel
4	180	0.015	18	channel
5	180	0.02	18	channel
6	180	0.025	18	channel
7	180	0.03	18	channel
8	180	0.025	1	channel
9	180	0.025	10	channel
10	180	0.025	15	channel
11	180	0.025	20	channel
12	180	0.025	22	channel
13	180	0.025	1	B. layer
14	180	0.025	10	B. layer
15	180	0.025	15	B. layer
16	180	0.025	18	B. layer

Table 2. The table lists all the cases of direct numerical simulations used in the current study along with $\tilde{V}Ri_\tau$ for each of the cases. $\tilde{V}Ri_\tau$ is a parameter which represents the level of self-stratification; and a higher value of $\tilde{V}Ri_\tau$ corresponds to a higher level self-stratification if Re_τ remains constant. Cases 1–12 correspond to the simulations for channel configuration and 13–16 correspond to the simulations for open-channel-like configuration. The corresponding sediment diffusivity profiles were quantified using Eq. (10), and the calculated parameters $K_{z\mu}$, $K_{z\sigma}$, $K_{z\gamma}$ have also been listed in the table. $K_{z\mu}$ parameterizes the mean sediment diffusivity, $K_{z\sigma}$ parameterizes the variance in the sediment diffusivity profile and $K_{z\gamma}$ parameterizes the skewness of the sediment diffusivity profile.

Case	$\tilde{V}Ri_\tau$	$K_{z\mu}$	$K_{z\sigma}$	$K_{z\gamma}$
2	0.090	0.0607	0.0231	−0.9897
3	0.180	0.0485	0.0212	−0.3191
4	0.270	0.0412	0.0231	0.3853
5	0.360	0.0426	0.0310	0.6639
6	0.450	0.0602	0.0516	0.3016
7	0.540	0.0610	0.0523	0.2976
8	0.025	0.0724	0.0272	−1.3216
9	0.250	0.0452	0.0272	0.5594
10	0.375	0.0618	0.0525	0.2931
11	0.500	0.0592	0.0511	0.31162
12	0.550	0.0582	0.0507	0.3255
13	0.025	0.1927	0.0882	−0.7991
14	0.25	0.1430	0.0716	−0.5649
15	0.375	0.1278	0.0676	−0.4417
16	0.450	0.1204	0.0653	−0.3882

sediment diffusivity profiles, three different parameters were defined and calculated for each of the sediment diffusivity profiles. The parameters (also referred to as *shape factors*) are mean sediment diffusivity $K_{z\mu}$, variance within the sediment diffusivity profile $K_{z\sigma}$ and skewness of the sediment diffusivity profile $K_{z\gamma}$. Three parameters were used in order to take care of ambiguity that may arise due to two different profiles having almost the same mean or/and variance. The three parameters have been defined below.

$$K_{z\mu} = \int_0^H \frac{K_z(z)}{H}\,\mathrm{d}z \tag{10a}$$

$$K_{z\sigma} = \left(\int_0^H \frac{(K_z - K_{z\mu})^2}{H}\,\mathrm{d}z \right)^{1/2} \tag{10b}$$

$$K_{z\gamma} = \frac{\int_0^H \frac{(K_z - K_{z\mu})^3}{H}\,\mathrm{d}z}{\left(\int_0^H \frac{(K_z - K_{z\mu})^2}{H}\,\mathrm{d}z \right)^{3/2}} \tag{10c}$$

In the equations defined above, H corresponds to the total depth of the flow.

3.1 Channel flow configuration

Among the twelve simulations done for the channel flow configuration, the first one was done without any suspended sed-

iment. A set of simulations (*set1*) was done for constant shear Richardson number but increasing $\tilde{V}$. This set is equivalent to the situation where the initial volume-averaged suspended sediment concentration is constant but the sediment particle size increases. The other set of simulations (*set2*) done for the same configuration is with a constant $\tilde{V}$ equal to 0.025 but increasing shear Richardson number. This set is equivalent to the situation where the particle size of sediment is constant but the initial volume-averaged suspended sediment concentration increases. For both *set1* and *set2* increase in $\tilde{V}$ or Ri_τ while keeping the other parameters constant results in an increase in the degree of stratification. Dutta (2012) showed that increase in either $\tilde{V}$ or Ri_τ increases the degree of self-stratification caused by suspended sediment and the extent to which a flow will stratify depends on the parameter $\tilde{V}Ri_\tau$. Cantero et al. (2012) made similar observations for turbidity currents.

In Figs. 2 and 3 the mean streamwise velocity, steady-state sediment concentration profile and normalized wall-normal turbulence intensity for *set1* and *set2* have been plotted. For both sets, the increase in degree of self-stratification leads to increase in bulk streamwise velocity of the flow. The flow also becomes asymmetric, with the velocity maximum getting skewed towards the channel bottom. One of the effects of the self-stratification is reduction of the bottom drag. The

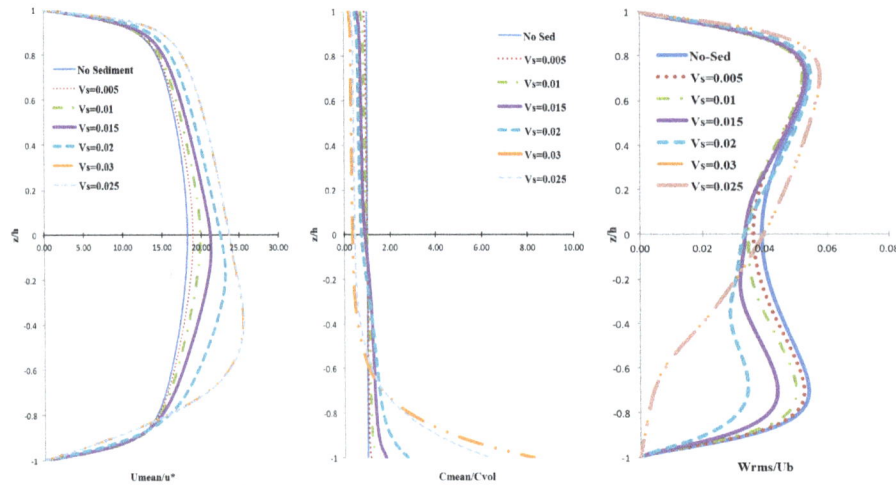

Figure 2. Results from the DNSs in a channel flow setting, for increasing $\tilde{V}$ and $Ri_\tau = 18$. The mean streamwise velocity and asymmetry of the flow increase with increase in $\tilde{V}$. Increase in $\tilde{V}$ increases the degree of self-stratification of the flow; this leads to increase in the sediment concentration gradient and higher amount of turbulence damping near the channel bottom. In the channel, normalized turbulence intensity (W_{rms}/U_b) is modulated, with a decrease in the lower half of the channel and slight increase in the upper half of the channel.

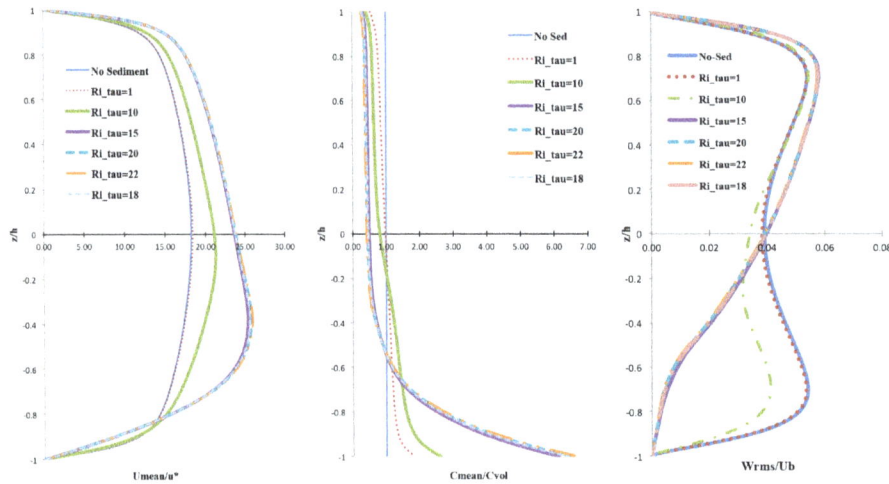

Figure 3. Results from the DNSs in a channel flow setting, for increasing Ri_τ and $\tilde{V} = 0.025$. The mean streamwise velocity and asymmetry of the flow increase with increase in Ri_τ. Increase in Ri_τ increases the degree of self-stratification of the flow; this leads to increase in the sediment concentration gradient and higher amount of turbulence damping near the channel bottom. In the channel, normalized turbulence intensity (W_{rms}/U_b) is modulated, with a decrease in the lower half of the channel and slight increase in the upper half of the channel.

decrease in drag results in an increase in the flow discharge in the channel, as the force driving the flow remains the same (the constant unit pressure gradient $\tilde{G}$). Even though the bulk streamwise velocity increases, turbulence intensity (W_{rms}) in the channel decreases, especially near the channel bottom. The reduction in bottom drag is connected to the reduction of turbulence activity near the bottom of the channel. Reduced turbulence activity in turn reduces the Reynolds stress, thus eventually reducing the total drag at the bottom. Flow in the channel at different levels of self-stratification can be clearly divided into two regimes: first regime (for $\tilde{V} Ri_\tau \leq 0.36$) in which the turbulence near the bottom of the channel is

damped but the flow in general is still turbulent. Second regime (for $\tilde{V} Ri_\tau \geq 0.45$) in which turbulence near the bottom of the channel is almost completely suppressed, but turbulence intensity in the upper half of the channel is slightly more than the case with no suspended sediment. Turbulent activity near the top wall is maintained due to lack of stratification in that region. The small increase in turbulent activity is due to the aforementioned increase in flow discharge. The steady-state sediment concentration profiles in Figs. 2 and 3 were used with Eq. (5) to calculate vertical sediment diffusivity profiles (Fig. 4) for all the cases in *set1* and *set2*. Sediment diffusivity profiles reflect the observable trends of the turbulence intensity profiles in Figs. 2 and 3. The aforementioned similarity is along the expected lines because the exact definition of vertical sediment diffusivity is nothing but a surrogate used to model the vertical sediment flux due to turbulence $(\overline{w'c'})$. So wherever turbulence is damped, K_z/Hu_* decreases and wherever turbulence increases K_z/Hu_* also increases. The effect of self-stratification on sediment diffusivity is similar to the effect it has on turbulent intensity. The calculated sediment diffusivity profiles (Fig. 4) were quantified using Eq. (10a–c), and the parameters $K_{z\mu}$, $K_{z\sigma}$ and $K_{z\gamma}$ have been listed in Table 2. The mean sediment diffusivity shows a counterintuitive trend, initially mean diffusivity ($K_{z\mu}$) decreases with increase in stratification ($\tilde{V}Ri_\tau$) but then it suddenly increases and then starts to decrease again (see Fig. 5). The sudden increase is associated with increase of sediment diffusivity in the upper part of the domain. The presence of a top wall which continues to pump turbulent energy into the system, even after the flow at the bottom has relaminarized, is responsible for the variation in the expected trend of decrease of mean sediment diffusivity with increase in stratification, which can be observed in Fig. 5 for the case of open-channel flow. $K_{z\mu}$ for the open-channel flow cases was obtained from DNS simulations that are discussed in the next section. The above-mentioned trend is also reflected in the parameter describing the variation of sediment diffusivity within each profile ($K_{z\sigma}$). The trends reflected by all the parameters clearly show the presence of two distinct regimes in the flow. Additionally, the normalized sediment concentration profiles of the simulated cases were plotted with the corresponding analytically obtained suspended sediment concentration profiles. The sediment concentration profiles were normalized using a reference concentration at a height of $b = 0.05H$, where H is the total depth of the channel. For the boundary-layer (open-channel-like) configuration, which will be discussed in the next section, the analytical sediment concentration is given by the well-known Rouse–Vanoni–Ippen suspended sediment distribution (Garcia, 2008):

$$\frac{\bar{c}}{\bar{c}_b} = \left[\frac{(H-z)/z}{(H-b)/b}\right]^{V/u^*\kappa}, \tag{11}$$

where $\bar{c}_b$ is the reference sediment concentration that is the sediment concentration at a height b, and b as defined above. H is the total depth of the flow and rest of the parameters are as previously defined. For the channel flow configuration, a similar relationship was not available. Thus a relationship for suspended sediment concentration in a channel was derived using Eq. (5), an approximate relationship for eddy viscosity in a pressure-driven-channel flow, and the appropriate boundary conditions (Schlichting and Gersten, 2000). The derived

relationship for sediment concentration is

$$\frac{\bar{c}}{\bar{c}_b} = \exp\left[-\frac{V}{u^*\kappa}\left\{\ln\left(\frac{z/(z-2H)}{b/(b-2H)}\right) + \sqrt{8}\right.\right. \tag{12}$$
$$\left.\left.\left(\tan^{-1}\left(\frac{\sqrt{2}(z-H)}{H}\right) - \tan^{-1}\left(\frac{\sqrt{2}(b-H)}{H}\right)\right)\right\}\right].$$

In the above relationship H is the channel half depth and the rest of the parameters are as earlier. Equation (12) was used along with the sediment concentration profiles from the DNS simulations to plot Fig. 6. The normalized sediment concentration profile predicated by Eq. (12) was found to match very well with the DNS-predicted normalized sediment concentration profiles for the cases having relatively small stratification ($\tilde{V}Ri_\tau < 0.1$). The performance of Eq. (12) deteriorated with increase in $\tilde{V}Ri_\tau$, so it clearly shows that the relationship is not able to accommodate the effect self-stratification might have on the suspended sediment concentration profiles, especially the one induced by increase in the concentration of suspended sediment.

3.2 Boundary-layer configuration

Four numerical simulations were done for the boundary-layer configuration. The boundary-layer configuration is similar to the open-channel configuration but not exactly the same. Like the open-channel configuration, a slip boundary condition is imposed at the top wall for the fluid phase. In Fig. 7, bulk streamwise velocity, steady-state sediment concentration profile and normalized turbulence intensity have been plotted. The four simulations for the open-channel-like configuration have the same particle fall velocity ($\tilde{V}$) but increasing shear Richardson number. Similar to the channel flow configuration, bulk streamwise velocity was found to increase with increase in shear Richardson number. Turbulence intensity was found to decrease with increase in shear Richardson number, and unlike the channel flow configuration where turbulence intensity decreases in the lower half of the channel and increases in the upper half; turbulence intensity was found to decrease throughout the channel. Though, the extent of damping in the upper half was found to be slightly more than the extent of damping in the lower half of the boundary layer. The steady-state sediment concentration profiles in Fig. 7 were used to calculate vertical sediment diffusivity profiles (Fig. 8) for the four cases simulated. Reflecting the trend shown by turbulence intensity, vertical sediment diffusivity was found to decrease with increase in level of self-stratification. The sediment diffusivity profiles were quantified using Eq. (10) and the parameters $K_{z\mu}$, $K_{z\sigma}$ and $K_{z\gamma}$ have been listed in Table 2. Unlike the channel flow cases where the trend was counterintuitive, the mean sediment diffusivity ($K_{z\mu}$) and variation ($K_{z\sigma}$) decreases with increase in stratification (see Fig. 5). Additionally, the normalized sediment concentration profiles of the simulated cases were plotted (see Fig. 8)

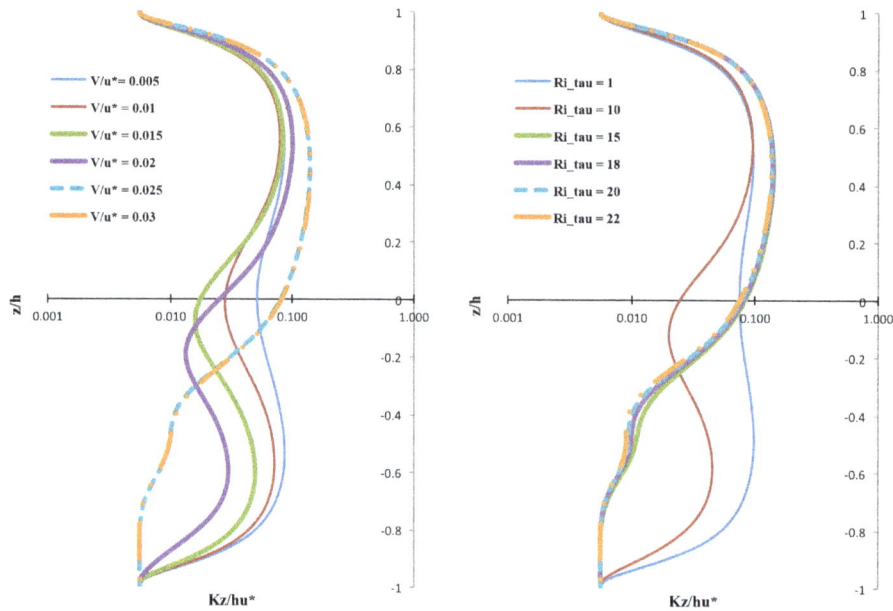

Figure 4. The trend for vertical sediment diffusivity mirrors the trend found for turbulence intensity. An increase in degree of self-stratification is found to decrease sediment diffusivity in the lower half of the channel and increase sediment diffusivity in the upper half of the channel. This is not completely unexpected because mixing of suspended sediment is primarily dependent on turbulence in the flow.

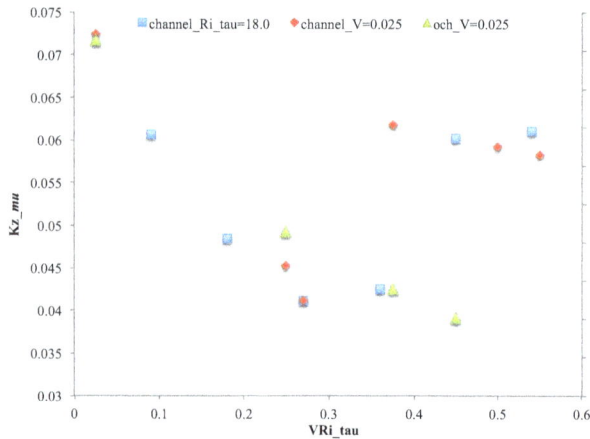

Figure 5. Mean sediment diffusivity ($K_{z\mu}$) calculated from all the DNS simulations has been plotted against the corresponding level of stratification ($\tilde{V} Ri_\tau$). $K_{z\mu}$ for the boundary-layer cases (open-channel-like configuration) has been plotted on the secondary y axis. $K_{z\mu}$ for the open-channel cases decrease monotonically with increase in stratification ($\tilde{V} Ri_\tau$). The aforementioned trend is not reflected in the channel flow configuration. $K_{z\mu}$ for the channel flow configuration has been plotted on the primary y axis, and show a counterintuitive trend of first decreasing, then increasing and then again decreasing with increase in $\tilde{V} Ri_\tau$.

with the corresponding analytically obtained suspended sediment concentration profile using the Rouse–Vanoni–Ippen relationship (Eq. 11). The normalized sediment concentration profile predicted by the Rouse–Vanoni–Ippen relation-

ship was found to match well with the DNS simulations for relatively lower stratification, and the difference increased appreciably with increase in stratification. In the next section the larger implication of the results presented in the previous sections will be discussed. Additionally, the trends observed in the DNS results will be compared with data from lab experiments and field measurements.

4   Discussion

In the preceding sections we saw how self-stratification due to suspended sediments can reduce sediment diffusivity in a flow. All the discussed results were on the basis of high-resolution numerical simulations, which were set up to exclusively capture the effect of self-stratification on sediment diffusivity. In order to vet our hypothesis further, we have compared our results with experimental observations of Ismail (1952) and Coleman (1986).

Ismail (1952) conducted a series of experiments in a closed rectangular channel. The aim of the experiments was to understand the transfer mechanism of turbulence and its interaction with suspended sediment. The rectangular closed-channel setup is similar to the DNS simulations for the channel flow configuration. Cases 74, 75, 76 and 78 from Ismail (1952) have been used in the current study. All the cases for which suspended sediment concentration profiles were available in Ismail (1952) had dunes at the bottom of the channel, whereas in the DNS simulations the walls were smooth. The presence of the dunes plays an important role, as they can enhance turbulence near the bottom of the channel

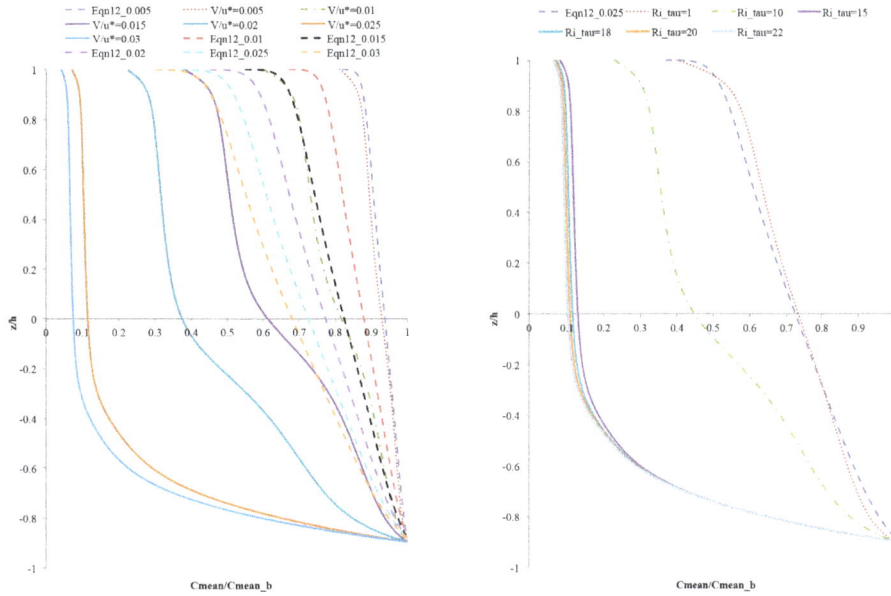

Figure 6. Normalized sediment concentration profiles from DNS simulations of the channel flow have been plotted with the sediment profile obtained using Eq. (12). The results from the DNS simulations matched very well with the derived relationship (Eq. 12) for relatively lower stratification ($\tilde{V} Ri_\tau < 0.1$).

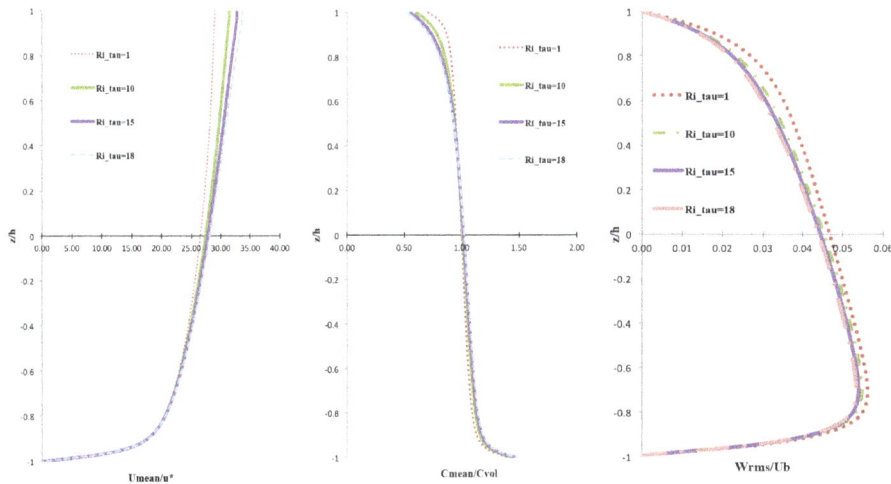

Figure 7. Results from the DNS simulations in a boundary-layer (open-channel-like) setting, for increasing Ri_τ and $\tilde{V} = 0.025$. Mean streamwise velocity of the flow increases with increase in Ri_τ. An increase in Ri_τ increases the degree of self-stratification of the flow; this leads to an increase in the sediment concentration gradient and higher amount of turbulence damping through out the domain. Normalized turbulence intensity (W_{rms}/U_b) is damped throughout the boundary layer but the level of suppression is slightly higher in the upper half.

(Ismail, 1952; Cellino and Graf, 2000). But as the comparison between the numerical and the experimental results is strictly qualitative, the effect of the dunes has been neglected. Mean streamwise velocity profiles for cases with (case 117) and without sediment (case 5) have also been reproduced from Ismail (1952), in order to compare the effect suspended sediment has on mean streamwise velocity. Figure 9a reflects the trend shown for streamwise velocity in the DNS results. Compared with the case without sediment, we clearly see in the case with sediment that streamwise velocity in the upper half of the channel increases whereas in the lower half decreases. Suspended sediment concentration profiles for cases 74–76 and 78 were used to calculate the sediment diffusivity profiles (Fig. 9b). All the cases used in the present study have been listed in Table 3; along with their corresponding shear Reynolds number, Ri_τ, $\tilde{V}$ and $\tilde{V} Ri_\tau$. Amongst the cases plotted in Fig. 9, case 74 has the highest level of self-stratification ($\tilde{V} Ri_\tau$) and case 78 the lowest. And moving from case 78 to 74, vertical sediment diffusivity decreases in the lower half of the channel and increases in upper half of the channel.

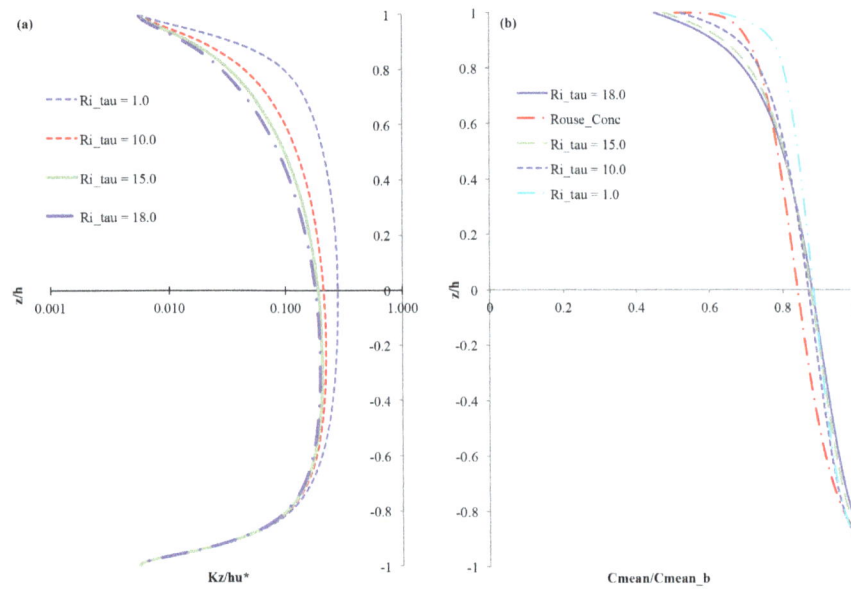

Figure 8. In (**a**), the trend for vertical sediment diffusivity mirrors the trend found for turbulence intensity. An increase in the degree of self-stratification is found to decrease sediment diffusivity in the boundary layer. The extent to which sediment diffusivity decreases in the upper half of the boundary layer is slightly higher than the lower half of the boundary layer. This is not completely unexpected because mixing of suspended sediment is primarily dependent on turbulence in the flow. In (**b**), normalized sediment concentration profiles from DNS simulations have been plotted along with the sediment profile obtained using the Rouse–Vanoni–Ippen relationship.

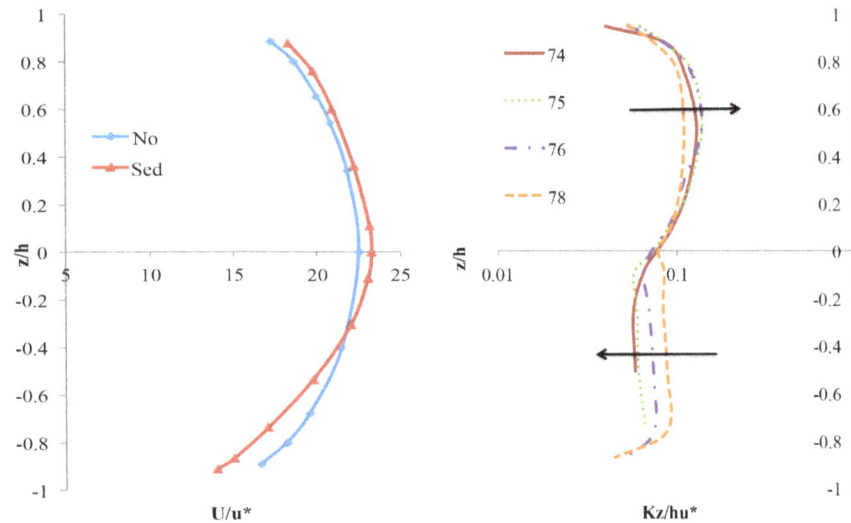

Figure 9. Streamwise velocity for cases 5 and 117 have been reproduced from Ismail (1952). Case 5 has no sediment in suspension and case 117 has suspended sediment. Vertical sediment diffusivity for cases 74–76, and 78 was calculated from sediment concentration profiles from experiments performed by Ismail (1952). With an increase in $\tilde{V}Ri_\tau$, vertical sediment diffusivity in the channel slightly increases in the top half of the channel and decreases in the lower half of the channel; this completely agrees with the DNS results.

The trend shown by sediment diffusivity calculated from the experimental data of Ismail (1952) concur with the trend observed in the DNS results (Fig. 4). The parameters $K_{z\mu}$, $K_{z\sigma}$ and $K_{z\gamma}$ were calculated using the sediment diffusivity profiles in Fig. 9 and Eq. (10) and have been listed in Table 4. The mean sediment diffusivity ($K_{z\mu}$) showed a trend similar to the one observed for the DNS results.

For comparing the boundary-layer (open-channel-like) case, suspended sediment concentration profiles published by Coleman (1986) were used. Coleman studied the effect of suspended sediment on the velocity distribution of an open-channel flow. The effect of suspended sediment on the streamwise velocity is similar to the effect observed in our DNS results (refer to Fig. 1 in Coleman, 1986). Coleman

Table 3. List of all the experiments of Ismail (1952) used in the present study.

Case	Re_τ	$\tilde{V} = V/u_*$	Ri_τ	$\tilde{V}Ri_\tau$
74	1409	0.25	3.629	0.90723
75	1546	0.197	4.426	0.87192
76	1978	0.176	3.192	0.56179
78	2698	0.133	1.729	0.22996
5	1768	0	0	0
117	2188	0.359	2.926	1.05043

Table 4. List of the parameters $K_{z\mu}$, $K_{z\sigma}$, and $K_{z\gamma}$, calculated from the sediment diffusivity profiles of the cases listed in Table 3.

Case	$\tilde{V}Ri_\tau$	$K_{z\mu}$	$K_{z\sigma}$	$K_{z\gamma}$
74	0.90723	0.0907	0.0276	−0.0819
75	0.87192	0.0893	0.0302	0.4050
76	0.56179	0.0896	0.0257	0.6515
78	0.22996	0.0898	0.0142	−0.9668

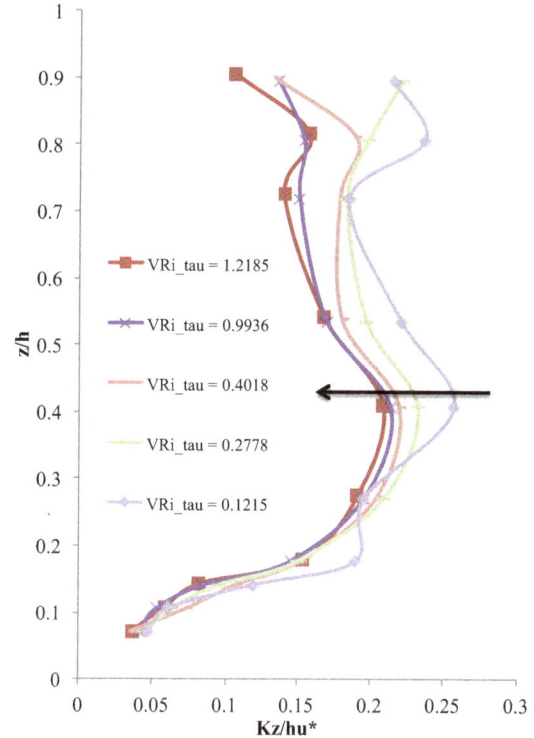

Figure 10. Vertical sediment diffusivity was calculated using sediment concentration profiles from experiments performed by Coleman (1986). With an increase in $\tilde{V}Ri_\tau$ (cases 1–5), the vertical sediment diffusivity in the open-channel flow decreases.

found the presence of suspended sediment slightly decreased the streamwise velocity near the bottom of the channel and slightly increased it in the upper half of the channel. An increase in streamwise velocity in the upper portion of the boundary layer is also consistent with observations by Barenblatt and Golitsyn (1974) for "mature dust storms". For the present study, we only used the suspended sediment concentration profiles for sediment with a diameter (D) of 0.210 mm. Coleman, in his experiments, used sediment of three different mean diameters, but the hydraulic condition (constant slope and discharge) was kept constant. For each sediment size, sediment was added to the flow till the amount of sediment suspended in the flow reached its maximum (also called the capacity condition). So, for a particular sediment size and flow condition, there are suspended sediment profiles for different net sediment concentrations. The sediment concentration profiles were used to calculate vertical sediment diffusivity profiles (Fig. 10). The cases used for the present study have been listed in Table 5, along with the corresponding $\tilde{V}$, Ri_τ and $\tilde{V}Ri_\tau$. $\tilde{V}Ri_\tau$ increases from case 1 to case 5. In Fig. 10 vertical sediment diffusivity decreases with increase in $\tilde{V}Ri_\tau$, especially if the difference between case 1 and case 5 is observed; though it is noticeable that the trend of decreasing sediment diffusivity with increase in $\tilde{V}Ri_\tau$ is not monotonic. This may be attributed to two or more competing mechanisms trying to influence the sediment diffusivity in opposite directions. For example, it is known that depending on the size of the sediment particles in suspension, suspended sediment can increase turbulent kinetic energy (Niño and Garcia, 1998) of the flow, which can then lead to higher sediment diffusivity; whereas increase in self-stratification tends to lower sediment diffusivity. Another factor that might be contributing to this slight incon-

sistency of the trend in the present case is irregularity of the $\tilde{V}$ values for different cases (refer to Table 5). Even though theoretically all the cases should have the same $\tilde{V}$, the experimental observations actually have some inconsistencies (we checked a copy of the actual data set of N. L. Coleman). As change in $\tilde{V}$ has an influence on the mixing length of the fluid (Nielsen and Teakle, 2004), these small inconsistencies in $\tilde{V}$ might be obfuscating the expected trend of decrease of sediment diffusivity with an increase in $\tilde{V}Ri_\tau$. Apart from the above-mentioned issue with the size of sediment in suspension, there are inherent uncertainties in estimation of sediment diffusivity, which stem from calculation of shear velocity u^* and correct measurement of the sediment concentration. The calculated values of u^* in each of the experiments fall within 2.5 % of each other, but this variation was not found to cause any substantial change to the calculated values of sediment diffusivity. The parameters $K_{z\mu}$, $K_{z\sigma}$ and $K_{z\gamma}$ were also calculated (Table 6); as expected, $K_{z\mu}$ was found to decrease with an increase in stratification. Whereas there was no appreciable change in $K_{z\sigma}$ and $K_{z\gamma}$.

A significant point to come out of the preceding discussions is that along with the particle settling velocity ($\tilde{V}$), suspended sediment concentration (Ri_τ) is an important parameter that influences the degree of self-stratification in a sediment suspension. Along with different mechanisms that

Table 5. List of all the experiments of Coleman (1986) used in our study.

$Re_\tau = 3700$		$D = 0.210$ mm	
	Ri_τ	$\tilde{V} = V/u_*$	$\tilde{V}\,Ri_\tau$
1	0.2005	0.606	0.1215
2	0.4583	0.606	0.2778
3	0.6631	0.606	0.4018
4	1.6615	0.598	0.9936
5	2.0140	0.605	1.2185

Table 6. List of the parameters $K_{z\mu}$, $K_{z\sigma}$, and $K_{z\gamma}$, calculated from the sediment diffusivity profiles of the cases listed in Table 5.

$Re_\tau = 3700$		$D = 0.210$ mm	
	$K_{z\mu}$	$K_{z\sigma}$	$K_{z\gamma}$
1	0.2009	0.5404	1.2555
2	0.1862	0.5404	1.2561
3	0.1748	0.5435	1.2565
4	0.1603	0.5403	1.2569
5	0.1560	0.5467	1.2571

increase sediment diffusivity with the increase in $\tilde{V}$ (van Rijn, 1984; Nielsen and Teakle, 2004) of the suspended sediment, the increase in stratification due to suspended sediment decreases sediment diffusivity. And this effect of self-stratification on sediment diffusivity can explain the anomaly in the expected trend of increasing sediment diffusivity with an increase of $\tilde{V}$ in Fig. 1.

5 Conclusions

In the present study sediment concentration profiles from direct numerical simulations of sediment-laden flow through a channel were used, to calculate sediment diffusivity profiles for the channel flow and boundary-layer configuration. The DNS simulations facilitated the evaluation of the effect sediment-induced self-stratification has on sediment diffusivity, without any other competing mechanism playing a role. For the channel flow configuration, an increase in stratification was found to decrease sediment diffusivity in the lower half of the channel and slightly increase sediment diffusivity in the upper half of the channel. For the boundary-layer (open-channel-like) configuration, throughout the channel, sediment diffusivity was found to decrease with the increase in stratification. Though the extent of suppression of sediment diffusivity in the lower half of the boundary layer is appreciably lower than rest of the boundary layer; along the expected lines, the sediment diffusivity profiles reflected the computed turbulence intensity profiles. Observations from the DNS results were vetted against experimental results from Ismail (1952) and Coleman (1986). Sed-

iment diffusivity profiles calculated using concentration profiles from Ismail's closed-channel experiments were found to be consistent with the DNS results. Sediment diffusivity profiles calculated using sediment concentration data from Coleman's (1986) experiments were more or less consistent with the DNS results, but the trend of decrease of sediment diffusivity with increase in stratification was erratic. For the present study only one set of Coleman's (1986) experiments was used; it would be interesting to repeat the calculations using data from the rest of Coleman's (1986) experiments and other similar experiments (Cellino, 1998). Through the current study, sediment-induced stratification has been put forth as a plausible explanation for the inconsistencies seen in the expected trends of sediment diffusivity profiles in Fig. 1. There are still other mechanisms (Niño and Garcia, 1998) whose effects on sediment diffusivity have to be eventually evaluated, because a better understanding of sediment diffusivity and the various factors it depends on will eventually help us to ascertain the suspended load in rivers/streams more accurately. At the end of the day, the interaction between suspended sediment and the ambient fluid is highly non-linear and will require further exploration to reveal more of its secrets.

Acknowledgements. We would like to thank Gary Parker, Carlos Pantano and Tzu-Hao Yeh for all the thought provoking and helpful discussions on the topic. This research was possible thanks to the Chester and Helen Siess Professorship in Civil Engineering at the University of Illinois. The support received by the junior author from a CSE Graduate Student Fellowship and the Ravindar K. and Kavita Kinra Fellowship in Civil and Environmental Engineering is gratefully acknowledged. We would also like to thank the two anonymous reviewers for their cogent and insightful remarks, which have contributed towards improving this work. Finally, we would like to thank handling associate editor Francois Metivier for his help during the whole process.

Edited by: F. Metivier

References

Anderson, A. G.: Distribution of suspended sediment in a natural stream, Trans. Amer. Geophys. Union, 23, 678–683, 1942.

Barenblatt, G. I. and Golitsyn, G. S.: Local structure of mature dust storms, J. Atmos. Sci., 31, 1917–1933, 1974.

Bennett, S. J., Bridge, J. S., and Best, J. L.: The fluid and sediment dynamics of upper-stage plane beds, J. Geophys. Res.-Oceans, 103, 1239–1274, 1998.

Cantero, M. I., Balachandar, S., and Garcia, M. H.: An Eulerian-Eulerian model for gravity currents driven by intertial paritcles, Int. J. Multiphas. Flow, 34, 484–501, 2008.

Cantero, M. I., Balachandar, S., Cantelli, A., Pirmez, C., and Parker, G.: Turbidity current with a roof: Direct numerical simulation of self-stratified turbulent channel flow driven by suspended sediment, J. Geophys. Res., 114, C03008, doi:10.1029/2008JC004978, 2009a.

Cantero, M. I., Balachandar, S., and Parker, G.: Direct Numerical Simulation of stratification effects in a sediment-laden turbulent channel flow, J. Turbul., 10, p. N27, 1–28, 2009b.

Cantero, M. I., Shringarpure, M., and Balachandar, S.: Towards a universal criteria for turbulence suppression in dilute turbidity currents with non-cohesive sediments, Geophys. Res. Lett., 39, L14603, doi:10.1029/2012GL052514, 2012.

Cellino, M.: Experimental study of suspension flow in open channels, Doctoral dissertation 1824, Ecole Polytechnique federale de Lausanne, Switzerland, 1998.

Cellino, M. and Graf, W. H.: Sediment-laden flow in open-channels under non-capacity and capacity conditions, J. Hydraul. Eng., 125, 455–462, 1999.

Cellino, M. and Graf, W. H.: Experiments on suspension flow in open channels with bed forms, J. Hydraul. Res., 38, 289–298, 2000.

Coleman, N. L.: Flume studies of the sediment transfer coefficient, Water Resour. Res., 6, 801–809, 1970.

Coleman, N. L.: Effects of suspended sediment on the open-channel velocity distribution, Water Resour. Res., 22, 1377–1384, 1986.

Dutta, S.: Effect of self-stratification on channel flows and boundary layers: a study using direct numerical simulations, MS thesis, University of Illinois at Urbana-Champaign, USA, 2012.

Ferry, J. and Balachandar, S.: A fast Eulerian method for disperse two-phase flow, Int. J. Multiphas. Flow, 27, 1199–1226, 2001.

Garcia, M. H. (Ed.): Manual 110. Sedimentation Engineering: processes, measurements, modelling and practice, ASCE, 46, Reston VA, USA, 2008.

Garcia, M. H. and Parker, G: Experiments on the entrainment of sediment into suspension by a dense bottom current, J. Geophys. Res., 98, 4793–4807, 1993.

Graf, W. H. and Cellino, M: Suspension flows in open channels; experimental study, J. Hydraul. Res., 40, 435–447, 2002.

Ismail, H. M.: Turbulent transfer mechanisms and suspended sediment in closed channels, T. Am. Soc. Civ. Eng., 117, 409–434, 1952.

McLean, S. R.: On the calculation of suspended load for non-cohesive sediments, J. Geophys. Res., 99, 5759–5770, 1992.

Mucha, P. J. and Brenner, M. P.: Diffusivities and front propagation in sedimentation, Phys. Fluids, 15, 1305–1313, 2003.

Nielsen, P.: Coastal bottom boundary layer and sediment transport, World Scientific, River Edge NJ, USA, 1992.

Nielsen, P. and Teakle, I. A. L.: Turbulent diffusion of momentum and suspended particles: A finite-mixing-length theory, Phys. Fluids, 16, 2342–2348, 2004.

Niño, Y. and Garcia, M. H.: On Engelund's analysis of turbulent energy and suspended load, J. Eng. Mech-ASCE, 124, 480–483, 1998.

Rouse, H.: Modern conceptions of the mechanics turbulence, T. Am. Soc. Civ. Eng., 102, 463–543, 1937.

Rouse, H.: Experiments on the mechanics of sediment suspension. Proceedings, 5th International Congress for Applied Mechanics, Vol. 55, 550–554, John Wiley & Sons, New York, 1938.

Schlichting, H. and Gersten, K.: Boundary Layer Theory, Springer-Verlag, 8th Edn., 2000.

Segre, P. N., Liu, F., Umbanhowar, P., and Weltz, D. A.: An effective gravitational temperature for sedimentation, Nature, 409, 594–597, 2001.

Shringarpure, M., Cantero, M. I., and Balachandar, S.: Dynamics of complete turbulence suppression in turbidity currents driven by monodisperse suspensions of sediment, J. Fluid Mech., 712, 384–417, 2012.

Smith, J. D. and McLean, S. R.: Spatially averaged flow over a wavy surface, J. Geophys. Res., 82, 1735–1746, 1977.

Vanoni, V. A.: Transportation of sediment by water, Trans. AGU, 3, 67–133, 1946.

Van Rijn, L. C.: Sediment Transport, part II: suspended load transport, J. Hydraul. Eng.-ASCE, 110, 1613–1641, 1984.

Wright, S. and Parker, G.: Density stratification effects in sand-bed river, J. Hydraul. Eng.-ASCE, 130, 783–795, 2004.

4

The role of hydrological transience in peatland pattern formation

P. J. Morris[1], A. J. Baird[2], and L. R. Belyea[3]

[1]Soil Research Centre, Department of Geography and Environmental Science, University of Reading, Reading, RG6 6DW, UK
[2]School of Geography, University of Leeds, Leeds, LS2 9JT, UK
[3]School of Geography, Queen Mary University of London, 327 Mile End Road, London, E1 4NS, UK

Correspondence to: P. J. Morris (p.j.morris@reading.ac.uk, paul.john.morris@gmail.com)

Abstract. The sloping flanks of peatlands are commonly patterned with non-random, contour-parallel stripes of distinct micro-habitats such as hummocks, lawns and hollows. Patterning seems to be governed by feedbacks among peatland hydrological processes, plant micro-succession, plant litter production and peat decomposition. An improved understanding of peatland patterning may provide important insights into broader aspects of the long-term development of peatlands and their likely response to future climate change.

We recreated a cellular simulation model from the literature, as well as three subtle variants of the model, to explore the controls on peatland patterning. Our models each consist of three submodels, which simulate: peatland water tables in a gridded landscape, micro-habitat dynamics in response to water-table depths, and changes in peat hydraulic properties.

We found that the strength and nature of simulated patterning was highly dependent on the degree to which water tables had reached a steady state in response to hydrological inputs. Contrary to previous studies, we found that under a true steady state the models predict largely unpatterned landscapes that cycle rapidly between contrasting dry and wet states, dominated by hummocks and hollows, respectively. Realistic patterning only developed when simulated water tables were still transient.

Literal interpretation of the degree of hydrological transience required for patterning suggests that the model should be discarded; however, the transient water tables appear to have inadvertently replicated an ecological memory effect that may be important to peatland patterning. Recently buried peat layers may remain hydrologically active despite no longer reflecting current vegetation patterns, thereby highlighting the potential importance of three-dimensional structural complexity in peatlands to understanding the two-dimensional surface-patterning phenomenon.

The models were highly sensitive to the assumed values of peat hydraulic properties, which we take to indicate that the models are missing an important negative feedback between peat decomposition and changes in peat hydraulic properties. Understanding peatland patterning likely requires the unification of cellular landscape models such as ours with cohort-based models of long-term peatland development.

1 Introduction

1.1 Background

The surface of northern peatlands often comprises a patchwork of distinct, small-scale (< 10 m; known as scale-level 1, or SL1 – Baird et al., 2009) micro-habitats, each with a char-acteristic vegetation type, micro-topographic relief, water-table and soil-moisture regimes, soil hydraulic properties, and soil biogeochemical regime (Ivanov, 1981; Alm et al., 1997; Belyea and Clymo, 2001; Rydin and Jeglum, 2006). These micro-habitats commonly aggregate into larger spatial structures at horizontal scales of tens to hundreds of metres (SL2), often forming landscapes composed of strongly directional, non-random patterns that may be linear and contour parallel, polygonal, or maze like (e.g. Aber et al., 2002; Eppinga et al., 2008; Korpela et al., 2009). Baird et al. (2009) demonstrated that the frequency distribution of water-table depths from across a peatland landscape depends not just on the proportion of the landscape covered by different micro-habitats but also on the pattern. In consequence, pattern may play an important role in determining peatland–atmosphere fluxes of greenhouse carbon gases (carbon dioxide and methane), which vary with water-table depth (Bubier et al., 1993, 1995; Roulet et al., 2007). Additionally, it seems likely that the development and maintenance of peatland patterning is governed by the same mechanisms that control peat accumulation, decomposition, and the development of soil hydraulic properties (Belyea and Clymo, 2001; Nungesser, 2003; Eppinga et al., 2009). An improved understanding of the mechanisms that control patterning may therefore reveal fundamental rules that govern broader aspects of peatland ecosystem and soil development at the landscape scale.

Observational (e.g. Foster and Fritz, 1987; Belyea and Clymo, 2001; Comas et al., 2005) and modelling (e.g. Nungesser, 2003; Swanson, 2007) studies have identified a variety of aspatial or one-dimensional (vertical only) feedback mechanisms that may help to explain directionless clumping of SL1 units into larger features. However, understanding the highly directional nature of patterning seen in many peatlands clearly requires an explicit consideration of spatial interactions. This problem lends itself naturally to investigation using 2- or 3-dimensional simulation models, in which directional transfers of water, energy and nutrients, and their effects on pattern, can be explored directly. Interplay between long- and short-range processes is a recurring theme in many apparently successful models of patterned landscapes, including peatlands (Rietkerk et al., 2004a, b; Eppinga et al., 2009) and other landscape types such as marshes (van de Koppel and Crain, 2006) and drylands (Lefever and Lejeune, 1997).

One current hypothesis on peatland pattern formation, the ponding mechanism, has been explored using cellular landscape models in a number of previous studies. The ponding mechanism consists of a pair of competing feedbacks between water-table depth, peatland micro-habitat succession, and peat hydraulic properties. Areas with deeper water tables are assumed to be more likely to support hummock vegetation, whereas areas with shallower water tables are assumed to be more likely to support hollow vegetation (cf. Rydin and Jeglum, 2006). Furthermore, hummock vegetation is assumed to produce near-surface peat that is less permeable than that produced in hollows (cf. Ivanov, 1981). Contrasting micro-habitat states (hummock, hollow) between adjacent SL1 units allow a positive feedback that reinforces local differences in water tables. Ponding occurs upslope of hummocks, leading to hollows there; while areas downslope of

Figure 1. Aerial photograph showing contour-parallel, striped patterning on a peatland complex in the James Bay lowlands, Ontario, Canada. The directions of slope and regional water flow are from the top right to the bottom left of the picture. Horizontal distance between tops of successive ridges is approximately 5 to 10 m. Image belongs to Brian Branfireun, reproduced here with kind permission.

hummocks are deprived of water, encouraging the development of hummocks. Conversely, rapid drainage occurs upslope of hollows, leading to hummocks there; while areas downslope of hollows are supplied readily with water, encouraging the development of hollows. This short-range positive feedback competes against a long-range negative feedback whereby water flow between the model's boundaries tends to homogenise local variations in water tables. Swanson and Grigal (1988), Couwenberg (2005) and Couwenberg and Joosten (2005) found that, when implemented in a numerical cellular landscape in which shallow groundwater flow is predominantly downslope, the ponding mechanism generated realistic-looking, contour-parallel stripes of hummocks and hollows at SL2 (we henceforth refer to these three papers, and the versions of the ponding model that they describe, collectively as SGCJ). However, Eppinga et al. (2009) came to different conclusions when using their own cellular model to explore the individual and combined effects of the ponding mechanism and two other feedbacks. They found that contour-parallel striped patterning such as that shown in Fig. 1 developed under several combinations of these feedbacks, but not when the ponding mechanism alone was used; that is, when using the ponding model in isolation they were unable to find a combination of parameters that generated anything other than a homogeneous, unpatterned landscape, seemingly in disagreement with the earlier studies. Consequently it is difficult to judge whether the ponding mechanism alone still represents a viable theory of peatland patterning.

Table 1. Glossary of algebraic terms, including default values where appropriate, for each of the four model versions.

Symbol	Description	Dimensions	Units	Model 1	Model 2	Model 3	Model 4
B	peatland surface height above arbitrary datum	L	m	0.22 to 4.20 m	0.22 to 4.20 m	0.48 to 4.17 m	0.48 to 4.17 m
Δt_e	hydrological submodel runtime (steady-state criterion)	T	h	50	50	50	17 520 (2 yr)
d	thickness of flow	L	m	–	auxiliary variable	auxiliary variable	auxiliary variable
H	water-table height above arbitrary datum	L	m	auxiliary variable	auxiliary variable	auxiliary variable	auxiliary variable
K	hydraulic conductivity	$L\,T^{-1}$	$m\,s^{-1}$	–	–	–	–
K_{ave}	depth-averaged hydraulic conductivity	$L\,T^{-1}$	$m\,s^{-1}$	–	–	–	–
K_{deep}	hydraulic conductivity of deep peat	$L\,T^{-1}$	$m\,s^{-1}$	–	0	1.25×10^{-5}	1.25×10^{-5}
K_{hol}	hydraulic conductivity of hollow peat	$L\,T^{-1}$	$m\,s^{-1}$	–	1.0×10^{-3}	1.0×10^{-3}	1.0×10^{-3}
K_{hum}	hydraulic conductivity of hummock peat	$L\,T^{-1}$	$m\,s^{-1}$	–	5.0×10^{-5}	5.0×10^{-5}	5.0×10^{-5}
p	probability of hummock formation	–	–	Eq. (3); Fig. 3	Eq. (3); Fig. 3	Eq. (3); Fig. 3	Eq. (3); Fig. 3
Q	hummock turnover rate	–	–	state variable	state variable	state variable	state variable
R	relative variance of hummocks per row	–	–	state variable	state variable	state variable	state variable
S	proportion of model landscape occupied by hummocks	–	–	state variable	state variable	state variable	state variable
θ	peat drainable porosity	–	–	0.3	0.3	0.3	0.3
T	peat transmissivity	$L^2\,T^{-1}$	$m^2\,s^{-1}$	–	–	–	–
T_{hol}	transmissivity of hollow cells	$L^2\,T^{-1}$	$m^2\,s^{-1}$	2.0×10^{-4}	–	–	–
T_{hum}	transmissivity of hummock cells	$L^2\,T^{-1}$	$m^2\,s^{-1}$	1.0×10^{-5}	–	–	–
U	net rainfall rate (precip. minus evapotran.)	$L\,T^{-1}$	$mm\,yr^{-1}$	0	0	400 (constant)	400 (time series)
x	spatial index (across-slope direction)	L	m	–	–	–	–
y	spatial index (along-slope direction)	L	m	–	–	–	–
Z	water-table depth below peat surface	L	m	auxiliary variable	auxiliary variable	auxiliary variable	auxiliary variable
Z_{final}	final water-table depth at end of developmental step	L	m	state variable	state variable	state variable	–
Z_{mean}	mean water-table depth during second half of dev. step	L	m	–	–	–	state variable

1.2   Aim and objectives

We recreated the SGCJ version of the ponding model to explore three characteristics that we suspected may have individually or collectively led to a Type-1 error in the SGCJ studies (i.e. causing a model to predict patterning despite the ponding mechanism being incapable of generating patterning in reality):

i. *Numerical implementation of shallow groundwater flow*: the transmissivity, T [dimensions of $L^2\,T^{-1}$] (see Table 1 for a glossary of algebraic terms), of any grid square in the original SGCJ models depends entirely on whether that square is currently a hummock or a hollow. Only two values of T are possible, one canonical value for hummocks, T_{hum}, and one for hollows, T_{hol}. A more realistic scheme would have been to assign canonical values of a more intrinsic property of peat such as saturated hydraulic conductivity, K_{hum} and K_{hol} [$L\,T^{-1}$], to hummocks and hollows, respectively, and to calculate transmissivity as the product of K and the thickness of flow in that square, thereby allowing for continuous variation in T across the model landscape. We extended the hydrological submodel from the SGCJ models in this way in order to remove any unrealistic constraints that the simplified numerical implementation of shallow groundwater flow may have placed upon overall model behaviour. We also wished to explore the effects of the absolute values of T or K (as appropriate) upon model behaviour.

ii. *Conceptual hydrogeological setting*: the original SGCJ models considered only a shallow layer of near-surface peat, and assumed that deeper peat was impermeable to groundwater flow. The top few decimetres of peat are usually the most permeable (e.g. Fraser et al., 2001; Clymo, 2004) and are therefore prone to the most rapid subsurface flow, although deeper peat is rarely truly impermeable. Indeed, a number of studies have indicated that drainage through deep peat layers may play an important role in peatland development and the ability of these ecosystems to self-organise (e.g. Ingram, 1982; Belyea and Baird, 2006; Morris et al., 2011). The assumption of impermeable peat below the uppermost few decimetres may have prevented the SGCJ models from representing a potentially important hydrological interaction between surficial hydraulic structures and deeper peat layers. Furthermore, in the original SGCJ models rainfall was not included; the only inputs of water to the models were from shallow groundwater flow and/or surface runon. In order to address these conceptual issues we experimented with the effects of incorporating a permeable lower layer and driving the model using simulated rainfall.

iii. *Dependence on hydrological transience*: the original SGCJ authors reported that strong striped patterning occurred in their models under "steady-state" hydrological conditions, whereby micro-habitat transitions only took place once the simulated water-table map had equilibrated with with the current distribution of hummocks and hollows (and so the spatial arrangement of transmissivity) (see Sect. 2 for full model description). The criterion used to determine steady state considers the proportion of cells in which the rate of water-table change [$L\,T^{-1}$] is less than a threshold rate. Steady-state conditions are deemed to occur when the proportion of cells below the threshold is greater than a proportion set by the model user. However, preliminary experiments (not reported here in full) with our own version of the ponding model indicated that the simulations reported in the earlier SGCJ studies may not have attained true hydrological steady state and that, curiously, patterning only developed under intermediate steady-state criteria where water tables were still transient. We chose to experiment with the effects of hydrological transience upon model patterning, either by manipulating the water-table steady-state criterion, or by driving the model with a real time series of rainfall data.

2 Models and methods

2.1 Overview

We began with a model that was as similar as possible to that employed by Couwenberg (2005), as far as the original model description allows. As well as this replica model (henceforth, Model 1) we created three additional models (Models 2, 3 and 4) with slightly altered routines. We designed Model 2 so as to address the numerical implementation issue identified in objective (i), while we designed Models 3 and 4 so as to address the conceptual issues identified in objective (ii) (see Sect. 1.2, above; and Table 2). Each of the four models may be thought to consist of three submodels that simulate: shallow saturated groundwater movements and the spatial distribution of water-table depths at SL1 (hydrological submodel); switches in micro-habitat type (ecological submodel); and changes in peat soil hydraulic properties (soil properties submodel). Model time progresses in developmental steps: during each developmental step the hydrological submodel is run until a predetermined steady-state criterion is met. The output from the hydrological submodel is a map of water-table depths within the model landscape. The ecological submodel then simulates a new micro-habitat map within the model landscape on the basis of the water-table map, such that every cell in the model landscape is assigned one of two binary micro-habitat types: hummock or hollow. Finally, the soil properties submodel uses the new micro-habitat map to reassign the spatial distribution of near-surface peat hydraulic properties. The new map of peat hydraulic properties is then used as an input to the hydrological model at the beginning of the next developmental step. As with the SGCJ studies, our models do not simulate explicitly the processes of peat formation or decomposition, and therefore do not incorporate a peat mass balance. Rather, the models represent plant community succession as a shifting mosaic of SL1 tiles superimposed onto a static peat landform.

2.2 Hydrological submodel

The hydrological submodel uses a modification of the DigiBog model of peatland saturated hydrology (Baird et al., 2012). We took the original DigiBog Fortran 95 code and added routines to represent the ecological and hydrophysical submodels. In the current study we deactivated DigiBog's peat accumulation, decomposition and hydrophysical subroutines described by Morris et al. (2012). We refer the reader to Baird et al. (2012) for a comprehensive description of DigiBog's governing equations and numerical implementation, although a few points are pertinent here. DigiBog represents a peatland as a grid of vertical peat columns; in plan each column is equivalent to one square grid cell in the SGCJ models, although DigiBog also allows for vertical variation in peat hydraulic properties. Horizontal saturated groundwater flow occurs between adjacent columns (four-square neighbourhood) at a rate equal to the product of the harmonic mean of inter-cell transmissivity (the product of depth-averaged hydraulic conductivity and the thickness of flow) and inter-cell hydraulic gradient, according to the Boussinesq equation (see also McWhorter and Sunada, 1977):

$$\frac{\partial H}{\partial t} = \frac{\partial}{\partial x}\left(\frac{K_{ave}(d)}{\theta}d\frac{\partial H}{\partial x}\right) + \frac{\partial}{\partial y}\left(\frac{K_{ave}(d)}{\theta}d\frac{\partial H}{\partial y}\right) + \frac{U}{\theta}, \quad (1)$$

where H is water-table height [L] above an arbitrary datum; t is simulated time [T]; x and y are across-slope and along-slope horizontal spatial dimensions, respectively [L]; K_{ave} is depth-averaged hydraulic conductivity [L T^{-1}]; θ is drainable porosity of the peat [dimensionless], assumed constant at 0.3; d is thickness of flow [L], defined as the height of the water table, H, minus the height of the impermeable base layer; and U is net rainfall [L T^{-1}]. Because U is added directly to the saturated zone it may be thought of as a black-box representation of total precipitation minus the sum of surface runoff, interception, evapotranspiration and unsaturated water-storage change.

The hydrological model is run for a predetermined length of simulated time in order to allow the water-table map to change in response to the boundary conditions and the current distribution of hydraulic properties. In Models 1 to 3, the length of time that the hydrological model is allowed to run, Δt_e [T], before an ecological transition takes place (Sect. 2.3) determines how close to a genuine steady state the simulated water tables are. Short equilibration times lead to highly transient water tables that are still changing rapidly and are far from being in equilibrium with the hydrological inputs or boundary conditions; the opposite is true of long runtimes. In Models 1, 2 and 3 we varied Δt_e between 1 h (highly transient water tables) and 10 000 h (approximately 417 days; highly steady water tables) to examine the effect of the steady-state criterion upon model behaviour. We performed only a single simulation with Model 4, with $\Delta t_e = 17\,520$ h (equal to two years of simulated time), and introduced hydrological transience by driving the model using a daily rainfall time series derived from observed data (Sect. 2.5).

We performed a simple test on Models 1, 2 and 3 in order to verify that increasing Δt_e led to increasing hydrological steadiness. We ran each model for a spin-up period of 25

Table 2. Summary of the four models and the objectives addressed by each.

Model	Transmissivity scheme	Deep peat layer	Net rainfall, U (precip. minus evapotran.)	Up-slope boundary condition	Aquifer shape	Objectives addressed
1	binary	impermeable	off	constant head, $Z = 0.025$ m	constant slope	iii
2	continuous	impermeable	off	constant head, $Z = 0.025$ m	constant slope	i, iii
3	continuous	$K_{deep} = 1.25 \times 10^{-5}$ m s^{-1}	400 mm yr^{-1} (constant)	impermeable (no-flow)	hemi-elliptical	i, ii, iii
4	continuous	$K_{deep} = 1.25 \times 10^{-5}$ m s^{-1}	400 mm yr^{-1} (time series)	impermeable (no-flow)	hemi-elliptical	iii

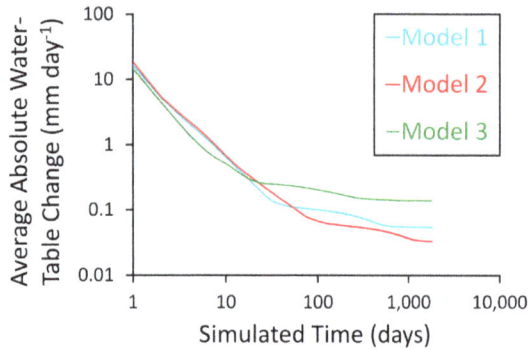

Figure 2. Time series of daily-average absolute water-table change per grid square during five-year test periods with Models 1, 2 and 3. See main text for full details. Note the logarithmic scales on both axes.

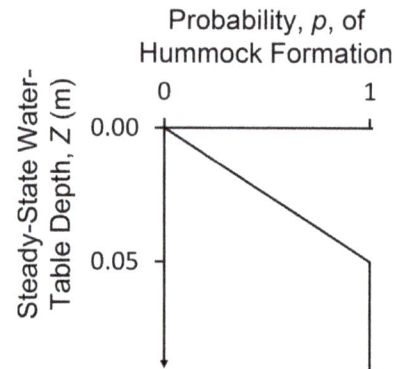

Figure 3. Graphical representation of the linear probability function used by the ecological submodel to assign hummock and hollow states to each model cell, based on water-table depth.

developmental steps with all default parameter values (see Table 1) and $\Delta t_e = 50$ h. We then allowed the hydrological submodel to continue for a further five years of simulated time, without any further changes in the spatial distributions of micro-habitats or peat properties. For each simulated day of the five-year test we calculated the daily sum of absolute water-table movements in each model grid square, and then averaged these absolute changes for all grid squares. In all three models the average rate of water-table change declined rapidly from initial values of between 14 and 19 mm day^{-1} to between 0.03 and 0.14 mm day^{-1} by the end of the five-year test period (Fig. 2). The tests indicate that water-table behaviour for all three models converges on steady state with increasing Δt_e.

The depth of the water table below the surface at any point in model time is given simply by

$$Z = B - H, \qquad (2)$$

where Z is water-table depth [L]; and B is the height of the peatland surface relative to the arbitrary datum [L]. We refer to the value of Z at the end of a developmental step as Z_{final}; and the time-averaged value of Z during the second half of a developmental step as Z_{mean}. These two metrics are used as inputs to the ecological submodel (Sect. 2.3).

2.3 Ecological submodel

In Models 1, 2 and 3 the probability, p [dimensionless], of any model cell being designated as a hummock during a given developmental step is a linear function of water-table depth, Z_{final}, at the end of the previous developmental step. When the water table is at the surface in any cell (i.e. when $Z_{final} = 0.00$ m), $p = 0$ (i.e. that cell is necessarily designated as a hollow for the following developmental step). The value of p increases linearly with increasing water-table depth up to $Z_{final} = 0.05$ m. When Z_{final} is equal to or greater than 0.05 m, the cell is automatically designated a hummock (i.e. $p = 1$). The relevant equations (see also Fig. 3) are

$$p = 0.0 \qquad \text{for } Z \leq 0.00 \text{ m}$$
$$p = 20Z \qquad \text{for } 0.00 \text{ m} < Z < 0.05 \text{ m}$$
$$p = 1.0 \qquad \text{for } Z \geq 0.05 \text{ m}, \qquad (3)$$

where Z is equal to either final water-table depth at the end of a developmental step, Z_{final} (Models 1, 2 and 3), or time-averaged water-table depth during the second half of a two-year developmental step, Z_{mean} (Model 4) (see below).

We refer to the upper 0.05 m of the soil profile as the transition zone. All cells that are not designated as hummocks are automatically designated as hollows. Swanson and Grigal (1988) provide a full description of micro-habitat transitions as simulated in our Models 1, 2 and 3. In Model 4 micro-habitat transitions are dealt with slightly differently. Rather than Z_{final}, Model 4 assigns hummocks and hollows based on each grid square's time-averaged water-table depth, Z_{mean} [L], during the second half of the previous 2 yr developmental step. The first 365 days of each developmental step are used to allow Model 4's simulated water tables to adjust to the newly updated distribution of peat hydraulic properties, but this adjustment period is not incorporated into Z_{mean}. This measure was intended to ensure that Z_{mean} in Model 4 contains no meaningful artefact of water-table geometries from earlier developmental steps. Along with a variable rainfall time series (see Sect. 2.5, below) the use of Z_{mean} provided a means of introducing transience to water-table behaviour in an arguably more realistic manner than the short equilibration times used in Models 1 to 3.

2.4 Soil properties submodel

Hummocks are assumed to produce peat that is less permeable than that produced by hollows. DigiBog's implementation of the Boussinesq equation uses depth-averaged saturated hydraulic conductivity, K_{ave}, and the thickness of flow, d, to calculate saturated groundwater flux between adjacent columns (Eq. 1). However, the original SGCJ models simply assigned a single value of transmissivity, T_{hum}, to hummocks and another, higher value, T_{hol}, to hollows. Model 1 uses the simple treatment of canonical transmissivities as per the original SGCJ models; thus, in the case of Model 1, Eq. (1) simplifies to

$$\frac{\partial H}{\partial t} = \frac{\partial}{\partial x}\left(\frac{T}{\theta}\frac{\partial H}{\partial x}\right) + \frac{\partial}{\partial y}\left(\frac{T}{\theta}\frac{\partial H}{\partial y}\right) + \frac{U}{\theta}. \quad (4)$$

We assumed default transmissivity values of $T_{\mathrm{hol}} = 2.0 \times 10^{-4}\,\mathrm{m^2\,s^{-1}}$ and $T_{\mathrm{hum}} = 1.0 \times 10^{-5}\,\mathrm{m^2\,s^{-1}}$, thereby preserving the T_{hol} to T_{hum} ratio of $20:1$ that led to strong patterning in the original SGCJ models. Models 2, 3 and 4 use a more sophisticated and realistic treatment allowed by Digi-Bog, whereby hummocks and hollows are assigned canonical values of hydraulic conductivity, K_{hum} and K_{hol}, respectively; T for each cell is recalculated during each iteration of the hydrological submodel as the product of K and the thickness of flow, d (height of water table above the model's impermeable base layer) (cf. Freeze and Cherry, 1979). In this way T is able to vary in a continuous manner based on water-table position, which is more realistic than the simple binary-T treatment of the original SGCJ models (Eq. 1). For Models 2, 3 and 4, we assumed default values of $K_{\mathrm{hol}} = 1.0 \times 10^{-3}\,\mathrm{m\,s^{-1}}$ and $K_{\mathrm{hum}} = 5.0 \times 10^{-5}\,\mathrm{m\,s^{-1}}$, thereby giving a K_{hol} to K_{hum} ratio of $20:1$. In models 1, 2 and 3 we also manipulated the default values of T or K by factors of between 0.05 and 20 in line with objective (i). In Model 1 we varied T_{hum} between 5.0×10^{-7} and $2.0 \times 10^{-4}\,\mathrm{m^2\,s^{-1}}$; and T_{hol} between 1.0×10^{-5} and $4.0 \times 10^{-3}\,\mathrm{m^2\,s^{-1}}$, whilst always maintaining a T_{hol} to T_{hum} ratio of $20:1$. Similarly, in Models 2 and 3 we varied K_{hum} between 2.5×10^{-6} and $1.0 \times 10^{-3}\,\mathrm{m\,s^{-1}}$; and K_{hol} between 5×10^{-5} and $2 \times 10^{-2}\,\mathrm{m\,s^{-1}}$, whilst maintaining a K_{hol} to K_{hum} ratio of $20:1$.

2.5 Model spatial and temporal domains; boundary conditions

We implemented all simulations in a 70 (across-slope, x direction) $\times 200$ (along-slope, y direction) grid of $1\,\mathrm{m} \times 1\,\mathrm{m}$ grid squares. In Models 1 and 2, the simulated peat aquifer overlaid a sloping impermeable base (replicating the assumption of an impermeable lower peat layer in the SGCJ models). Both the peat surface and the impermeable base had a constant slope of $1:50$; the permeable upper peat had a uniform thickness of 0.2 m. In Models 3 and 4 we assumed the thick, lower peat layer is also permeable with its own hydraulic conductivity, K_{deep}, meaning that K for each cell is depth-averaged to account for the vertical transition in K between the upper and lower layers. Baird et al. (2012) provide a full description of DigiBog's calculation of K_{ave} and inter-cell T. We used the groundwater mound equation (Ingram, 1982) to calculate the dimensions of a deep peat layer that is hemi-elliptical in cross section, has a uniform K of $1.25 \times 10^{-5}\,\mathrm{m\,s^{-1}}$, is 200 m from central axis to the margin, is underlain by a flat impermeable base, and receives a net water input (from an implied upper peat layer) of $155\,\mathrm{mm\,yr^{-1}}$. The resulting deep peat layer was 3.97 m thick along its central axis, curving elliptically down to 0.28 m at the peatland's margin. Within DigiBog, this deep layer was overlain by a surficial, more permeable layer, which, like Models 1 and 2, was 0.2 m thick.

We allowed all simulations to run for 100 developmental steps; the initial condition consisted of randomly generated water tables, between 0.0 and 0.05 m below the peat surface, in each cell. The initial micro-habitat and soil-properties maps were based on this random initial water-table map in the usual manner described above.

In all simulations we set the lateral (along-slope) boundaries to impermeable (Neumann zero-flux condition) in order to simplify the representation of the simulated peatland within its local hydrogeological setting. In the simulations with an impermeable lower peat layer (Models 1 and 2) both the upslope and downslope boundaries were set to constant water-table (Dirichlet) conditions of 0.025 m below the surface such that the overall hydraulic gradient along the model was equal to the topographic gradient and the gradient of the impermeable peat layer. As with the original SGCJ models, Models 1 and 2 received no rainfall; all inputs of water in these models came from the upper boundary condition, representing shallow groundwater influx and/or surface runon. In this way Models 1 and 2 might be thought to be more representative of throughflow fens than of raised bogs. In Models 3 and 4 we replaced the constant-head condition at the up-slope boundary with a no-flow (Neumann zero flow) condition to represent the drainage divide along the crest of a raised bog in a manner similar to Morris et al. (2012) (see also Ingram, 1982), but retained the constant water-table condition at the lower boundary to represent shallow drainage outflow, to a lag stream, for example. Water-table levels in

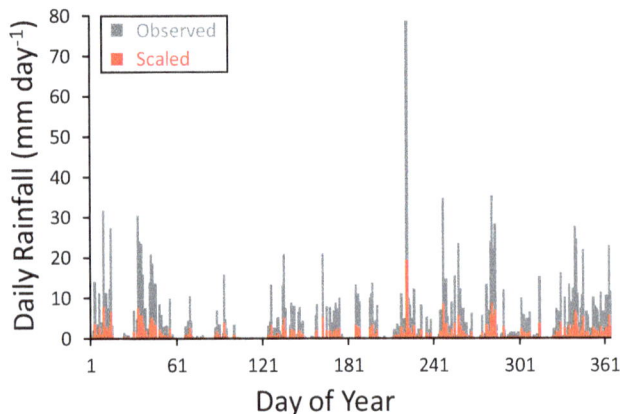

Figure 4. 365-day time series of daily observed rainfall, and its conversion to net rainfall, U, after scaling by a factor of 0.245. See main text for full details.

Model 3 were maintained by a constant simulated net rainfall rate of $U = 400\,\mathrm{mm\,yr^{-1}}$ (U is assumed to be net of evapotranspiration, interception and runoff; hence, its low value). We drove Model 4 using a daily time series of U. We took 365 days of daily rainfall data from close to Malham Tarn Moss, a raised bog in North Yorkshire, United Kingdom, for the calendar year 2011. The rain gauge recorded precipitation on 294 days of that year, with an annual total of 1633 mm of precipitation. Maximum daily precipitation for the year was 78.6 mm on 10 August. We multiplied each daily rainfall total in the time series by approximately 0.245 so as to give an annual total of 400 mm of net precipitation, equal to the constant U assumed in Model 3, whilst maintaining a plausible temporal variation (Fig. 4). During each 2 yr developmental step in Model 4 we cycled twice through the 365-day rainfall series. By using the same year of rainfall data repeatedly in this way we were able to induce water-table transience without the potentially complicating effects of inter-annual variability in rainfall patterns. The thick peat deposit represented in Models 3 and 4, and the fact that their groundwater mounds were maintained by simulated net rainfall, means that they might be thought to be most representative of ombrotrophic raised bogs.

2.6   Analysis

Previous studies (e.g. Andreasen et al., 2001; Eppinga et al., 2009) have demonstrated that the human eye is a powerful tool for assessing pattern strength in landscape models, and we used a visual appraisal of the two-dimensional micro-habitat maps produced by the four models as our primary means of estimating simulated pattern strength. In addition we calculated the relative variance, R [dimensionless], of hummocks per across-slope row of cells, following the method of Swanson and Grigal (1988), as an objective and reproducible metric of pattern strength. The value of R in-

creases with pattern strength: values less than 2 indicate unordered landscapes without discernible patterning; values of R greater than 4 represent highly ordered landscapes with clear, strong patterning. We also calculated the proportion, S [dimensionless], of model cells occupied by hummocks during each developmental step. Finally, we calculated model turnover rate, Q [dimensionless], defined as the proportion of model cells that undergo a transition from hummock to hollow, or vice versa, in each developmental step.

3   Results

3.1   Hydrological transience

In Models 1, 2 and 3, the strength of across-slope, striped patterning at SL2 initially increased strongly with increasing hydrological steadiness, from an apparently entirely unordered, random mixture of hummocks and hollows at SL1 when $\Delta t_e = 1\,\mathrm{h}$, to peak "strengths" at around $\Delta t_e = 100\,\mathrm{h}$ (for Models 1 and 2) or 50 h (Model 3). This increase in patterning strength is evident from both a visual appraisal of the final micro-habitat maps (Fig. 5), and an increase in the values of relative variance, R, with increasing Δt_e (Fig. 6a, b, c). For values of Δt_e greater than 50 h (Models 1 and 2) and 20 h (Model 3), an increase in the spatial scale of the simulated SL2 stripes is apparent with increasing Δt_e (i.e. the stripes became broader in the y (along-slope) direction). Increasing Δt_e also led to SL2 stripes whose upslope and downslope edges were increasingly straight and sharply defined (Fig. 5). For values of Δt_e greater than 100 h (Models 1 and 2) or 50 h (Model 3), the simulated SL2 stripes had broadened to the point of appearing to be "over-developed", such that they no longer resembled realistic peatland striped patterning (cf. Fig. 1).

As well as governing the spatial configuration of simulated patterning, the value of Δt_e also affected the temporal dynamics of Models 1, 2 and 3. For low values of Δt_e, turnover rates of model cells were high: when $\Delta t_e = 1\,\mathrm{h}$, approximately half of all cells would undergo a transition from designation as a hummock to a hollow (or vice versa) during each developmental step, providing further indication of a random landscape. Increasing Δt_e led to a reduction in turnover rates in Models 1 and 2 as SL2 structures began to stabilise, to a minimum of approximately 40 % of cells per developmental step when $\Delta t_e = 100\,\mathrm{h}$ (Fig. 6d, e). Increasing Δt_e had little effect on turnover rates in Model 3 until $\Delta t_e = 1000$ or 10 000 h (Fig. 6f). The temporal trends in SL1 turnover rate and proportional hummock coverage also indicate what appears to be a limit cycle in the behaviour of all models under highly stringent hydrological steady-state criteria. Particularly when $\Delta t_e = 17\,520$ (Model 4 only) or 10 000 h, and to a lesser extent when $\Delta t_e = 1000\,\mathrm{h}$, the models exhibited a cyclical behaviour in time whereby the simulated landscape would alternate on an approximately regular cycle of between four and ten developmental steps between two contrasting states:

Model 1

Model 2

Model 3

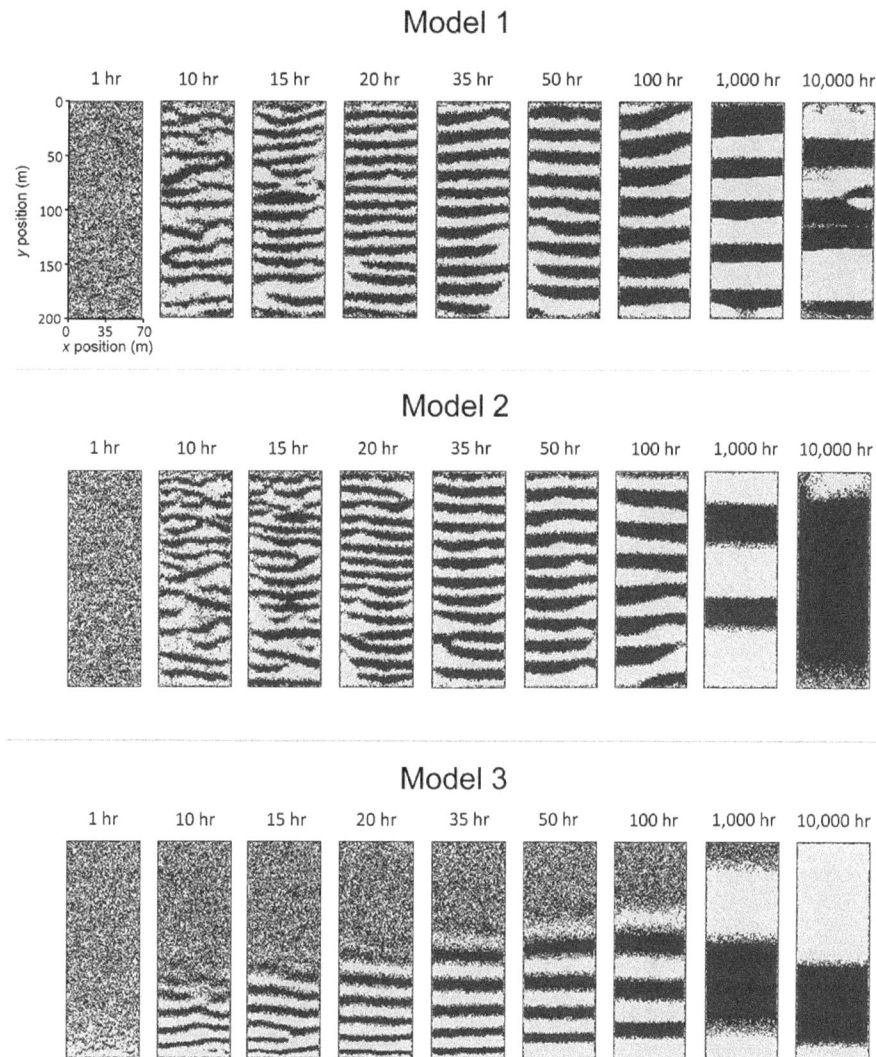

Figure 5. Final micro-habitat maps after 100 developmental steps, showing effects of increasing Δt_e on patterning in typical simulations with Model 1 (top panel), Model 2 (middle panel) and Model 3 (bottom panel). The value of Δt_e assumed for each simulation is indicated immediately above the map. Light pixels represent hummocks, dark pixels represent hollows. Low values of y represent upslope locations ($y = 0$ is the upslope boundary); high values of y represent downslope locations ($y = 200$ is the downslope boundary); as such, groundwater flow is generally down the page. All maps have the same horizontal (x, y) scale as that shown for the upper-leftmost map.

a dry landscape with deeper water tables, dominated by hummocks; and a wet landscape with water tables near the bog surface, dominated by hollows (Fig. 6). The contrast between the wet and dry states was particularly pronounced in Model 2, which cycled between approximately 15 % and 95 % hummock coverage when $\Delta t_e = 10\,000$ h (Fig. 6h). The single simulation with Model 4 ($\Delta t_e = 17\,520$ h) behaved in a very similar manner to Models 1, 2 and 3 when $\Delta t_e = 10\,000$ h, predicting nearly homogeneous, unpatterned landscapes (Fig. 7) that cycled rapidly between being dominated by hummocks and hollows (Fig. 6f, i).

In all simulations that generated striped SL2 patterning, those SL2 units migrated consistently downslope (not

shown) in the same manner as that previously reported in the published accounts of the SGCJ models.

3.2 Simplified calculation of transmissivity

Any artefacts introduced to model behaviour by the simple, binary treatment of peat permeability employed in the original SGCJ models (see objective i, above) would have been evident as differences in behaviour between Model 1 (simple, binary transmissivity scheme, based solely on hummock/hollow designation) and Model 2 (continuous treatment of transmissivity as product of peat hydraulic conductivity and thickness of flow). Models 1 and 2 behaved in a very similar manner to one another, in terms of their response

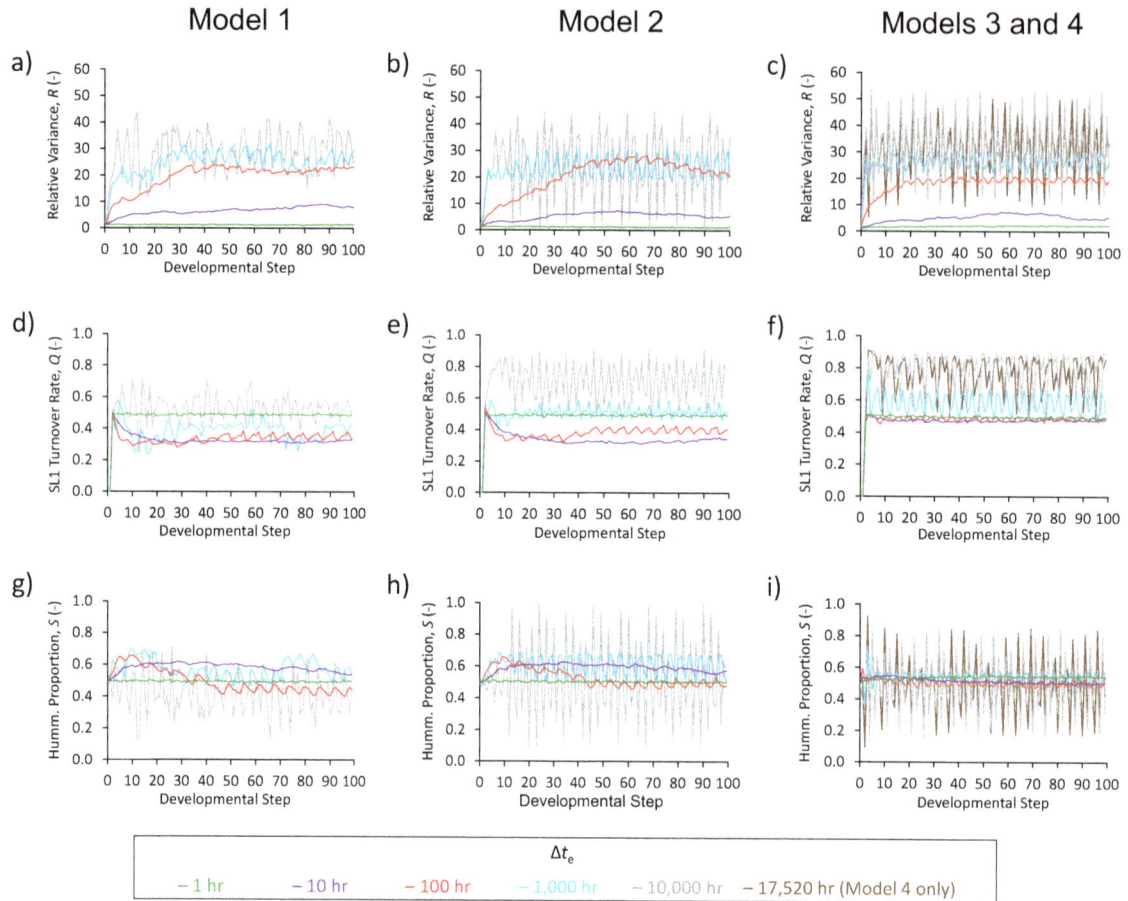

Figure 6. Influence of hydrological equilibration time, Δt_e, over temporal development of relative variance of hummocks per model row, R (top row); SL1 turnover rate, Q (middle row); and proportion of model landscape occupied by hummocks, S (bottom row); in Models 1 (left column), 2 (middle column), 3 and 4 (right column).

to both changing values of Δt_e and changing absolute values of peat permeability. Both models developed realistic looking, contour-parallel SL2 stripes over the entire model domain for intermediate values of Δt_e. In both models the striped patterning was relatively weak and discontinuous for $\Delta t_e = 15$ and 20 h; the patterns were stronger and continuous when Δt_e was between 35 and 100 h; the SL2 stripes became unrealistically broad at $\Delta t_e = 1000$ h, and were mainly absent when $\Delta t_e = 10\,000$ h (Fig. 5). Temporal patterns of summary metrics (SL1 turnover rate, hummock proportion, relative variance of hummocks per row) were qualitatively and quantitatively similar between models 1 and 2, and again responded similarly to changes in Δt_e (Fig. 6) and the ratio of T_{hum} to T_{hol} or K_{hum} to K_{hol} (not shown).

3.3 Hydrogeological setting

Any artefacts introduced to model behaviour by the simplifying assumptions made in the original SGCJ models about the hydrogeological setting of the simulated peatland (see objective ii, above) would have been evident as differences in

behaviour between Model 2 (impermeable deep peat layer; constant-head upslope boundary condition; zero rainfall addition; constant slope of peatland surface) and Model 3 (permeable deep peat layer; no-flow upslope boundary; constant net rainfall rate of 400 mm yr^{-1}; hemi-elliptical aquifer shape). Model 2 simulations that generated patterning did so over the entire model domain, although the same was not true of Model 3. Striped patterning at SL2 developed in Model 3 only in the downslope portion of the model domain. Furthermore, the area of the model domain that exhibited patterning extended upslope with increasing values of Δt_e. For instance, when $\Delta t_e = 10$ h, Model 3 only produced striped patterning within 60 m or so of the downslope boundary, but when $\Delta t_e = 100$ h the patterned area covered the majority of the model domain and extended as far as approximately 140 m from the downslope boundary (Fig. 5). Clear, continuous striped patterning that extended all the way across the across-slope direction (x direction) of the model domain developed in Model 3 at much lower values of Δt_e than in Model 2. Continuous, closely spaced SL2 stripes were evident in the downslope area of Model 3 when $\Delta t_e = 20$ h,

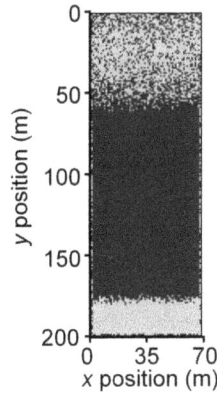

Figure 7. Final micro-habitat map after 100 developmental steps from Model 4, with $\Delta t_e = 17520$ h (2 yr). Light pixels represent hummocks, dark pixels represent hollows. Low values of y represent upslope locations ($y = 0$ is the upslope boundary); high values of y represent downslope locations ($y = 200$ is the downslope boundary); as such, groundwater flow is generally down the page.

although sharply defined, continuous SL2 stripes in Model 2 did not develop for values of Δt_e less than 35 h. Additionally, Model 3 behaved differently from Model 2 in terms of its responses to changes in the absolute values of peat hydraulic conductivity (see below).

3.4 Manipulation of peat properties

Modest changes in the absolute values of T or K brought about large changes in the nature and strength of simulated patterning in Models 1, 2 and 3, although the nature of these responses varied between models. In Models 1 and 2, the lowest values of T or K led to largely unpatterned landscapes, dominated by apparently random distributions of hummocks and hollows, with either a complete absence of patterning (×0.05 treatment) or very weak, discontinuous patterning (×0.1 treatment) (Fig. 8). In both Models 1 and 2, pattern strength increased with increasing values of T or K up to the default combination. For combinations of T and K greater than the default values, Models 1 and 2 both predicted an increase in the spatial scale of SL2 stripes in a manner similar to the effect of increasing Δt_e; for the ×10 and ×20 treatments Model 1 became almost a uniform landscape of hummocks with only small SL2 groupings of hollows. The behaviour of Model 3 was quite different. Only the default and the ×0.2 treatments generated any kind of contour-parallel stripes; all other treatments produced near-uniform, unpatterned landscapes composed almost entirely of either hollows (for the ×0.05 and ×0.1 treatments) or hummocks (for the ×5, ×10 and ×20 treatments).

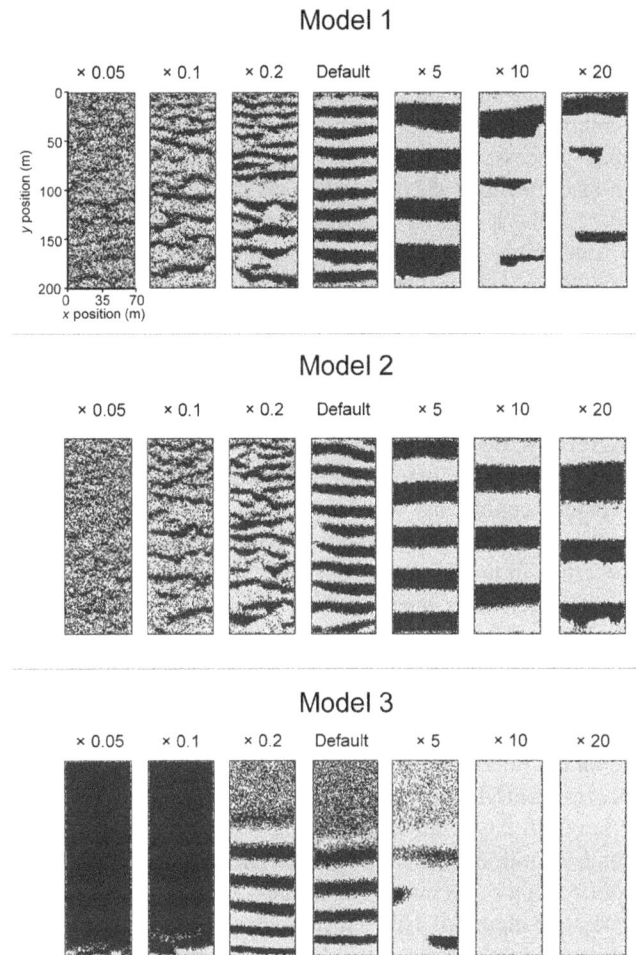

Figure 8. Final micro-habitat maps after 100 developmental steps, showing effects of different combinations of T_{hum} and T_{hol} (in the case of Model 1, top panel) or K_{hum} and K_{hol} (Model 2, middle panel; Model 3, bottom panel) on pattern geometries in typical model runs. The values of T or K used in each simulation are indicated immediately above the maps, and are expressed as a factor of the default values. See main text for full explanation. Light pixels represent hummocks, dark pixels represent hollows. Low values of y represent upslope locations ($y = 0$ is the upslope boundary); high values of y represent downslope locations ($y = 200$ is the downslope boundary); as such, groundwater flow is generally down the page. All maps have the same horizontal (x, y) scale as that shown for the upper-leftmost map.

4 Discussion

4.1 Hydrological transience and ecological memory

The striking similarity between the patterns generated by our Models 1, 2 and 3 at intermediate values of Δt_e and the striped patterns reported by the previous SGCJ authors, as well as the absence of patterning in our models under more stringent steady-state hydrological conditions, are highly suggestive that the simulations reported in the previous SGCJ studies had not attained genuine steady-state hydrological conditions. The dependence of realistic patterning in our models on intermediate equilibration times can be reconciled with the currently popular theory that landscape patterning commonly arises from interplay between long- and short-range processes (e.g. Rietkerk and van de Koppel, 2008; see also Turing, 1952). The strength of the long-range negative feedback (which acts to homogenise local differences in water-table position in response to distant hydrological boundary conditions) appears to increase as the hydrological equilibration time is lengthened. When equilibration times are very short each grid square's water-table behaviour is influenced only by its immediate neighbours, and long-range influences are unable to propagate across the model domain. Conversely, the longest equilibration times cause the simulated water-table map to be dominated by its long-range boundary conditions, which, under steady state, eliminate much of the short-range effects of contrasting peat hydraulic properties in neighbouring SL1 and SL2 units. Without additional feedbacks such as those explored by Eppinga et al. (2009), it is only at intermediate values of Δt_e that the model strikes the balance of long- and short-range feedbacks seemingly required for patterning.

If the absolute values of Δt_e are taken literally then the model predicts that realistic patterning only occurs if micro-succession at SL1 operates on unrealistically short timescales of hours to days, rather than years. Moreover, Model 4, in which we introduced water-table transience not through short equilibration times but through a daily rainfall series, did not generate patterning. This indicates clearly that it is the short equilibration times used in Models 1, 2 and 3, rather than simply non-steady water tables (which are also present in Model 4) that caused patterning in our simulations. It is perhaps initially tempting to discard the model in light of its reliance on hydrological transience.

However, an alternative and more abstract interpretation of the models' dynamics allows one to conceptualise the length of a developmental step simply as a measure of hydrological steadiness under which succession takes place. With this in mind, we can think of succession as occurring over a period of years, during which the degree of hydrological steadiness is determined by Δt_e. Therefore, Δt_e is no longer a literal time period during which plant community succession occurs but merely a means of representing hydrological steadiness. For lower values of Δt_e, the water tables across the models' domains are not in full equilibrium with the current distribution of hummocks and hollows. As such, those water-table maps reflect the distribution of hummocks and hollows during not only the current developmental step but also partly the previous developmental step. It may be that the formation of peatland patterning relies on some ecological memory effect (cf. Peterson, 2002), such as the influence of recently buried peat that no longer reflects current positions of hummocks and hollows but which remains hydrologically important due to its shallow depth. Despite Digi-

Bog allowing for 3-dimensional variation in peat properties, the ponding model is in essence a 2-dimensional model insofar as the new micro-habitat map during each developmental step supersedes the previous map entirely; the modelled system retains no memory of previous SL1 units or their associated soil properties. As such, our models neglect some of the structural complexity that peatlands exhibit in three spatial dimensions (e.g. Barber, 1981). Our simulations with non-conservative hydrological steady-state criteria (particularly $\Delta t_e = 50$ and $100\,h$) implicitly contain a type of ecological memory effect, because the influence of previous patterns of surface vegetation and shallow soil properties are expressed in the model's hydrological behaviour; this memory effect decreases in strength with increasingly stringent hydrological steady-state criteria. Our results indicate that this memory effect is important to the formation of patterning in the ponding model and should be investigated directly in future studies, perhaps via the use of cohort-based peat accumulation models (e.g. Frolking et al., 2010; Morris et al., 2011, 2012).

4.2 Peat hydraulic properties

Models 1, 2 and 3 were all highly sensitive to the absolute values of T or K, and patterning only occurred within a narrow region of parameter space. Even in the ×0.05 and ×20 treatments, the values of T or K were well within reported ranges; the fact that these parameterisations failed to produce patterning suggests that a full suite of relevant processes and feedbacks has not been represented. Particularly in Model 3, in which model water-table levels are maintained by the simulated addition of rainfall, even modest changes in the absolute values of K_{hol} and K_{hum} caused simulations to "run away" to either wet or dry end-member states. In simulations with higher values of hydraulic conductivity Model 3 drained rapidly to the downslope boundary, causing water tables to fall below the transition zone and giving rise to uniformly dry simulated landscapes dominated by hummocks. Conversely, the simulations with lower values of hydraulic conductivity caused Model 3 to drain so slowly that water tables rose to the surface of all columns, leading to uniformly wet landscapes dominated by hollows. The same effect was not evident in Models 1 and 2 because the constant-head condition at the upslope boundary maintained water-table levels within the transition zone. Nonetheless, manipulating T and K in Models 1 and 2, respectively, still resulted in landscapes devoid of realistic patterning by strengthening the long-range negative feedback relative to the short-range positive feedback.

The high sensitivity to the absolute values of T and K may be taken to suggest that our models are lacking at least one important negative feedback, namely that between peat decomposition and changes in peat hydraulic properties. For example, saturated hydraulic conductivity is known to decrease strongly with increasing time-integrated decomposition of peat (e.g. Boelter, 1969; 1972; Grover and Baldock, 2013). Areas with deep water tables are prone to more rapid decomposition and more rapid collapse of pore spaces. The resultant reduction in hydraulic conductivity in turn causes peat to retain water more readily and causes decay rates to fall, stabilising system behaviour (see Belyea, 2009; and Morris et al., 2012). This concept is partially represented in the ponding model by lower values of T or K in hummocks (where peat spends more time under oxic conditions, and hence peat at the water table is more degraded and less permeable) than in hollows (where peat has a shorter residence time under oxic conditions, producing better-preserved and more permeable peat at the depth of the water table). However, in the ponding model only two K values are possible, K_{hum} and K_{hol}, meaning that hydraulic conductivity cannot vary as a continuous function of decomposition; the ponding model's representation of this relationship may therefore be overly constrained. Modelling studies by Morris et al. (2011) and Swindles et al. (2012) have indicated that a continuous relationship between peat decomposition and hydraulic conductivity may be highly important to the ability of peatlands to self-organise and to maintain homeostatic water-table behaviour. The representation of this negative feedback within models of peatland patterning such as ours has the potential to stabilise model behaviour by allowing simulated water tables and peat permeability to self-organise, thereby reducing model sensitivity to small changes in soil hydraulic parameter values. This would require the expansion of the ponding model so as to include routines that describe litter production and decomposition, and continuous changes in peat hydraulic conductivity. The inclusion of these processes in patterning models would also help to address the question raised above of ecological memory effects, and suggests the need to unify models of peatland surface patterning such as those considered here, and cohort models of long-term peatland development (e.g. Frolking et al., 2010; Morris et al., 2011, 2012).

4.3 Improved model hydrology

Our alterations to the ponding model's hydrological basis compared to previously published versions had little impact on model behaviour, evidenced by the fact that Models 1, 2 and 3 all behaved in qualitatively similar manners, including their response to hydrological transience. The previous SGCJ authors reported that patterning is stronger and forms more readily as the slope angle of the model domain, and associated hydraulic gradients, are increased. The curved cross-sectional shape of the bog in Model 3 means that slope angle near the downslope boundary is greater than the uniform 1 : 50 slope in Models 1 and 2. This appears to have allowed weak striping to develop at the downslope end of Model 3 even under hydrological conditions that were so transient as to prevent patterning in Models 1 and 2. Increasing hydrological steadiness in Model 3 then allowed patterning to

spread upslope onto increasingly shallow slopes. The differences in behaviour between Models 2 and 3 can therefore be explained by the curved aquifer shape, reflecting the slope angle effect reported by previous authors. In particular, the nature of SL2 patterning in Model 3 (Fig. 5) is strikingly similar to that seen in the simulations with domed aquifers reported by Couwenberg and Joosen (2005). We are therefore left to deduce that the permeable deep peat layer in Model 3 had little independent effect, and that the generation of patterning by the ponding mechanism is not dependent on either hydrological interaction with, or isolation from, deeper peat layers.

The simple binary treatment of hummock and hollow transmissivity used in the earlier SGCJ studies also appears to have produced no artefact in the behaviour of those models (or in our Model 1) compared to our more realistic Boussinesq treatment (in Models 2 and 3). Nonetheless, it is important to recognise that neither the simplified treatment of transmissivity nor the assumption of an impermeable deep peat layer, despite both being questionable assumptions in themselves, were responsible for the model's reliance on the hydrological steady-state criterion and the absolute values of peat hydraulic properties, nor the prediction of downslope migration of SL2 stripes. As such we are confident that these behaviours (reliance on hydrological transience; high sensitivity to absolute values of transmissivity; downslope pattern migration) are genuine facets of the ponding model and are not artefacts of numerical implementation.

4.4 Pattern migration

The downslope movement of SL2 stripes in the ponding model is an emergent behaviour that should be treated as a testable hypothesis against which the ponding model could, in part, be tested (cf. Grimm et al., 2005). We are aware of very little direct evidence as to the long-term stationarity or otherwise of peatland patterning, likely because the long timescales involved preclude direct observation. Koutaniemi (1999) found that SL2 units in a Finnish aapa mire expanded and contracted on a seasonal basis due to frost heave, but he saw no evidence of consistent downslope migration of any of these features over a 21 yr period. Foster and Wright (1990) analysed the age of pool-bottom sediments from a Swedish raised bog. They also found no evidence of migration over thousands of years. Rather, they concluded that the age of pools was determined largely by the timing of the lateral expansion of the bog, and that once formed the pools remained stationary. However, Kettridge et al. (2012) used data from a combination of ground-penetrating radar and soil cores to investigate the below-ground physical properties of peat soil in a raised bog in Wales. They found subsurface layers that dipped consistently towards the bog's margins, which they interpreted as evidence of down-slope migration of SL1 or SL2 units during the bog's development. More observational work is clearly required to ascertain whether or not the findings of Kettridge et al. (2012) hold in the general case, but it appears that the ponding model's prediction of the downslope migration of SL2 stripes cannot be used to falsify the model at this stage.

5 Summary and conclusions

Our numerical experiments suggest strongly that simulations reported in previous studies had not attained true hydrological steady-state conditions; under true steady state patterning does not occur. Realistic, striped patterns only form when micro-habitat transitions occur in response to highly transient water-table patterns. The equilibration times required by the hydrological submodel to attain the level of transience necessary for patterning are of the order of hours to days: these timescales are largely meaningless in terms of vegetation dynamics. The models' reliance on the hydrological steady-state criterion indicates that ecological memory may play a role in pattern formation. Although ecological memory is not represented explicitly (or deliberately) in our models, the transient water tables appear to have inadvertently replicated such an effect.

We increased incrementally the sophistication of the models' numerical implementation and the realism of their conceptual bases, chiefly by representing a permeable deep peat layer; improving the models' calculation of transmissivity; and introducing a plausible temporal variation in rainfall. In doing so we removed a number of simplifying assumptions made in previous studies. However, these improvements had little effect on the models' behaviours, indicating that the sensitivity to hydrological steady state is a genuine facet of the models and is not an artefact of numerical implementation.

The models appear to be unrealistically sensitive to the absolute values of the parameters used to represent peat permeability. We interpret this sensitivity as indicating that the models are missing a negative feedback between peat decomposition and hydraulic conductivity, which previous studies have shown is important to the ability of peatlands to self-organise.

Peatland structures and processes exhibit complexity in three spatial dimensions. A logical next step for patterning research would be to combine 2-dimensional (horizontal only) cellular landscape models such as those presented here with (mostly 1-dimensional; vertical only) peatland development models that provide a more detailed and realistic representation of peat formation, decomposition and dynamic changes in soil hydraulic properties. This would allow a direct process-based exploration of both the mechanism involved in ecological memory, and a continuous relationship between peat decomposition and hydraulic properties.

Acknowledgements. This research was funded in part by a Queen Mary University of London PhD studentship awarded to Paul Morris. We are grateful to Brian Branfireun (University of Western Ontario) for permission to use the photograph in Fig. 1, and to The Field Studies Council at Malham, in particular Robin Sutton, for providing us with the data from Malham Tarn Weather Station. We are grateful to the associate editor for helpful dialogue, and to three referees for insightful and constructuive comments on an earlier version of the manuscript.

Edited by: F. Metivier

References

Aber, J. S., Aaviskoo, K., Karofeld, E., and Aber, S. W.: Patterns in Estonian bogs as depicted in color kite aerial photographs, Suo, 53, 1–15, 2002.

Alm, J., Talanov, A., Saarnio, S., Silvola, J., Ikkonen, E., Aaltonen, H., Nykänen, H., and Martikainen, P. J.: Reconstruction of the carbon balance for microsites in a boreal oligotrophic pine fen, Finland, Oecologia, 110, 423–431, doi:10.1007/s004420050177, 1997.

Andreasen, J. K., O'Neill, R. V., Noss, R., and Slosser, N. C.: Considerations for the development of a terrestrial index of ecological integrity, Ecol. Indic., 1, 21–35, doi:10.1016/S1470-160X(01)00007-3, 2001.

Baird, A. J., Belyea, L. R., and Morris, P. J.: Upscaling of peatland-atmosphere fluxes of methane: small-scale heterogeneity in process rates and the pitfalls of "bucket-and-slab" models, in: Carbon Cycling in Northern Peatlands, Geophysical Monograph Series, 184, edited by: Baird, A. J., Belyea, L. R., Comas, X., Reeve, A., and Slater, L., American Geophysical Union, Washington, DC, 37–53, doi:10.1029/2008GM000826, 2009.

Baird, A. J., Morris, P. J., and Belyea, L. R.: The DigiBog peatland development model 1: Rationale, conceptual model, and hydrological basis, Ecohydrology, 5, 242–255, doi:10.1002/eco.230, 2012.

Barber, K. E.: Peat Stratigraphy and Climate Change: A Palaeoecological Test of the Theory of Cyclic Peat Bog Regeneration, Balkema, Rotterdam, Netherlands, 1981.

Belyea, L. R.: Nonlinear dynamics of peatlands and potential feedbacks on the climate system, in: Carbon Cycling in Northern Peatlands, Geophysical Monograph Series, 184, edited by: Baird, A. J., Belyea, L. R., Comas, X., Reeve, A., and Slater, L., American Geophysical Union, Washington, DC, 5–18, doi:10.1029/2008GM000829, 2009.

Belyea, L. R. and Baird, A. J.: Beyond "the limits to peat bog growth": Cross-scale feedback in peatland development, Ecol. Monogr., 76, 299–322, doi:10.1890/0012-9615(2006)076[0299:BTLTPB]2.0.CO;2, 2006.

Belyea, L. R. and Clymo, R. S.: Feedback control of the rate of peat formation, P. Roy. Soc. Lond. B, 268, 1315–1321, doi:10.1098/rspb.2001.1665, 2001.

Boelter, D. H.: Physical properties of peats as related to degree of decomposition, Soil Sci. Soc. Am. Pro., 33, 606–609, 1969.

Boelter, D. H.: Methods for analysing the hydrological characteristics of organic soils in marsh-ridden areas, in: Proceedings of the IAHS Symposium on the Hydrology of Marsh Ridden Areas, Minsk, IAHS/UNESCO, Paris, 161–169, 1972.

Bubier, J., Costello, A., Moore, T. R., Roulet, N. T., and Savage, K.: Microtopography and methane flux in boreal peatlands, northern Ontario, Canada, Can. J. Botany, 71, 1056–1063, doi:10.1139/b93-122, 1993.

Bubier, J. L., Moore, T. R., Bellisario, L., Comer, N. T., and Grill, P. M.: Ecological controls on methane emissions from a northern peatland complex in the zone of discontinuous permafrost, Manitoba, Canada, Global Biogeochem. Cy., 9, 455–470, doi:10.1029/95GB02379, 1995.

Clymo, R. S.: Hydraulic conductivity of peat at Ellergower Moss, Scotland, Hydrol. Process., 18, 261–274, doi:10.1002/hyp.1374, 2004.

Comas, X., Slater, L., and Reeve, A.: Stratigraphic controls on pool formation in a domed bog inferred from ground penetrating radar (GPR), J. Hydrol., 315, 40–51, doi:10.1016/j.jhydrol.2005.04.020, 2005.

Couwenberg, J.: A simulation model of mire patterning – revisited, Ecography, 28, 653–661, doi:10.1111/j.2005.0906-7590.04265.x, 2005.

Couwenberg, J. and Joosten, H.: Self-organization in raised bog patterning: the origin of microtope zonation and mesotope diversity, J. Ecol., 93, 1238–1248, doi:10.1111/j.1365-2745.2005.01035.x, 2005.

Eppinga, M. B., Rietkerk, M., Borren, W., Lapshina, E. D., Bleuten, W., and Wassen, M. J.: Regular surface patterning of peatlands: Confronting theory with field data, Ecosystems, 11, 520–536, doi:10.1007/s10021-008-9138-z, 2008.

Eppinga, M. B., de Ruiter, P. C., Wassen, M. J., and Rietkerk, M.: Nutrients and hydrology indicate the driving mechanisms of peatland surface patterning, The American Naturalist, 173, 803–818, doi:10.1086/598487, 2009.

Foster, D. R. and Fritz, S. C.: Mire development, pool formation and landscape processes on patterned fens in Dalarna, central Sweden, J. Ecol., 75, 409–437, doi:10.2307/2260426, 1987.

Foster, D. R. and Wright, H. E.: Role of Ecosystem Development and Climate Change in Bog Formation in Central Sweden, Ecology, 71, 450–463, doi:10.2307/1940300, 1990.

Fraser, C. J. D., Roulet, N. T., and Laffleur, M.: Groundwater flow patterns in a large peatland, J. Hydrol., 246, 142–154, doi:10.1016/S0022-1694(01)00362-6, 2001.

Freeze, R. A. and Cherry, J. A.: Groundwater. Prentice Hall, Englewood Cliffs, New Jersey, United States, 1979.

Frolking, S., Roulet, N. T., Tuittila, E., Bubier, J. L., Quillet, A., Talbot, J., and Richard, P. J. H.: A new model of Holocene peatland net primary production, decomposition, water balance, and peat accumulation, Earth Syst. Dynam., 1, 1–21, doi:10.5194/esd-1-1-2010, 2010.

Grimm, V., Revilla, E., Berger, U., Jeltsch, F., Mooij, W. M., Railsback, S. F., Thulke, H., Weiner, J., Wiegand, T., and DeAngelis, D. L.: Pattern-oriented modeling of agent-based complex systems: lessons from ecology, Science, 310, 987–991, doi:10.1126/science.1116681, 2005.

Grover, S. P. P. and Baldock, J. A.: The link between peat hydrology and decomposition: Beyond von Post, J. Hydrol., 479, 130–138, doi:10.1016/j.jhydrol.2012.11.049, 2013.

Ingram, H. A. P.: Size and shape in raised mire ecosystems: a geophysical model, Nature, 297, 300–303, doi:10.1038/297300a0, 1982.

Ivanov, K. E.: Water Movement in Mirelands (translated from Russian by Thompson, A. and Ingram, H. A. P.), Academic Press, London, 1981.

Kettridge, N., Binley, A., Comas, X., Cassidy, N. J., Baird, A. J., Harris, A., van der Kruk, J., Strack, M., Milner, A. M., and Waddington, J. M.: Do peatland microforms move through time? Examining the developmental history of a patterned peatland using ground-penetrating radar, J. Geophys. Res., 117, G03030, doi:10.1029/2011JG001876, 2012.

Korpela, I., Koskinen, M., Vasander, H., Hopolainen, M., and Minkkinen, K.: Airborne small-footprint discrete-return LiDAR data in the assessment of boreal mire surface patterns, vegetation, and habitats, Forest Ecol. Manag., 258, 1549–1566, doi:10.1016/j.foreco.2009.07.007, 2009.

Koutaniemi, L.: Twenty-one years of string movements on the Liippasuo aapa mire, Finland, Boreas, 28, 521–530, doi:10.1111/j.1502-3885.1999.tb00238.x, 1999.

Lefever, R. and Lejeune, O.: On the origin of tiger bush, B. Math. Biol., 59, 263–294, doi:10.1007/BF02462004, 1997.

McWhorter, D. B. and Sunada, D. K.: Ground-water Hydrology and Hydraulics. Water Resources Publications, Fort Collins, Colorado, United States, 1977.

Morris, P. J., Belyea, L. R., and Baird, A. J.: Ecohydrological feedbacks in peatland development: A theoretical modelling study, J. Ecol., 99, 1190–1201, doi:10.1111/j.1365-2745.2011.01842.x, 2011.

Morris, P. J., Baird, A. J., and Belyea, L. R.: The DigiBog peatland development model 2: Ecohydrological simulations in 2D, Ecohydrology, 5, 256–268, doi:10.1002/eco.229, 2012.

Nungesser, M. K.: Modelling microtopography in boreal peatlands: Hummocks and hollows, Ecol. Model., 165, 175–207, doi:10.1016/S0304-3800(03)00067-X, 2003.

Peterson, G. D.: Contagious disturbance, ecological memory, and the emergence of landscape pattern, Ecosystems, 5, 329–338, doi:10.1007/s10021-001-0077-1, 2002.

Rietkerk, M. G. and van de Koppel, J.: Regular pattern formation in real ecosystems, Trends Ecol. Evol., 23, 169–175, doi:10.1016/j.tree.2007.10.013, 2008.

Rietkerk, M. G., Dekker, S. C., de Ruiter, P. C., and van de Koppel, J.: Self-organized patchiness and catastrophic shifts in ecosystems, Science, 305, 1926–1929, doi:10.1126/science.1101867, 2004a.

Rietkerk, M. G., Dekker, S. C., Wassen, M. J., and Verkroost, A. W. M.: A putative mechanism for Bog Patterning, The American Naturalist, 163, 699–708, doi:10.1086/383065, 2004b.

Roulet, N. T., Lafleur, P. M., Richard, P. J. H., Moore, T. R., Humphreys, E. R., and Bubier, J.: Contemporary carbon balance and late Holocene carbon accumulation in a northern peatland, Glob. Change Biol., 13, 397–411, doi:10.1111/j.1365-2486.2006.01292.x, 2007.

Rydin, H. and Jeglum, J. K.: The Biology of Peatlands, Oxford University Press, Oxford, UK, 2006.

Swindles, G. T., Morris, P. J., Baird, A. J., Blaauw, M., and Plunkett, G.: Ecohydrological feedbacks confound peat-based climate reconstructions, Geophys. Res. Lett., 39, L11401, doi:10.1029/2012GL051500, 2012.

Swanson, D. K. and Grigal, D. F.: A simulation model of mire patterning, Oikos, 53, 309–314, 1988.

Swanson, D. K.: Interaction of mire microtopography, water supply, and peat accumulation in boreal mires, Suo, 58, 37–47, 2007.

Turing, A.: The chemical basis of morphogenesis, Philos. T. R. Soc. Lon. B, 237, 37–72, doi:10.1016/S0092-8240(05)80008-4, 1952.

van de Koppel, J. and Crain, C. M.: Scale-dependent inhibition drives regular tussock spacing in a freshwater marsh, The American Naturalist, 168, 136–147, doi:10.1086/508671, 2006.

5

Rapid marine deglaciation: asynchronous retreat dynamics between the Irish Sea Ice Stream and terrestrial outlet glaciers

H. Patton[1], A. Hubbard[1], T. Bradwell[2], N. F. Glasser[1], M. J. Hambrey[1], and C. D. Clark[3]

[1]Institute of Geography and Earth Sciences, Aberystwyth University, Aberystwyth, SY23 3DB, UK
[2]British Geological Survey, Murchison House, West Mains Road, Edinburgh, EH9 3LA, UK
[3]Department of Geography, University of Sheffield, Sheffield, S10 2TN, UK

Correspondence to: H. Patton (henrypatton@gmail.com)

Abstract. Understanding the retreat behaviour of past marine-based ice sheets provides vital context for accurate assessments of the present stability and long-term response of contemporary polar ice sheets to climate and oceanic warming. Here new multibeam swath bathymetry data and sedimentological analysis are combined with high resolution ice-sheet modelling to reveal complex landform assemblages and process dynamics associated with deglaciation of the Celtic ice sheet within the Irish Sea Basin. Our reconstruction indicates a non-linear relationship between the rapidly receding Irish Sea Ice Stream and the retreat of outlet glaciers draining the adjacent, terrestrially based ice cap centred over Wales. Retreat of Welsh ice was episodic; superimposed over low-order oscillations of its margin are asynchronous outlet readvances driven by catchment-wide mass balance variations that are amplified through migration of the ice cap's main ice divide. Formation of large, linear ridges which extend at least 12.5 km offshore (locally known as sarns) and which dominate the regional bathymetry are attributed to repeated frontal and medial morainic deposition associated with the readvancing phases of these outlet glaciers. Our study provides new insight into ice-sheet extent, dynamics and non-linear retreat across a major palaeo-ice stream confluence zone, and has ramifications for the interpretation of recent fluctuations observed by satellites over short timescales across marine sectors of the Greenland and Antarctic ice sheets.

1 Introduction

The mass balance and stability of ice sheets is strongly determined by the dynamic behaviour of fast-flowing ice streams and outlet glaciers, as it is through these rapid conveyors of mass that the majority of ice flux and ultimate loss to calving and melt occur. However, our understanding of long-term stability of contemporary ice sheets is fundamentally hampered by the slow thermodynamic response and evolution of the ice-sheet system, leading to large uncertainties in the prediction of ice-sheet responses on timescales longer than the satellite-derived observational record (Hindmarsh, 1995; van der Veen, 2004; IPCC WG1, 2007; Bamber et al., 2007). Reconstruction of the complex deglaciation of palaeo-ice

sheets therefore offers an opportunity to explore dynamic behaviour and interactions, providing critical insight and context to potential centennial- to millennial-scale responses of our present-day polar ice sheets and their contribution to global sea-level rise (e.g. Calov et al., 2002; Knutz et al., 2007; Greenwood and Clark, 2009; Hubbard et al., 2009).

Across the British–Irish continental shelf, high-resolution, marine based, geophysical data sets have helped to shift consensus on the maximum reconstructed extent of the last ice sheet centred over Britain and Ireland to a version that was largely marine-influenced, characterised by high dynamism and advanced asynchronously across much of the continental shelf (Bowen et al., 2002; Bradwell et al., 2008; Hubbard et al., 2009; Chiverrell and Thomas, 2010; Clark et al., 2012;

Cofaigh et al., 2012). However, the Irish Sea Basin, host to the largest ice stream of the ice sheet, has received relatively little attention, and inferences on its advance, rapid recession, and interaction with adjacent ice accumulation centres have largely relied on sedimentological interpretations and cosmogenic isotope exposure ages taken from coastal sections (Eyles and McCabe, 1989; Huddart, 1991; Merritt and Auton, 2000; Glasser et al., 2001; Ó Cofaigh and Evans, 2001; Evans and Cofaigh, 2003; Patton and Hambrey, 2009; Van Landeghem et al., 2009; Chiverrell et al., 2013).

In this paper we present new high-resolution, multibeam echo sounder data from the eastern margin of the Irish Sea Basin that reveal submarine glacial landforms close to the former confluence zone of the Irish Sea Ice Stream and the terrestrially based Welsh Ice Cap (Fig. 1). Geomorphological mapping, in combination with previous sedimentological interpretations and numerically modelled output, are used to propose a regional reconstruction of complex ice dynamics and retreat that is driven, in part, by a response to climate variations, and also to internal flow reorganisations. Rather than treating the empirical evidence and modelling as separate exercises (or one tested against the other), here we pioneer a new approach that uses both to yield a modelling-informed empirical reconstruction of ice-sheet history.

Insight concerning the retreat of the Welsh Ice Cap, particularly so for its location at the peripheral margin of the glacierised area of Celtic Britain, is seen as a useful analogue for the future response of contemporary terrestrially based ice caps, such as those in Svalbard (Moholdt et al., 2010), Canada (Burgess and Sharp, 2004) and Iceland (Magnússon et al., 2005; Bradwell et al., 2013). Predictions of eustatic sea-level rise in the 21st century indicate glaciers such as these will be significant contributors (e.g. Meier et al., 2007). The bathymetric data also elucidate more details of the enigmatic "sarns" (large gravel ridges) that extend into Cardigan Bay, the origins of which are still equivocal.

1.1 Geological context

The stark geological differences between the Irish Sea Basin and Welsh hinterland allow for relatively easy discrimination of clast provenances within glacial deposits. The Irish Sea Basin is characterised largely by Mesozoic sandstones, mudstones and limestone, while volcanic and metamorphic rocks of Neoproterozoic to Ordovician age dominate the Llŷn Peninsula in NW Wales. The Welsh mountains are composed largely of Cambro-Ordovician volcanic and clastic sedimentary rocks, flanked by extensive Silurian mudstones and sandstones.

The Llŷn Peninsula marks an important glaciological zone of confluence between the former Irish Sea Ice Stream and Welsh Ice Cap in the eastern Irish Sea Basin, where a distinction can be made between glacial sediments to the west deposited exclusively by Irish Sea ice flowing southwards (e.g. Porth Oer), and those to the east deposited solely by

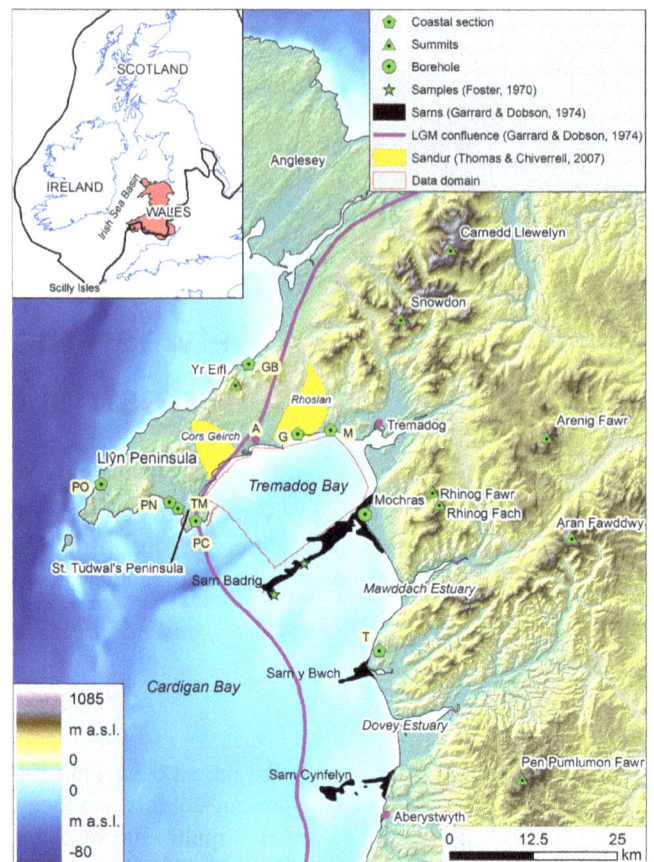

Figure 1. The terrestrially based Welsh Ice Cap, a semi-independent accumulation centre on the southern periphery of glaciation in Celtic Britain (inset box). The study area of Tremadog Bay in NW Wales, delimited in red, is bounded to the north by the Llŷn Peninsula, and Sarn Badrig to the south. Coastal sedimentological sections: Porth Oer (PO), Porth Neigwl (PN), Tai Morfa (TM), Porth Ceiriad (PC), Gwydir Bay (GB), Glanllynnau (G), Morannedd (M), and Tonfanau (T). Towns: Abererch (A). Relief is displayed using NEXTMap Britain (Intermap Technologies) data onshore, and British Geological Survey (BGS) DigBath250 bathymetry data offshore.

the Welsh Ice Cap flowing west (e.g. Morannedd) (Campbell and Bowen, 1989; Fig. 1). The exposure of Welsh-sourced till at Porth Ceiriad places an important constraint on the maximum eastward limit of Irish Sea ice on the Llŷn Peninsula (Whittow and Ball, 1970), emphasising St. Tudwal's Peninsula as an important confluence zone between Welsh and Irish Sea ice masses.

For coastal exposures in this vicinity, the history of deposition has been complicated by numerous interpretations derived over the last century, particularly in terms of the number of Irish Sea and Welsh glacial phases (even glaciations), their relative timings, and the limits of glaciation associated with each (Jehu, 1909; Smith and George, 1961; Synge, 1964; Saunders, 1968a, b; Whittow and Ball, 1970; Mitchell, 1972; Bowen, 1973a, b; Boulton, 1977). More recent and extensive stratigraphic description, mapping and interpretation (Thomas et al., 1998; Thomas and Chiverrell, 2007) across the Llŷn Peninsula have since established a lithostratigraphical framework, recording a set of sediment–landform assemblages that reflect rapidly changing erosional and depositional environments during ice interaction and deglaciation within a single glaciation.

An important 2 km section at Tonfanau has shown Irish Sea-sourced diamicton to overlie Welsh till (Patton and Hambrey, 2009). Offshore in Cardigan Bay, numerous boreholes and geophysical surveys have identified a limit of Welsh-sourced till broadly perpendicular to the western end of the three sarns (Garrard and Dobson, 1974; Fig. 1).

1.2 Data collection

High-resolution bathymetric data covering $270 \, km^2$ of Tremadog Bay were collected using a multibeam echo-sounder during 2006 as part of the UK Civil Hydrography Programme co-ordinated by the Maritime and Coastguard Agency. A Kongsberg 3002D 300 kHz multibeam was used for data collection. Sound velocity data were also collected from a hull-mounted AML Smart sound velocity sensor. Attitude (heave, pitch and roll), heading and positional data were collected using an Applanix POS MV motion sensor, Ashtech GSR2200, Siemens GSM and Trimble DSU4100 Dynamic Global Positional Systems. All data were compiled on board using Kongsberg Seafloor Information Systems software. Post-acquisition data processing to remove data artefacts and prepare basic mean bathymetric and CUBE (Combined Uncertainty Bathymetric Evaluation) surfaces was conducted by the UK Hydrographic Office. The British Geological Survey carried out subsequent data processing and visualisation using Caris HIPS and SIPS, and Fledermaus software. Surface models (grids) and geotiffs of the xyz data were produced at a horizontal resolution of 7 m. To describe the geomorphological features, a variety of different data formats were used, including acoustic backscatter (amplitude of return, essentially a proxy for seabed substrate physical properties, such as hardness and roughness), hill-shaded bathymetric models, and xyz elevation data. Digital mapping of glacial landforms was carried out using ArcMap GIS software.

2 Geomorphological mapping results

Tremadog Bay is dominated by a generally uniform NE-SW trending shallow depression, bounded to the north by a large rocky platform extending from the Llŷn Peninsula. Maximum water depths recorded in the multibeam data reach 24.3 m below present-day sea level approximately at the midpoint in the basin. A number of landforms are clearly visible on the seabed (Fig. 2):

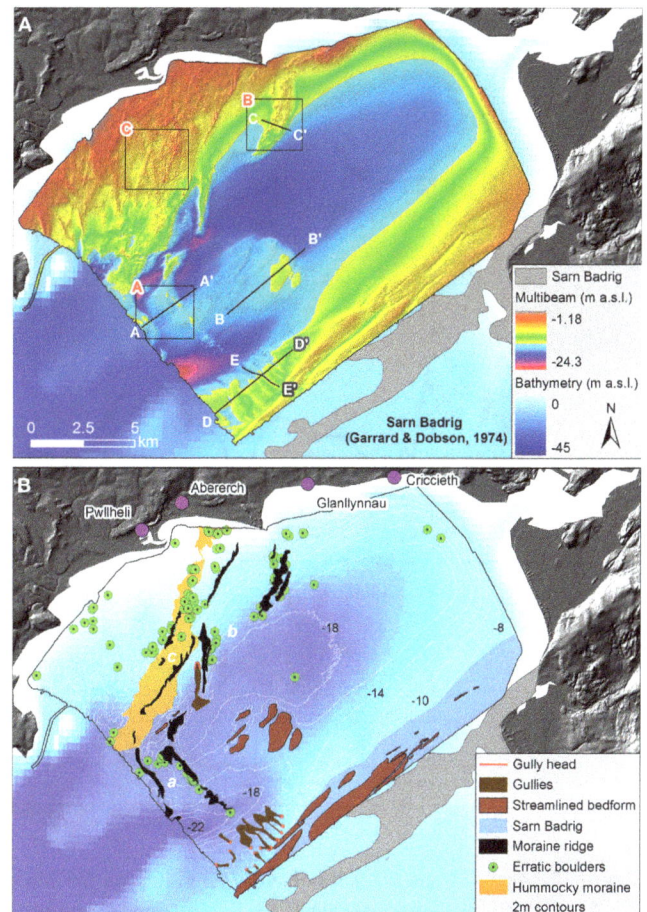

Figure 2. (A) Hillshaded high-resolution bathymetry of Tremadog Bay with locations of the transects in Fig. 3 identified. Inset boxes relate to close views of the data in Fig. 4. **(B)** Glacial landforms and features in Tremadog Bay are mapped from multibeam echo-sounder data. Interpreted moraines have been grouped according to suggested mechanisms of deposition. The grey outline of Sarn Badrig is defined from original survey results by Garrard and Dobson (1974). Contains Maritime and Coastguard Agency bathymetry data © Crown Copyright.

2.1 Recessional moraines

Moraines in the mapped site are divided into 3 sets (a–c) based on their distinctive morphologies and locations (Fig. 2).

2.1.1 Moraines: (a)

Description. To the SW of the domain lie a number of curvilinear ridges, parallel to each other and running perpendicular to Sarn Badrig and the Llŷn Peninsula, although near the margins of the bay they turn towards the coastline following the bathymetric contours. Near Sarn Badrig, the most landward moraine changes develops a relatively strong sinuous form. Their sharp crest lines are generally gently arcuate, trending NE at their northern end. The ridges vary in

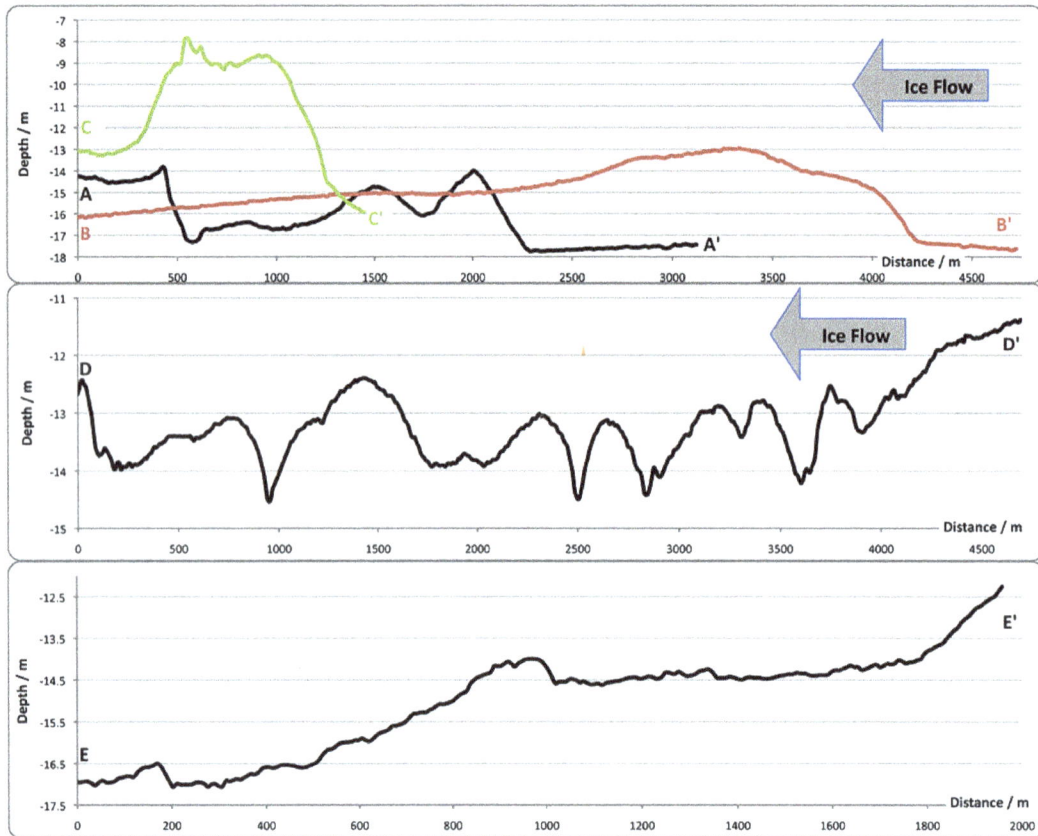

Figure 3. Vertical transects across landforms visible on the seabed of Tremadog Bay, taken from Fig. 2a. Landforms are interpreted as (A–A′) glaciotectonic push moraines, (B–B′) streamlined bedforms, (C–C′) frontal dump moraine, (D–D′) transverse profile of the gullies on Sarn Badrig, and (E–E′) along-profile of one gully.

height from 3 to 5 m above the sea floor on their stoss side. In cross-section their profiles are asymmetrical, with the lee slopes 2–3 m higher (Fig. 3a). Associated with the ridges are a number of boulders recognisable against the subdued relief of the seabed, the majority of which are found on the distal (western) slopes of the ridges (Fig. 4a).

Interpretation. On the basis of their morphology, distribution, and setting, the curvilinear features are interpreted as end moraines associated with ice flow from the Welsh Ice Cap. Their arcuate, steep-crested, asymmetrical, and sometimes sinuous form is typical of contemporary glaciotectonic push moraines that develop during a period of positive mass balance or glacier surge (Boulton, 1986; Boulton et al., 1999; Bennett, 2001; Ottesen and Dowdeswell, 2006; Bennett and Glasser, 2009), and have been previously described in a number of other submarine environments (e.g. Ottesen and Dowdeswell, 2006; Ottesen et al., 2007; Nygård et al., 2008). The fact that these ridges are found well within the Welsh drift limit identified by Garrard and Dobson (1974) and east of exposures of Welsh till on the Llŷn Peninsula at Porth Ceiriad (Whittow and Ball, 1970) indicates that these ridges do not mark the maximum extent of Welsh glaciation,

and must have been deposited following separation from Irish Sea ice during overall ice-sheet retreat.

2.1.2   Moraines: (b)

Description. Northeast from set (a) towards the Llŷn Peninsula are two sets of ridges slightly different in character; ridge-lines are generally less pronounced, more symmetrical (Fig. 3c), and in places appear to have undergone rotational sliding (Fig. 4b). As before, a number of boulders can be found close by, and in areas where the ridges have slumped, they can also be found on the proximal (eastern) side. The positions of these ridges may to be linked with bedrock rises around the margins of Tremadog Bay.

Interpretation. Based on their position, form, nested character, and association with numerous large boulders, these ridges are interpreted as ice-marginal moraines (e.g. Small, 1983; Benn and Evans, 2010). Structurally, the ice-proximal parts of ice-marginal moraines tend to be complex because of widespread collapse and reworking following the removal of ice support and melt out of buried ice (e.g. Benn and Owen, 2002).

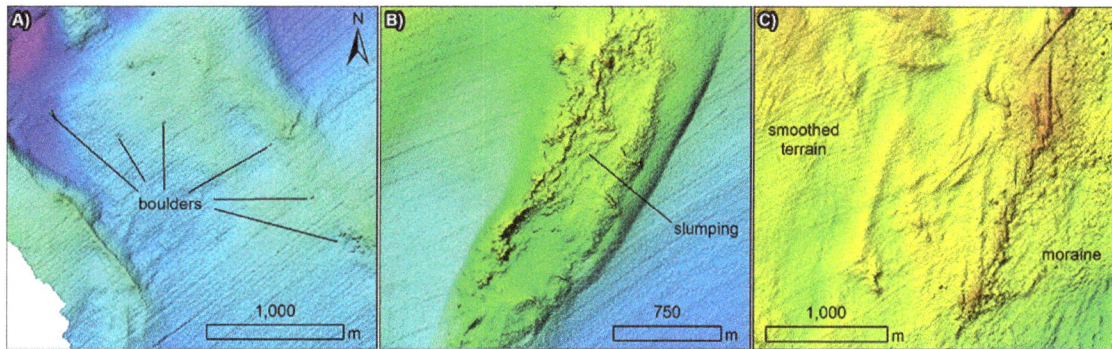

Figure 4. Magnified views of the Tremadog Bay seabed. Locations of each box are given in Fig. 2a. (**A**) Moraines and boulders on the western edge, (**B**) slumped moraines, and (**C**) contrasts between the smoothed terrain and moraine deposits to the north. Contains Maritime and Coastguard Agency bathymetry data © Crown Copyright.

2.1.3 Moraines: (c)

Description. To the northwest of the study area, running southwestwards from the coastline near Abererch, small (< 2 m), yet extensive, linear ridges can be found associated with numerous large scattered boulders. The general topography is composed of subdued hummocky relief with an irregular surface morphology, characterised by meandering ridges and collections of mounds, up to ~ 40 m wide and with no preferred orientation. Further west, the seabed dramatically changes character, becoming smooth and dissected by a number of large gullies.

Interpretation. One possibility is that the irregular and chaotic topography represented here is the remnant of a *controlled* hummocky moraine, with the pronounced transverse elements inherited from concentrated debris bands within the downwasting Tremadog outlet glacier (Boulton, 1972; Evans, 2009). Genetically, hummocky moraine has been used to encompass landforms with a variety of origins, including ice stagnation and englacial thrusting (cf. Hambrey et al., 1997, 1999). However, in its more restrictive sense it is used to refer to moraines deposited during the melt out of debris-mantled glaciers (Sharp, 1985; Benn and Evans, 1998). Without detailed glaciotectonic and sedimentological analyses, the term "hummocky moraine" is used here purely in its purely descriptively sense. An adjacent coastal exposure at Glanllynau, previously interpreted by Boulton (1977) as the result of stagnating ice-cored ridges, could form an important onshore equivalent of this offshore feature.

2.2 Streamlined bedforms

Description. In the centre of Tremadog Bay lie a number of smooth, partially oval-shaped hills (Fig. 2b). They are 3–5 m high and 2–3 km long with length-to-width ratios less than 50 : 1. The landforms are not associated with any obvious exposed bedrock features, and their long axes are all aligned perpendicular to the end moraines interpreted above. Similar elongated features are also present along the southern edge

of the domain, overprinting the northern flank of the sarn and trending the same direction as the bedforms in the centre of the bay. The elongation ratios (length/width) of these features vary from 7 : 1 to 12 : 1.

Interpretation. Although not on the scale of mega-scale glacial lineations (e.g. Clark, 1993), without further knowledge of the subsurface components of these features they are interpreted using the non-genetic term, streamlined bedforms (cf. Ó Cofaigh et al., 2002; King et al., 2009). Their long profile – a high, blunt stoss end pointing in the upstream direction and a more gently sloping, pointed end (tail) facing down-ice (Fig. 4b) – indicates principal ice flow to be associated with Welsh-sourced ice emanating from Snowdonia and flowing into Tremadog Bay towards the southwest (Glasser and Bennett, 2004).

2.3 Sarn Badrig

Description. The southern edge of the data domain encroaches onto Sarn Badrig, which is an obvious topographical ridge extending at least 12.5 km to the SW. This substantial seabed feature divides the northern basin of Cardigan Bay (Tremadog) from the mid-basin (Mawddach), and even at its western end the sea floor in places is still only < 7.5 m below sea level. Our new mapping from the bathymetric data indicate that the sarn is > 4 km wide, almost double its previously mapped width.

Interpretation. The superimposed streamlined bedforms, combined with the high percentage of clasts it contains from the Welsh hinterland (Foster, 1970), strongly indicates a glacial origin for this broad gravel ridge. Considering the position of all three sarns in Cardigan Bay at the interfluve of major onshore valleys in West Wales (Fig. 5b), Sarn Badrig is interpreted as a large "ice-stream" interaction medial moraine – the merging of two lateral moraines (Eyles and Rogerson, 1978a). This type of moraine forms at the intersection of confluent valley glaciers below or close to the equilibrium line, and their morphological expression

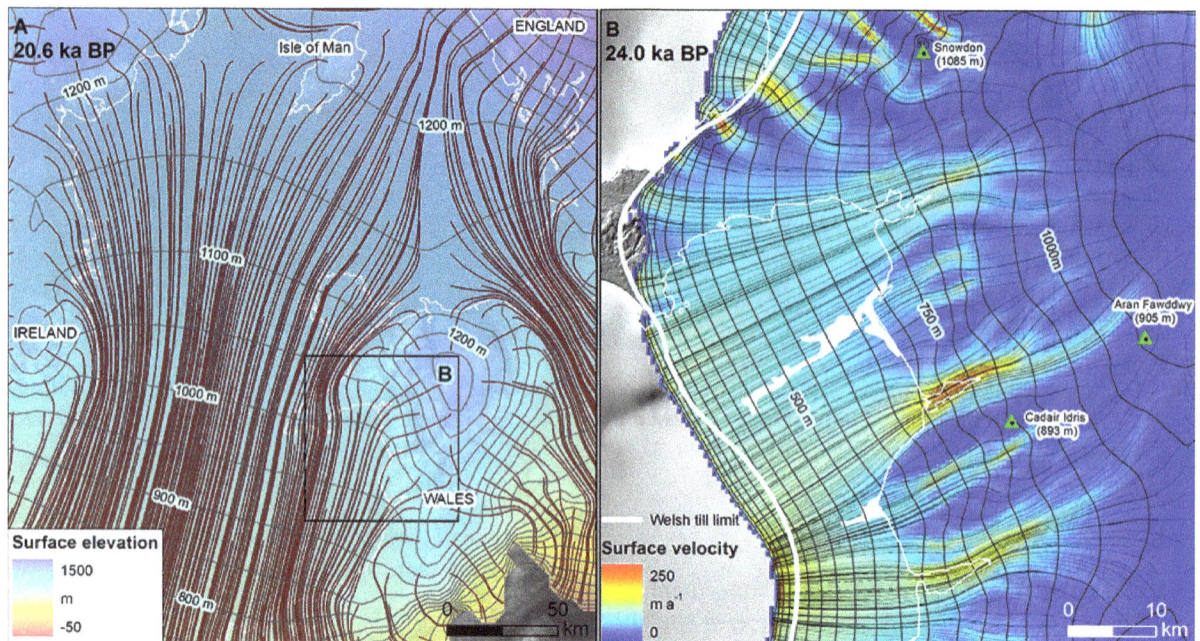

Figure 5. (A) Numerically modelled flowlines in the Irish Sea Basin and Welsh domain during the Last Glacial Maximum (LGM). Data are taken from the median experiment E109b8 by Hubbard et al. (2009). **(B)** Flowlines of the independently modelled Welsh Ice Cap by Patton et al. (2013a). The limit of Welsh till in Cardigan Bay was delimited by Garrard and Dobson (1974).

is influenced by the amount of available debris, with large amounts of englacial and pre-convergence subglacial debris encouraging long-lived, high-relief moraine ridges. Lateral compression between the fast-flowing glaciers forces the debris into a longitudinal septum (Smiraglia, 1989), with complex surface forms also prone to develop if the two glaciers have different velocities (Eyles and Rogerson, 1978a, b). We propose that its present submarine position accounts for much of the fines being removed, probably by a combination of wave washing and marine currents, leaving a predominantly gravel ridge, visible in places at low tide. Based on Sarn Badrig's relatively large volume ($\sim 1\,\text{km}^3$) and proximity to the main accumulation centres of the Welsh Ice Cap, it is speculated that the three sarns in Cardigan Bay are enduring features that have been reworked and added to over the course of numerous glaciations during the Pleistocene. This is supported by boreholes indicating "Welsh drift" over 100 m deep in the centre of Tremadog Bay (Garrard and Dobson, 1974), attesting to long-lived and/or rapid rates of glacigenic sediment deposition in this area.

2.4 Parallel ridges and linear depressions

Description. On the western flank of Sarn Badrig, approximately 10 superimposed linear depressions and parallel ridges occur orientated predominantly NW–SW. They are ~ 200–400 m wide and ~ 2 m deep, and contain along-profile undulations of < 0.5 m. Gradients along the floor of the depressions are very shallow, sometimes dropping only ~ 4 m

over several kilometres. Morphological transects reveal a stepped profile, with a steep section on their northwestern end (Fig. 3e). The depressions and associated ridges show a distinct increase in spacing in a down-ice direction (Fig. 3d).

Interpretation. One possible interpretation is that these depressions are erosional gullies, possibly formed through a combination of (or solely by) debris flows, ice-marginal discharge, or ice-cored moraine slumping (e.g. Bennett et al., 2000). Whether these gullies represent subaerial or submarine erosion is equivocal; iceberg pits, scour marks and De Geer moraines offshore from north Wales (> 36 m present water depth) indicate that deglaciation of grounded Irish Sea Basin ice occurred in the presence of a proglacial water body (Van Landeghem et al., 2009). However, given that present water depths in Tremadog Bay range from 8 to 22 m, and many coastal sections around the central and southern Irish Sea Basin have been shown to be deposited terrestrially (e.g. Thomas et al., 1998; Glasser et al., 2001; Lambeck and Purcell, 2001), we suggest that subaerial processes are more probable. A possible source for proglacial meltwater discharge on this flank of the sarn is the frontal margin of the adjacent Mawddach Glacier. With Tremadog Bay free of ice, a readvance of the adjacent Mawddach outlet lobe as far as the southern flank of Sarn Badrig could provide sufficient meltwater to have initiated erosion of these gullies. A more extensive multibeam survey and further mapping of the whole sarn would be required to test this hypothesis.

An alternative interpretation is that the intervening subparallel ridges are constructional glacigenic features. In light

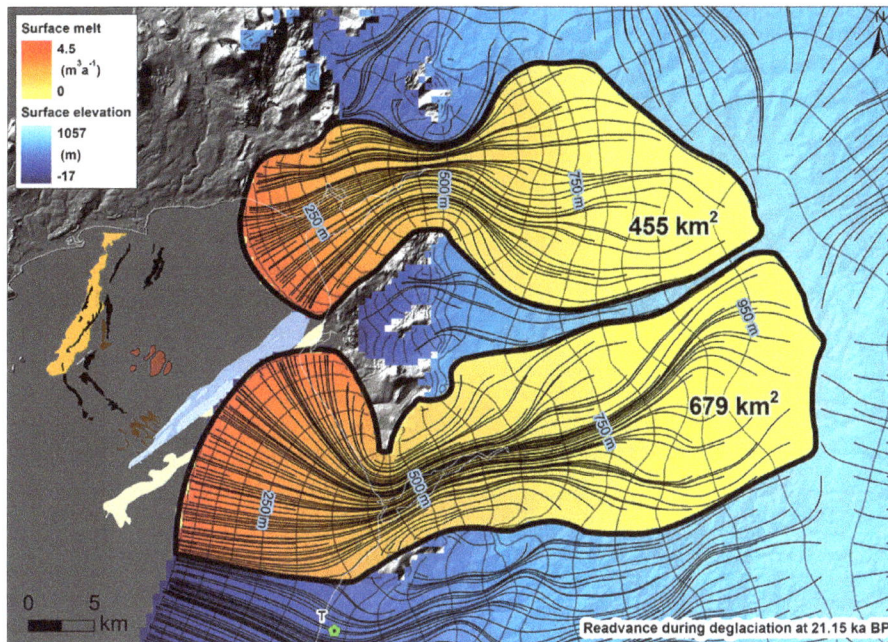

Figure 6. State of the modelled Welsh Ice Cap while readvancing during deglaciation at 21.15 ka BP. Areal extents of the Tremadog (northern) and Mawddach (southern) Glacier catchments are highlighted. A coastal section at Tonfanau (T) records Irish Sea-sourced diamicton overlying a till of distinct Welsh origin.

of only geomorphological data, several possibilities for their formation exist: (1) closely spaced recessional moraines, laid down at the retreating margin of a large (probably land-terminating) outlet glacier. Numerous analogous landforms currently exist around the margins of the present-day ice sheet and large ice caps in Greenland and Iceland (Evans and Twigg, 2002; Forman et al., 2007). (2) They could also be low-amplitude Rogen (ribbed) moraines, formed by sub-glacial deformation under a partially thawed or warm-based thermal regime (Lundqvist, 1989; Möller, 2006). The long axes of these short ridges align transverse to ice flow, consistent with the main Welsh ice-flow direction from Snowdonia; however their morphology is subdued and not typical of these landforms. (3) The third formation possibility is polygenetic ridges, possibly formed by ice overriding and reshaping/modifying pre-existing ice-marginal moraines. Some workers have proposed this model for sites of closely spaced transverse ridges where sedimentary structures reveal complex (polyphase) formation histories (e.g. Möller, 2006). Without knowledge of the sediment facies or structures within these submarine landforms in Tremadog Bay, their precise origin remains speculative.

3 Comparison with modelled output

The major benefit of numerical modelling is in its ability to describe ice masses holistically and time-transgressively. Geomorphological mapping commonly records isochronous events such as the maximum extension of the glacier ter-

minus, or time-indefinite processes such as ice streaming, with little indication of glaciodynamics occurring through a glacial cycle. When viewed in combination with modelled data, however, insights concerning landscape evolution can be more objectively assessed.

High-resolution modelled output used for comparison here is taken from experiments carried out by Patton et al. (2013a, b), who modelled the independent Welsh Ice Cap during the last glacial cycle. The optimal experiment referred to in this paper, E397, was derived using a 3-D thermomechanical model and ensemble methodology, whereby the sensitivities of key individual parameters in the model were examined through the systematic perturbation of their values. "Optimal" experiments were chosen based on how well the modelled data matched key empirical constraints (e.g. margin positions, inferred retreat dates), whilst still within a realistic parameter space.

Key characteristics for the experiment E397 include a maximum temperature suppression at the Last Glacial Maximum (LGM) of 11.85 °C, combined with a precipitation reduction of $\sim 1/3$rd of present-day values. Moraine limits to the east of the domain were matched by applying an enhanced west–east precipitation gradient (rain shadow effect). Post LGM, temperature and precipitation suppressions were moderately relaxed, although still kept scaled with the GISP2 climate curve. Model time slices from the northwest sector of Wales are shown in Figs. 5–7.

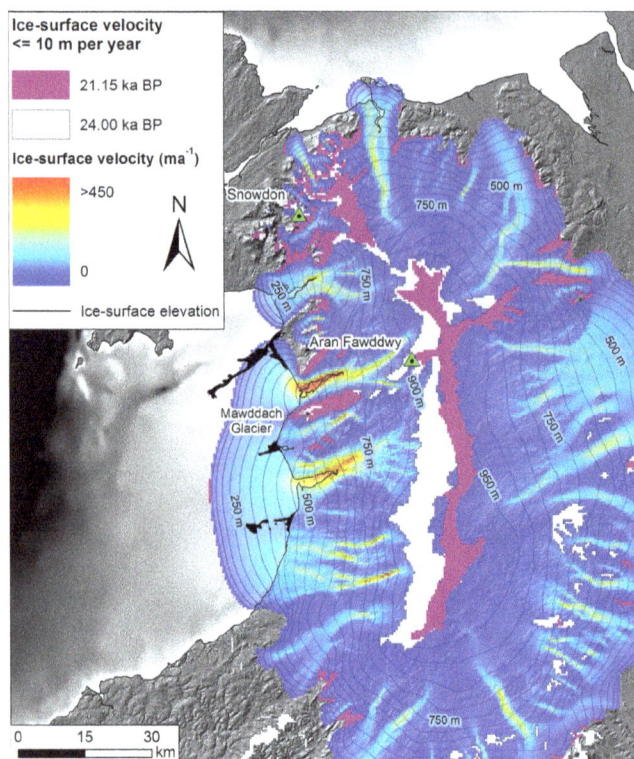

Figure 7. Modelled eastward migration of the central ice divide between 24.0 and 22.3 ka BP significantly affects the asymmetric response of the Welsh Ice Cap to renewed positive mass balance during a short-lived climatic downturn in the GISP2 forcing curve. At 22.3 ka BP the ice divide covering Snowdonia remains static, whilst further south around Aran Fawddwy the upper reaches of the Mawddach outlet glacier catchment have notably increased.

3.1   Ice advance

During advance of the modelled Welsh Ice Cap, conditions conducive to fast-flowing outlet glaciers are strongly modulated by oscillations within the GISP2 record of climate forcing, and triggered by transitions to a relatively warm climate (Patton et al., 2013a, b). It is envisaged that these "purge" or fluctuation events would have led to the formation of the streamlined bedforms described in the data domain above. Figure 5b shows the position of the modelled Welsh Ice Cap 500 yr before its glacial maximum. Of particular note from this time slice is the position of the ice front on the Llŷn Peninsula and in Cardigan Bay compared with the limit identified by Garrard and Dobson (1974) from numerous borehole observations. Also of interest are the striking ~100 m high crags just north of Tremadog, which appear to exert some control on ice flow, close to the transition zone between fast and slow ice-surface velocities of the Tremadog Glacier. The position of the sarns at the interfluves of outlet glaciers entering Cardigan Bay, as well as their orientation trending parallel to the direction of modelled flowlines, further supports the hypothesis that they have been streamlined

by Welsh-sourced ice, and at least in part, represent inter-stream or medial moraine deposits.

It was during this time period that the Irish Sea Ice Stream was advancing through the Irish Sea Basin (Scourse, 1991; McCarroll et al., 2010; Chiverrell et al., 2013), coalescing with the Welsh Ice Cap (Fig. 5a). The confluence of these two glaciers would have led to high localised shear strain rates, with the flow units of differing velocities being forced along a path of convergence, inferred here as the Welsh "drift" limit mapped by Garrard and Dobson (1974). Direct mapping of present-day ice-sheet systems reveal ubiquitous longitudinal surface lineations called flow stripes or flow lines (Casassa et al., 1991; Casassa and Brecher, 1993; Fahnestock et al., 2000), which are sometimes interpreted as longitudinal foliation (Reynolds and Hambrey, 1988; Casassa et al., 1991; Casassa and Brecher, 1993; Hambrey and Dowdeswell, 1994; Glasser and Scambos, 2008; Glasser et al., 2011). Where glaciers converge, larger flow units tend to "pinch out" these structures where they meet smaller tributary glaciers (e.g. Glasser and Gudmundsson, 2012). Based on these contemporary analogues, it is speculated here that the western ends of the sarns would reflect this converging flow pattern, showing strong deflection southwards. Numerically modelled flow lines indicate that the Welsh Ice Cap was large enough to deflect advancing Irish Sea Basin ice onto the Llŷn Peninsula and away from the present-day west Wales coastline (Fig. 5a). This result supports observations from sedimentological sections along the Llŷn Peninsula that Welsh ice dominated in Tremadog Bay and eastern Cardigan Bay, leaving St. Tudwal's Peninsula as a key convergence point for both ice masses. However, the absence of the horizontal shear stress within the numerical model's first-order solution of the ice-flow equations means that the high localised shear strain rates predicted above are not replicated in this reconstruction (cf. Hubbard et al., 2009). More expansive submarine mapping of Cardigan Bay would also be needed to fully unlock the key ice dynamics in this region.

3.2   Ice retreat

The history and chronology of the Irish Sea Ice Stream has been relatively well constrained using numerous absolute dating techniques and Bayesian modelling, with retreat starting around 24.0–23.3 ka from the Isles of Scilly (cf. Chiverrell et al., 2013). Cosmogenic nuclide ages between 22.5–21.2 ka from the Llŷn Peninsula slightly predate general thinning of the Welsh Ice Cap and the exposure of mountain summits in mid-Wales between ca. 20–17 ka (Glasser et al., 2012). The general retreat pattern is complicated, however, by observations of repeated minor advances. Minor fluctuations of the Irish Sea Ice Stream have been recorded from Anglesey (Thomas and Chiverrell, 2007) and southeast Ireland (Thomas and Chiverrell, 2011), and the sharp-crested end moraines (set a) identified in this study north of Sarn Badrig also strongly suggest similar small-scale, ice-marginal oscillations of the Welsh Ice Cap.

Superimposed on these relatively low-order fluctuations are the more complex high-magnitude and non-linear response of the former ice masses to internal and external drivers. Bayesian modelling has given strong indication of a general slowdown in the rate of retreat of the Irish Sea Ice Stream once it reached the narrowing of the Irish Sea Basin between Wales and Ireland, with topographic confinement, "sticky spots", and change in bed slope all cited as probable reasons (Chiverrell et al., 2013). Similarly, new ice-sheet modelling of the Welsh Ice Cap has indicated deglaciation was punctuated by a phase of major, yet asymmetric readvance along its western margins ca. 21.15 ka BP (Fig. 6). This minor readvance captured in the model experiment coincides with a known Dansgaard–Oeschger event in the GISP2 climate record (Dansgaard et al., 1993; Grootes et al., 1993) – a short-lived cold episode that punctuated the last glaciation in the North Atlantic region. Analyses of the ice-rafted debris record from the British–Irish continental shelf indicate that such millennial-scale variability of palaeo ice-sheet dynamics in this region was a common phenomenon, with numerous, large flux events occurring during Marine Isotope Stages 3 and 2, in phase with known Dansgaard–Oeschger events (Scourse et al., 2009).

Model output reveals that the asymmetric response of outlet glaciers entering Cardigan Bay at this time is largely driven by the amplification of mass-balance variations between adjacent catchments, enhanced by eastward migration of the central ice-divide of the ice cap (Fig. 7). Similar, though more complex, dynamics in both spatial and temporal domains associated with ice divide migration have been inferred from the nearby Irish Ice Cap (Greenwood and Clark, 2009).

This dynamically-forced readvance episode captured in the model experiment raises another possibility for the formation of Sarn Badrig in particular: it is a composite feature, also representing morainic deposition at the frontal margin of the Mawddach outlet glacier. Figure 6 illustrates this variation; once Tremadog Bay is free of ice, the Mawddach Glacier is left unconstrained on its northern side and is thus free to advance across low-lying ground as far as Sarn Badrig. Although the relatively larger volume of Sarn Badrig could be a function of preservation, it may also reflect this composite development during readvances of the Welsh Ice Cap. A borehole at Mochras Point at the head of Sarn Badrig discovered a large number of different "drift" units up to 77 m below OD, some of which have been speculated to be of pre-Devensian age (e.g. O'Sullivan, 1971), lending support to repeated deliveries of glacial material from the Welsh hinterland. Unfortunately no conclusive ages have been found for these sediments (cf. Herbert-Smith, 1971). The interpretation, however, is complicated by the presence of an extensive Irish Sea-sourced diamicton overlying Welsh till at Tonfanau (Patton and Hambrey, 2009). The presence of well-preserved

Welsh-sourced glacial landforms in Tremadog Bay provides evidence that Irish Sea Basin ice must have impinged upon the coastline at Tonfanau prior to regional deglaciation, possibly when the Welsh Ice Cap and Irish Sea Ice Stream were confluent at the glacial maximum. Irish Sea Basin-sourced erratics deposited at Tonfanau may then have been subsequently incorporated and reworked into the upper till during readvance of the Mawddach outlet glacier (Fig. 6). This hypothesis may also account for the small proportion of Irish Sea erratics within Sarn Badrig (Foster, 1970).

In light of this hypothesis, rising sea level is not considered to be a driver for asynchronous retreat between adjacent outlet glaciers in Cardigan Bay since the similarity and shallow depth of the seabed around west Wales precludes any potential differences in calving rates or in the timing of retreat (Fig. 1). Also, despite initial and rapid eustatically forced retreat of Irish Sea Basin ice from the Celtic Sea (Scourse and Furze, 2001; Scourse et al., 2009; Chiverrell et al., 2013), coastal sections around the central and southern Irish Sea basin have consistently shown deglaciation occurred under terrestrial conditions (cf. McCarroll, 2001), in contrast to earlier interpretations of a glaciomarine model for deglaciation (Eyles and McCabe, 1989).

4   Sequence of events

Through interpretation of landform mapping, ice-sheet modelling and previously described sedimentological sections, the following sequence of events at the coalescent margin between the Irish Sea Ice Stream and Welsh Ice Cap is suggested for the Last Glacial Maximum (Fig. 8):

1. With the onset of widespread glaciation, Welsh-sourced ice rapidly inundated near-shore areas from proximal accumulation areas in the Welsh hinterland (Fig. 8a). Meanwhile further north, ice centres in Scotland, northern England and Northern Ireland were also growing, feeding ice into the Irish Sea Basin and initiating ice drainage, which would eventually stream as far south as the Scilly Isles (Scourse et al., 1990; Scourse, 1991; Hiemstra et al., 2006). In assuming a glacial origin for the sarns, these landforms were likely to have been present in Cardigan Bay throughout much of the Quaternary (Fig. 8a).

2. During the LGM, Welsh and Irish Sea ice coalesced; the topography of the Llŷn Peninsula, in combination with the rapidly expanding Welsh Ice Cap, acted as a major obstacle to south-flowing ice, leaving Tremadog Bay dominated by Welsh-sourced ice (Fig. 8b). As the Tremadog glacier reached maximum mass turnover, sediments at its base were being reworked, producing the streamlined bedforms observed within the bay and along the northern sarn edge. Abundant Irish Sea erratics exposed at Tonfanau indicate the Irish Sea Ice

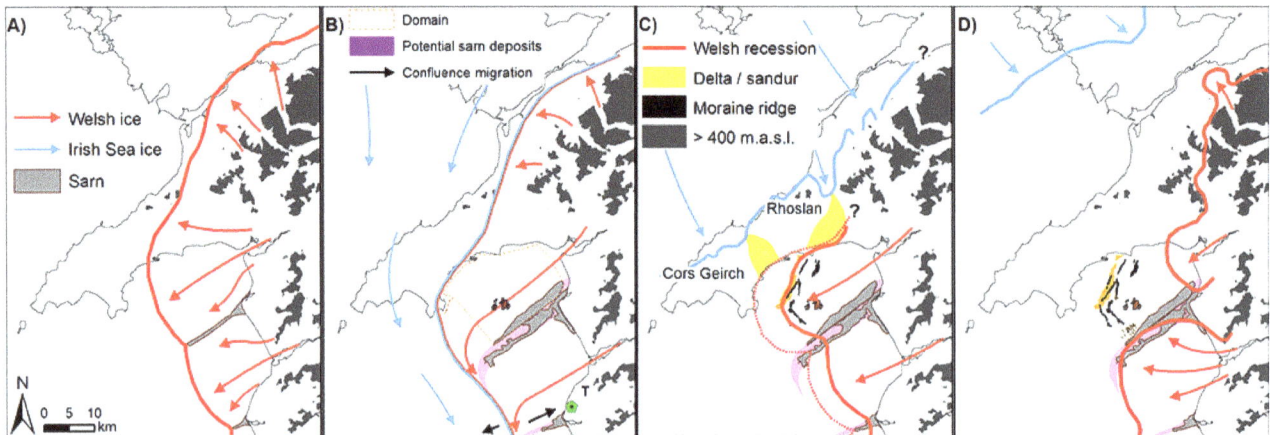

Figure 8. Proposed schematic sequence of events during the Last Glacial Maximum (LGM) in NW Wales (Welsh Ice: red; Irish Sea ice: blue): **(A)** Prior to advance of the Irish Sea Ice Stream, Welsh ice inundated the near-shore areas of Cardigan Bay. **(B)** At the LGM, Irish Sea and Welsh ice coalesced, the boundary defined by offshore borehole records (Garrard and Dobson, 1974) and geomorphological mapping on the Llŷn Peninsula (Thomas et al., 1998). Irish Sea erratics at Tonfanau (T) suggest possible impingement here after advance of Welsh ice (Patton and Hambrey, 2009). **(C)** During deglaciation, Welsh Ice still dominated Cardigan Bay, helping create the lacustrine delta terraces at Cors Geirch (Thomas et al., 1998). **(D)** Modelled output indicates a probable readvance during deglaciation of the Welsh Ice Cap ca. 21.15 ka BP, enhanced by migration of the central ice divide and subsequent enlargement of the Mawddach Glacier catchment.

Stream impinged along the coastline here after a period of Welsh advance (Patton and Hambrey, 2009), possibly as a result of migration of the confluence zone between the two ice masses. Convergence of Welsh and Irish Sea ice flow units along the Irish Sea "drift" limit would probably have extended and deflected sarn-deposition southeastwards.

3. While the Irish Sea Ice Stream rapidly retreated from its unstable maximum extent with the onset of widespread deglaciation ~ 24.0–23.3 ka (Scourse et al., 2009; Chiverrell et al., 2013), Welsh glaciers still dominated NE Cardigan Bay (Fig. 8c). Evidence for this includes the Cors Geirch terraces on the Llŷn Peninsula, which are the remnants of a glacial lake fed by Irish Sea Basin ice from the north, but dammed by Welsh-sourced ice to the south (Matley, 1936; Thomas and Chiverrell, 2007).

4. As deglaciation continued, moraines were laid down in Tremadog Bay during minor or seasonal readvances. The larger catchment of the Mawddach basin, coupled with a shift in the central ice divide, probably drove a minor readvance of ice from mid-Wales extending as far north as the ridge of Sarn Badrig during the Dansgaard–Oeschger event ~ 21.15 ka BP. In contrast, ice in Tremadog Bay retreated back to the present-day coastline at this time, during its overall recession towards the N Wales mountains (Fig. 8d).

5. With rapid marine transgression in Cardigan Bay following deglaciation, the sarns were submerged and were probably washed clean of fine glacial sediment by cur-

rents within the intertidal zone. The present-day geomorphology on the shallow sea floor of Tremadog Bay (Fig. 2b) has been subsequently preserved beneath wave-base level.

5 Conclusions

New multibeam echo-sounder data collected from the eastern margin of the Irish Sea Basin reveal insights concerning glacier dynamics close to the confluence zone of the marine-influenced Irish Sea Ice Stream and terrestrially based Welsh Ice Cap. For the first time, definitive glacial landforms associated with Welsh ice flowing offshore are presented. Through a combined approach using landform mapping, sedimentological interpretations and ice-sheet modelling, a regional reconstruction for complex flow and deglaciation for the eastern Irish Sea Basin is proposed. Superimposed on the low-order, small-scale oscillations of both ice masses, a general pattern of non-linear retreat is suggested for the Welsh Ice Cap, with a major asymmetric readvance attributed to ice-divide migration amplifying mass-balance variations between adjacent catchments. Our findings support others that show asymmetric and asynchronous marginal behaviour to be an emerging characteristic feature of the ice sheet centred over Britain and Ireland, in response to a range of internal and external drivers. Such higher-order dynamism underscores the importance for understanding ice-stream processes and dynamics at the scale of individual drainage basins – not only to more accurately reconstruct palaeo ice-sheet dynamics, but also to predict future centennial- to millennial-scale changes of polar ice masses.

Modelled and geomorphological data indicate that the three sarns (large gravel ridges) in Cardigan Bay are composite features of frontal- and medial-moraine deposits that have been reworked through repeated advances of the Welsh outlet glaciers. Further sedimentological and bathymetric data collection in and around Cardigan Bay will serve to test the proposed reconstruction of events in this region.

Acknowledgements. This study was supported by a Joint Studentship (2K08/E108) provided by the British Geological Survey and Aberystwyth University via the BGS-University Funding Initiative, as well as a grant from the Climate Change Consortium of Wales. Swath bathymetry data used in this study was provided courtesy of the Maritime & Coastguard Agency's UK Civil Hydrography Programme. Rhys Cooper is thanked for processing the multibeam data, and Nick Golledge for insightful comments on an earlier version of the MS. James Scourse, Ola Fredin and Greg Hancock are also thanked for their constructive reviews. T. Bradwell acknowledges support from the MAREMAP programme and publishes with permission of the Executive Director, BGS (NERC).

Edited by: G. Hancock

References

Bamber, J. L., Alley, R. B., and Joughin, I.: Rapid response of modern day ice sheets to external forcing, Earth Planet. Sc. Lett., 257, 1–13, 2007.

Benn, D. I. and Evans, D. J. A.: Glaciers and Glaciation, Arnold, London, 1998.

Benn, D. I. and Evans, D. J. A.: Glaciers and Glaciation, Hodder Education, London, 2010.

Benn, D. I. and Owen, L. A.: Himalayan glacial sedimentary environments: a framework for reconstructing and dating the former extent of glaciers in high mountains, Quatern. Int., 97–98, 3–25, 2002.

Bennett, M. P.: The morphology, structural evolution and significance of push moraines, Earth-Sci. Rev., 53, 197–236, 2001.

Bennett, M. R. and Glasser, N. F.: Glacial Geology: ice sheets and landforms, John Wiley and Sons, Chichester, 2009.

Bennett, M. R., Huddart, D., Glasser, N. F., and Hambrey, M. J.: Resedimentation of debris on an ice-cored lateral moraine in the high-Arctic (Kongsvegen, Svalbard), Geomorphology, 35, 21–40, 2000.

Boulton, G. S.: Modern arctic glaciers as depositional models for former ice sheets, J. Geol. Soc. Lond., 128, 361–393, 1972.

Boulton, G. S.: A multiple till sequence formed by a late-Devensian Welsh ice cap: Glanllynau, Gwynedd, Cambria, 4, 10–31, 1977.

Boulton, G. S.: Push-moraines and glacier-contact fans in marine and terrestrial environments, Sedimentology, 33, 677–698, 1986.

Boulton, G. S., Van der Meer, J. J. M., Beets, D. J., Hart, J. K., and Ruegg, G. H. J.: The sedimentary and structural evolution of a recent push moraine complex: Holmstrombreen, Spitsbergen, Quaternary Sci. Rev., 18, 339–371, 1999.

Bowen, D., Phillips, F., McCabe, A., Knutz, P., and Sykes, G.: New data for the Last Glacial Maximum in Great Britain and Ireland, Quaternary Sci. Rev., 21, 89–101, 2002.

Bowen, D. Q.: The Pleistocene history of Wales and the borderland, Geol. J., 8, 207–224, 1973a.

Bowen, D. Q.: The Pleistocene succession of the Irish Sea, Proceedings of the Geologists' Association, 84, 249–272, 1973.

Bradwell, T., Stoker, M. S., Golledge, N. R., Wilson, C. K., Merritt, J. W., Long, D., Everest, J. D., Hestvik, O. B., Stevenson, A. G., Hubbard, A. L., Finlayson, A. G., and Mathers, H. E.: The northern sector of the last British Ice Sheet: Maximum extent and demise, Earth-Sci. Rev., 88, 207–226, 2008.

Bradwell, T., Sigurðsson, O., and Everest, J.: Recent, very rapid retreat of a temperate glacier in SE Iceland, Boreas, 42, 959–973, 2013.

Burgess, D. O. and Sharp, M. J.: Recent changes in areal extent of the Devon Ice Cap, Nunavut, Canada, Arct. Antarct. Alp. Res., 36, 261–271, 2004.

Calov, R., Ganopolski, A., Petoukhov, V., Claussen, M., and Greve, R.: Large-scale instabilities of the Laurentide ice sheet simulated in a fully coupled climate-system model, Geophys. Res. Lett., 29, 2216, doi:10.1029/2002GL016078, 2002.

Campbell, S. and Bowen, D. Q. (Eds.): The Quaternary of Wales, Geological Conservation Review, Nature Conservancy Council, Peterborough, 1989.

Casassa, G. and Brecher, H. H.: Relief and decay of flow stripes on Byrd Glacier, Antarctica, Ann. Glaciol., 17, 255–261, 1993.

Casassa, G., Jezek, K. C., Turner, J., and Whillans, I. M.: Relict flow stripes on the Ross Ice Shelf, Ann. Glaciol., 15, 132–138, 1991.

Chiverrell, R. C. and Thomas, G. S. P.: Extent and timing of the Last Glacial Maximum (LGM) in Britain and Ireland: a review, J. Quaternary Sci., 25, 535–549, 2010.

Chiverrell, R. C., Thrasher, I. M., Thomas, G. S. P., Lang, A., Scourse, J. D., Van Landeghem, K. J. J., McCarroll, D., Clark, C. D., Ó Cofaigh, C., Evans, D. J. A., and Ballantyne, C. K.: Bayesian modelling the retreat of the Irish Sea Ice Stream, J. Quaternary Sci., 28, 200–209, 2013.

Clark, C. D.: Mega-scale glacial lineations and cross-cutting ice-flow landforms, Earth Surf. Proc. Land., 18, 1–29, 1993.

Clark, C. D., Hughes, A. L. C., Greenwood, S. L., Jordan, C., and Sejrup, H. P.: Pattern and timing of retreat of the last British-Irish Ice Sheet, Quaternary Sci. Rev., 44, 112–146, 2012.

Cofaigh, C. Ó., Dunlop, P., and Benetti, S.: Marine geophysical evidence for Late Pleistocene ice sheet extent and recession off northwest Ireland, Quaternary Sci. Rev., 44, 147–159, 2012.

Dansgaard, W., Johnsen, S. J., Clausen, H. B., Dahl-Jensen, D., Gundestrup, N. S., Hammer, C. U., Hvidberg, C. S., Steffensen, J. P., Sveinbjörnsdottir, A. E., Jouzel, J., and Bond, G.: Evidence for general instability of past climate from a 250-kyr ice-core record, Nature, 364, 218–220, 1993.

Evans, D. and Cofaigh, C. Ó.: Depositional evidence for marginal oscillations of the Irish Sea ice stream in southeast Ireland during the last glaciation, Boreas, 32, 76–101, 2003.

Evans, D. J. A.: Controlled moraines: origins, characteristics and palaeoglaciological implications, Quaternary Sci. Rev., 28, 183–208, 2009.

Evans, D. J. A. and Twigg, D. R.: The active temperate glacial landsystem: a model based on Breiðamerkurjökull and Fjallsjökull, Iceland, Quaternary Sci. Rev., 21, 2143–2177, 2002.

Eyles, N. and McCabe, A. M.: The Late Devensian (< 22,000 BP) Irish Sea Basin: The sedimentary record of a collapsed ice-sheet margin, Quaternary Sci. Rev., 8, 307–351, 1989.

Eyles, N. and Rogerson, R. J.: A framework for the investigation of medial moraine formation: Austerdalsbreen, Norway, and Berendon Glacier, British Columbia, Canada, J. Glaciol., 20, 99–113, 1978a.

Eyles, N. and Rogerson, R. J.: Sedimentology of medial moraines on Berendon Glacier, British-Columbia, Canada – implications for debris transport in a glacierized basin, Geol. Soc. Am. Bull., 89, 1688–1693, 1978b.

Fahnestock, M. A., Scambos, T. A., Bindschadler, R. A., and Kvaran, G.: A millennium of variable ice flow recorded by the Ross Ice Shelf, Antarctica, J. Glaciol., 46, 652–664, 2000.

Forman, S. L., Marín, L., Van Der Veen, C., Tremper, C., and Csatho, B.: Little Ice Age and neoglacial landforms at the Inland Ice margin, Isunguata Sermia, Kangerlussuaq, west Greenland, Boreas, 36, 341–351, 2007.

Foster, H. D.: Sarn Badrig, a submarine moraine in Cardigan Bay, north Wales, Z. Geomorphol., 14, 473–486, 1970.

Garrard, R. A. and Dobson, M. R.: The nature and maximum extent of glacial sediments off the west coast of Wales, Mar. Geol., 16, 31–44, 1974.

Glasser, N. F. and Bennett, M. R.: Glacial erosional landforms: origins and significance for palaeoglaciology, Prog. Phys. Geogr., 28, 43–75, 2004.

Glasser, N. F. and Gudmundsson, G. H.: Longitudinal surface structures (flowstripes) on Antarctic glaciers, The Cryosphere, 6, 383–391, doi:10.5194/tc-6-383-2012, 2012.

Glasser, N. F. and Scambos, T. A.: A structural glaciological analysis of the 2002 Larsen B Ice Shelf collapse, J. Glaciol., 54, 3–16, 2008.

Glasser, N. F., Hambrey, M. J., Huddart, D., Gonzalez, S., Crawford, K. R., and Maltman, A. J.: Terrestrial glacial sedimentation on the eastern margin of the Irish Sea basin: Thurstaston, Wirral, Proceedings of the Geologists' Association The Geologists' Association, 112, 131–146, 2001.

Glasser, N. F., Hughes, P. D., Fenton, C., Schnabel, C., and Rother, H.: 10Be and 26Al exposure-age dating of bedrock surfaces on the Aran ridge, Wales: evidence for a thick Welsh Ice Cap at the Last Glacial Maximum, J. Quaternary Sci., 27, 97–104, 2012.

Glasser, N. F., Scambos, T. A., Bohlander, J., Truffer, M., Pettit, E., and Davies, B. J.: From ice-shelf tributary to tidewater glacier: continued rapid recession, acceleration and thinning of Röhss Glacier following the 1995 collapse of the Prince Gustav Ice Shelf, Antarctic Peninsula, J. Glaciol., 57, 397–406, 2011.

Greenwood, S. L. and Clark, C. D.: Reconstructing the last Irish Ice Sheet 2: a geomorphologically-driven model of ice sheet growth, retreat and dynamics, Quaternary Sci. Rev., 28, 3101–3123, 2009.

Grootes, P. M., Stuiver, M., White, J. W. C., Johnsen, S., and Jouzel, J.: Comparison of oxygen-isotope records from the GISP2 and GRIP Greenland ice cores, Nature, 366, 552–554, 1993.

Hambrey, M. J. and Dowdeswell, J. A.: Flow regime of the Lambert Glacier-Amery Ice Shelf system, Antarctica: structural evidence from Landsat imagery, Ann. Glaciol., 20, 401–406, 1994.

Hambrey, M. J., Bennett, M. R., Dowdeswell, J. A., Glasser, N. F., and Huddart, D.: Debris entrainment and transfer in polythermal valley glaciers, J. Glaciol., 45, 69–86, 1999.

Hambrey, M. J., Huddart, D., Bennett, M. R., and Glasser, N. F.: Genesis of "hummocky moraines" by thrusting in glacier ice: Ev-

idence from Svalbard and Britain, J. Geol. Soc., 154, 623–632, 1997.

Herbert-Smith, M.: Palynology of the Tertiary and Pleistocene deposits of the Llanbedr (Mochras Farm) borehole, in: The Llanbedr (Mochras Farm) borehole, edited by: Woodland, A. W., Institute of Geological Sciences Report 71/18, 115 pp., 1971.

Hiemstra, J. F., Evans, D. J. a., Scourse, J. D., McCarroll, D., Furze, M. F. A., and Rhodes, E.: New evidence for a grounded Irish Sea glaciation of the Isles of Scilly, UK, Quaternary Sci. Rev., 25, 299–309, 2006.

Hubbard, A., Bradwell, T., Golledge, N., Hall, A., Patton, H., Sugden, D., Cooper, R., and Stoker, M.: Dynamic cycles, ice streams and their impact on the extent, chronology and deglaciation of the British–Irish ice sheet, Quaternary Sci. Rev., 28, 758–776, 2009.

Huddart, D.: The glacial history and deposits of the north and west Cumberland lowlands, in: Glacial Deposits of Great Britain and Ireland, edited by: Rose, J., Gibbard, P., and Ehlers, J., Balkema, Rotterdam, 157–167, 1991.

IPCC WG1: Climate Change 2007: The Physical Science Basis. Contribution of Working Group 1 to the Fourth Assessment Report of the Intergovernmental Panel on Climate Change, edited by: Solomon, S., Qin, D., Manning, M., Chen, Z., Marquis, M., Averyt, K. B., Tignor, M., and Miller, H. L., Cambridge University Press, 2007.

Jehu, T. J.: The glacial deposits of western Caernarvonshire, T. Roy. Soc. Edin., 47, 17–56, 1909.

King, E. C., Hindmarsh, R. C. A., and Stokes, C. R.: Formation of mega-scale glacial lineations observed beneath a West Antarctic ice stream, Nat. Geosci., 2, 585–588, 2009.

Knutz, P. C., Zahn, R., and Hall, I. R.: Centennial-scale variability of the British Ice Sheet: Implications for climate forcing and Atlantic meridional overturning circulation during the last deglaciation, Paleoceanography, 22, PA1207, doi:10.1029/2006PA001298, 2007.

Lambeck, K. and Purcell, A. P.: Sea-level change in the Irish Sea since the Last Glacial Maximum: constraints from isostatic modelling, J. Quaternary Sci., 16, 497–506, doi:10.1002/jqs.638, 2001.

Lundqvist, J.: Rogen (ribbed) moraine – identification and possible origin, Sediment. Geol., 62, 281–292, 1989.

Magnússon, E., Björnsson, H., Dall, J., and Pálsson, F.: Volume changes of Vatnajökull ice cap, Iceland, due to surface mass balance, ice flow, and subglacial melting at geothermal areas, Geophys. Res. Lett., 32, L05504, doi:10.1029/2004GL021615, 2005.

Matley, C. A.: A 50-foot coastal terrace and other lateglacial phenomena in the Lleyn Peninsula, Proceedings of the Geologists' Association, 47, 221–223, 1936.

McCarroll, D.: Deglaciation of the Irish Sea Basin: a critique of the glaciomarine hypothesis, J. Quaternary Sci., 16, 393–404, 2001.

McCarroll, D., Stone, J. O., Ballantyne, C. K., Scourse, J. D., Fifield, L. K., Evans, D. J. A., and Hiemstra, J. F.: Exposure-age constraints on the extent, timing and rate of retreat of the last Irish Sea ice stream, Quaternary Sci. Rev., 29, 1844–1852, 2010.

Meier, M. F., Dyurgerov, M. B., Rick, U. K., O'Neel, S., Pfeffer, W. T., Anderson, R. S., Anderson, S. P., and Glazovsky, A. F.: Glaciers dominate eustatic sea-level rise in the 21st century, Science, 317, 1064–1067, 2007.

Merritt, J. W. and Auton, C. A.: An outline of the lithostratigraphy and depositional history of Quaternary deposits in the Sellafield district, west Cumbria, P. Yorks. Geol. Soc., 53, 129–154, 2000.

Mitchell, G.: The Pleistocene history of the Irish Sea: second approximation, Proceedings of the Royal Society of Dublin, 4, 181–199, 1972.

Moholdt, G., Hagen, J. O., Eiken, T., and Schuler, T. V.: Geometric changes and mass balance of the Austfonna ice cap, Svalbard, The Cryosphere, 4, 21–34, doi:10.5194/tc-4-21-2010, 2010.

Möller, P.: Rogen moraine: an example of glacial reshaping of pre-existing landforms, Quaternary Sci. Rev., 25, 362–389, 2006.

Morgan, V. I., Jacka, T. H., Akerman, G. J., and Clarke, A. L.: Outlet glacier and mass budget studies in Enderby, Kemp, and Mac, Robsertson Lands Antarctica, Ann. Glaciol., 3, 204–210, 1982.

Nygård, A., Sejrup, H. P., Haflidason, H., Cecchi, M., and Ottesen, D.: Deglaciation history of the southwestern Fennoscandian Ice Sheet between 15 and 13 14C ka BP, Boreas, 33, 1–17, 2008.

Ó Cofaigh, C. and Evans, D. J. A.: Sedimentary evidence for deforming bed conditions associated with a grounded Irish Sea glacier, southern Ireland, J. Quaternary Sci., 16, 435–454, 2001.

Ó Cofaigh, C., Pudsey, C. J., Dowdeswell, J. A., and Morris, P.: Evolution of subglacial bedforms along a paleo-ice stream, Antarctic Peninsula continental shelf, Geophys. Res. Lett., 29, 1199, doi:10.1029/2002.GL014488, 2002.

O'Sullivan, K. N.: Log of the Llanbedr (Mochras Farm) borehole: Recent, Pleistocene and Tertiary, in: The Llanbedr (Mochras Farm) borehole, edited by: Woodland, A. W., Institute of Geological Sciences Report 71/18, 115 pp., 1971.

Ottesen, D. and Dowdeswell, J. A.: Assemblages of submarine landforms produced by tidewater glaciers in Svalbard, J. Geophys. Res., 111, F01016, doi:10.1111/j.1502-3885.2007.tb01251.x, 2006.

Ottesen, D., Dowdeswell, J. A., Landvik, J. Y., and Mienert, J.: Dynamics of the Late Weichselian ice sheet on Svalbard inferred from high-resolution sea-floor morphology, Boreas, 36, 286–306, 2007.

Patton, H. and Hambrey, M. J.: Ice-marginal sedimentation associated with the Late Devensian Welsh Ice Cap and the Irish Sea Ice Stream: Tonfanau, West Wales, Proceedings of the Geologists' Association, 120, 256–274, 2009.

Patton, H., Hubbard, A., Glasser, N. F., Bradwell, T., and Golledge, N. R.: The last Welsh Ice Cap: Part 1 – Modelling its evolution, sensitivity and associated climate, Boreas, 42, 471–490, 2013a.

Patton, H., Hubbard, A., Glasser, N. F., Bradwell, T., and Golledge, N. R.: The last Welsh Ice Cap: Part 2 – Dynamics of a topographically controlled ice cap, Boreas, 42, 491–510, 2013b.

Reynolds, J. M. and Hambrey, M. J.: The structural glaciology of George VI Ice Shelf, Antarctic Peninsula, Brit. Antarct. Surv. B., 79, 79–95, 1988.

Saunders, G. E.: Reappraisal of glacial drainage phenomena in the Lleyn peninsula, Proceedings of the Geologists' Association, 79, 305–324, 1968a.

Saunders, G. E.: A fabric analysis of the ground moraine deposits of the Lleyn peninsula of south-west Caernarvonshire, Geol. J., 6, 105–118, 1968b.

Scourse, J. D.: Late Pleistocene stratigraphy and palaeobotany of the Isles of Scilly, Philos. T. R. Soc. B, 334, 405–448, 1991.

Scourse, J. D. and Furze, M. F. A.: A critical review of the glaciomarine model for Irish sea deglaciation: evidence from southern Britain, the Celtic shelf and adjacent continental slope, J. Quaternary Sci., 16, 419–434, 2001.

Scourse, J. D., Austin, W. E. N., Bateman, M. D., Catt, J. A., Evans, C. D. R., Robinson, J. E., and Young, J. R.: Sedimentology and micropalaeontology of glacimarine sediments from the central and southwestern Celtic Sea, Geological Society, London, 329–347, doi:10.1144/GSL.SP.1990.053.01.19, 1990.

Scourse, J. D., Haapaniemi, A. I., Colmenero-Hidalgo, E., Peck, V. L., Hall, I. R., Austin, W. E. N., Knutz, P. C., and Zahn, R.: Growth, dynamics and deglaciation of the last British–Irish ice sheet: the deep-sea ice-rafted detritus record, Quaternary Sci. Rev., 28, 3066–3084, 2009.

Sharp, M.: Sedimentation and stratigraphy at Eyjabakkajökull – an Icelandic surging glacier, Quaternary Res., 24, 268–284, 1985.

Small, R. J.: Lateral Moraines of Glacier de Tsidjiore-Nouve – form, development, and implications, J. Glaciol., 29, 250–259, 1983.

Smiraglia, C.: The medial moraines of Ghiacciaio Dei Forni, Valtellina, Italy – morphology and sedimentology, J. Glaciol., 35, 81–84, 1989.

Smith, B. and George, T. N.: North Wales. British Regional Geology, HMSO, London, 97 pp., 1961.

Synge, F. M.: The glacial succession in Caernarvonshire, Proceedings of the Geologists' Association, 75, 431–444, 1964.

Thomas, G. S. P. and Chiverrell, R. C.: Structural and depositional evidence for repeated ice-marginal oscillation along the eastern margin of the Late Devensian Irish Sea Ice Stream, Quaternary Sci. Rev., 26, 2375–2405, 2007.

Thomas, G. S. P. and Chiverrell, R. C.: Styles of structural deformation and syn-tectonic sedimentation around the margins of the late Devensian Irish Sea Ice Stream: the Isle of Man, Llyn Peninsula and County Wexford, in: Glacitectonics: A Field Guide, edited by: Phillips, E., Lee, J. R., and Evans, H. M., Quaternary Research Association, Pontypool, 59–78, 2011.

Thomas, G. S. P., Chester, D. K., and Crimes, P.: The Late Devensian glaciation of the eastern Lleyn Peninsula, North Wales: evidence for terrestrial depositional environments, J. Quaternary Sci., 13, 255–270, 1998.

van der Veen, C. J.: Polar ice sheets and global sea level: how well can we predict the future?, Global Planet. Change, 32, 165–194, 2002.

Van Landeghem, K. J. J., Wheeler, A. J., and Mitchell, N. C.: Seafloor evidence for palaeo-ice streaming and calving of the grounded Irish Sea Ice Stream: Implications for the interpretation of its final deglaciation phase, Boreas, 38, 119–131, 2009.

Whittow, J. B. and Ball, D. F.: North-west Wales, in: The glaciations of Wales and adjoining areas, edited by: Lewis, C. A., Longman, London, 21–58, 1970.

6

A reduced-complexity model for river delta formation – Part 1: Modeling deltas with channel dynamics

M. Liang[1,*], V. R. Voller[1], and C. Paola[2]

[1]Department of Civil, Environmental, and Geo-Engineering, National Center for Earth Surface Dynamics, Saint Anthony Falls Laboratory, University of Minnesota, Twin Cities, Minneapolis, Minnesota, USA

[2]Department of Geology and Geophysics, National Center for Earth Surface Dynamics, Saint Anthony Falls Laboratory, University of Minnesota, Twin Cities, Minneapolis, Minnesota, USA

[*]now at: Department of Civil, Architectural and Environmental Engineering and Center for Research in Water Resources, The University of Texas at Austin, Austin, Texas, USA

Correspondence to: M. Liang (manliang@utexas.edu)

Abstract. In this work we develop a reduced-complexity model (RCM) for river delta formation (referred to as DeltaRCM in the following). It is a rule-based cellular morphodynamic model, in contrast to reductionist models based on detailed computational fluid dynamics. The basic framework of this model (DeltaRCM) consists of stochastic parcel-based cellular routing schemes for water and sediment and a set of phenomenological rules for sediment deposition and erosion. The outputs of the model include a depth-averaged flow field, water surface elevation and bed topography that evolve in time. Results show that DeltaRCM is able (1) to resolve a wide range of channel dynamics – including elongation, bifurcation, avulsion and migration – and (2) to produce a variety of deltas such as alluvial fan deltas and deltas with multiple orders of bifurcations. We also demonstrate a simple stratigraphy recording component which tracks the distribution of coarse and fine materials and the age of the deposits. Essential processes that must be included in reduced-complexity delta models include a depth-averaged flow field that guides sediment transport a nontrivial water surface profile that accounts for backwater effects at least in the main channels, both bedload and suspended sediment transport, and topographic steering of sediment transport.

1 Introduction

Home to hundreds of millions of people, major coastal cities and infrastructure, immensely productive wetlands, and some of the most compelling and diverse landscapes on Earth – yet low-lying and vulnerable to storms and rising sea levels – deltas are emerging as among the most critical environments in a changing world (Syvitski et al., 2009). They are also immensely complex. The science of deltas comprises, in roughly equal parts, geomorphology, ecology, hydrology, organic and microbial geochemistry, and human dynamics. The physical dynamics alone would present a formidable challenge, even if they were restricted to just turbulent flow interacting with sand; but most natural deltas involve major additional complications such as fine-grained cohesive sediment (mud) and strong, two-way interactions with biota.

A fundamental debate is developing across the sciences as to the best way to model and understand such complexity (e.g., Murray, 2003; Overeem et al., 2005; Paola and Leeder, 2011; Paola et al., 2011; Hajek and Wolinsky, 2012). Should we try to capture everything, creating models that simulate the processes in as much detail as current knowledge and computing power allow, or should we simplify, even at the risk of losing connections with reality? Modeling of deltas in recent years has produced excellent examples of both approaches, which we review below. Our aim here is to present a model that resides in the middle ground between detailed simulation and abstract simplification. We use a method based on weighted random walks, where the random walks are constrained by rules based on a hybrid of simplified governing equations for fluid motion and phenomenological representation of sediment transport processes. With suitable rules, DeltaRCM (reduced-complexity model for river delta formation) is able to produce delta morphologies that compare well with those produced by more complex models such as Delft3D and with the morphology of deltas in the field. We believe that the availability of abundant computing power strengthens rather than weakens the case for so-called reduced-complexity models such as the one we propose here. Understanding – as opposed to simulating – complex natural phenomena requires a spectrum of approaches and a clear understanding of the advantages and disadvantages of each.

The paper begins with a review (Sect. 2) of current approaches to modeling deltas, emphasizing previous reduced-complexity models. The detailed implementation of our model is presented in Sect. 3, and results from it in Sects. 4 and 5. In Sect. 6 we discuss the meaning of the model results to date. Conclusions are provided in Sect. 7.

2 Modeling river delta formation

As with any morphodynamic model, the most direct delta formation model would solve the governing equations for water flow and sediment particles based on first principles, i.e., the conservation of mass and momentum or energy, in detail, given all the necessary initial and boundary conditions. However, this is still not practical, not only because of limits of computational power, but also because of the potential error accumulation in such complex "full physics" models (Hajek and Wolinsky, 2012). Existing models for delta formation cover a wide range of scales and complexity (Fagherazzi and Overeem, 2007; Paola et al. 2011).

On the simple side, models based on spatially averaged delta surface topography can predict average delta dynamics, such as laterally averaged surface profile, position of the shoreline, and position of the alluvial–bedrock transition (Parker et al., 2008; Kim et al., 2009; Lorenzo-Trueba et al., 2013). These models do not attempt to provide detailed structure, such as topography and channel networks. On the more complex side, to date, the most inclusive physics-based delta formation model is Delft3D, which solves a depth-integrated version of the Reynolds-averaged Navier–Stokes equations (shallow water equations) with a turbulence closure term for horizontal Reynolds stresses, and coupled with empirical sediment transport formulas based on bed shear stress (Lesser et al., 2004; Edmonds and Slingerland, 2007). Delft3D can resolve deltaic processes from smaller, engineering scales such as river mouth-bar formation and bifurcation (Edmonds and Slingerland, 2007) to larger, geological scales such as the whole delta morphodynamics controlled by sediment cohesion (Edmonds and Slingerland,

2009), waves, tides and antecedent stratigraphy (Geleynse et al., 2010). Delft3D is widely considered the best high-resolution delta model available to the research community, and its utility is greatly enhanced by the release of an open-source version in 2012. In the middle ground of the model complexity spectrum are the so-called reduced-complexity models (RCMs). These models feature descriptive constructions and intuitive simplifications over the hierarchy of natural processes, in contrast to highly detailed but computationally complex models such as Delft3D, while still evolving the topography and channel network without simplifying to the degree of spatially averaged models. The most common form of models in this category is a rule-based cellular routing scheme, such as the braided river model by Murray and Paola (1994, 1997) and some of the early erosional-landscape models (e.g., Willgoose et al., 1991). In terms of channel-resolving delta formation models, an excellent example is found in Seybold et al. (2007, 2009, 2010). In their model, the water flux field is calculated on a lattice grid via a set of simplified hydrodynamic equations which are equivalent to a diffusive-wave form of the shallow water equations with constant diffusivity. A few other examples of delta-related channel-resolving RCMs include an avulsive delta building model by Sun et al. (2002) and a channel-floodplain co-evolution delta building model, AquaTellUS, by Overeem et al. (2005).

RCMs are less computationally intensive than CFD (computational fluid dynamics)-based high-fidelity models yet still produce morphodynamic features at system scales, such as stream braiding and floodplain aggradation. While computational efficiency is often considered the reason for developing RCMs, their most important advantage is the flexible rule-based framework which allows for direct translation of phenomenological observations into the model (as opposed to hoping that they will emerge given a sufficiently detailed description of the underlying mechanics). The challenges of building a RCM for delta formation are the following: (i) the low topographic slope of the majority of river deltas (10^{-4}–10^{-5}) does not provide a strong guide for topographic flow routing, which is a key component in many RCMs for geomorphodynamic systems; (ii) the low slope together with relatively deep, slow channel flow creates a low-Froude-number environment such that the flow senses downstream information over relatively long distances, making it difficult to design localized rules which are essential for RCMs; (iii) the self-organized distributary channel network includes loops that further complicate flow routing; and (iv) many river deltas have suspended load and wash load as a primary sediment input component, which make sediment routing more complex than in a bedload-only system. In addition, the low-Froude-number flow condition implies, as the Froude number tends to zero, a "rigid-lid" condition in which the shape of the free surface is nearly flat. This condition potentially offers computational advantages as the flow depth can be estimated from a fixed surface elevation (usually sea level or a simple function using backwater equations) and bed elevation, but is almost decoupled from the bed topography.

In this work, we present a RCM delta model using the "weighted random walk" method. The basic goal is to develop a model that includes just enough of the dynamics to tackle the main problems listed above. To be more specific, we seek complexity-reduction in the following aspects: (i) the solution of water surface elevation, (ii) the flow momentum balance, and (iii) the criteria for sediment deposition and erosion. A detailed model description is given in the next section, followed by results and comparisons with experimental and field deltas, along with the results of more detailed delta models, and then a discussion of the strengths and weaknesses of our model approach.

3   Model construction

DeltaRCM has two components: a cellular flow routing scheme as the hydrodynamic component, and a set of sediment transport rules as the morphodynamic component. The model uses a lattice of square cells for its domain, where water and sediment flux are routed in a cell-by-cell fashion. The model evolves in time by updating the depth-averaged flow field, water surface elevation, sediment flux, and bed elevation at each time step.

3.1   Model setup

The physical setting of our delta formation model is simplified to a rectangular basin of constant water depth (h_B) with a short inlet channel on one side (Fig. 1). At the inlet we assume a constant water discharge Q_{w0} (m^3 s^{-1}) and sediment discharge Q_{s0} (m^3 s^{-1}). The boundary with the inlet channel is a wall boundary such that no water or sediment crosses. The other three boundaries are ocean boundaries with the boundary condition of a fixed sea level, H_{SL}.

For water and sediment routing, we first define a set of global parameters that remain constant for each model run: (1) a reference water depth h_0, i.e., a representative flow depth for the system, and (2) a reference slope S_0, which is a representative overall water surface slope of the system. For example, for a lowland river delta, a typical value of h_0 is from a few meters to tens of meters, with S_0 on the order of 10^{-4} to 10^{-5}, while for an experimental fan delta, a typical value of h_0 is tens of millimeters and S_0 on the order of 10^{-2}. The values are not precise but rather represent scale values, and may require trial and error to validate for each specific system. The depth of the inlet channel is set at h_0 and the inlet flow velocity is calculated as $U_0 = \frac{Q_{w0}}{h_0 W}$, which will be referred to as a reference velocity of the system. W is the inlet channel width, specified for each model run.

The domain is shown in Fig. 2, with cell size δ_c, a value that depends on the target scale of the model run; e.g., in the results section we use 50 m for a field-scale delta and 2 cm for a laboratory-scale fan delta. The total number of cells along

the dip direction (from the inlet, into the basin) is N_x and the number of cells along the strike direction (perpendicular to the inlet, across the basin) is N_y. Typically, N_x and N_y are both on the order of a hundred, with N_y being roughly twice as large as N_x to allow for a semicircular delta growth. The inlet has a width of N_0 cells. Typically, N_0 is around 5. The primary quantities associated with each cell include (i) water unit discharge vector $\boldsymbol{q}_w = (q_x, q_y)$, (ii) water surface elevation H, and (iii) bed elevation η. These primary quantities are updated at each time step. Other useful quantities such as velocity vector $\boldsymbol{u} = (u_x, u_y)$ and water depth h can be easily calculated from the primary quantities by $h = H - \eta$ and $\boldsymbol{u} = \frac{\boldsymbol{q}_w}{h}$.

Two types of parcels that carry a water or sediment attribute are routed through the domain. A time step is defined by the addition of n_w water parcels and n_s sediment parcels. This is done through a sequence of water parcels carrying an equal fraction of the total input water discharge during a time step followed by sediment parcels carrying an equal fraction of the total input sediment discharge during a time step.

Within each model run, the size of the time step Δt is constant. As is often the case in numerical modeling, the choice of Δt is a balance between computation efficiency and model stability. In each time step, the total amount of sediment added to the domain is measured by $\Delta V_s = Q_{s0} \Delta t$. A smaller ΔV_s means less change to the topography and allows the cellular routing scheme to perform better with a more consistent terrain but, obviously it will take more steps to build the delta to a certain size. Here we introduce a reference volume,

$$V_0 = h_0 \delta_c^2, \qquad (1)$$

which is the volume of a channel inlet cell from the bed to water surface. If we assume that channels on the delta self-organize in scale with the reference depth h_0, this reference volume gives a good measurement of the characteristic topographic change on the growing delta. Currently we set the time step size so that the sediment volume added in each time step satisfies

$$\Delta V_s = 0.1 N_0^2 V_0. \qquad (2)$$

Therefore, time step size is given by

$$\Delta t = \frac{0.1 N_0^2 V_0}{Q_{s0}}. \qquad (3)$$

3.2   Model operation

The operations can best be understood by describing the processes in a single time step. There are four distinct phases: (1) the addition and routing of the water; (2) updating of the water surface elevation; (3) routing the sediment parcels and updating the bed elevation through deposition and erosion; and (4) updating of the routing direction, a vector field that

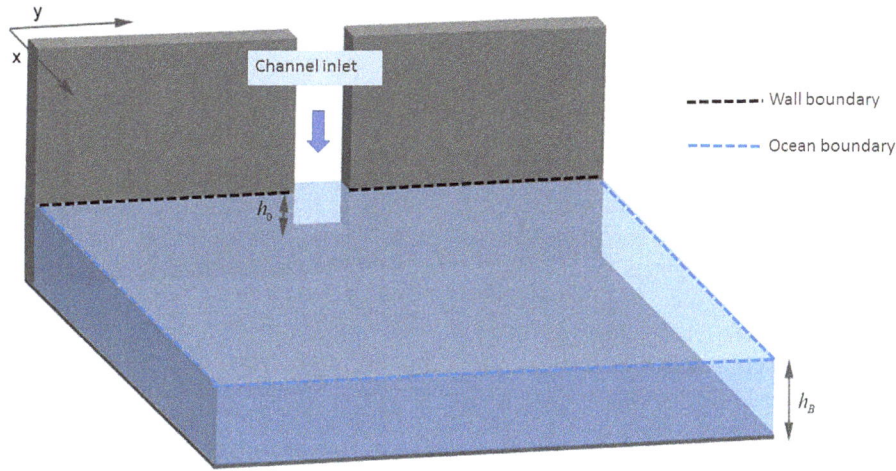

Figure 1. Illustration of the basin, boundaries and inlet channel.

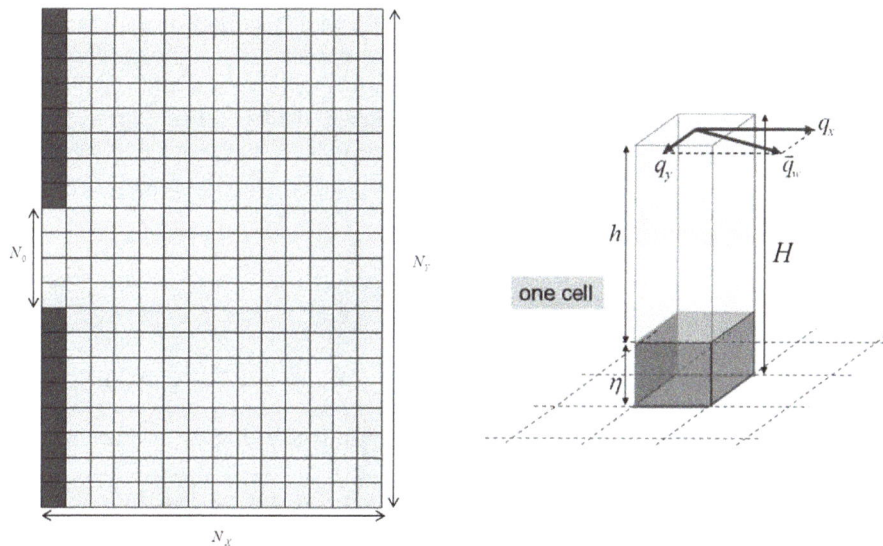

Figure 2. Diagram of the lattice grid and the primary values at each cell (water unit discharge, water surface elevation and bed elevation). Note that the total number of cells is reduced for the illustration.

determines the direction of flow through each cell in the domain. Each of these phases is described in turn.

To prepare, we divide the upstream water discharge (Q_{w0}) and the total sediment input volume during a time step (ΔV_s) into parcels. Typically, we use $n_w = 2000$ water parcels and each water parcel carries an equal amount of discharge:

$$Q_{p\_water} = \frac{Q_{w0}}{n_w}. \tag{4}$$

Likewise, we use $n_s = 2000$ sediment parcels and each sediment parcel carries an equal amount of sediment volume:

$$V_{p\_sed} = \frac{\Delta V_s}{n_s}. \tag{5}$$

3.2.1 Phase 1: water routing

At the start of a time step we assume that we have a delta with known shape and topography, i.e., at each cell we have a value of the water surface elevation H, bed elevation η, and water depth (difference between the water surface elevation and the bed elevation) h. We also have, at each cell, a unit vector F, referred to as the routing direction, which indicates the average downstream direction of flow through that cell. If the current time step is the first step in the model run, the routing directions are all in line with the inlet channel.

For the purpose of routing water, we define a binary cell state: 0 – dry, 1 – wet. This is done by doing a sweep through the domain and marking cells with a water depth larger than a small threshold value h_{dry} as wet cells. This threshold value is typically a fraction (10 %) of the characteristic depth scale of the environment of interest or 0.1 m, whichever is smaller. For example, for a natural delta, h_{dry} is typically 0.1 m, while for an experimental delta in laboratory, h_{dry} is typically a few millimeters which is 10 % of the characteristic flow depth.

The process in the first part of the time step requires us to route, in turn, each of the water parcels through the domain. When the parcel is at a given cell, a decision is needed indicating to which of the eight neighbor cells it will move to. We achieve this by using a so-called weighted random walk where the movement is dictated by a predefined probability distribution between the eight neighbor cells. The specification of the probability distribution is as follows.

At a given cell, first we calculate the routing weights for the eight neighbor cells. With the local routing direction F specified, the routing weights are determined by two factors: (i) the angle between the relative direction of the neighbor cell i and the routing direction, which we will estimate using a dot product method that we describe below; and (ii) the resistance to the flow from each neighbor cell i. In this model we calculate the routing weight for neighbor cell i as

$$w_i = \frac{\frac{1}{R_i}\max\left(0, F \cdot d_i\right)}{\Delta_i}, \qquad (6)$$

where resistance R_i is estimated as an inverse function of local water depth h_i,

$$R_i = \frac{1}{h_i^{\theta}}. \qquad (7)$$

For the current version of flow routing, the exponent θ is set to 1, hence, leading to the following relationship of the routing weight:

$$w_i = \frac{h_i \max\left(0, F \cdot d_i\right)}{\Delta_i}. \qquad (8)$$

The cellular direction vector, d_i, is a unit vector pointing to neighbor i from the given cell. Finally, Δ_i is the cellular distance: 1 for cells in main compass directions and $\sqrt{2}$ for corner cells (Fig. 3).

The weights above are calculated only for the wet neighbor cells of the given channel cell. All dry neighbor cells take a weight value of 0. At each channel cell we can then calculate routing probabilities p_i:

$$p_i = \frac{w_i}{\sum\limits_{\text{nb}=1}^{8} w_{\text{nb}}}, \quad i = 1, 2, \dots, 8. \qquad (9)$$

To obtain a discharge vector at each cell based on the motion of visiting water parcels, our starting point is to construct, for each visiting parcel, an average vector of the input and output vectors (Fig. 4). So the result is, for each channel cell, a set (size N_{visit}) of vectors, each expressing the average path of a visiting parcel through that cell. A summation of this set of

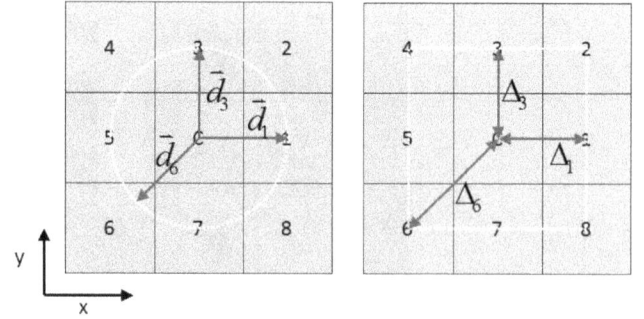

Figure 3. Definition of cellular direction d_i and cellular distance Δ_i. For example, $d_1 = (1, 0)$, $d_6 = (-\frac{1}{\sqrt{2}}, -\frac{1}{\sqrt{2}})$, $\Delta_1 = 1$, $\Delta_6 = \sqrt{2}$.

vectors provides, after appropriate normalization, a representative direction for water parcels through the cell. In this way, a vector with this direction and a magnitude of $N_{\text{visit}} Q_{\text{p\_water}}$ can be regarded as a discharge vector for the cell, Q_{cell}.

Then, for the purpose of later sediment transport, we need to estimate the local flow unit discharge and velocity. To do this we take the cell discharge vector ($\text{m}^3\,\text{s}^{-1}$) and divide it by the cell size δ_c to obtain a unit water discharge vector ($\text{m}^2\,\text{s}^{-1}$):

$$q_{\text{w}} = \frac{Q_{\text{cell}}}{\delta_c}. \qquad (10)$$

3.2.2 Phase 2: water surface calculation

Water surface elevation is essential in this model not only because it participates in the calculation of flow depth but, even more importantly, because the gradient of water surface plays a major role in determining the routing probabilities, w_i (Eq. 8), of water parcels.

In this reduced-complexity model, our goal is to obtain a sufficiently accurate surface profile without solving the full 2-D hydrodynamic equations. We propose a method that uses a finite-difference scheme along the movement path of individual water parcels, analogous to the simplified surface solver developed by Rinaldo et al. (1999).

To start with the simplest formulation, we assume that water surface slope along a channel streamline can be approximated by the reference slope S_0, and in the ocean the water surface slope is always zero. With the downstream water surface boundary condition $H = H_{\text{SL}}$, ideally along any given streamline, we can reconstruct the surface profile with a simple finite-difference calculation. In the model, however, instead of tracing a flow streamline, we take advantage of the walking path of water parcels, which can be considered as an approximation to the flow streamlines. The difference between the water-parcel paths (the "zigzag" version of streamlines) and the real flow streamlines is illustrated in Fig. 5. In the following we explain how to construct a water surface

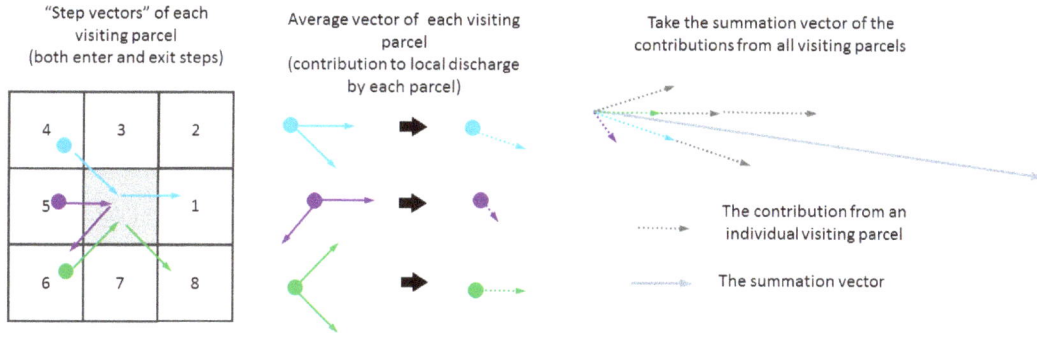

Figure 4. Calculation of the direction of the cell-representative discharge vector. The representative discharge vector takes the direction of the summation vector of all contributions from each visiting water parcel, and for N_{visit}-visiting parcels its magnitude is $N_{\text{visit}} Q_{\text{p\_water}}$.

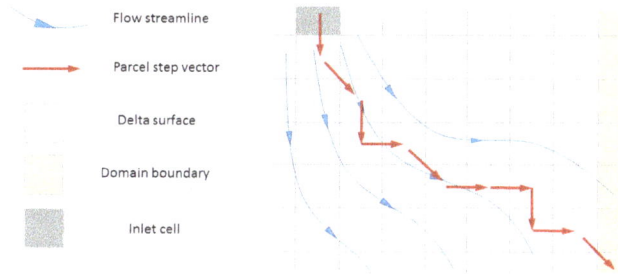

Figure 5. A diagram showing the path of one individual water parcel compared to smooth flow streamlines.

profile along a water-parcel path with a given reference slope S_0.

First, we need to locate the part of the path that is on the delta surface, as the part in ocean is considered flat. In general, a water-parcel path starts at one of the inlet cells, moves from one cell to an adjacent cell, and ends at one of the downstream ocean boundary cells. We distinguish the cells along the path on the delta surface and the cells in the open ocean by checking two values at each cell such that either a cell is on the delta, or a cell is in the ocean if both of the following criteria are met:

1. local bed elevation η is lower than a threshold value η_{shore} (set to $\eta_{\text{shore}} = H_{\text{SL}} - 0.9\,h_{\text{ref}}$);

2. local flow speed $|\boldsymbol{u}|$ is smaller than a threshold value U_{shore} (set to $U_{\text{shore}} = 0.5\,U_{\text{ref}}$).

With a given water-parcel path, the calculation starts from the end of the path and goes backward towards the inlet. For the kth cell in the direction of calculation,

– if cell k is in the ocean, $H|_k = H_{\text{SL}}$;

– if cell k is on the delta, $H|_k = H|_{k-1} - \Delta \delta_c (\boldsymbol{q}_{\text{w}}|\boldsymbol{q}_{\text{w}}| \cdot \boldsymbol{d}|_k) S_0$, where Δ is the cellular distance between the kth and $(k-1)$th cell, δ_c is cell size, and $\boldsymbol{d}|_k$ is the parcel step vector from cell k to cell $k - 1$.

The purpose of the term $(\boldsymbol{q}_{\text{w}}|\boldsymbol{q}_{\text{w}}| \cdot \boldsymbol{d}|_k)$ is to take into account the angle between the parcel path and the streamline.

This calculation gives the surface profile along the path of an individual water parcel and is repeated for all water-parcel paths. There are two additional situations to be taken care of.

1. If a cell is visited by multiple water parcels, all the values from each visiting path are recorded and an average value is taken from these stored values in the end to obtain a single value for water surface elevation at each cell.

2. If a cell is not visited by any water parcels, its water surface elevation retains the old value (from the previous time step).

This newly calculated surface profile is recorded as H^{temp}. We then apply a diffuser to smooth the calculated surface profile, which is typically spiky due to the 1-D stepwise method of calculation. The diffusion is applied as

$$H^{\text{smooth}} = (1 - \varepsilon)H^{\text{temp}} + 0.125\varepsilon \sum_{\text{nb}=1}^{8} H_{\text{nb}}. \tag{11}$$

We have used a diffusivity of $\varepsilon = 0.1$ and applied the smoothing calculation in Eq. (11) 10 times in each time step. This number is selected by checking samples of the resulting surface profile until no obvious spikes appear. We will discuss more in detail how sensitive the results are along with other features in calculating the free surface.

In the end, the water surface elevation is updated with an underrelaxation scheme for numerical stability:

$$H^{\text{new}} = (1 - \varpi)H^{\text{old}} + \varpi H^{\text{smooth}}. \tag{12}$$

The underrelaxation coefficient ϖ is set to 0.1, which allows the surface profile to transit slowly and smoothly from one time step to another, avoiding numerical instability.

To ensure conservation of water mass, the unit discharge field remains the same within one time step. Therefore, as the water surface elevation is updated, only water flow depth and velocity are adjusted accordingly.

3.2.3 Phase 3: sediment transport and bed topography update

Now, both the flow field, $\boldsymbol{q}_\mathrm{w}$, and water surface elevation, H, are updated. These two variables will remain constant until the next time step. To calculate the changes to the topography in a time step, we propose two sets of rules for the transport, deposition and erosion of sediment. The first set describes the routing of the sediment parcels, and the second set describes the rate of deposition and erosion as the exchange of sediment volume between sediment parcels and the bed. The rules are phenomenological and the goal is to build them via our understanding of macroscopic behavior rather than via fine-scale physical interactions between the fluid, sediment and bed. To this end, we distinguish two types of sediment that have different behaviors in the model:

- coarse sediment, referred to as "sand", is noncohesive, and transported as bedload;

- fine sediment, referred to as "mud", is cohesive, and transported as suspended load.

A sediment parcel is either a "sand" parcel or a "mud" parcel. At the beginning of each run, an input parameter f_sand gives the portion of sand in the total upstream sediment input. Therefore, a total number of $f_\mathrm{sand} n_\mathrm{s}$ parcels are designated as sand parcels and a total number of $(1 - f_\mathrm{sand}) n_\mathrm{s}$ parcels are designated as sand parcels for each time step.

3.2.4 Routing of the sediment parcels

For routing sediment parcels we use the same weighted random walk method as for the routing of water parcels (Eq. 6) with two modifications:

1. The routing direction $\boldsymbol{F}$ in Eq. (6) is replaced with the newly calculated water discharge vector $\boldsymbol{q}_\mathrm{w}$ at the given cell (from Phase 1 above), assuming that sediment parcels move with the water flow.

2. Transport resistance for sediment maintains the inverse function of flow depth but has different exponents. The idea is that sediment flux tends to concentrate in the lower portion of the water column and therefore it is more likely to follow topographically low areas. For now we use an exponent $\theta = 2$ for sand parcels (bedload) which is twice the value for water, and $\theta = 1$ for mud parcels (suspended load) which is equal to the value for water. The physical reason for the values chosen is that the distribution of the concentration of coarse material is skewed towards the lower portion of the water column and the distribution of fine material is more evenly distributed throughout the water column.

Thus, the routing weights for sediment parcels are

$$w_i = \frac{h_i^2 \max\left(0, \boldsymbol{q}_\mathrm{w} \cdot \boldsymbol{d}_i\right)}{\Delta_i} \text{ for sand parcels, and} \tag{13}$$

$$w_i = \frac{h_i \max\left(0, \boldsymbol{q}_\mathrm{w} \cdot \boldsymbol{d}_i\right)}{\Delta_i} \text{ for mud parcels.} \tag{14}$$

And routing probabilities are calculated as

$$p_i = \frac{w_i}{\sum_{\mathrm{nb}=1}^{8} w_\mathrm{nb}}, \ i = 1, 2, \ldots, 8. \tag{15}$$

3.2.5 The rate of deposition and erosion

Sediment parcels are routed sequentially in a weighted random walk fashion according to the probabilities calculated with Eqs. (13), (14) and (15). The change to the bed topography is obtained by the exchange of sediment volume between the moving parcel and the local bed at each cell along the path – during deposition a sediment parcel loses part of its volume and this volume is added to the bed, and vice versa for erosion. We use simple phenomenological rules to decide (i) where deposition or erosion happens and (ii) how much volume should be exchanged between the sediment parcel and the bed. The rules for sand and mud parcels are different.

For convenience of description, we refer to the initial volume of each sediment parcel $V_\mathrm{p\_sed}$ as the "reference sediment parcel volume", and the remaining volume during the walking process of a sediment parcel as the "residual sediment parcel volume", $V_\mathrm{p\_res}$. The amount of deposition at each cell by an individual parcel is referred to as $V_\mathrm{p\_dep}$. The amount of erosion at each cell by an individual parcel is referred to as $V_\mathrm{p\_ero}$. The detailed rules are as follows.

For the deposition from a sand parcel we do the following:

- At each cell in the domain, we calculate a "transport capacity" for sand flux, $q_\mathrm{s\_cap}$, as the maximum flux per unit width, which is a nonlinear function of local flow velocity U_loc. The scaling between sediment flux and flow velocity takes the form of the Meyer-Peter and Müller (1948) formula,

$$q_\mathrm{s\_cap} = q_\mathrm{s0} \frac{U_\mathrm{loc}^3}{U_0^3}, \tag{16}$$

where q_s0 is calculated by dividing the upstream sand flux input by the inlet channel width:

$$q_\mathrm{s0} = \frac{f_\mathrm{sand} Q_\mathrm{s0}}{N_0 \delta_\mathrm{c}}. \tag{17}$$

- Similar to the calculation of water discharge, as the sand parcels are routed sequentially, we track the accumulated total sand flux, $q_\mathrm{s\_loc}$, which increases with each visiting bedload parcel:

$$q'_\mathrm{s\_loc} = q_\mathrm{s\_loc} + \frac{V_\mathrm{p\_res}}{\delta_\mathrm{c} \Delta t}. \tag{18}$$

– Deposition occurs if a sand parcel visits a cell that has an accumulated local sand flux exceeding the transport capacity:

$$V_{p\_dep} = V_{p\_res} \text{ if } q_{s\_loc} > q_{s\_cap}, \tag{19a}$$

$$V_{p\_dep} = 0 \text{ if } q_{s\_loc} \leq q_{s\_cap}. \tag{19b}$$

For deposition from a mud parcel we do the following:

– Deposition occurs if a mud parcel visits a cell that has a local flow velocity U_{loc} smaller than a threshold velocity, U_{dep}. The amount of deposition is proportional to the residual sediment volume of the mud parcel as well as the relative difference between the squares of U_{loc} and U_{dep}, a simplified representation of standard empirical laws for fine-sediment deposition (van Rijn, 1984):

$$V_{p\_dep} = V_{p\_res} \frac{U_{dep}^3 - U_{loc}^3}{U_{dep}^3} \text{ if } U_{loc} < U_{dep}, \tag{20a}$$

$$V_{p\_dep} = 0 \text{ if } U_{loc} \geq U_{dep}. \tag{20b}$$

– U_{dep} is set to $U_{dep} = 0.3\,U_{ref}$. The idea is that the finer the grain size, the slower the flow it requires to settle.

For the erosion by both types of sediment parcels, we do the following:

– Erosion occurs if local flow velocity magnitude is larger than a threshold value, U_{ero}, that differs for sand and mud parcels (García and Parker, 1991):

$$V_{p\_ero} = V_{p\_sed} \frac{U_{loc}^3 - U_{ero}^3}{U_{ero}^3} \text{ if } U_{loc} > U_{ero}, \tag{21a}$$

$$V_{p\_ero} = 0 \text{ if } U_{loc} \leq U_{ero}. \tag{21b}$$

– For a sand parcel, $U_{ero} = 1.05\,U_{ref}$.

– For a mud parcel, $U_{ero} = 1.5\,U_{ref}$.

For volume exchange between sediment parcel and the bed:

– At each step, the volume of the sediment parcel is updated as

$$V'_{p\_res} = V_{p\_res} - V_{p\_dep} + V_{p\_ero}. \tag{22}$$

– The elevation of the local bed is updated as:

$$\eta' = \eta + \frac{V_{p\_dep}}{\delta_c^2} - \frac{V_{p\_ero}}{\delta_c^2}. \tag{23}$$

– The local flow velocity and flow depth are updated in accordance with each event of deposition or erosion: $h' = H - \eta'$ and $\boldsymbol{u}' = \frac{\boldsymbol{q}_w}{h}$.

Note that in this setup a parcel can only take sediment of its own category (e.g., sand or mud), and the volume is equal to the total volume entrained. Therefore, in the erosion process, only the total sediment mass is preserved rather than the individual category of sand or mud. Given that deltas are predominantly depositional environments this method provides a reasonable conservation of sediment. We note, however, that if our approach is to be extended to model environments that involve strong erosion over mixed sand/mud beds our treatment will need modification to allow each parcel to carry multiple sediment categories.

The reason for updating local flow depth and velocity immediately after each event of deposition and erosion is to avoid excess change to the bed. Similarly, we add an extra control on the rate of change to the bed by limiting the amount of deposition and erosion by a sediment parcel so that the change to local depth is less than 25 %, so that the change to local flow velocity is less than 33 %. For example, if local flow depth is 4 m, then the maximum deposition or erosion by a single sediment parcel is limited to 1 m change to the bed.

After all sediment parcels finish their random walk, to take into account the influence of topographical slope on sediment flux in an approximation of the Bagnold–Ikeda expressions (García, 2008), we apply a topographic diffuser that assumes the diffusive flux is proportional to local sand (bedload) flux and topographical slope:

$$q_{s\_diff} = \alpha |\nabla \eta| q_{s\_loc}, \tag{24}$$

where α is a scaling coefficient, by default set to 0.1, and $|\nabla \eta|$ is bed slope. The total change to the bed elevation by the topographic diffuser is obtained by summing up the inbound and outbound diffusive fluxes at each cell over the time period Δt. This topographic diffusion also introduces lateral erosion by allowing sediment on the bank to be removed and added to the channels. This lateral erosion gives channels the mobility to migrate or even to meander. Examples are shown in the results section.

3.2.6 Phase 4: update routing direction

Before moving to the next time step, we need to update the routing direction: a unit vector at each cell indicating the downstream direction for routing water parcels. In this last phase of the time step, at each cell we calculate the updated value of the unit water discharge vector $\boldsymbol{q}_w$, water surface elevation H, bed elevation η, water depth h, etc.

To achieve this, we combine two physical processes dictating the flow direction: (i) at an instant in time flow has a tendency to continue in the same direction as the direction at the previous instant due to inertia, and (ii) in the absence of any other drivers the flow goes downslope which in our case is indicated by the water-surface slope rather than bed slope.

Table 1. List of model constants and parameters.

Parameter	Values and rationale
α	Coefficient of topographic diffusion, set to 0.1. This parameter controls the cross-slope sediment flux as well as bank erosion. The magnitude of 10^{-1} comes from the portion of bedload that is steered by bed slope.
γ	Partitioning coefficient between routing direction by inertia and routing direction by water surface gradient. This parameter essentially controls how much water spread laterally (caused by cross-channel component of water surface gradient) and is usually a small value (on the order of 10^{-2}).
ε	Coefficient for water surface diffusion, set to a small value of 0.1 to ensure stability.
θ	Depth dependence in routing water and sediment parcels. The value is set to 1 for water parcels and mud parcels, and 2 for sand parcels. The higher this value is the more skewed in the routing probabilities towards cells with larger depth value.
U_{dep}	Threshold velocity for sediment deposition. Currently, it only applies to mud parcels and is set to 30 % of the reference velocity U_0. The smaller this value is the longer a mud parcel can travel before losing all its mud volume.
U_{ero}	Threshold velocity for sediment erosion. The value is set to $1.05 \cdot U_0$ for sand parcels and $1.5 \cdot U_0$ for mud parcels. The higher this value is the more difficult it is to erode the bed.
h_{dry}	Threshold depth for a cell to be considered "dry" and turned off from flow routing. The value is user defined and should be estimated depending on the physical environment. We suggest 1–10 % of the characteristic flow depth. In the model runs presented in this paper a value of 0.1 m is used for field scale, which comes from the observation in Wax Lake Delta, LA; and 0.002 m for experimental scale, which comes from the observation of delta basin experiments in the lab.

First, we calculate a unit vector from the downstream direction based on the previous time step:

$$F_{int} = \frac{q_{w,old}}{|q_{w,old}|}. \tag{25}$$

Then, we calculate a unit vector from the water surface gradient (from the previous time step):

$$F_{sfc} = \frac{\nabla H_{old}}{|\nabla H_{old}|}. \tag{26}$$

Then, a linear combination of the two vectors is taken with a partitioning coefficient γ:

$$F^* = \gamma F_{sfc} + (1-\gamma) F_{int} \text{ and } F = \frac{F^*}{|F^*|}. \tag{27}$$

The value of γ is set to a small number, typically 0.05 in the runs reported here.

By implementing the method described in this section, we have achieved our goal of complexity reduction: (i) the construction of the water surface via 1-D profiles captures the overall trend of water surface gradients without solving the full hydrodynamic equations; (ii) the flow momentum balance is relaxed, e.g., the effect of flow inertia is considered only in the form of direction rather than magnitude; and (iii) the criteria for sediment deposition and erosion are in the very basic form of a nonlinear relation between sediment carrying capacity and flow velocity. Key constants and parameters that do not vary in our tests are listed in Table 1. In the next section, we will show that when implemented in our DeltaRCM model these reduced-complexity constructions predict delta growth characteristics and channel dynamics that are comparable to those of high-fidelity modeling and field observations.

4 Model results

In this section we present various morphological features produced by DeltaRCM with different domain setup and input parameters. All simulations assume no effects from wave or tidal energy, i.e., the delta is a classic river-dominated delta (Galloway, 1975). We investigate (1) the effects of input sediment composition and (2) the model's ability to simulate deltas at field and laboratory scales. The former has been studied via field observation (Orton and Reading, 1993) and numerical simulation (Edmonds and Slingerland, 2009). The latter is based on the availability of data from experimental deltas; also, we believe that if a model can handle both field and experimental scales, it could potentially inform the interpretations and connections of both. Furthermore, we demonstrate DeltaRCM as a tool for hypothesis testing through study of the effects of the receiving basin depth.

As discussed above, two types of sediment are routed through the system: coarse (sand) and fine (mud). The ratio of the numbers of these two types of parcels at the inlet gives the ratio of sand and mud coming into the system. To set the physical scale of the simulation, domain and grid size are adjusted by changing cell size and physical input parameters, such as total input water and sediment discharge, and also global parameters such as the reference energy slope.

The input parameters (Table 2) include

1. the portion of sand in the upstream sediment input, f_{sand};

2. global parameters; i.e., the reference flow depth h_0, basin depth h_B, and the reference slope, S_0;

3. total discharge Q_{w0} and Q_{s0}.

Strictly speaking, the choice of the reference slope S_0 is dependent on the sand : mud ratio as well as the scale of the physical setting. In our model runs for field scale we use 3×10^{-4} for purely sandy deltas, 1×10^{-4} on purely muddy deltas and a linear combination of the two for mixed deltas; for laboratory scale, we use values on the order of 10^{-2} for S_0. The magnitude of the reference slope is scaled with the ratio of bedload and water fluxes that come from the inlet channel, such that $S_0 \sim Q_{s0\_bed}/Q_{w0}$.

4.1 Effects of input coarse/fine sediment ratio

In this group, the domain is a lattice grid of 120 by 60 square cells. Cell size is taken to be 50 m. The channel inlet is five-cells wide (250 m), with a reference flow depth of $h_0 = 5$ m. The total water discharge is 1250 m^3 s^{-1}. The total sediment discharge is 0.1 % by volume, which is 1.25 m^3 s^{-1}. We use a time step calculated from Eq. (3) of 25 000 s (~ 7 h). Both water and sediment discharge stay constant and we assume they represent channel-forming conditions.

We show three model runs in Fig. 6 with the portion of sand in the upstream sediment discharge set to 25, 50, and 75 %. The resultant deltas differ systematically based on the input mud fraction in the following characteristics, which are consistent with those found in the investigation on the effects of sediment cohesion by Edmonds and Slingerland (2009).

- On a sandy delta the channels are relatively shallow and mobile, without well-defined levees. Flow is less confined. There are large areas of sheet flow. The shoreline is smooth and the delta grows roughly in a semicircular shape.

Table 2. List of delta model runs and parameter values.

Run	f_{sand}	S_0	Q_{w0} (m^3 s^{-1})	h_0 (m)	h_B (m)
1	0.9	2.8×10^{-4}	1250	5	5
2	0.5	2.0×10^{-4}	1250	5	5
3	0.1	1.2×10^{-4}	1250	5	5
4	0.3	1.6×10^{-4}	1250	5	5
5	1.0	1.0×10^{-2}	0.0006	0.02	0.02
6	0.3	1.6×10^{-4}	1250	5	2.5
7	0.3	1.6×10^{-4}	1250	5	10
8	1.0	2.0×10^{-2}	0.0006	0.02	0.02

- On a muddy delta, channels are deeper and stable, with well-defined levees. Channels tend to elongate. The shoreline is rugose, and deltas build in different directions by switching lobes.

- The contrast in the model-predicted roughness between a sandy and muddy delta is illustrated in Fig. 7, where plots of the time variation of the ratio of number of cells on the shoreline to average delta radius (measured in number of cells) is presented. In these calculations, the shoreline is defined using the opening-angle method (OAM) developed by Shaw et al. (2008), employing an elevation threshold of -1 m and an opening-angle threshold of 30°. Also note that the calculation of the roughness ratio in Fig. 7 is made across the range of time intervals where the predicted delta consists of several lobes but has not yet filled the calculation domain.

4.2 Experimental fan deltas

Laboratory experiments, numerical modeling and field observation are three important approaches of understanding the formation of deltas. Because we would like to test our model across as wide a scale range as possible, we include experimental deltas at laboratory scales. To do this, we change the domain to a lattice grid of 90 by 180 cells with a cell size of 0.02 m. The inlet channel is still five-cells wide but has a flow depth of 0.02 m and a water discharge of 0.6 L s^{-1}. Basin water depth is 0.02 m. The reference slope is set at 0.02. The time step is estimated at 1.67 s. Sediment input is considered to be coarse-grained only. These conditions are representative of laboratory experiments such as those reported by Reitz and Jerolmack (2012).

In Fig. 8a–f we show a time series of the resultant deltas during one avulsion cycle. These plots reveal the key characteristics of an alluvial fan delta, in which a few active channels quickly switch (avulse) to build a semicircular shape with a relatively smooth shoreline (Reitz and Jerolmack, 2012). To evaluate the details of this channel-switching process, we calculate the wet fraction of delta surface that is covered by active channels (defined by cells that have a flow

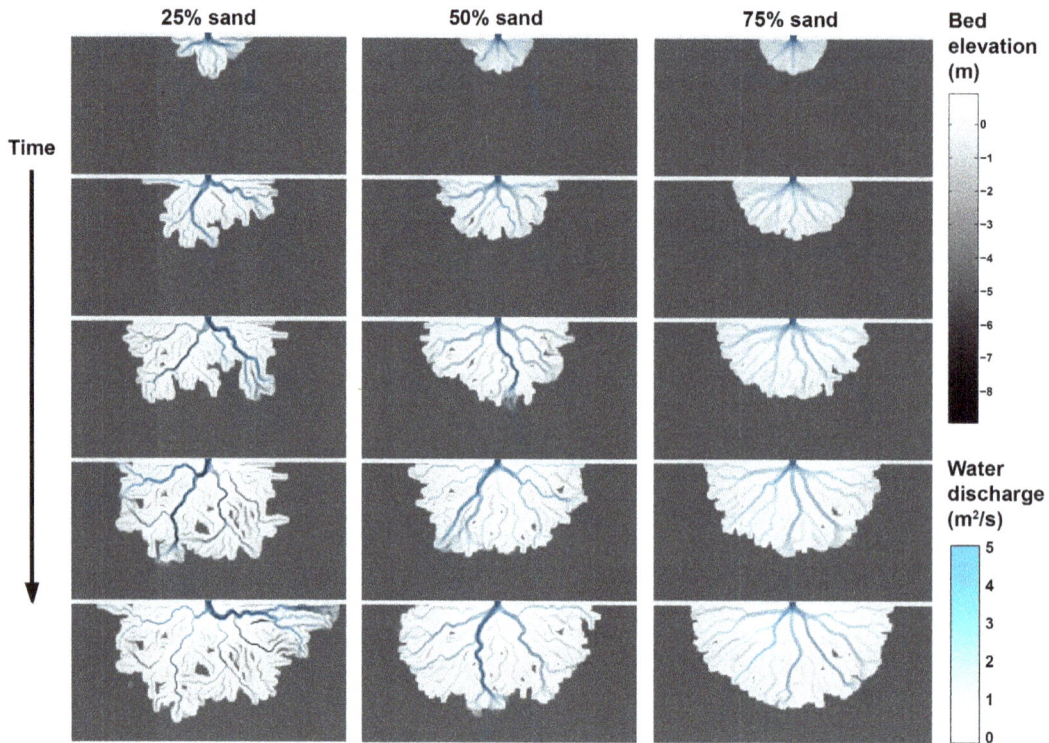

Figure 6. Time series of delta formation with different ratios of sand and mud flux (runs 1, 2 and 3). The time interval between rows is roughly 200 days of delta building time with continuous bank-full discharge.

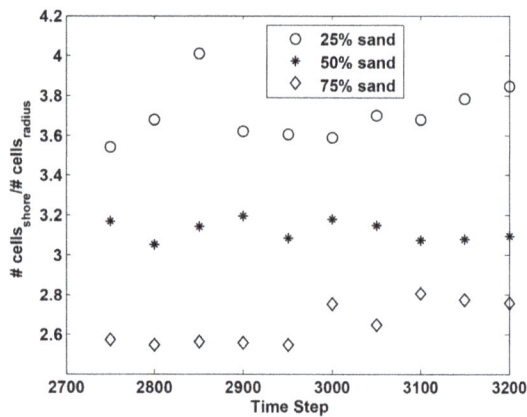

Figure 7. Comparing shoreline roughness between simulated deltas with input sand fractions of 25, 50, and 75 %. Shoreline roughness here is measured by the ratio between (i) the number of cells in the domain that contain a piece of shoreline of the simulated delta, and (ii) the average radius of the delta toposet in number of cells.

velocity greater than 50 % of the characteristic flow velocity) and plot it against time (Fig. 8g). Each avulsion event can be identified by a sudden drop of the wet fraction followed by a relatively slow rise caused by backfilling and flooding. An avulsion timescale estimated from this plot is in the range of 5–10 min, a value that is of the same order as the laboratory observations made by Reitz and Jerolmack (2012).

4.3 Effects of basin depth

It has been suggested that the accommodation – the space that a delta can grow into – plays an important role in the architecture and behavior of a growing delta (e.g., Paola, 2000; Heller et al., 2001). However, for the case of river deltas with very low-Froude-number flow, it is still unclear how the depth of the basin affects the overall morphology of the delta. Storms et al. (2007) use Delft3D to model initial delta formation from a river effluent discharging constant flow and sediment loads into shallow and deep receiving basins under homopycnal conditions; they show that the shallow basin delta is dominated by mouth-bar bifurcations and a shoaling channel network, and exhibits significant stratigraphic complexity and subaerial development, while the deep basin delta is dominated by unstable bifurcations, levee breaches and avulsions (Storms et al., 2007). The authors suggest that the shallow basin case resembles the Wax Lake Delta. In our model runs 6 and 7, we test scenarios with the same inlet channel conditions and discharge, but different basin depths. In run 6, the receiving basin depth is half of the reference depth (defined by the inlet channel which is supposed to be at equilibrium state in terms of sediment transport), while in run 7, the receiving basin depth is double the reference depth. In Fig. 9 we show that our results yield similar behaviors to the ones modeled by Storms et al. (2007) using Delft3D. For the shallow basin the morphological development is very close

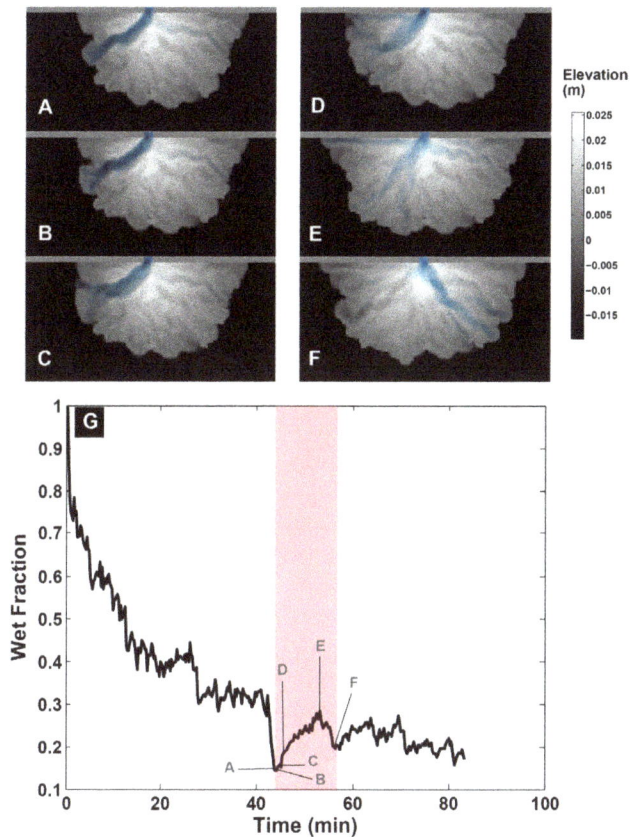

Figure 8. The series of images matches the avulsion cycles observed in physical experiments (Reitz and Jerolmack, 2012): (a) channelizes, (b) pushes out the shoreline (only deposits at the channel mouth), (c) flares out locally to establish a semicircular lobe (deposits minilobes around original channel mouth by local avulsions and sheet flow), (d) backfills (channel widens), (e) floods, and (f) channelizes again. Note that the time interval between (d) and (e) is about 4 times longer than any other pair of consecutive frames.

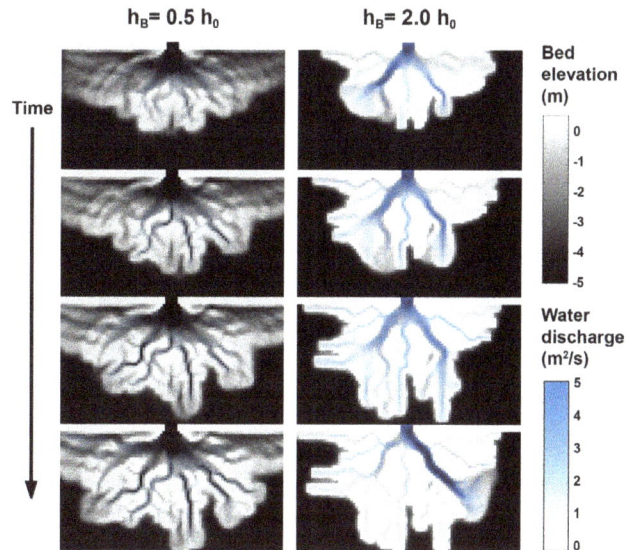

Figure 9. Two model runs (runs 6 and 7) with different basin depths and everything else the same. The shallow basin delta is dominated by more frequent bifurcations while the deep basin delta is dominated by few channels with more avulsions.

to the description of Storms et al., while the deep basin delta has similar outcomes but the middle ground bar and avulsion over the levee are not as clear in the RCM results.

The differences between a shallow and deep receiving basin, according to our model results, are the following:

- Channels will still try to maintain the same unit power of transporting sediment by maintaining a certain cross-sectional geometry with levees on the side and erosion or deposition on the bottom.

- In general, a distributary channel network shoals up and channels are stable at shallower depths going seaward. With a shallow basin the amount of work is reduced. Also, the narrow space promotes the splitting of flow which enhances the growth of a distributary network.

- A deep basin increases the timescale of establishing a stable channel and, therefore, introduces stronger com-

petition among channels by allowing larger differences to develop.

- The total number of active channels is higher in the shallow basin case, with about 5–6 channels, as compared to 1–3 channels in the deep-basin case.

Finally, we note two interesting emergent features from our model that have also been observed in the field at Wax Lake Delta by Shaw (2013) and Shaw et al. (2013) (Fig. 10). First, the channels in the shallow basin delta are initially erosional, and carve into the basin bottom. This is consistent with the observations at the Wax Lake Delta (Shaw et al., 2013). Second, the channel network on this delta develops "tributary" subnetworks on islands (highlighted in Fig. 10), which collect flow both from tie channels directly connected to the main channel network and from sheet flow topping the levees into the islands. As to whether this subnetwork is erosional or depositional, Shaw (2013) points out that at least the channels comprising it are likely not favorable for deposition. In our model results, we notice the following process that might explain the situation.

1. The subnetwork mainly collects fine sediment from the main channel network, which requires a much slower flow to settle.

2. As the tributary subnetwork joins into bigger trunk channels, the ability of the flow to carry sediment increases.

3. Finally, at the downstream end of the network, where the trunk channel collecting water coming out of the island meets the open water, the sorting of the sediment

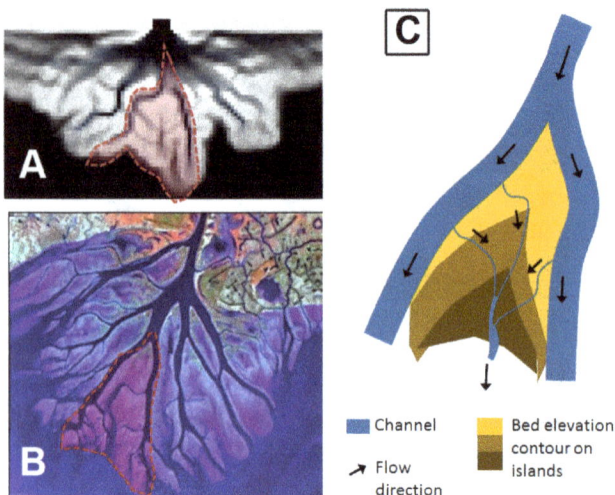

Figure 10. Flow features on the island of a delta formed in a shallow basin. **(a)** Model result from run 6, where basin depth (2.5 m) is only half of the inlet channel depth (5 m); **(b)** Wax Lake Delta, where basin depth (< 5 m) is much lower than the inlet channel depth (> 20 m); **(c)** schematic drawing showing the "tributary" flow feature on the island (Shaw et al., 2013) observed both in the field (Shaw, 2013) and in our numerical model results.

deposited is very similar to a normal channel that has a coarser bar-like structure at the mouth.

5   Recording of stratigraphy

A delta writes (and rewrites) its own autobiography by building a sedimentary record from deposition and erosion. These sedimentary records allow us understand the past and to use delta deposits to reconstruct their range of natural behavior. Therefore, the ability to record stratigraphy in a delta formation model enables us to directly investigate the connection between surface and subsurface processes. In this model, we have two methods that track the stratigraphy of model-produced deltas: the first method tracks the distribution of coarse and fine sediment by recording the percentage of sand in each deposition event; and the second method tracks the age of the deposit by labeling each deposition event with the time that its sediment enters the domain from the inlet channel.

To track stratigraphy each cell in the domain is viewed as a storage column (shown in Fig. 2), and the volume below the bed surface is further divided into thin layers of an equal thickness (these layers are visible especially in Figs. 11 and 12). The thickness is chosen to be about a thousandth of the reference depth, although it can be set to different values to allow for different vertical resolutions. Each layer is recorded with a value associated with it – at present it is either the percentage of sand (a value between 0 and 1) or the age of the deposit (represented by the number of time step).

For example, if a cell has net deposition, the volume it received from passing parcels will fill up as many layers as needed above the previous bed surface, and all values associated with these layers are set to the ratio between the volume of sand deposited and the total volume of sediment deposited during this time step. If a cell has net erosion, the bed surface will be lowered and all values associated with the layers above the new bed surface will be erased (by resetting these values to −1 in the code).

Here we present two examples. (1) We take a sample run of a field-scale delta and 30 % sediment input (run 4). In Fig. 11, we show a stratigraphic slice in the dip direction along the center line of the inlet channel. In Fig. 12, we show the time series of the stratigraphic slice in the strike direction about 20 cells (1 km in this case) away from the inlet channel. In both figures white represents pure sand and dark blue represents pure mud, with mixed deposits represented by linear combinations of the two endmembers. Generally speaking, coarse sediment (sand) can be found in channel belts and mouth bars, while fine sediment (mud) can be found in distal regions such as the bottom set of the delta, on the floodplain or in abandoned channels. (2) In Fig. 13 we show a sample model run for laboratory conditions (run 8). Note the evolution of the area pointed to by the yellow arrow. The series of images shows the deposition sequence from an individual avulsion event.

6   Discussion

One of the themes running through this paper is that even in the framework of a reduced-complexity delta model there are a number of important details that must be modeled fairly accurately to achieve even qualitatively correct model results. These include a reasonably accurate representation of the water surface and the inclusion of suspended sediment deposition and entrainment. To demonstrate the importance of the water surface we switch off this component in routing water parcels, i.e., we set the partitioning coefficient (γ) to zero. In this case, the delta is completely dominated by inertia and as a result a single elongated channel extends without avulsion or bifurcation (Fig. 14a). (Note that this is not the same behavior as setting the input sediment to contain 0 % sand which exhibits multiple elongated channels – see Fig. 14b.) By contrast, the effect of deposition and entrainment of fine-grained sediment in DeltaRCM is illustrated by removing the suspended sediment load from the calculation. In such a case, the predicted channels are highly mobile and levees separating channels and floodplains are absent; i.e., we arrive at a delta formation with no stable channel networks, the characteristics of an alluvial fan (Fig. 14c). The importance of the water surface and suspended sediment is also well illustrated in the previous RCM delta model developed by Seybold et al. (2007, 2009), where a reduced-complexity water surface and depth calculation, along with a treatment of cohesive and

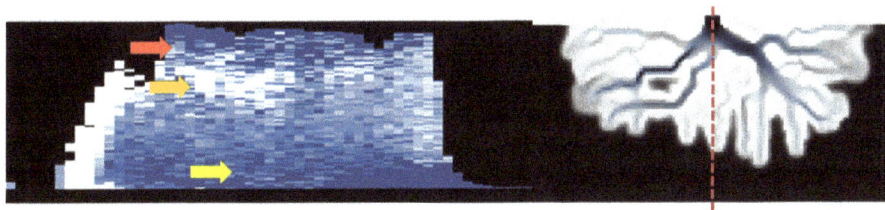

Figure 11. Stratigraphy slice in the dip direction of run 4 (30 % sand input). Note the layering of coarse and fine grains over time. Yellow arrow points to the bottom layer that accumulates fine grains at the bottom set of the delta; orange arrow points to the coarse grain layer deposited by channels that used to be active at that location; the two together show the classic "coarsening-up" pattern in stratigraphy. The red arrow points to the fine grains deposited after the channels are abandoned.

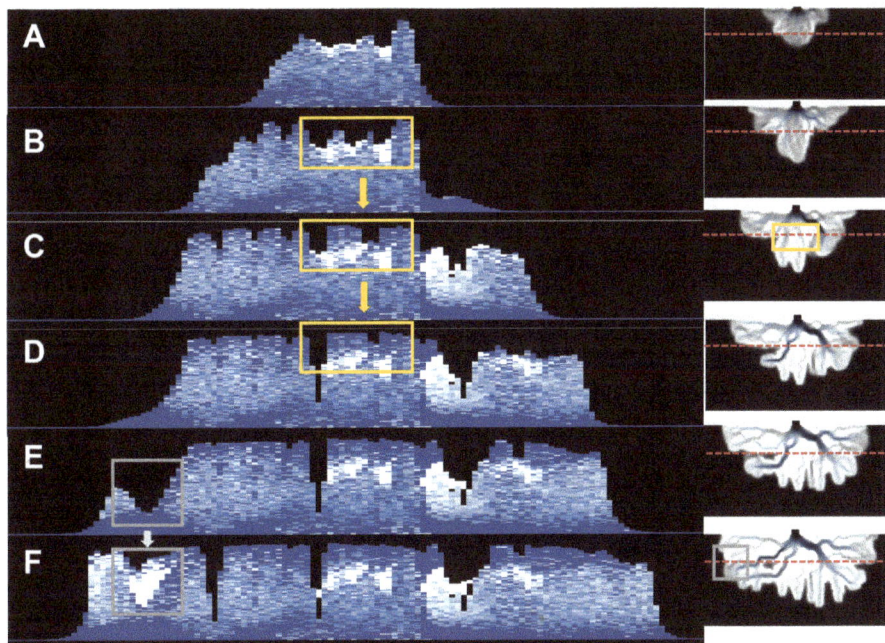

Figure 12. Time series of the stratigraphic slice in the strike direction about 20 cells (1 km) away from the inlet channel. Note that between (**b**), (**c**) and (**d**), in the yellow box, the abandoned channel belts are covered by muddy floodplain deposits. Also note that between (**e**) and (**f**), in the light gray box, a mouth bar quickly deposits a significant amount of sand.

noncohesive sediment behaviors through a flow strength and flow velocity terms, respectively, was able to build both elongated bird-foot and multichannel fan deltas. Part 2 of this work further explores the hydrodynamic mechanism of the water surface and investigates the feedback between the flow solver and the sediment transport processes in determining channel bifurcations.

The need for accurate representation of some of the physical details in DeltaRCM is quite striking compared to the success of even fairly radical reduced-complexity approaches in modeling other morphodynamic environments such as erosional landscapes (e.g., Willgoose et al., 1991), braided rivers (e.g., Murray and Paola, 1994) and eolian bedforms (e.g., Werner, 1995). So why is it that deltas seem to require more attention to detail? Can we learn anything from this experi-

ence that might help us better understand what systems are most and least amenable to reduced-complexity approaches?

Since deltas and drainage basins share dendritic channel patterns – one is a distributary network while the other is a tributary network – we first look at the differences between these two systems. In modeling the evolution of drainage tributary networks, even highly simplified relations for water flux and sediment transport yield quite reasonable drainage networks and elevation changes in the long-term evolution of catchments (e.g., Willgoose et al., 1991). The equation describing the evolution of land elevation in Willgoose et al. (1991) includes two transport processes: fluvial transport and diffusive transport. The former is dependent on the discharge and the slope in the steepest downhill direction, and the latter is dependent on slope and diffusivity. Relations of similar simplicity cannot be easily applied to modeling

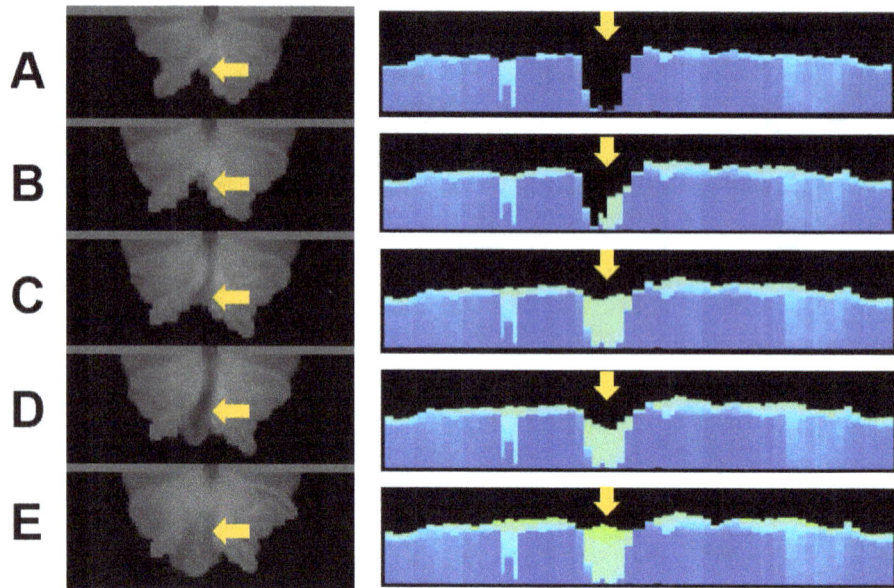

Figure 13. Time series of a delta produced by DeltaRCM with laboratory settings, and stratigraphy slices in the strike direction about 20 cells (0.4 m) away from channel inlet. Note the evolution of the area pointed to by the yellow arrow. (**a**) A concave shoreline – empty space in stratigraphy; (**b**) channel begins to receive water and sediment – deposition begins; (**c**) more water and sediment switch to the channel – space is filled-up quickly; (**d**) full avulsion completed – a channel is established by water eroding existing deposits; (**e**) backfilling causes flooding and the channel loses its advantage – the channel is refilled and there is a discontinuity in deposition age (yellowish green in the upper portion and bluish green in the lower portion).

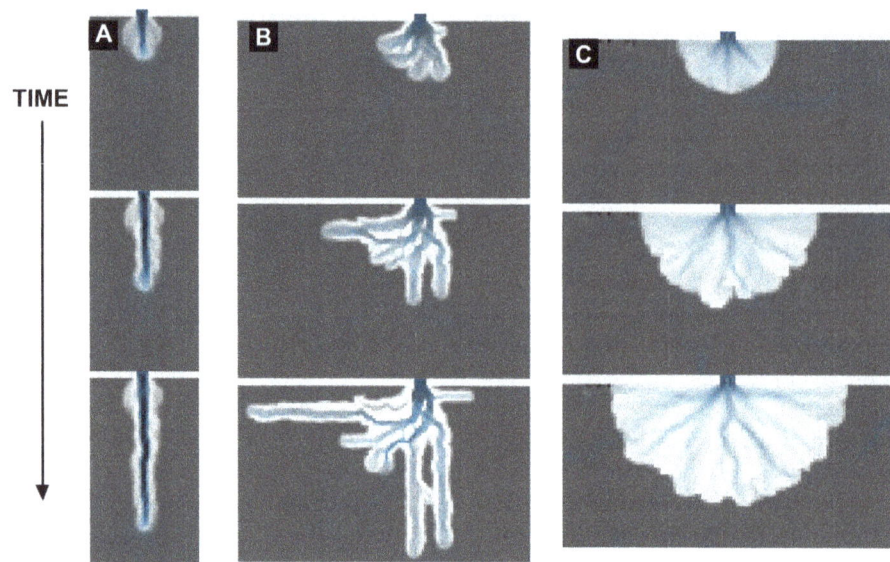

Figure 14. Effects of model parameters. (**a**) An elongated channel is formed with 50 % sand input by switching off the influence of water surface in routing water flow (i.e., setting parameter γ to zero). (**b**) Multiple elongated channels are formed with 0 % sand (100 % mud) input without modifying any parameter values. (**c**) A fan delta is formed with 100 % sand input which shows that a stable channel network with levees cannot be achieved with only bedload.

deltas because deltas are low-gradient environments where the transport direction and capacity are to some extent decoupled from bed elevation and slope. To be more specific, (1) bed slope in low-gradient environments is often uncorrelated with flow direction and strength; for example, bed slope points opposite to the direction of flow where channels shoal up towards the shoreline; (2) the water surface, which dominates local flow routing, is largely independent of bed topography; (3) the typical low-Froude-number flow in low-gradient deltaic environments creates strong backwater effects that imply strong nonlocality in flow and sediment flux control (Lamb et al., 2012; Nittrouer et al., 2011) – meaning that downstream conditions control upstream flow dynamics (Hoyal and Sheets, 2009); and (4) river mouth and shore processes such as waves and tides also control the overall morphology of deltas, providing additional process complexity.

According to Werner (1995), for a nonlinear and dissipative system, considerable simplification can be applied if the system exhibits the following two properties: (1) it has a finite number of steady states as "attractors", and (2) it has macroscopic emergent behaviors that are self-organized and consistent with, but decoupled, from microscopic physics. If we compare drainage networks with deltas, the former exhibits a strong generic pattern and scale-invariant properties expressed in generalizations such as Horton's laws (Horton, 1945). In contrast, the networks on deltas have many varieties, responding to a wide range of processes; no universal geometry applies to them all. Regarding model complexity, the lack of universality in the system pattern indicates the requirement for a more detailed, system-specific approach in modeling them.

So, is the low gradient the main cause of the modeling difficulty, making deltas more "unforgiving" than erosional landscapes in terms of the accuracy of hydrodynamic calculation? For cellular models that use explicit flow routing schemes, the complexity level rises as factors other than topographic slope alone determine water and sediment routing. It also increases with nonlocality in the broad sense of the sensitivity of dynamics at one point to conditions far away in the system. Other contributing factors such as water surface gradient and flow inertia weigh in as the overall topographic gradient decreases. For example, dune fields may have very low to zero average topographic slope, but they have locally high steepness meaning that, as in erosional landscapes, the sediment dynamics are dominated by bed topography. In deltas, however, the controlling factor is the relatively subtle water surface topography, therefore simple descriptions relating sediment deposition and erosion to e.g., local elevation and slope give realistic dune field dynamics but do not work in deltas.

Can we be more systematic about evaluating the amount of detail needed to model a geomorphodynamic system? This is an important fundamental question in morphodynamic modeling, and we do not pretend to resolve it here. But our experience with DeltaRCM suggests the following guidelines as a starting point.

- For gravity-driven systems, the overall gradient of the landform is one important index in the sense that in high-gradient systems the gradient alone is enough to route the flow.

- A closely related indicator is the wetted area fraction in the sense that a combination of low wetted fraction and high topographic gradient is the limit in which steepest-path methods (Passalacqua et al., 2010) are sufficient to determine the flow path, without the need for simulation of the flow details.

- Froude number (Fr): as Fr tends to unity, the backwater length tends to zero (Cui and Parker, 1997), so the simplification of a local normal-flow assumption provides a satisfactory means of accounting for momentum balance in the flow.

- For systematic behaviors on scales greater than the backwater length scale, in-channel-scale hydrodynamic details can be resolved at much lower complexity; this applies for example to avulsion models that use single-cell-wide threads to represent channel belts (Jerolmack and Paola, 2007).

- Whether the system to be modeled exhibits a strong generic pattern or scale-invariant (e.g., fractal) properties, the lack of universal patterns in a dynamic system is an indicator of sensitivity to local detail.

We see the potential of this type of modeling as analogous to that of laboratory experiments, which can also provide useful insight despite not capturing all the details of complex natural systems (Paola et al., 2009). The strength of RCMs is to serve as (1) exploratory models that allow for direct representation of phenomenological observation; (2) a tool to identify those aspects of large-scale system behavior that are not sensitive to the details of smaller-scale processes; and (3) a framework for hybrid modeling in which higher-resolution model results can be integrated where precise description of smaller-scale processes is needed even for larger-scale dynamics.

7 Conclusions

In this paper we have introduced a new reduced-complexity model (RCM) for river delta formation. Key techniques include that (1) water and sediment fluxes are represented as parcels and routed through the domain in a Lagrangian point of view; (2) the movements of parcels are based on a probability field calculated from rules abstracting the governing physics; (3) deposition and erosion are achieved by exchanging the volume of passing sediment parcels and bed sediment columns, and the condition for this exchange depends on a set of rules that distinguish bedload and suspended load; (4) bed sediment columns record the composition of coarse and fine material in layers; (5) a topographic diffusion process takes into account cross-slope sediment transport and bank erosion. By varying input conditions such as the ratio of coarse and fine sediment, reference slope, and dimensions of the domain, the simulated deltas yield a range of different behaviors that compare well to higher-fidelity model results and observations of field and experiment deltas.

We find that the relatively simple cellular representation of water and sediment transport is able to replicate delta morphology at the scale of channel dynamics, including the emergent channel network with channel extension, bifurcation and avulsion. Here, we summarize the basic components needed for a RCM to produce major static and dynamic features of river deltas:

- a depth-averaged flow field that guides sediment transport

- a nontrivial water surface profile that accounts for backwater effects at least in the main channels

- representation of both bedload and suspended load

- topographic steering of sediment transport.

Even at the RCM level of modeling, the following items still require a physically consistent treatment:

- the instability at channel mouths that creates bars and subsequent bifurcation

- the variation in water surface profile associated with lobe extension that causes channel avulsion

- water surface slope along channel sides which creates flooding onto the floodplain.

Appendix A

Table A1. List of notations.

Symbol	Definition and unit	Symbol	Definition and unit
α	Topographic diffusion coefficient ($m^2\,s^{-1}$)	n_s	Number of sediment parcels (–)
γ	Partitioning parameter for routing by inertia and by free surface (–)	P_i	Routing probability (–)
Δ_i	Cellular distance (–)	$\boldsymbol{Q}_{cell}$	Total discharge at a cell ($m^3\,s^{-1}$)
δ_c	Grid size (m)	Q_{w0}	Total water discharge from inlet channel ($m^3\,s^{-1}$)
ε	Diffusion coefficient for water surface smoothing (–)	Q_{s0}	Total sediment discharge from inlet channel ($m^3\,s^{-1}$)
η	Bed/land elevation (m)	Q_{p_water}	Discharge represented by a water parcel ($m^3\,s^{-1}$)
η_{shore}	Threshold bed elevation for marking shoreline (m)	q_{s_cap}	Sediment flux capacity at a cell ($m^2\,s^{-1}$)
θ	Exponent of depth dependence (–)	q_{s_diff}	Diffusive sediment flux at a cell ($m^2\,s^{-1}$)
ϖ	Underrelaxation coefficient for water surface (–)	q_{s_loc}	Local coarse sediment flux at a cell ($m^2\,s^{-1}$)
$\boldsymbol{d}_i$	Cellular unit direction (–)	$q_w = (q_x, q_y)$	Water unit discharge vector ($m^2\,s^{-1}$)
F	Routing direction (–)	R_i	Flow resistance (–)
$\boldsymbol{F}_{int}$	Routing direction by inertia (–)	S_0	Reference slope (–)
$\boldsymbol{F}_{sfc}$	Routing direction by water surface (–)	Δt	Time step (s)
f_{sand}	Fraction of sand (–)	U_0	Reference velocity ($m\,s^{-1}$)
H	Water surface elevation (m)	U_{dep}, U_{ero}	Threshold velocity for deposition and erosion ($m\,s^{-1}$)
H_{SL}	Sea level (m)	U_{loc}	Local velocity at a cell ($m\,s^{-1}$)
H^{smooth}	Smoothed water surface elevation (m)	U_{shore}	Threshold velocity for marking shoreline ($m\,s^{-1}$)
H^{temp}	Temporary water surface solution (m)	$\boldsymbol{u} = (u_x, u_y)$	Flow velocity vector ($m\,s^{-1}$)
h	Water depth (m)	V_{p_sed}	Initial volume of a sediment parcel (m^3)
h_0	Reference water depth (m)	V_{p_dep}	Volume removed from a sediment parcel by deposition (m^3)
h_B	Basin water depth (m)	V_{p_ero}	Volume added to a sediment parcel by erosion (m^3)
h_{dry}	Threshold water depth for dry land (m)	V_{p_res}	Remaining volume of a sediment parcel (m^3)
N_{visit}	Number of water-parcel visits at a cell (–)	V_0	Reference volume (m^3)
N_x, N_y	Number of cells in the x and y directions of the computational domain (–)	ΔV_s	Total volume of sediment input at each time step (m^3)
N_0	Number of cells across inlet channel (–)	W	Width of inlet channel (m)
n_w	Number of water parcels (–)	w_i	Routing weights (–)

Acknowledgements. This work was supported by the National Science Foundation via the National Center for Earth-surface Dynamics (NCED) under agreement EAR-0120914 and EAR-1246761. This work also received support from the National Science Foundation via grant FESD/EAR-1135427 and from ExxonMobil Upstream Research Company. The authors thank P. Passalacqua, D. A. Edmonds, N. Geleynse, and J. Martin for discussions and comments. The authors also thank S. Castelltort, R. Slingerland and A. Ashton for their insightful comments and reviews.

Edited by: S. Castelltort

,

References

Cui, Y. and Parker, G.: A quasi-normal simulation of aggradation and downstream fining with shock fitting, Int. J. Sediment Res., 12, 68–82, 1997.

Edmonds, D. A. and Slingerland, R. L.: Mechanics of river mouth bar formation: implications for the morphodynamics of delta distributary networks, J. Geophys. Res., 112, F02034, doi:10.1029/2006JF000574, 2007.

Edmonds, D. A. and Slingerland, R. L.: Significant effect of sediment cohesion on delta morphology, Nat. Geosci., 3, 105–109, 2009.

Fagherazzi, S. and Overeem, I.: Models of deltaic and inner continental shelf landform evolution, Annu. Rev. Earth Pl. Sc., 35, 685–715, 2007.

García, M. and Parker, G.: Entrainment of bed sediment into suspension, J. Hydraul. Eng., 117, 414–435, 1991.

García, M. H.: Sedimentation Engineering: Processes, Measurements, Modeling, and Practice, American Society of Civil Engineers, Reston, VA, 1132 pp., 2008.

Galloway, W. E.: Process framework for describing the morphologic and stratigraphic evolution of deltaic depositional systems, in: Deltas, Models for Exploration, Houston Geological Society, Houston, TX, 87–98, 1975.

Geleynse, N., Storms, J. E. A., Stive, M. J. F., Jagers, H. R. A., and Walstra, D. J. R.: Modeling of a mixed-load fluvio-deltaic system, Geophys. Res. Lett., 37, L05402, doi:10.1029/2009GL042000, 2010.

Hajek, E. A. and Wolinsky, M. A.: Simplified process modeling of river avulsion and alluvial architecture: connecting models and field data, Sediment. Geol., 257–260, 1–30, 2012.

Heller, P. L., Paola, C., Hwang, I.-G., John, B., and Steel, R.: Geomorphology and sequence stratigraphy due to slow and rapid base-level changes in an experimental subsiding basin (xes 96-1), AAPG Bull., 85, 817–838, 2001.

Horton, R. E.: Erosional development of streams and their drainage basins, Geol. Soc. Am. Bull., 56, 275–370, 1945.

Hoyal, D. C. and Sheets, B. A.: Morphodynamic evolution of experimental cohesive deltas, J. Geophys. Res., 110, F02009, doi:10.1029/2007JF000882, 2009.

Jerolmack, D. J. and Paola, C.: Complexity in a cellular model of river avulsion, Geomorphology, 91, 259–270, 2007.

Kim, W., Mohrig, D., Twilley, R., Paola, C., and Parker, G.: Is it feasible to build new land in the Mississippi River Delta?, EOS Trans. Am. Geophys. Un., 90, 373–374, 2009.

Lamb, M. P., Nittrouer, J. A., Mohrig, D., and Shaw, J.: Backwater and river-plume controls on scour upstream of river mouths: implications for fluvio-deltaic morphodynamics, J. Geophys. Res., 117, F01002, doi:10.1029/2011JF002079, 2012.

Lesser, G. R., Roelvink, J. A., van Kester, J. A. T. M., and Stelling, G. S.: Development and validation of a three-dimensional morphological model, Coast. Eng., 51, 883–915, 2004.

Lorenzo-Trueba, J., Voller, V. R., and Paola, C.: A geometric model for the dynamics of a fluvially dominated deltaic system under base-level change, Comput. Geosci., 53, 39–47, 2013.

Meyer-Peter, E. and Müller, R.: Formulas for bed-load transport, in: Proceedings of the 2nd Meeting of IAHSR, Stockholm, Sweden, 39–64, 1948.

Murray, A. B.: Contrasting the goals, strategies, and predictions associated with simplified numerical models and detailed simulations, Geophys. Monogr. Ser., 135, 151–165, 2003.

Murray, A. B. and Paola, C.: A cellular model of braided rivers, Nature, 371, 54–57, 1994.

Murray, A. B. and Paola, C.: Properties of a cellular braided-stream model, Earth Surf. Proc. Land., 22, 1001–1025, 1997.

Nittrouer, J. A., Mohrig, D., Allison, M. A., and Peyret, A. P. B.: The lowermost Mississippi River: a mixed bedrock-alluvial channel, Sedimentology, 58, 1914–1934, 2011.

Orton, G. J. and Reading, H. G.: Variability of deltaic processes in terms of sediment supply, with particular emphasis on grain size, Sedimentology, 40, 75–512, 1993.

Overeem, I., Syvitski, J. P. M., and Hutton, E. W. H.: Three-dimensional numerical modeling of deltas, in: River Deltas: Concepts, Models and Examples, edited by: Bhattacharya, J. P. and Giosan, L., SEPM Spec. Publ. 83, SEPM, Tulsa, OK, 13–30, 2005.

Paola, C.: Quantitative models of sedimentary basin filling, Sedimentology, 47, 121–178, 2000.

Paola, C. and Leeder, M.: Environmental dynamics: simplicity versus complexity, Nature, 469, 38–39, 2011.

Paola, C., Straub, K., Mohrig, D., and Reinhardt, L.: The "unreasonable effectiveness" of stratigraphic and geomorphic experiments, Earth-Sci. Rev., 97, 1–43, 2009.

Paola, C., Twilley, R. R., Edmonds, D. A., Kim, W., Mohrig, D., Parker, G., Viparelli, E., and Voller, V. R.: Natural processes in delta restoration: application to the Mississippi Delta, Annu. Rev. Mar. Sci., 3, 67–91, 2011.

Parker, G., Muto, T., Akamatsu, Y., Dietrich, W. E., and Lauer, J.: Unravelling the conundrum of river response to rising sea-level from laboratory to field, Part I: Laboratory experiments, Sedimentology, 55, 1643–1655, 2008.

Passalacqua, P., Do Trung, T., Foufoula-Georgiou, E., Sapiro, G., and Dietrich, W. E.: A geometric framework for channel network extraction from lidar: Nonlinear diffusion and geodesic paths, J. Geophys. Res., 115, F01002, doi:10.1029/2009JF001254, 2010.

Reitz, M. D. and Jerolmack, D. J.: Experimental alluvial fan evolution: Channel dynamics, slope controls, and shoreline growth, J. Geophys. Res., 117, F02021, doi:10.1029/2011JF002261, 2012.

Rinaldo, A., Fagherazzi, S., Lanzoni, S., Marani, M., and Dietrich, W. E.: Tidal networks: Landscape-forming discharges and studies in empirical geomorphic relationships, Water Resour. Res., 35, 3919–3929, 1999.

Seybold, H., Andrade Jr., J. S., and Herrmann, H. J.: Modeling river delta formation, P. Natl. Acad. Sci. USA, 104, 16804–16809, 2007.

Seybold, H. J., Molnar, P., Singer, H. M., Andrade, J. S., Herrmann, H. J., and Kinzelbach, W.: Simulation of birdfoot delta formation with application to the Mississippi Delta, J. Geophys. Res., 114, F03012, doi:10.1029/2009JF001248, 2009.

Seybold, H. J., Molnar, P., Akca, D., Doumi, M., Cavalcanti Tavares, M., Shinbrot, T., Andrade, J. S., Kinzelbach, W., and Herrmann, H. J.: Topography of inland deltas: observations, modeling, and experiments, Geophys. Res. Lett., 37, L08402, doi:10.1029/2009GL041605, 2010.

Shaw, J. B.: The kinematics of distributary channels on the Wax Lake Delta, coastal Louisiana, USA, PhD dissertation, University of Texas at Austin, Austin, TX, 2013.

Shaw, J. B., Wolinsky, M. A., Paola, C., and Voller, V. R.: An image-based method for shoreline mapping on complex coasts, Geophys. Res. Lett., 35, L12405, doi:10.1029/2008GL033963, 2008.

Shaw, J. B., Mohrig, D., and Whitman, S. K.: The morphology and evolution of channels on the Wax Lake Delta, Louisiana, USA, J. Geophys. Res.-Earth, 118, 1–23, 2013.

Storms, J. E. A., Stive, M. J. F., Roelvink, D. (J.) A., and Walstra, D. J.: Initial Morphologic and Stratigraphic Delta Evolution Related to Buoyant River Plumes, Coastal Sediment'07, American Society of Civil Engineers, New Orleans, Louisiana, 736–748, 2007.

Sun, T., Paola, C., Parker, G., and Meakin, P.: Fluvial fan deltas: linking channel processes with large-scale morphodynamics, Water Resour. Res., 38, 26-1–26-10, 2002.

Syvitski, J. P. M., Kettner, A. J., Overeem, I., Hutton, E. W. H., Hannon, M. T., Brakenridge, G. R., Day, J., Vörösmarty, C., Saito, Y., Giosan, L., and Nicholls, R. J.: Sinking deltas due to human activities, Nat. Geosci., 2, 681–686, 2009.

Van Rijn, L. C.: Sediment transport II: Suspended load transport, J. Hydraul. Eng., 110, 1431–1456, 1984.

Werner, B. T.: Eolian dunes: computer simulations and attractor interpretation, Geology, 23, 1107–1110, 1995.

Willgoose, G., Bras, R. L., and Rodriguez-Iturbe, I.: A coupled channel network growth and hillslope evolution model: 1. Theory, Water Resour. Res., 27, 1671–1684, 1991.

A reduced-complexity model for river delta formation – Part 2: Assessment of the flow routing scheme

M. Liang[1,2]**, N. Geleynse**[2,*]**, D. A. Edmonds**[3]**, and P. Passalacqua**[1]

[1]Department of Civil, Architectural and Environmental Engineering and Center for Research in Water Resources, The University of Texas at Austin, Austin, Texas, USA
[2]Department of Geological Sciences, The University of Texas at Austin, Austin, Texas, USA
[3]Department of Geological Sciences, Indiana University, Bloomington, Indiana, USA
[*]now at: ARCADIS, Water and Environment Division, Zwolle, the Netherlands

Correspondence to: M. Liang (manliang@utexas.edu)

Abstract. In a companion paper (Liang et al., 2015) we introduced a reduced-complexity model (RCM) for river delta formation, developed using a parcel-based "weighted random walk" method for routing water and sediment flux. This model (referred to as DeltaRCM) consists of a flow routing scheme as the hydrodynamic component (referred to as FlowRCM) and a set of sediment transport rules as the morphodynamic component. In this work, we assess the performance of FlowRCM via a series of hydrodynamic tests by comparing the model outputs to Delft3D and theoretical predictions. These tests are designed to reveal the capability of FlowRCM to resolve flow field features that are critical to delta dynamics at the level of channel processes. In particular, we focus on (1) backwater profile, (2) flow around a mouth bar, (3) flow through a single bifurcation, and (4) flow through a distributary channel network. We show that while the simple rules are not able to reproduce all fine-scale flow structures, FlowRCM captures flow field features that are essential to deltaic processes such as bifurcations and avulsions, the partitioning of flux between channels and inundated islands, and the instability of flux distribution at channel mouths which is responsible for mouth-bar growth. Finally, we discuss advantages and limitations of FlowRCM and identify environments most suitable for it.

1 Introduction

Flow routing plays a fundamental role in geomorphologic modeling. Although all models resolving coupled fluid flow and sediment transport are simplified to a certain degree, reduced-complexity models (RCMs) gain their name in comparison to more detailed reductionism models, sometimes referred to as "high-fidelity" models, typically based on rigorous computational fluid dynamics (CFD) solutions (e.g., Lane et al., 2002; Lesser et al., 2004; Nicholas, 2013; Duan and Julien, 2010; Siviglia et al., 2013). Reduced-complexity (RC) flow routing schemes are typically in the form of cellular models, in which topography is represented by a lattice of cells with elevation information and wherein flux is calculated by cellular rules abstracting governing physics (Nicholas, 2005; Murray, 2007; Paola et al., 2011). Due to the highly rule-based nature and extensive parameterization, RC flow routing schemes are often tailored to the characteristics of the target environment, and the processes that are simplified and represented are carefully selected. For steep terrains such as drainage networks and alluvial fans, routing schemes are typically based on topographic slopes alone. Examples include single-direction methods such as the steepest descent method (O'Callaghan and Mark, 1984), and the multidirection method (MFD) (Freeman, 1991). The successive flow-routing method (Pelletier, 2008) modifies the MFD method by accounting for the effects of flow depth on flow pathways in an iterative fashion. The method can handle divergent flow with flooding, although it is still primarily based on topographic slope. For flow routing in fluvial channels and floodplain inundation, flow depth and water surface slope have to be taken into account (Bates et al., 2010). Routing schemes for these environments are usually based on equations such as Manning's to convert discharge to flow depth and obtain the local water surface slope and thus the flux between neighboring cells. Models in this category include LISFLOOD-FP (Bates and de Roo, 2000; Bates et al., 2010), CAESAR (Coulthard et al., 2002; Van De Wiel et al., 2007) and the ones proposed by Parsons and Fonstad (2007), Nicholas (2010) and Larsen and Harvey (2011).

Despite these efforts, there is a relatively small number of flow routing schemes developed for river deltas, which is surprising given the importance of deltaic environments for people and resources, and their increasingly recognized vulnerability to natural and anthropogenic disturbance (Syvitski et al., 2009). Existing RC flow routing schemes for channel-resolving river delta formation are the model proposed by Seybold et al. (2007) and the hydrodynamic component (referred to as FlowRCM) of the parcel-based delta formation model (DeltaRCM) developed by Liang et al. (2015). The challenges of routing flow in a deltaic environment include (i) low topographic slopes (typically 10^{-4}–10^{-5}), (ii) low-Froude-number flow which exhibits strong backwater effects making it difficult to design localized rules essential for RC routing schemes, and (iii) distributary channel networks with strong spatial variation in flow directions and loops which further complicate flow routing. Although methods such as LISFLOOD-FP, CAESAR and the one proposed by Nicholas et al. (2012) have shown good performance in low-gradient environments, they would likely need to be modified to account for flow characteristics of deltaic environments (Liang et al., 2015).

In this paper we provide a comprehensive assessment of our reduced complexity flow routing model FlowRCM, testing for the first time the performance of a RC model for deltas. We see this assessment as a test of model plausibility (Hardy et al., 2003), or the credibility of the processes represented in the model, rather than a "validation" of the model as often performed in numerical modeling. Some argue that validation of numerical models in Earth science is impossible and model confirmation by the demonstration of agreement between observation and prediction is inherently partial (Oreskes et al., 1994). Still, approaches at different levels have been defined and applied (Martin and Church, 2004) to compare a simulated landscape and a real landscape in terms of (i) behavior of governing processes (e.g., Tucker and Bras, 1998), (ii) qualitative consistency (e.g., Howard, 1997), (iii) full quantitative comparison (e.g., Ferguson et al., 2001), and (iv) statistical properties (e.g., Willgoose, 1994). Validating RCMs poses difficulties for several reasons as outlined below. First of all, RCMs in general put emphasis on "explanation" rather than "prediction" (Murray, 2007). Therefore, the validation for a RCM is different in the sense that evaluation of the representation of the classical physics in full is not the subject of the validation. Rather, the validation focuses on the evaluation of the correct representation of the behavior and physical structures of the system under consideration. Due to the wide spectrum of purpose and complexity of RCMs, standards for model validation are still poorly defined. Statistical methods offer quantitative metrics for validation, but the identification of the most revealing and discriminating statistics of a certain system is a challenge in itself. Examples of on-going efforts in developing better metrics for the quantitative description of river deltas include the work by Wolinsky et al. (2010), Edmonds et al. (2011), and Passalacqua et al. (2013). Second, a large portion of RCMs are built for geological timescales, such as models for studying channel avulsions and alluvial architecture (e.g., Jerolmack and Paola, 2007; Karssenberg and Bridge, 2008). Validation of such models requires intensive stratigraphic data which are not easily available. This issue has been addressed by many researchers, calling for a combined effort of model development and field observations (Overeem et al., 2005; Hajek and Wolinsky, 2012). Third, for cellular models, model validation (whether the model correctly represents physical processes) and verification (whether the numerical solutions are in agreement with the given equations) are sometimes mixed. For example, achieving a grid-independent solution is inherently more difficult than for other types of numerical models as the grid structure represents an implicit element of the process parameterization (Nicholas, 2005). After recent improvement of RC modeling techniques, especially in flow routing schemes, it is possible to develop RCMs that are capable of predicting flow patterns with accuracy comparable to CFD-based models (Nicholas et al., 2012). In the validation work by Nicholas et al. (2012), an intercomparison of a relatively simple RCM with two CFD models shows that all the three models are able to replicate patterns of depth-averaged velocity and unit discharge evident in acoustic Doppler current profiler (ADCP) cross-sectional surveys.

We design a series of tests for the RC flow routing scheme – referred to as FlowRCM – of the delta formation model introduced in Part 1 (Liang et al., 2015). With the proposed tests we aim to examine to what extent FlowRCM is able to reproduce hydrodynamic details, in comparison to higher-fidelity CFD-based models such as Delft3D, which have been shown to accurately model hydrodynamics (Lesser et al., 2004). The goal of our analysis is not to match Delft3D simulations, but to identify in retrospective similarities and differences in model results, and to identify what hydrodynamic features are the controlling factors in delta morphodynamics. The tests are designed in a way that each case represents a critical hydrodynamic process essential to delta morphology at the scale of channel dynamics: (1) backwater profile, (2) flow around a mouth bar, (3) flow through a single bifurcation, and (4) flow through a distributary channel network.

The paper is organized as follows. In Sect. 2, we briefly describe the key steps in FlowRCM, based on a parcel-based "weighted random walk" method. In Sect. 3, we present the design, setup and results of the test cases and their relevance

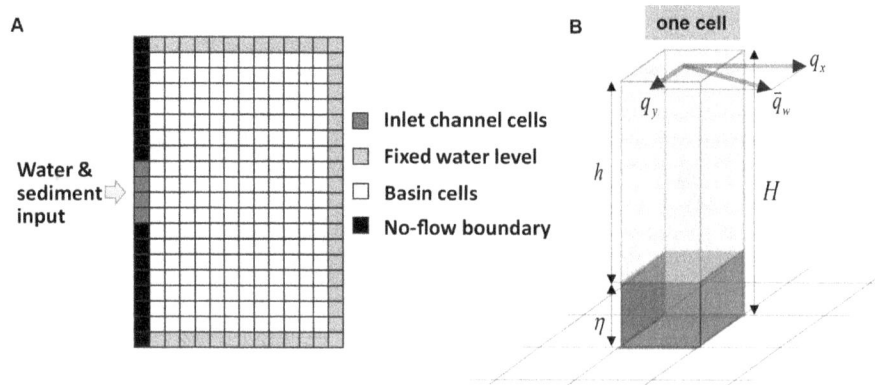

Figure 1. Basic setup of the model. (**a**) The calculation domain is represented by a lattice of square cells (shown as a diagram). (**b**) Primary values associated with each cell include flow depth, water surface elevation, bed elevation and flow unit discharge vector.

to delta formation. We discuss our results and ideas for future research in Sect. 4. Finally, we present our conclusions in Sect. 5.

2   Review of the proposed flow routing scheme

Here we give a brief review of our reduced-complexity flow routing scheme FlowRCM, focusing on the key steps and parameterizations (for a complete description refer to Part 1; Liang et al., 2015). In this section, we assume a nondeformable topography (i.e., no morphodynamic processes), over which the flow routing scheme resolves the depth-averaged flow field and the water surface profile given appropriate initial and boundary conditions.

The topography is represented by a lattice of square cells. At each cell, quantities such as bed elevation η, water surface elevation ∇H, and water unit discharge ∇H are recorded (Fig. 1). Water depth h can be calculated by taking the difference between water surface elevation and bed elevation (with H and η both relative to a reference datum). Water flux is represented by a large number of small water parcels, typically hundreds to a few thousands. With a larger number of parcels, the probability-based routing scheme is less affected by extreme events, but requires longer computational time. We found that the magnitude of 1000 works best in terms of efficiency. The solution is reached in an iterative fashion: first, parcels are routed individually from cell to cell based on a probability field, then the flow field and water surface elevation are updated. This process is repeated until the calculated flow field and water surface elevation converge. Convergence is defined as a dynamic equilibrium of flux distribution, which is typically achieved in 500–1000 iteration steps. We do not use the classic convergence criterion which is based on the difference between consecutive iterations due to two factors: (1) the probabilistic nature of the routing scheme introduces considerable noise between iterations, (2) there is an "oscillation" behavior caused by the feedback between underrelaxed water surface update and wa-

ter flux update. An example of this oscillation is shown and discussed in the next section (Test 4a).

The key component in our flow routing scheme is calculating the probabilities for routing water parcels, based on rules abstracting the governing physics – hence the name weighted random walk. As the detailed procedure is explained in Part 1, here we only revisit the main idea. The likelihood of a neighbor cell receiving a water parcel is determined by a combination of local quantities including water surface gradient, the direction of flow inertia, and flow depth at the cell.

The steps for calculating the routing probability field are described below.

2.1   Step 1: define a routing direction at each cell

The routing direction (F) is essentially an estimate of local downstream direction for the purpose of directing the water flux. In our model, it is a combination of the estimated local water surface gradient and direction of flow inertia:

$$F^* = \gamma F_{\text{sfc}} + (1 - \gamma) F_{\text{int}}, \tag{1}$$

$$F = \frac{F^*}{|F^*|}, \tag{2}$$

where $F_{\text{int}} = q|q|$ is a unit vector indicating inertia, and $F_{\text{sfc}} = \frac{\nabla H}{|\nabla H|}$ is a unit vector indicating surface gradient. Both q and ∇H take value from the latest iteration step. γ is a dimensionless coefficient, which is set to 0.05 in our runs if not indicated otherwise (Liang et al., 2015).

The value of γ affects the results of FlowRCM by controlling how sensitive the flow's response is to the water surface slope. For example, in the case of a distributary network, the higher concentration of water flux in channels will result in a slightly higher water surface profile than that of the surrounding floodplain, which causes a lateral surface gradient pointing away from the channels to the floodplains. This mechanism controls how much flow "escapes" from channels and spreads onto the surrounding floodplain. A guideline for the choice of the value of γ for a range of environments, including steeper terrains such as alluvial fans, can be obtained by expressing γ as a function of the characteristic slope and flow velocity of the environment, and grid size:

$$\gamma_0 = \frac{g \delta_c S_0}{U_0^2}, \tag{3}$$

where g is the gravitational acceleration, δ_c is the grid size, S_0 is the characteristic slope of the system, and U_0 is the characteristic flow velocity. The characteristic slope can be estimated by the average topographic or water surface slope of the system, and the characteristic flow velocity can be estimated as the mean velocity of the major channels. Equation (3) directly takes into account the steepness of the environment as well as the grid size. Using Wax Lake delta (WLD) – a subdelta of the Mississippi River delta system – as an example, the characteristic slope is 5×10^{-5}, grid size is 60 m (see Test 4a) and the characteristic flow velocity from field surveys is $0.5–1\,\mathrm{m\,s^{-1}}$ depending on the flooding stage, resulting in a range of γ values from 0.03 to 0.12. The choice of 0.05 is within this range. A quantitative analysis of the effects of γ on the modeling results can be found in Test 4a.

2.2 Step 2: calculate relative routing weights for the neighbors at each cell

With the routing direction $\boldsymbol{F}$ specified, the relative routing weights (w_i) for neighbors around each cell (eight neighbors in the case of square lattice setup) are determined as follows:

$$w_i = \frac{h_i \max(0, \boldsymbol{F} \cdot \boldsymbol{d}_i)}{\Delta_i}, \quad i = 1, \dots, 8. \tag{4}$$

The cellular direction vector, $\boldsymbol{d}_i$, is a unit vector pointing to neighbor i from the given cell, Δ_i is the cellular distance, taking a value of 1 for cells in main compass directions and a value of $\sqrt{2}$ for corner cells, and h_i is water depth of neighbor i.

2.3 Step 3: calculate routing probabilities (p_i)

The weights above are then processed according to the wet–dry status of cells. The model considers cells with flow depth greater than a threshold value, typically 1–5 % of the characteristic flow depth of the system, as "wet" cells, and the opposite as "dry" cells. The weights for dry cells are then converted to the value of 0. The routing probabilities (p_i) are calculated as

$$p_i = \frac{w_i}{\sum\limits_{\mathrm{nb}=1}^{8} w_{\mathrm{nb}}}, \quad i = 1, \dots, 8, \tag{5}$$

where "nb" is the numbering of neighbors around a given cell (1–8 for a 3-by-3 square grid). With the routing probabilities calculated, water parcels are released one by one from

the upstream inlet cells and follow a weighted random walk based on the probability field. The cumulative movements of parcels are summed in terms of vectors at each cell to obtain an estimation of flow unit discharge. The direction is given by the average walking direction of the passing parcels and the magnitude is given by the summation of the fluxes carried by these parcels.

The calculation of the water surface profile is performed with a 1-D scheme along the water parcel paths, rather than the solution of a system of partial differential equations. The basic assumption of this approach is that the water surface profile along a streamline can be approximated by a 1-D equation. In the morphodynamic results in Part 1, this 1-D equation takes the simplest form that the water surface slope on the delta equals a constant value – the "reference slope". For the hydrodynamic tests in this work, we formulate the 1-D equation such that (i) it satisfies the backwater equation if the local Froude number is low ($Fr^2 \le 0.5$) and (ii) the water surface slope (S) is equal to the friction slope (S_f) if the Froude number is high ($Fr^2 > 0.5$):

$$\frac{\partial h}{\partial l} = \frac{S - S_f}{1 - Fr^2}, \quad \text{if } Fr^2 \le 0.5, \tag{6}$$

$$\frac{\partial H}{\partial l} = S_f, \quad \text{if } Fr^2 > 0.5, \tag{7}$$

where C_f is the coefficient of friction, $S_f = C_f Fr^2 = C_f \frac{U^2}{gh}$ is the friction slope and l is the distance along an arbitrary flow streamline.

This calculation is done along each water parcel path via a finite difference scheme. An average value is taken for cells belonging to multiple paths. The obtained surface profile is then smoothed with numerical diffusion to remove bumps and ditches caused by the gaps between different 1-D paths. In addition, an underrelaxation is applied for numerical stability between iterations (see Part 1 for details). This method assumes that along the streamline the flow is 1-D so that Eqs. (6) and (7) apply. However, the combination of a large number of flow streamlines covering many flow paths essentially constitutes a 2-D surface. This surface calculation does not work for strongly varying 2-D flows, but in the tests below we will show that this highly simplified surface calculation is sufficient to reproduce large-scale flow partitioning, a very important application for deltaic distributary networks.

Using this parcel-based flow routing method, flow continuity is always satisfied because (1) there is no gain or loss of the water flux represented by each water parcel, and (2) at each cell in the domain the number of parcels coming in is always equal to the number of parcels leaving that cell. Equations (6) and (7) not only use the streamlines identified from the new unit discharge field, but also use the value of discharge itself to update the water surface profile. During this process, the model assumes that the discharge at each cell remains the same, while the depth and velocity are updated according to the new water surface elevation. Therefore flow continuity remains satisfied.

FlowRCM is written in Matlab and typically runs in 1–2 h on a personal laptop for simulations in the tests presented in the next section. The running speed could be reduced by an order of magnitude if the code were written in more efficient languages (such as C) and optimized.

3 Hydrodynamic test cases

The hydrodynamic test cases are designed to analyze two important aspects in the RCM. First, we target the "reduced-complexity" features in our flow routing model, i.e., the rules and parameterizations that represent governing physics; second, we want to capture critical hydrodynamic processes affecting the overall delta morphology. The hydrodynamics of river deltas involve a hierarchy of processes occurring at a wide range of scales, from flow structures behind individual ripples on the bed to channel avulsions that change the distribution of water flux across the entire delta surface. We focus on hydrodynamic processes at the channel scale, which are essential to creating, maintaining and modifying a distributary channel network. The proposed tests are a (1) backwater profile in a straight channel, (2) flow around a mouth bar, (3) partitioning of flow at a single bifurcation, and (4) partitioning of flow in a distributary channel network with submerged islands. FlowRCM results are compared to Delft3D outputs and theoretical predictions. The description and results of each test case are given below. All Delft3D simulations were done in depth-averaged mode, with a time step of 30 s and a constant horizontal eddy viscosity ($1\,\mathrm{m^2\,s^{-1}}$).

3.1 Test 1: backwater profile in a straight channel

River deltas have subtle topography and water surface gradients and flow is typically subcritical. Nevertheless, the gradient in the surface profile is essential to water motion in deltaic environments (Edmonds and Slingerland, 2008). As mentioned in the previous section, FlowRCM estimates the water surface based on the assumption of a 1-D surface profile along flow streamlines. As the movement of water parcels on the grid is in both x and y directions, usually in a "zigzag" fashion and not necessarily along flow streamlines, the calculation of water surface along the flow paths involves projecting each step a parcel makes to the estimated streamline. Thus, in the first test we use a straight channel of constant width with no variation in cross-stream direction, whose surface can be described by the 1-D backwater equation. The model performance is compared to the theoretical backwater profile.

The domain is 2000 m wide and 15 000 m long with a rectangular cross section (Fig. 2a). The bed has a constant slope $S = 10^{-3}$ and a constant friction coefficient $C_f = 0.01$. A constant discharge of $2 \times 10^4\,\mathrm{m^3\,s^{-1}}$ is fed into the upstream inlet. The downstream outlet has a fixed water surface elevation as boundary condition. The discharge is chosen so that the flow remains subcritical in normal flow conditions.

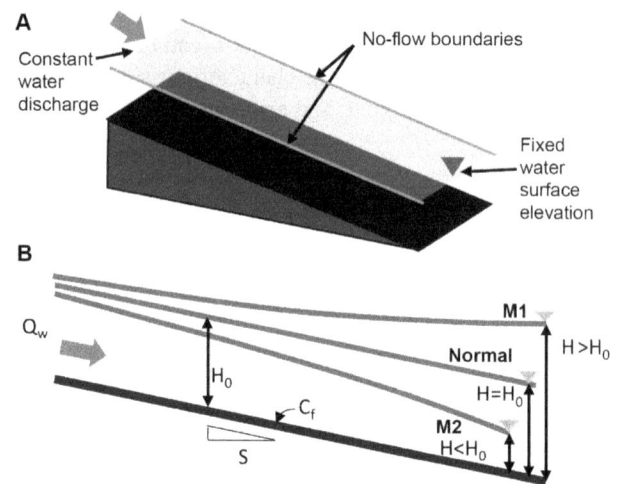

Figure 2. The setup of test 1 (straight channel test). (a) The channel has a rectangular cross section, constant bed slope and friction coefficient. A constant discharge is fed into the upstream inlet. A fixed water surface elevation boundary condition is assigned at the downstream outlet. (b) Three backwater profiles (M1, normal, and M2) are shown in the bottom diagram where downstream water surface elevation is respectively above, equal to, or below that of normal flow.

By varying the downstream water surface elevation relative to that of the normal flow, different backwater profiles can be achieved, such as M1, M2 and normal flow (Chow, 1959) (Fig. 2b). As the flow does not have cross-stream variation, the surface profile can be resolved using the 1-D backwater equation (Eq. 6).

For the FlowRCM calculation, we use a grid of 20×150 cells, with cell size of 100 m. The water surface elevation calculated from FlowRCM is averaged in the cross-stream direction and compared to the numerical solution of Eq. (6).

The test results show that in all three scenarios FlowRCM is able to successfully reproduce the backwater profile solution (Fig. 3). The predicted longitudinal water surface profile (averaged in the cross-stream direction), in fact, matches the solution of the 1-D backwater equation. The simple scheme used in FlowRCM is thus able to produce realistic water surface profiles.

3.2 Test 2: flow around a mouth bar

The formation of channel bifurcations is essential to the formation of the distributary channel network on river deltas (Edmonds and Slingerland, 2007; Kleinhans et al., 2013). Therefore, the ability of a numerical model to represent the bifurcation process is of great importance. As shown by the work of Edmonds and Slingerland (2007), mouth-bar development is critical to channel bifurcation. Particularly, one key feature in this process is the transition from flow acceleration to deceleration over the top of the mouth bar as

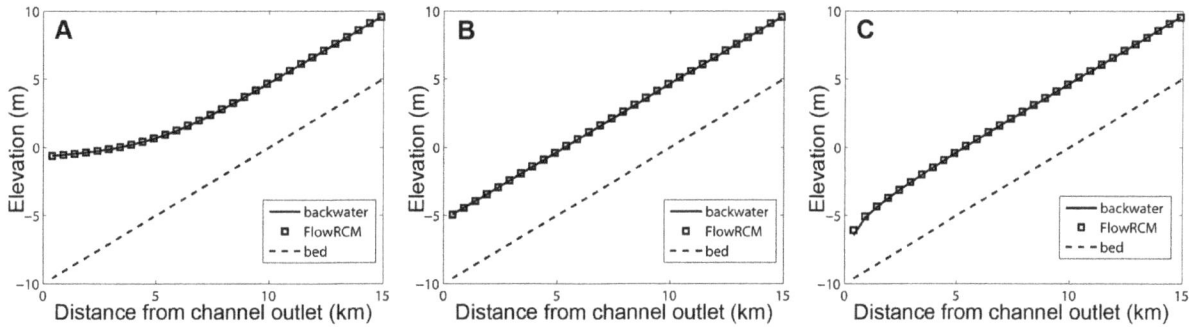

Figure 3. Results from test 1 (straight channel test). The water surface profile in the longitudinal direction calculated by FlowRCM (averaging the 2-D profile in the cross-stream direction) is compared to the numerical solution of the 1-D backwater equation in three scenarios: **(a)** M1 curve, **(b)** normal flow, and **(c)** M2 curve.

the bar height grows and the bar top approaches the water surface (Fig. 4). Here we design a similar test with variable bar heights. The mouth-bar topography is represented by a Gaussian-shaped bump in a straight channel. The bump does not deform and its height is varied from 10 to 99 % of the normal flow depth.

The domain is a straight channel section, 100 m wide and 200 m long with rectangular cross section. The bed has a constant slope $S = 2.5 \times 10^{-3}$ and a constant friction coefficient $C_f = 0.1$. A constant water discharge of $50\,\mathrm{m}^3\,\mathrm{s}^{-1}$ is fed into the upstream inlet. The downstream outlet has a fixed water surface elevation as downstream boundary condition. The Gaussian-shaped bump has a diameter equal to approximately one third of the channel width. The side walls of the channel have a no-flow boundary condition. Both FlowRCM and Delft3D use a 40×80 grid with cell size of 25 m.

We compare the outputs from FlowRCM and Delft3D in terms of water surface elevation and flow velocity, focusing on (i) the location of "hot spots" of high and low velocity (ii) the deformation of the water surface in proximity of the bump, and (iii) the transition of flow velocity right over the top of the bump as the bump height increases from 10 to 99 % of the normal flow depth.

The maximum flow velocity over the top of the bump occurs for a bump height of around 60 % of the flow depth (Edmonds and Slingerland, 2007). Therefore, we compare the outputs of FlowRCM and Delft3D for bump heights equal to 20, 60, and 90 % of the flow depth (Fig. 5). The results show that (i) both FlowRCM and Delft3D reproduce the transition from flow acceleration to deceleration over the top of the bump as the height of the bump increases (Figs. 5, 6); (ii) both models show that higher velocity occurs in a "bow shape" over the bump and curved into the downstream direction, while lower velocity occurs behind the bump and slightly in front of the bump (Fig. 5); and (iii) both models show that the water surface is superelevated in front of the bump and is drawn down behind the bump. FlowRCM thus captures the key hydrodynamic features, although their magnitude and shape do not exactly match the results from

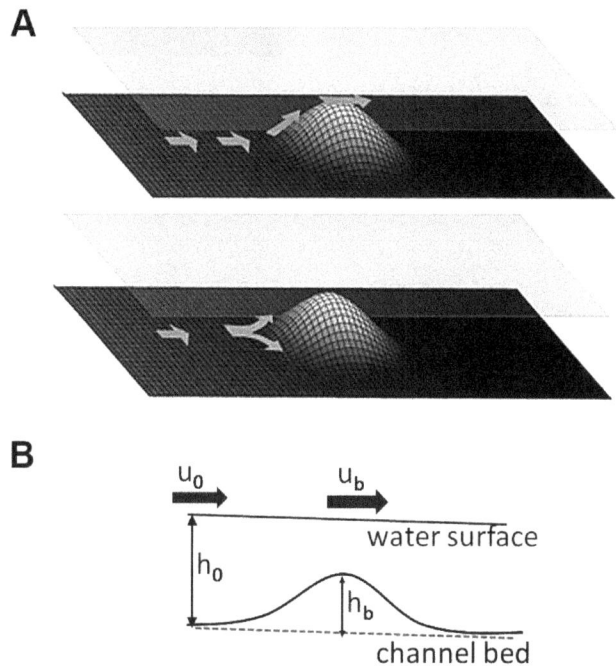

Figure 4. Setup of test 2 (flow around a mouth bar). A smooth Gaussian-shape bump is placed in a straight channel. **(a)** Two scenarios are observed: (i) acceleration over the bump or, (ii) deceleration over the bump and diversion around the bump. **(b)** Sketch of the key variables: normal flow depth (h_0) and velocity (u_0), bump top velocity (u_b), and bump height (h_b).

Delft3D (Fig. 5). The irregularities in the FlowRCM output are caused by the "randomness" inherent in the weighted random walk scheme. As seen in Figs. 4 and 5 such irregularities do not affect the ability of the model to reproduce key hydrodynamic features.

3.3 Test 3: flow through a single bifurcation

Bifurcations control the partitioning of water and sediment in the deltaic system and affect the stability of the whole

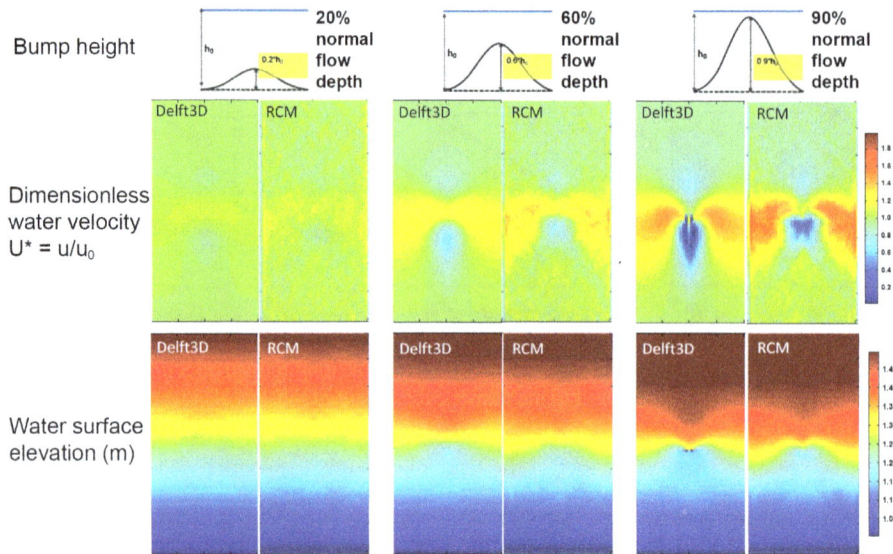

Figure 5. Contour plots of water velocity magnitude (scaled by the normal flow velocity $u_0 = 0.5\,\mathrm{m\,s^{-1}}$) and water surface elevation for three cases (bump heights at 20, 60 and 90 % of normal flow depth). Flow direction is from top to bottom. Notice the deformation of the water surface and the development of a low velocity region in front of and behind the bump.

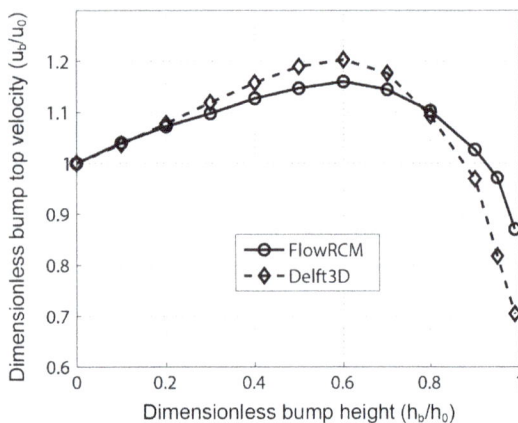

Figure 6. Comparison of dimensionless bump top velocity (u_b/u_0) as a function of dimensionless bump height (h_b/h_0) predicted by FlowRCM and Delft3D. Both models predict an initially rising and then falling flow velocity as the dimensionless bump height goes to 1. FlowRCM also captures the turning point where rising changes to falling (~ 60 % flow depth). Flow velocity is scaled by the normal flow velocity magnitude, and bump height by normal flow depth.

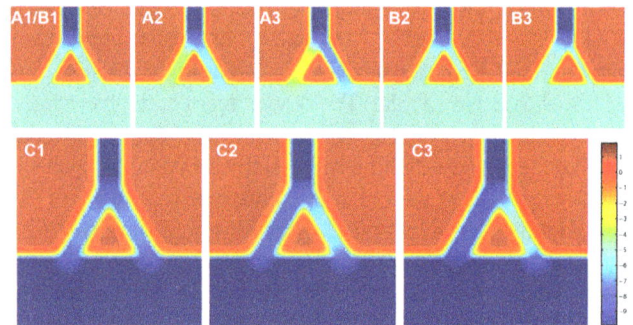

Figure 7. Channel geometries for test 3 (single bifurcation test). Group A (A1, A2 and A3) and group B (B1, B2 and B3) have a smaller domain and input discharge. Within group A the ratio of depth between the two downstream branches is varied while keeping width constant. Within group B the ratio of width between the two downstream branches is varied while keeping depth constant. Within group C (C1, C2 and C3) width and depth are varied in the same proportion while keeping the summation of cross-sectional area constant.

network (Edmonds and Slingerland, 2008; Edmonds et al., 2010; Kleinhans et al., 2008, 2013). We use a simple and idealized bifurcation topography to test the response of FlowRCM to bifurcation asymmetry, e.g., changes in width and/or depth ratio of the two downstream branches.

There are three groups of tests, referred to as group A, B and C. In all groups the main channel splits into two branches that open at an angle of $60°$, and both branches enter a basin with constant depth (Fig. 7). Channel banks are no-flow

boundaries so that the water flux stays in the channel and there is no flooding onto the bank. For all tests, FlowRCM and Delft3D use a 100 by 100 grid. The tests in groups A and B have smaller domain sizes of 2500 m by 2500 m with 25 m grid size, and total input discharge of 3000 $\mathrm{m^3\,s^{-1}}$. The tests in group C have larger domain sizes of 5000 m by 5000 m with 50 m grid size, and total input discharge of 5000 $\mathrm{m^3\,s^{-1}}$. In group A we explore the effect of asymmetry in depth alone, keeping the two branches at the same width; in group B the effect of asymmetry in width alone, keeping the two branches at same depth; and in group C the effect of

Table 1. Discharge asymmetry and upstream surface elevation results from FlowRCM and Delft3D for all single bifurcation runs.

Run	Discharge asymmetry (%)			Upstream surface elevation (m)		
	FlowRCM	Delft3-D	Difference	FlowRCM	Delft3-D	Difference
A1	0.16 %	0.00 %	+0.16 %	0.3226	0.2067	+0.1159
A2	20.82 %	24.04 %	−3.23 %	0.3160	0.2048	+0.1112
A3	41.04 %	47.69 %	−6.65 %	0.3156	0.2016	+0.1140
B1	0.16 %	0.00 %	+0.16 %	0.3226	0.2067	+0.1159
B2	21.59 %	21.17 %	+0.42 %	0.3230	0.2068	+0.1162
B3	40.76 %	42.78 %	−2.02 %	0.3301	0.2075	+0.1226
C1	0.21 %	0.01 %	+0.19 %	0.1472	0.1091	+0.0381
C2	28.22 %	27.84 %	+0.37 %	0.1417	0.1057	+0.0360
C3	51.49 %	52.66 %	−1.17 %	0.1286	0.0963	+0.0323

asymmetry in cross-sectional area, by keeping the summation of the cross-sectional areas constant and all channels at the same width-to-depth ratio. Detailed channel geometries are shown in Fig. 7.

The output of FlowRCM is compared to that of Delft3D in terms of (i) the spatial pattern of flow velocity and water surface elevation and (ii) the ratio of fluxes between the two branches.

To evaluate the effect of channel geometry asymmetry on flow partitioning between bifurcations, we calculate two values from each test result for both FlowRCM and Delft3D, namely the asymmetry of discharge ψ_Q and the asymmetry of cross-sectional area ψ_A:

$$\psi_Q = \frac{|Q_L - Q_R|}{Q_L + Q_R}, \tag{8}$$

$$\psi_A = \frac{|A_L - A_R|}{A_L + A_R}, \tag{9}$$

where Q_L and Q_R are the water discharge in the left and right branches respectively; A_L and A_R are the cross-sectional area (calculated with bed elevation and resolved water surface elevation) of the left and right branches respectively.

The asymmetry values defined in Eqs. (8) and (9) are plotted in Fig. 8. FlowRCM captures the same trend of flow discharge partitioning predicted by Delft3D: the amount of water flux into the two branches is proportional to the cross-sectional area of the branches. This means that despite the variation in the ratio of depth and/or width, the flow tends to have the same mean velocity in both branches. Also, both FlowRCM and Delft3D predict a water surface gradient in the shallower branch that is significantly higher than the one in the deeper branch, consistent with field observations by Edmonds and Slingerland (2008). An example of flow field and water surface elevation (test C3) is plotted in Fig. 9. Also, within each test group both FlowRCM and Delft3D predict the same behavior of upstream water surface elevation in response to the discharge asymmetry: in groups A and C the upstream water surface elevation decreases as discharge

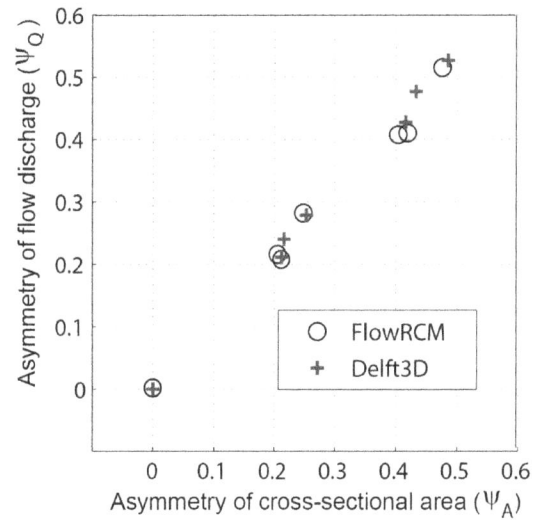

Figure 8. Comparison of discharge asymmetry as a function of cross-sectional area asymmetry predicted by FlowRCM and Delft3D. Both models result in asymmetry of discharge proportional to the asymmetry of cross-sectional area.

asymmetry increases; in group B the upstream water surface elevation increases as discharge asymmetry increases.

Qualitative differences in the observed flow plan-view pattern produced by FlowRCM and Delft3D emerge due to what we call "local effects". In the results of FlowRCM there is significant concentration of flux right upstream of the island, where flow bifurcates directly against the tip of the island (in this case, the no-flow condition at the boundary of the islands does not allow any water flux to escape the channels) (Fig. 9, dashed square region). The water surface elevation calculated by FlowRCM does not show superelevation right at the tip of the island as in the Delft3D results; instead, there is a drawdown. We believe that the superelevation creates a gradient that causes the flow to divert around the island much further upstream.

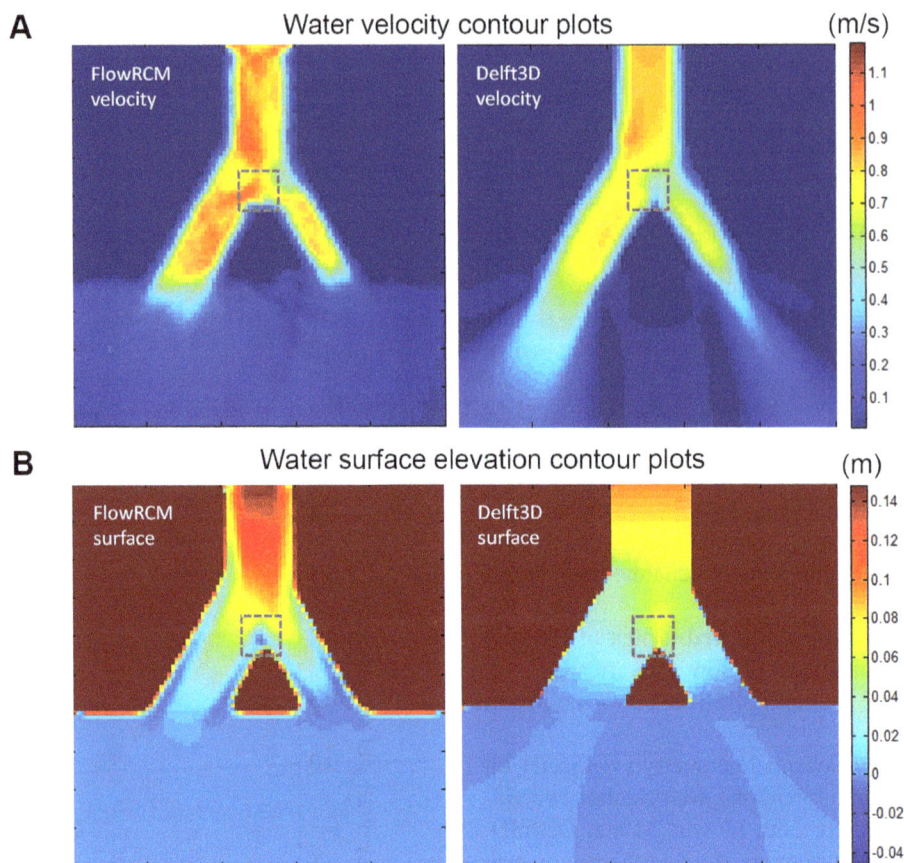

Figure 9. Results from run C3: (**a**) water velocity and (**b**) water surface elevation. FlowRCM and Delft3D behave differently at the upstream side of the triangle island (marked by the dashed gray square): while Delft3D exhibits a low velocity zone with a high water surface elevation, FlowRCM exhibits a high velocity zone and a low water surface elevation.

In quantitative terms, this single-bifurcation test shows that FlowRCM predicts a discharge asymmetry within 7 % of difference compared to Delft3D, but tends to overpredict surface elevation by up to 50 % (Table 1). As discussed in the companion paper, the inclusion of water surface is required but its accuracy can be relaxed when modeling deltas. The error in the calculation of the surface elevation is due to the fact that FlowRCM calculates the discharge vector at each cell by (1) summing up the volume of all passing water parcels at each time step to obtain the magnitude of the discharge vector, and (2) doing a vector summation of the moving directions of all passing water parcels (each will be a unit direction pointing to one of the eight neighboring cells) to obtain the direction of the discharge vector. By doing so, in the area where there is flux convergence, some cross-stream flux caused by the parcels traveling almost sideways to the main flow direction will be counted as contributing to the total downstream flux as well, causing an overestimation of the water surface elevation. An alternative method would be directly doing a vector summation of the discharge vector represented by each passing parcel; however, in this way, discharge and water surface elevation would be underpredicted.

Despite this effect, the parcel-based Lagrangian approach is able to predict flow partitioning at a bifurcation and offers more flexibility in flow routing in complex terrains with a low wet-to-dry ratio and frequent changes in wetted area.

3.4   Test 4: flow through a distributary channel network

We now evaluate the performance of FlowRCM at the scale of a complete deltaic distributary channel network. We are also interested in assessing whether the "inaccuracy" of the flow field at finer scales accumulates as the flow propagates through the whole channel network. The topography setups used to run FlowRCM and Delft3D are the synthetic topography of a natural river delta (Test 4a) and a DeltaRCM-simulated topography (Test 4b). By introducing a whole distributary network, which has no constrains on lateral flux exchange between channel and islands, we essentially test the behavior of FlowRCM in a transition from confined to unconfined flow. Both tests 4a and 4b use a distributary network that features flow spreading from a single feeding channel which is strictly confined by nonpenetration high walls to an approximately 90° or 180° open space. Water not only flows

Table 2. Effect of the parameter γ on the flux distribution between channels and islands. Detailed cross-section locations are given in Fig. 14.

γ	$Q^1_{\text{channel}}/Q^2_{\text{island}}$
0.025	1.2554
0.05	1.1417
0.075	1.0161
0.1	0.9713

$^1 \; Q_{\text{channel}} = Q_2 + Q_4 + Q_6 + Q_8$
$^2 \; Q_{\text{island}} = Q_1 + Q_3 + Q_5 + Q_7 + Q_9$

through the channel network but also onto the islands. Therefore the whole delta surface, except for the upstream tips of a few islands closer to the apex, is inundated.

3.4.1 Test 4a: synthetic Wax Lake delta (WLD) topography

The WLD is a modern river-dominated delta in the coast of Louisiana, part of the Mississippi River delta system (Roberts et al., 1980; Wellner et al., 2005; Shaw et al., 2013). We construct a synthetic bathymetry (Fig. 10) from satellite images and bathymetry measurements along nine transects (USACE, 1999). We use an image processing software (Adobe Photoshop CS6) to integrate the plan view of islands and channels from the satellite imagery and obtain a smooth profile between neighboring transects. FlowRCM and Delft3D are given the same initial and boundary conditions: an upstream inlet channel discharge of $2490 \, \text{m}^3 \, \text{s}^{-1}$ and a water surface elevation at the downstream boundary of 0 m mean sea level. The friction coefficient is set to 0.01. Both FlowRCM and Delft3D simulations for this test case use a 200 by 200 grid with a cell size of 60 m.

The results show that the two models predict similar flow distribution among channels and islands with "hot spots" of higher flow velocity occurring at the same locations (Fig. 11). Delft3D shows a more "diffused" velocity map while FlowRCM gives a more "noisy" map. The flow pattern on the islands predicted by FlowRCM seems to respond to island topography, exhibiting a converging stream in the low elevation area towards the lower center of the islands (Fig. 11). In Delft3D results, flow is more evenly distributed across each island, thus not revealing much of the island's topographic detail (although higher-resolution Delft3D modeling would, while constraints exist in its drying–flooding algorithm). The depth-averaged mode and the choice of horizontal eddy viscosity might also add to the relatively smooth outputs from Delft3D. The "local effects" in FlowRCM discussed in the previous section cause the appearance of a high velocity zone right along the tip of the islands rather than a low velocity zone caused by flow diversion starting upstream of the island as in the Delft3D results (Fig. 12).

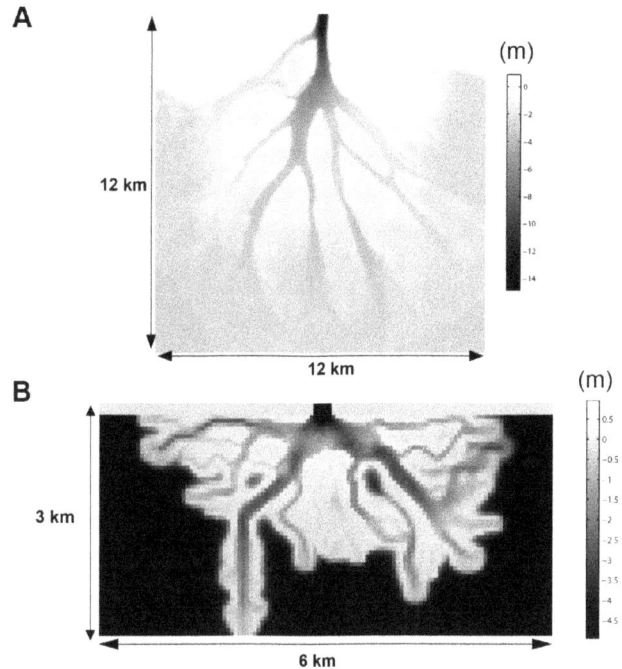

Figure 10. Topography used in tests 4a and 4b (flow through a distributary channel network). **(a)** Synthetic topography constructed for WLD; **(b)** a DeltaRCM-generated delta topography.

Our reduced-complexity water surface calculation method requires a large underrelaxation to achieve smooth transition in the surface profile. This underrelaxation causes a delay between the response of the water discharge distribution to changes in water surface slope and, therefore, an oscillation in the cross-stream direction discharge distribution over 10–20 iterations. By averaging over several iterations (e.g., the last 100 of a total 500 iterations) stable results are achieved, resulting in what we call a dynamic equilibrium. An example of this oscillation on the synthetic WLD topography is shown in Fig. 13.

To quantitatively evaluate the effect of the parameter γ, the setup of test 4a is used with values of γ from 0.025 to 0.1 while keeping all other parameters unchanged. We measure the ratio of the discharge coming out of four main channels and the discharge coming out of the downstream side of the islands (Fig. 14). From the results listed in Table 2, it can be seen that with a higher value of γ, more flux is routed through the islands. In order to compare the flux distribution obtained with FlowRCM (for a value of the parameter γ equal to 0.05) to results from Delft3D, we selected two transects located at different distances from the apex of the delta (Fig. 15). The closer transect, A, is divided into 10 sections and the farther transect, B, is divided into 14 sections. The fraction of discharge at each section is calculated by the discharge crossing that section normalized by the total discharge crossing the whole transect. Results from FlowRCM and Delft3D are plotted for each transect. There is a close match between

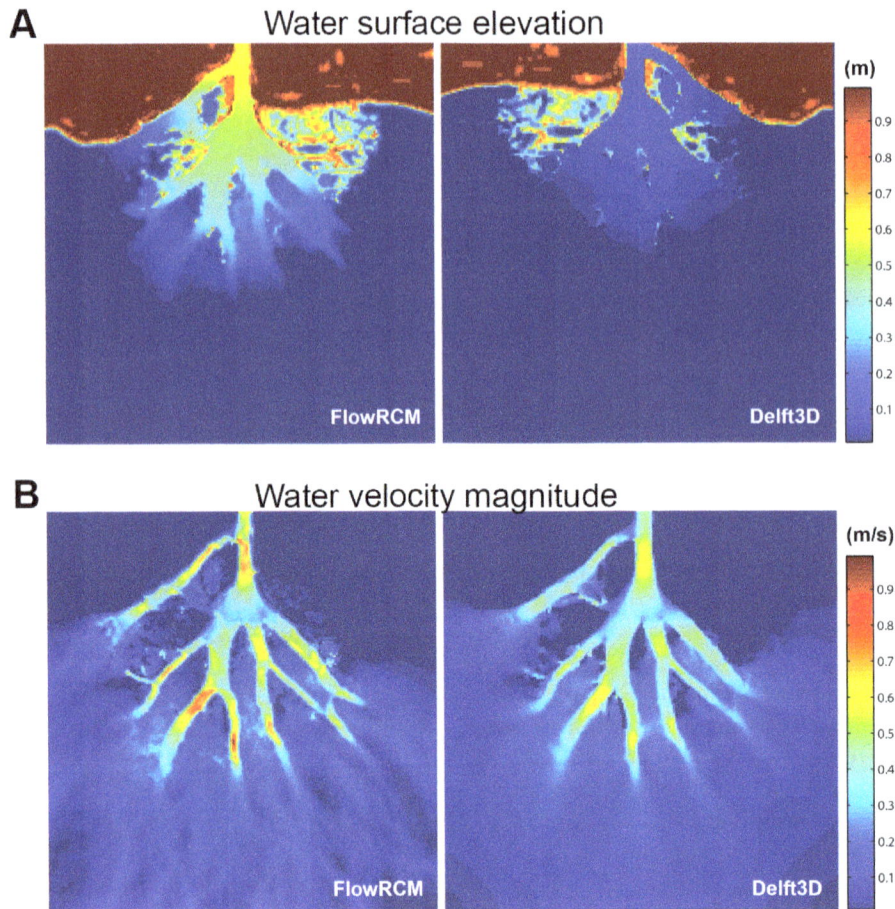

Figure 11. Results from test 4a (flow over synthetic bathymetry of WLD). **(a)** FlowRCM predicts a significantly higher gradient, while the water surface distribution is similar to the one predicted by Delft3D. **(b)** Velocity contour maps from model results; FlowRCM and Delft3D give similar flow distributions and predict "hot zones" where water velocity is significantly higher in channels.

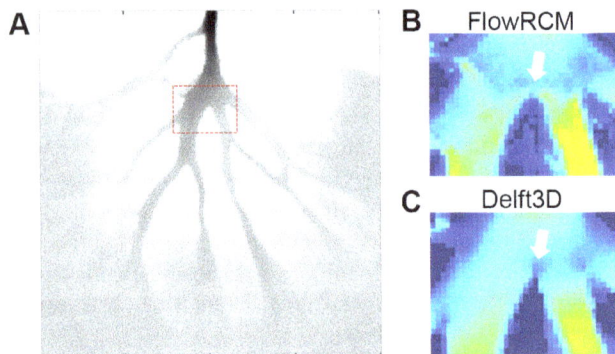

Figure 12. (a) The red dashed square marks one of the regions where FlowRCM displays strong local effects; **(b)** detailed view from both model results showing the "local effects" in FlowRCM, not present in the Delft3D simulation.

FlowRCM and Delft3D throughout transect A, while the results differ when flow reaches transect B. The difference between transects A and B is due to the lateral flux normal to the main channel direction. FlowRCM has significantly less flux coming out of the very eastern and western sides of the delta ($< 5\%$), compared to Delft3D ($> 10\%$), resulting in a more concentrated flux distribution among the main channels and islands. Field surveyed data are needed to understand which model reproduces more closely the behavior of WLD.

3.4.2 Test 4b: bed topography generated by DeltaRCM

For this test, the delta topography is a snapshot of the mixed grain size delta simulated by DeltaRCM. The delta is composed of 70 % fine grain and 30 % coarse grain, and has a very well-defined distributary channel network (Liang et al., 2015) (Fig. 10). The rectangular domain is 6 km wide and 3 km long. The inlet channel has a discharge of $1250\,\mathrm{m}^3\,\mathrm{s}^{-1}$, and the delta is formed in a basin with a constant depth of 5 m. Both FlowRCM and Delft3D use a 60 by 120 grid with cell size of 50 m.

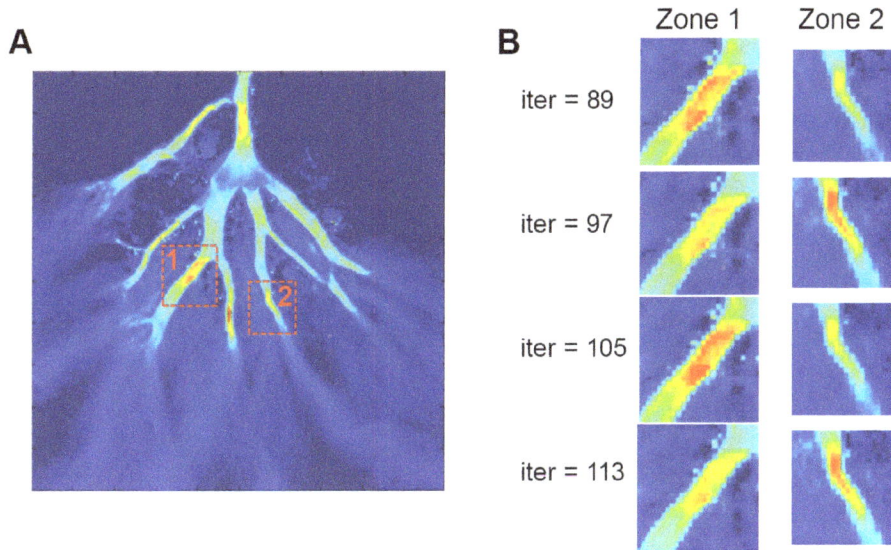

Figure 13. The dynamic balance between water surface gradient and flux distribution. (**a**) The dashed squares show focus areas enlarged on the right. (**b**) Oscillations of flow velocity between iterations in the focus areas.

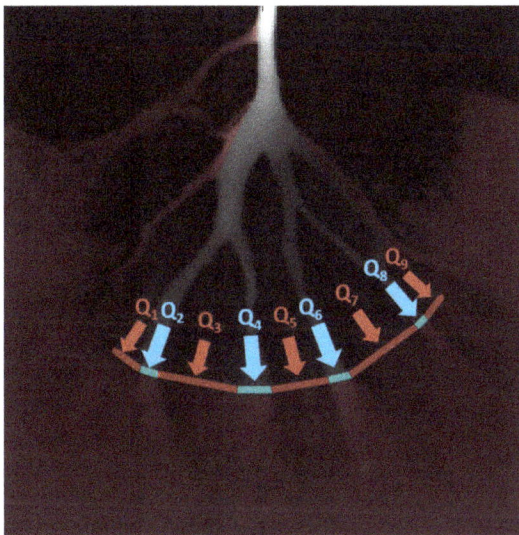

Figure 14. Setup used for analyzing the effect of the parameter γ on modeling results. The ratio of the discharge coming out of four main channels (Q_2, Q_4, Q_6, Q_8) and the discharge coming out of the downstream side of the islands (Q_1, Q_3, Q_5, Q_7, Q_9) is measured to quantify how much flow is routed through the island as a function of the parameter γ.

These results show that, analogously to the synthetic WLD topography test, FlowRCM and Delft3D predict a similar flow distribution through the network, e.g., the partitioning among channels and islands (Fig. 16). Moreover, the water surface profile calculated by FlowRCM is smoother than the one obtained from Delft3D, as FlowRCM is able to handle discontinuities in the domain represented by frequent alternations of "wet" and "dry" cells. In this test "local effects"

are not pronounced. A possible reason is that this simulated topography has already adapted to the features of the flow field calculated by FlowRCM, such that at the tip of the island local topography creates a smooth transition from deep channels to shallow islands that diverts the flow gradually. At the same time, since islands are mostly submerged and can be inundated, the calculation of the free surface allows for a more accurate surface construction compared to using a no-flow boundary condition at the edge of the islands.

4 Discussion

In this work, we evaluated the performance of FlowRCM in terms of its ability to reproduce plausible hydrodynamics of deltaic systems. We showed the performance of FlowRCM in several setups that mimic hydrodynamic scenarios characteristic of river deltas. The computed flow fields were compared to theoretical predictions and to numerical simulation results obtained with a higher-fidelity hydrodynamic model (Delft3D), previously validated for similar purposes (Lesser et al., 2004). From the comparison to analytical solutions and Delft3D, we address the goal stated in our introduction: what level of hydrodynamic physics does our FlowRCM capture? In terms of water surface profile modeling, FlowRCM is able to replicate (i) the backwater profile along a simple straight channel matching the solution of the theoretical backwater equation; (ii) the local surface deformation around submerged bumps, including a superelevated region upstream of the bump and a drawdown region downstream of the bump; (iii) the difference in surface gradients between two competing bifurcation branches; and (iv) the outward surface gradient from channels to floodplains. In terms of flow discharge and velocity field, FlowRCM is able to repli-

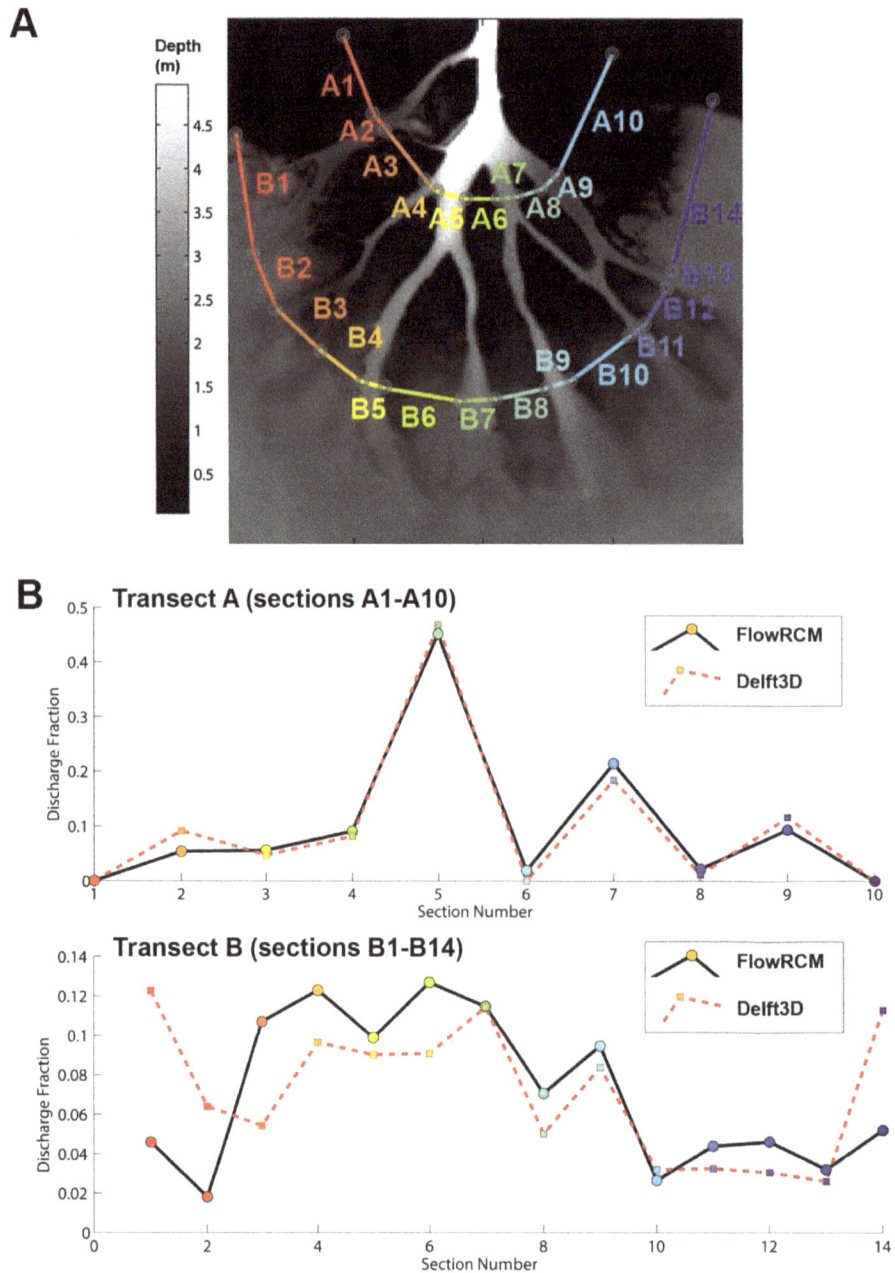

Figure 15. Spatial distribution of water discharge at two transects on the synthetic WLD topography. (**a**) Transect A is divided into 10 sections and transect B is divided into 14 sections. The fraction of discharge at each section is calculated by the discharge across that section normalized by the total discharge across the whole transect. (**b**) Results of FlowRCM and Delft3D are plotted for each transect.

cate (i) the nonlinear response of flow velocity to the height of the submerged bump structure, (ii) the discharge partitioning between two competing bifurcation branches in a single bifurcation, and (iii) the discharge partitioning through a whole distributary network of channels and floodplains. The results also show that FlowRCM falls short with respect to Delft3D in terms of detailed hydrodynamic features. For example, in the single bifurcation test, FlowRCM results show strong "local effects" resulting into the flow splitting right in

front of the island. Also, occasional irregularities appear due to the probabilistic approach used for routing water parcels.

Overall, FlowRCM performs well at the multichannel network scale, where spatial flux distribution matters more than detailed channel flow structure. Furthermore, for single-channel-scale cases, such as submerged smooth obstacles, FlowRCM is able to predict regions of high and low velocities, and regions of water surface superelevation and drawdown, although the exact shape of the regions or the values of

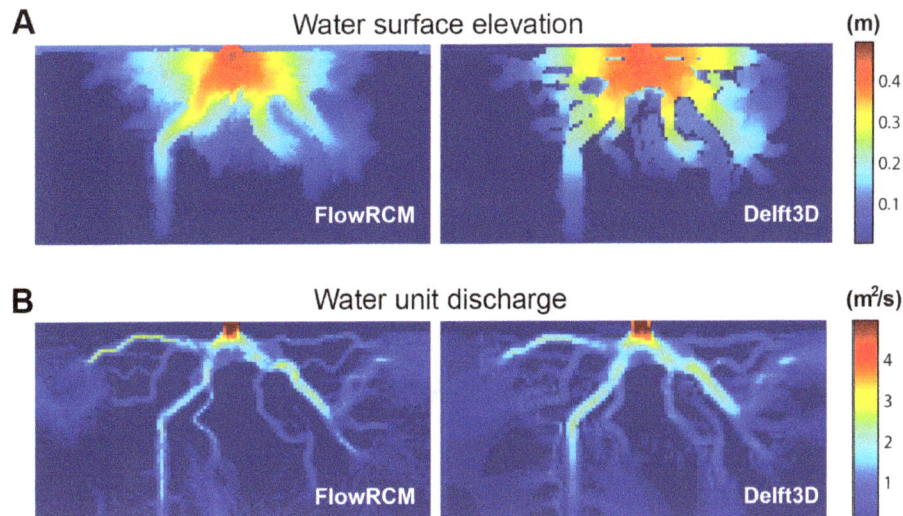

Figure 16. Results of test 4b (flow over a DeltaRCM-generated delta topography): (a) water surface elevation and (b) water unit discharge. The results obtained with DeltaRCM are compared to the results from Delft3D using the same topography and input conditions. The results show that (1) both models yield similar discharge distribution throughout the topography, and (2) both models yield similar water surface profiles in magnitude, although DeltaRCM shows a more consistent surface profile and more pronounced gradient across the channel into the floodplains.

the velocity and water surface elevation may be more qualitatively rather than quantitatively correct.

While FlowRCM has been mainly applied in low-gradient environments where flow is typically subcritical, the routing method can be applied to high-gradient and supercritical systems, where topographic slope itself is a reasonable representation of water surface slope and is sufficient for routing fluxes. The laboratory-scale alluvial fan simulation in Part 1 is an example of the application of FlowRCM to high-gradient environments. The key parameters for switching between these environments are the slope/inertia partitioning parameter γ for calculating routing probabilities and the 1-D profile equation for calculating the water surface elevation.

One characteristic of rule-based RCMs is the flexibility in making changes to one specific rule in the model, resulting in the opportunity to isolate processes, understand their effects and the behavior of the model, and potentially guide future model improvements. For example, FlowRCM does not resolve the water surface profile based on 2-D hydrodynamic equations. The water surface plays a role in distributing water fluxes through a feedback mechanism: the convergence of water flux causes water surface to rise up along the flow path, introducing a positive surface gradient pointing away from the cells with high flux and diverting flux sideways. This process is done iteratively and the delay in the responses between water surface and water flux causes an oscillation in the model outputs (example from test 4a in Fig. 13). At the same time, this iterative feedback mechanism allows the water flux to adjust beyond the conditions of the immediate neighbors, thus reducing local effects. A similar iterative

feedback mechanism can be found in the row-by-row depth-based iteration method developed by Nicholas (2010).

FlowRCM does have dependency on grid resolution, as each cell only sees its immediate neighbors and the flux is routed using cells as units rather than physical distances. This is widely considered a fundamental property of cellular approaches (Nicholas, 2005), and has been shown in a number of cellular routing models (e.g., Doeschl-Wilson and Ashmore, 2005). The dependency on grid resolution could potentially be removed by introducing a relationship between RC model process parameterization and spatial resolution (Nicholas et al., 2012).

The goal of FlowRCM is not to achieve an accurate solution of the water surface elevation, but to reproduce morphodynamic features obtained so far only by high-fidelity models. As discussed in the companion paper (Liang et al., 2015), the inclusion of the water surface calculation is required but its accuracy can be relaxed when modeling deltas. While our work addresses the plausibility of the model's hydrodynamic results, it is interesting to investigate whether the ability of a flow routing scheme to correctly predict hydrodynamics controls its ability to produce plausible morphodynamic features. We can think, for example, about the process of bifurcation formation. In DeltaRCM a channel mouth goes through a series of stages (Fig. 17): (i) flow expands and sediment deposits with reduced flow velocity, which creates a wide lunate bar with noisy surface; (ii) the irregular topography of the bar forms "ridges" and "troughs" where troughs attract more water flux than ridges; (iii) the enhanced flow rate in troughs prevents sediment from settling while ridges continue to experience deposition; and (iv) the pro-

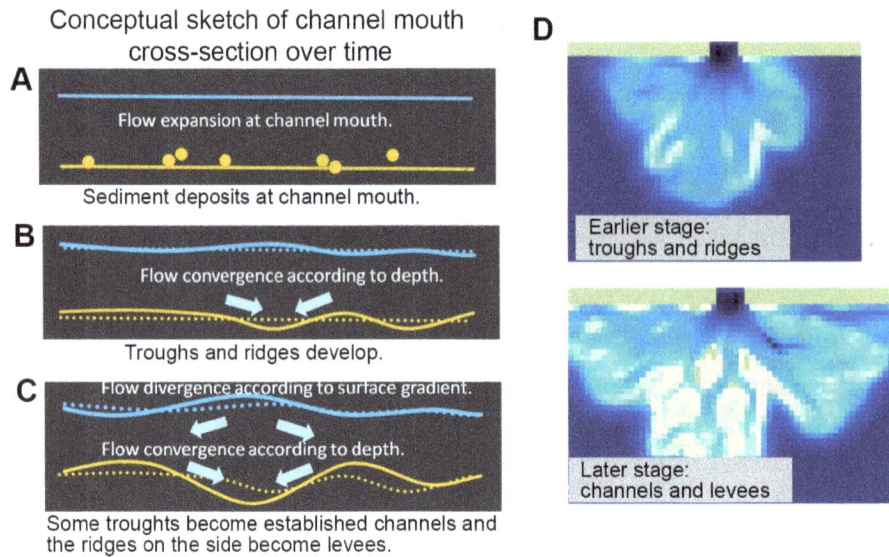

Figure 17. Illustration of the mechanics of bifurcation formation in the DeltaRCM: **(a)** flow expands and sediment deposits with reduced flow velocity, which creates a wide lunate bar with noisy surface; **(b)** the irregular topography of the bar forms "ridges" and "troughs" where troughs attract more water flux than ridges, the enhanced flow rate in troughs prevents sediment from settling while ridges continue to experience deposition; **(c)** the process reaches equilibrium by a feedback mechanism between water surface elevation and water flux. The few paths established from the troughs outcompete the others. **(d)** A sample simulation from DeltaRCM shows two stages of the conceptual processes described above: the earlier stage with troughs and ridges, and the later stage with channels and levees.

cess reaches equilibrium through the feedback mechanism between water surface elevation and water flux. The few paths established from the troughs outcompete the others. Interestingly, this process not only produces bifurcations, but also trifurcations. Accordingly, DeltaRCM suggests that distributary channel networks are not necessarily solely built by mouth-bar-induced flow bifurcation (Edmonds and Slingerland, 2007; Mariotti et al., 2013).

Another example of the translation from modeled hydrodynamic features to modeled morphodynamic features is the sensitivity of cellular flow routing to topographic details, such as flow acceleration and changes in flow orientation over short distances (Nicholas et al., 2012), which are caused by local effects. In test 4a, the results from Delft3D look more "diffusive" than FlowRCM. Local effects, when combined with the Lagrangian representation of fluxes as parcels, may offer a richer flow pattern in response to small changes on the floodplain. Although this observation requires further investigation, based on our results RC flow routing schemes seem to be less constrained by complex flow boundaries caused by wet–dry partitioning than higher-fidelity models based on partial differential equations.

CFD has the advantage of being able to resolve 3-D details such as velocity gradients normal to the bed and turbulence structures. The need for computational efficiency, uncertainty in initial and boundary conditions and lack of detailed information on parameterization for direct simulation often prompt the use of simpler modeling approaches. For example, depth-averaged models with appropriate parame-

terizations of 3-D features (e.g., spiral flow in bends) can be used for realistic sediment transport modeling (e.g., Falcini and Jerolmack, 2010). Simple rule-based RCMs are able to resolve hydrodynamic details previously believed to be produced by CFD exclusively (Wolfram, 2002). Such models share many common characteristics as the lattice Boltzmann method for fluid flow (which is considered a class of CFD). As for sediment transport, the detailed 3-D structure flow may be important to many morphodynamic features, which cannot be captured by simple depth-averaged flow routing methods. Identifying these delicate linkages between hydrodynamics and morphodynamics requires the combined effort of CFD modelers and RC modelers.

FlowRCM shares key features with existing RC flow routing schemes (Bates et al., 2010; Nicholas et al., 2012; Pelletier, 2008; Larson and Harvey, 2011), but differs in several ways. First, in order to model low-gradient environments, flow routing should be based on water surface slope, rather than on topographic slope. Some models assume a constant slope throughout the calculation region (Larson and Harvey, 2011) and others almost reproduce the exact solution of water surface profiles in benchmark cases (Bates et al., 2010). FlowRCM calculates the water surface elevation with the purpose of determining the direction of the water flux, without aiming at an exact solution. The underrelaxation of this condition helps us understand to what extent large-scale flow partitioning in deltaic environment is sensitive to hydrodynamic details. Second, FlowRCM does not require predefined maximum flow depth or minimum slope for its iterations as in successive MFD iterations (Pelletier, 2008). Third, FlowRCM uses a Lagrangian approach to route fluxes. The routing direction is calculated iteratively, eliminating the need of the traditional "row-by-row" discharge-splitting method as in Nicholas et al. (2012), while handling the multidirectionality typical of distributary networks. An intercomparison between flow routing schemes in a deltaic environment will be the subject of future research.

5 Conclusions

In this work we have applied a series of numerical tests to validate the reduced-complexity flow routing scheme, FlowRCM, which is introduced in Part 1 (Liang et al., 2015) as the hydrodynamic component of our RC delta formation model. We selected key hydrodynamic processes essential to channel processes in deltas and designed numerical tests as "benchmarking" cases for this specific environment. We compared the output from FlowRCM with the output from a higher-fidelity hydrodynamic model, Delft3D, which is based on rigorous CFD solutions. We also used theoretical solutions to demonstrate FlowRCM's ability to reproduce hydrodynamic details.

The results show that FlowRCM is able to reproduce most of the flow and water surface features of interest. Overall, it captures (1) the trend of water surface from upstream to downstream and from channels to floodplain, and (2) flow partitioning corresponding to complex bed topography such as flow divergence and convergence around obstacles. Furthermore, the routing scheme is able to produce morphodynamic features such as mouth bars, bifurcations and levee formation. The responsible process for these morphodynamic features is an instability-feedback mechanism resulting from the coupling of the hydrodynamic component and the sediment transport rules.

This work suggests an assessment framework for RCMs that include explicit flow routing schemes. The key ideas include (1) designing test cases to evaluate the performance of the routing scheme in producing features related to the processes of interest, (2) identifying the effects of model rules and connecting them to model output individually, and (3) connecting the hydrodynamic to the morphodynamic performance to evaluate the flow routing scheme's ability to model morphodynamics accurately.

We suggest that FlowRCM is appropriate for modeling environments with multichannel networks where morphodynamic features can be produced from the estimation of channel-to-channel and channel-to-island/floodplain flow partitioning. More detailed prediction of in-channel flow patterns such as spatial distribution of high/low velocity and surface deformation could also be achieved but FlowRCM has not been designed for small-scale engineering applications.

Acknowledgements. M. Liang acknowledges support from National Science Foundation via the National Center for Earth-surface Dynamics (NCED) under agreements EAR-1246761. The authors also acknowledge support from the National Science Foundation (FESD/OCE-1135427, CAREER/EAR-1350336 and GSS/BCS-1063228 to P. Passalacqua, and OCE-1061380 to D. A. Edmonds). The authors also thank A. Ashton and the anonymous reviewer for their insightful opinions.

Edited by: S. Castelltort

References

Bates, P. D. and De Roo, A. P. J.: A simple raster-based model for flood inundation simulation, J. Hydrol., 236, 54–77, 2000.

Bates, P. D., Horritt, M. S., and Fewtrell, T. J.: A simple inertial formulation of the shallow water equations for efficient two-dimensional flood inundation modelling, J. Hydrol., 387, 33–45, 2010.

Chow, V. T.: Open-channel hydraulics, in: Open-Channel Hydraulics, McGraw-Hill, New York, 1959.

Coulthard, T. J., Macklin, M. G., and Kirkby, M. J.: A cellular model of Holocene upland river basin and alluvial fan evolution, Earth Surf. Proc. Land., 27, 269–288, 2002.

Doeschl-Wilson, A. B. and Ashmore, P. E.: Assessing a numerical cellular braided-stream model with a physical model, Earth Surf. Proc. Land., 30, 519–540, doi:10.1002/esp.1146, 2005.

Duan, J. G. and Julien, P. Y.: Numerical simulation of meandering evolution, J. Hydrol., 391, 34–46, 2010.

Edmonds, D. A. and Slingerland, R. L.: Mechanics of river mouth bar formation: implications for the morphodynamics of delta distributary networks, J. Geophys. Res., 112, F02034, doi:10.1029/2006JF000574, 2007.

Edmonds, D. A. and Slingerland, R. L.: Stability of delta distributary networks and their bifurcations, Water Resour. Res., 44, W09426, doi:10.1029/2008WR006992, 2008.

Edmonds, D. A., Slingerland, R., Best, J., Parsons, D., and Smith, N.: Response of riverdominated delta channel networks to permanent changes in river discharge, Geophys. Res. Lett., 37, L12404, doi:10.1029/2010GL043269, 2010.

Edmonds, D. A., Paola, C., Hoyal, D. C. J. D., and Sheets, B. A.: Quantitative metrics that describe river deltas and their channel networks, J. Geophys. Res., 116, F04022, doi:10.1029/2010JF001955, 2011.

Falcini, F. and Jerolmack, D. J.: A potential vorticity theory for the formation of elongate channels in river deltas and lakes, J. Geophys. Res., 115, F04038, doi:10.1029/2010JF001802, 2010.

Ferguson, R. I., Church, M., and Weatherly, H.: Fluvial aggradation in Vedder River: testing a one-dimensional sedimentation model, Water Resour. Res., 37, 3331–3347, 2001.

Freeman, T. G.: Calculating catchment area with divergent flow based on a rectangular grid, Comput. Geosci., 17, 413–422, 1991.

Hajek, E. A. and Wolinsky, M. A.: Simplified process modeling of river avulsion and alluvial architecture: connecting models and field data, Sediment. Geol., 257, 1–30, 2012.

Hardy, R. J., Lane, S. N., Ferguson, R. I., and Parsons, D. R.: Assessing the credibility of a series of computational fluid dynamic simulations of open channel flow, Hydrol. Process., 17, 1539–1560, 2003.

Howard, A. D.: Badland morphology and evolution: interpretation using a simulation model, Earth Surf. Proc. Land., 22, 211–227, 1997.

Jerolmack, D. J. and Paola, C.: Complexity in a cellular model of river avulsion, Geomorphology, 91, 259–270, 2007.

Karssenberg, D. and Bridge, J. S.: A three-dimensional numerical model of sediment transport, erosion and deposition within a network of channel belts, floodplain and hill slope: extrinsic and intrinsic controls on floodplain dynamics and alluvial architecture, Sedimentology, 55, 1717–1745, 2008.

Kleinhans, M. G., Jagers, H. R. A., Mosselman, E., and Sloff, C. J.: Bifurcation dynamics and avulsion duration in meandering rivers by one-dimensional and three-dimensional models, Water Resour. Res., 44, W08454, doi:10.1029/2007WR005912, 2008.

Kleinhans, M. G., Ferguson, R. I., Lane, S. N., and Hardy, R. J.: Splitting rivers at their seams: bifurcations and avulsion, Earth Surf. Proc. Land., 38, 47–61, 2013.

Lane, S. N., Hardy, R. J., Elliott, L., and Ingham, D. B.: High-resolution numerical modelling of three-dimensional flows over complex river bed topography, Hydrol. Process., 16, 2261–2272, 2002.

Larsen, L. G. and Harvey, J. W.: Modeling of hydroecological feedbacks predicts distinct classes of landscape pattern, process, and restoration potential in shallow aquatic ecosystems, Geomorphology, 126, 279–296, 2011.

Lesser, G. R., Roelvink, J. A., van Kester, J. A. T. M., and Stelling, G. S.: Development and validation of a three-dimensional morphological model, Coast. Eng., 51, 883–915, 2004.

Liang, M., Voller, V. R., and Paola, C.: A reduced-complexity model for river delta formation – Part 1: Modeling deltas with channel dynamics, Earth Surf. Dynam., 3, 67–86, doi:10.5194/esurf-3-67-2015, 2015.

Mariotti, G., Falcini, F. , Geleynse, N., Guala, M., Sun, T., and Fagherazzi, S.: Sediment eddy diffusivity in meandering turbulent jets: Implications for levee formation at river mouths, J. Geophys. Res.-Earth, 118, 1908–1920, doi:10.1002/jgrf.20134, 2013.

Martin, Y. and Church, M.: Numerical modelling of landscape evolution: geomorphological perspectives, Prog. Phys. Geogr., 28, 317–339, 2004.

Murray, A. B.: Reducing model complexity for explanation and prediction, Geomorphology, 90, 178–191, 2007.

Nicholas, A. P.: Cellular modelling in fluvial geomorphology, Earth Surf. Proc. Land., 30, 645–649, 2005.

Nicholas, A. P.: Reduced-complexity modeling of free bar morphodynamics in alluvial channels, J. Geophys. Res.-Earth, 115, F04021, doi:10.1029/2010JF001774, 2010.

Nicholas, A. P.: Modelling the continuum of river channel patterns, Earth Surf. Proc. Land., 38, 1187–1196, 2013.

Nicholas, A. P., Sandbach, S. D., Ashworth, P. J., Amsler, M. L., Best, J. L., Hardy, R. J., Lane, S. N., Orfeo, O., Parsons, D. R., Reesink, A. J. H., Sambrook Smith, G. H., and Szupiany, R. N.: Modelling hydrodynamics in the Rio Paraná, Argentina: an evaluation and intercomparison of reduced-complexity and physics based models applied to a large sand-bed river, Geomorphology, 169–170, 192–211, 2012.

O'Callaghan, J. F. and Mark, D. M.: The extraction of drainage network from digital elevation data, Comput. Vis. Graph. Image Process., 28, 323–344, 1984.

Oreskes, N., Shrader-Frechette, K., and Belitz, K.: Verification, validation, and confirmation of numerical models in the earth sciences, Science, 263, 641–646, 1994.

Overeem, I., Syvitski, J. P. M., and Hutton, E. W. H.: Three-dimensional numerical modeling of deltas, in: River Deltas: Concepts, Models and Examples, edited by: Bhattacharya, J. P. and Giosan, L., SEPM Spec. Publ., 83, SEPM, Tulsa, 13–30, 2005.

Paola, C., Twilley, R. R., Edmonds, D. A., Kim, W., Mohrig, D., Parker, G., Viparelli, E., and Voller, V. R.: Natural processes in delta restoration: application to the Mississippi Delta, Ann. Rev. Mar. Sci., 3, 67–91, 2011.

Parsons, J. A. and Fonstad, M. A.: A cellular automata model of surface water flow, Hydrol. Process., 21, 2189–2195, doi:10.1002/hyp.6587, 2007.

Passalacqua, P., Lanzoni, S., Paola, C., and Rinaldo, A.: Geomorphic signatures of deltaic processes and vegetation: the Ganges-Brahmaputra-Jamuna case study, J. Geophys. Res.-Earth, 118, 1838–1849, 2013.

Pelletier, J. D.: Quantitative Modeling of Earth Surface Processes, Cambridge U. Press, Cambridge, 2008.

Roberts, H., Adams, R., and Cunningham, R.: Evolution of sand-dominant subaerial phase, Atchafalaya Delta, Louisiana, AAPG Bull., 64, 264–279, 1980.

Seybold, H., Andrade Jr., J. S., and Herrmann, H. J.: Modeling river delta formation, P. Natl. Acad. Sci. USA, 104, 16804–16809, 2007.

Shaw, J. B., Mohrig, D., and Whitman, S. K.: The morphology and evolution of channels on the Wax Lake Delta, Louisiana, USA, J. Geophys. Res.-Earth, 118, 1–23, 2013.

Siviglia, A., Stecca, G., Vanzo, D., Zolezzi, G., Toro, E. F., and Tubino, M.: Numerical modelling of two-dimensional morphodynamics with applications to river bars and bifurcations, Adv. Water Resour., 52, 243–260, 2013.

Syvitski, J. P. M., Kettner, A. J., Overeem, I., Hutton, E. W. H., Hannon, M. T., Brakenridge, G. R., Day, J., Vörösmarty, C., Saito, Y., Giosan, L., and Nicholls, R. J.: Sinking deltas due to human activities, Nat. Geosci., 2, 681–686, 2009.

Tucker, G. E. and Bras, R. L.: Hillslope processes, drainage density, and landscape morphology, Water Resour. Res., 34, 2751–2764, 1998.

USACE – US Army Corps of Engineers: Atchafalaya River hydrographic survey, 1998–1999: Old River to Atchafalaya Bay including main channel and distributaries, The District, New Orleans, LA, 1999.

Van De Wiel, M. J., Coulthard, T. J., Macklin, M. G., and Lewin, J.: Embedding reach-scale fluvial dynamics within the CAESAR cellular automaton landscape evolution model. Geomorphology, 90, 283–301, doi:10.1016/j.geomorph.2006.10.024, 2007.

Wellner, R., Beaubouef, R., Van Wagoner, J., Roberts, H., and Sun, T.: Jet-plume depositional bodies – the primary building blocks of Wax Lake delta, Gulf Coast Assoc. Geol. Soc. Trans., 55, 867–909, 2005.

Willgoose, G.: A statistic for testing the elevation characteristics of landscape simulation models, J. Geophys. Res.-Sol. Ea., 99, 13987–13996, 1994.

Wolfram, S.: A new kind of science, Wolfram Media, Champaign, IL, 2002.

Wolinsky, M. A., Edmonds, D. A., Martin, J., and Paola, C.: Delta allometry: growth laws for river deltas, Geophys. Res. Lett., 37, L21403, doi:10.1029/2010GL044592, 2010.

8

Data-driven components in a model of inner-shelf sorted bedforms: a new hybrid model

E. B. Goldstein[1], G. Coco[2], A. B. Murray[1], and M. O. Green[3]

[1]Division of Earth and Ocean Sciences, Nicholas School of the Environment, Center for Nonlinear and Complex Systems, Duke University, P.O. Box 90227, Durham, NC 27708, USA
[2]Environmental Hydraulics Institute, "IH Cantabria", c/Isabel Torres no. 15, Universidad de Cantabria, 39011 Santander, Spain
[3]National Institute of Water and Atmospheric Research (NIWA), P.O. Box 11-115, Hamilton, New Zealand

Correspondence to: E. B. Goldstein (evan.goldstein@duke.edu)

Abstract. Numerical models rely on the parameterization of processes that often lack a deterministic description. In this contribution we demonstrate the applicability of using machine learning, a class of optimization tools from the discipline of computer science, to develop parameterizations when extensive data sets exist. We develop a new predictor for near-bed suspended sediment reference concentration under unbroken waves using genetic programming, a machine learning technique. We demonstrate that this newly developed parameterization performs as well or better than existing empirical predictors, depending on the chosen error metric. We add this new predictor into an established model for inner-shelf sorted bedforms. Additionally we incorporate a previously reported machine-learning-derived predictor for oscillatory flow ripples into the sorted bedform model. This new "hybrid" sorted bedform model, whereby machine learning components are integrated into a numerical model, demonstrates a method of incorporating observational data (filtered through a machine learning algorithm) directly into a numerical model. Results suggest that the new hybrid model is able to capture dynamics previously absent from the model – specifically, two observed pattern modes of sorted bedforms. Lastly we discuss the challenge of integrating data-driven components into morphodynamic models and the future of hybrid modeling.

1 Introduction

Parameterizations become necessary in morphodynamic models when processes cannot be described entirely from conservation laws. This is often the case with descriptions of sediment transport, where the mechanics are multidimensional and highly nonlinear (e.g., have thresholds). Parameterizations are often developed through the collection and processing of experimental data. This results in formulas that, because they have been developed through inductive methods, are subject to many caveats: constraints regarding the applicable forcing conditions or the appropriate setting for use. The inaccuracy of individual predictors has significant consequences in nonlinear morphodynamic models because errors accumulate as inaccuracy is (1) propagated through the nonlinear pieces of the model (e.g., Bolaños et al., 2012) and (2) propagated in time (e.g., Pape et al., 2010).

Some prediction schemes may perform well only in specific settings or under specific hydrodynamic conditions (Cacchione et al., 2008; Bolaños et al., 2012). This is an example of locally optimal predictors, performing well with a single set of data but not necessarily transferable to other settings (both physical locations and hydrodynamic conditions). The existence of many locally optimal predictors (each developed from its own data set) leads to the problem of selecting the appropriate predictor for a morphodynamic model. One solution to this difficulty is to sidestep it entirely and instead develop globally optimal predictors from multi-setting data sets that encompass wide ranges of forcing conditions and independent variables. The hope is that differences in locally optimal solutions may be attributed to an independent variable that may become apparent when building a single, unified globally optimal model.

The construction of globally optimal predictors is difficult because large multi-setting data sets with nonlinear relationships and multiple independent variables are difficult to visualize and interpret. Traditional techniques for developing successful parameterizations include converting multidimensional data sets into low-dimensional spaces and then fitting a curve. However, collapsing data into combined parameters may inherently bias the resultant predictor and may obscure subtle relationships in the data. One method to detect relationships in large, nonlinear, multidimensional data sets is machine learning (ML), a class of computational optimization routines. A range of ML techniques have previously been used successfully to develop data-driven parameterizations: artificial neural networks (ANN) have been used to parameterize alongshore suspended sediment transport in the surf zone (van Maanen et al., 2010), sediment suspension in the surf zone (Yoon et al., 2013), and near-bed reference concentration (Oehler et al., 2012). Boosted regression trees (BRT) have been used to parameterize suspended sediment reference concentration (Oehler et al., 2012), and genetic programming techniques have been used to develop predictions of wave-generated ripple geometry (Goldstein et al., 2013), roughness in vegetated flows (Baptist et al., 2007), and fluvial sediment transport (Kitsikoudis et al., 2013). Aside from small-scale process descriptions, data-driven approaches have also been used as stand-alone morphodynamic models (Pape et al., 2007, 2010) and to calibrate model parameters (Knaapen and Hulscher, 2002, 2003; Ruessink, 2005).

In this contribution we focus on the data-driven prediction of near-bed reference concentration under unbroken waves. As the bottom boundary condition for calculating suspended sediment transport, reducing error is of paramount importance for accurate predictions of total suspended sediment load. Several parameterizations already exist, notably Nielsen (1986) and Lee et al. (2004). Recent work by Oehler et al. (2012) demonstrated the ability of ML predictors to outperform traditional empirical prediction schemes for reference concentration (i.e., Lee et al., 2004; Nielsen, 1986). The BRT and ANN model developed by Oehler et al. (2012) is an accurate predictor of reference concentration, but the predictor is not smooth, physically interpretable, or economical in length; all problems when attempting to incorporate the results into a morphodynamic model. Here we use genetic programming (GP) to develop a smooth and physically interpretable parameterization of near-bed reference concentration. GP is a population-based optimization technique where the population is composed of individual predictors (Koza, 1992). Using evolutionary principles (e.g., crossover, mutation) to develop new solutions, the functional form of the pre-

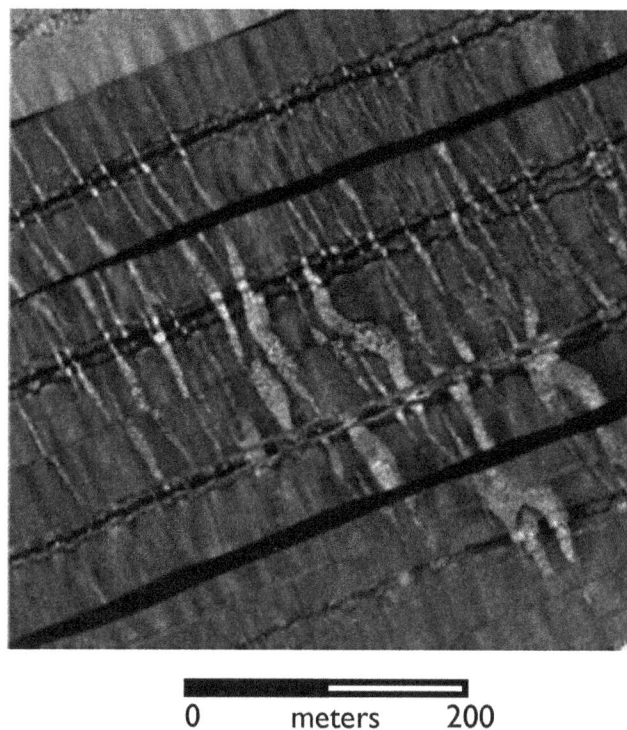

0 meters 200

Figure 1. Sorted bedforms present in ~ 5 m of water off the coast of Tairua Beach, New Zealand (Coco et al., 2007a). White areas are composed of coarse sediment, while dark areas are floored by fine sediment. Shoreline is towards the bottom of the panel.

dictor and the location and presence of the variables within a given predictor are adjusted and optimized to find a globally optimum solution.

The development of a new near-bed suspended sediment reference concentration predictor using GP is the first objective of this work. The second objective is to incorporate this new predictor (and a previously developed predictor for ripple geometry, built with GP) into a previously developed model of inner-shelf sorted bedforms (Coco et al., 2007a) to develop a "hybrid" numerical model (Krasnopolsky and Fox-Rabinovitz, 2006), where data-driven components are combined with widely accepted formulas for hydrodynamics and sediment transport. Previous examples of the hybrid approach are found in studies of shoreline change (Karunarathna and Reeve, 2013), hydrology (Corzo et al., 2009), and the atmospheric and climate system (Krasnopolsky and Fox-Rabinovitz, 2006).

Spatially extensive (kilometer scale) patches of segregated coarse and fine-grained sediment (Fig. 1) with only slight bathymetric relief (centimeter to meter scale) relative to bedform pattern wavelength (10 m–km) are present on many continental shelf systems (Coco et al., 2007b). Unlike most bedforms that develop solely as an interaction between bathymetry and flow, recent work implicates a sorting feedback as the mechanism for the development of inner-shelf

Table 1. Summary of experiments used in this study.

Study	Mean water depth (m)	Sediment grain size (mm)	Sampling rate (Hz)	Burst duration (min)	Sorted bedform field?
(Green, 1999; Green and Black, 1999)	7	0.23	4–5	10–17.06	No
(Green et al., 2004; Trembanis et al., 2004)	15	0.22	1	15	Yes
(Green et al., 2004; Trembanis et al., 2004)	22	0.22	4	8.5	Yes
(Green et al., 2004; Trembanis et al., 2004)	22	0.75	1	15	Yes
(Vincent and Green, 1999)	25	0.33	4	10	No
(Green and MacDonald, 2001)	1.7	0.15	4–5	4.267–5	No

"sorted bedforms" (Murray and Thieler, 2004; Coco et al., 2007a, b). The sorting feedback is initiated by wave-generated ripples whose size is a function of seabed composition and hydrodynamic forcing conditions (e.g., Cummings et al., 2009). Regions covered with fine sediment support smaller wave-generated ripples than areas mantled by coarse sediment. Strong turbulence above the large wave ripples on coarse domains enhances the erosion of fine material from the bed (and also functions as a barrier to the deposition of suspended fine sediment). Near-bottom currents lead to the advection of suspended fine material and the preferential settling of suspended fine sediment in areas where the seabed is composed of predominantly fine sediment with small wave ripples (and correspondingly less turbulence induced by the smaller features). Through self-organization this local sorting feedback leads to spatially extensive features. The numerical model of Coco et al. (2007a) indicates that the sorting feedback operates in a wide range of forcing conditions (Coco et al., 2007b).

Sorted bedforms show several configurations that we divide into two distinct end-member patterns typified by the location of the coarse domain, either in the trough of the bedform or on the flanks of the bedforms (appearance on both the updrift and/or downdrift are possible; e.g., Goff et al., 2005; Ferrini and Flood, 2005). We note that within an individual sorted bedform field the pattern configuration can change (Thieler et al., 2014; Ferrini and Flood, 2005). Previous work with the finite-amplitude models by Murray and Thieler (2004) and Coco et al. (2007a) showed the presence of coarse domains solely on the downdrift flank of bedforms. While Coco et al. (2007b) did show the potential for coarse domains to occur in the trough of bedforms, this configuration was highly path dependent (i.e., the result of a high wave event that is preceding and followed by smaller waves). Van Oyen et al. (2010, 2011), through linear stability analysis, showed the presence of two pattern modes in the initial infinitesimal-amplitude instability that correspond to these two distinct configurations. However Van Oyen et al. (2010, 2011) showed that each pattern mode is the result of separate feedback mechanisms, where coarse domains present in troughs occurred as the result of a flow–bathymetry feed-

back, while coarse domains present on bedform flanks is the result of the previously described sorting feedback (refereed to as the "roughness" feedback by Van Oyen et al. (2010, 2011).

With the goal of presenting a new hybrid model, we first describe the development of the near-bed suspended sediment reference concentration predictor from the large data set of Green and colleagues (Green, 1996, 1999; Green and Black, 1999; Vincent and Green, 1999; Green and MacDonald, 2001; Green et al., 2004; Trembanis et al., 2004). We then outline the sorted bedform model and the modifications to incorporate the new data-driven components. This new model is meant as an update to the Coco et al. (2007a) model. The new predictors in the hybrid model are more accurate and better performing than the formulations used in the Coco et al. (2007a) model. Finally, we present a novel experiment with the new hybrid model to show autogenic behaviors that were not present in the Coco et al. (2007a) model (i.e., the appearance of two pattern configurations solely from a sorting feedback) and discuss advantages and disadvantages of this data-driven approach. This paper does not attempt to quantitatively compare the new hybrid model against older modeling efforts: instead we offer this new model as a refinement to the previous model that is additionally able to capture new dynamics.

2 GP methods

2.1 Data set

Figure 2 shows the multi-setting field data set composed of 1748 individual measurements from 6 separate field experiments at different locations in New Zealand. We briefly summarize the experiments below and in Table 1; a detailed summary of each experiment and the specific methodology used to determine the near-bed suspended sediment reference concentration (C_0; g L^{-1}), significant near-bed orbital velocity (U_{sig}; m s^{-1}), wave orbital diameter at the bed (d_0; m), mean grain size (d_{50}; m), and mean spectral wave period at the bed (T_{mean}; s) is available in the associated references. A single experiment (Green and Black, 1999; Green, 1999) collected 127 measurements seaward of the surf zone with mean water

Table 2. Solutions for reference concentration.

Solution	Complexity	MSE
$C_0 = 0.182$	1	0.070
$C_0 = U_{\text{sig}}^2$	2	0.057
$C_0 = 0.637 U_{\text{sig}}$	3	0.056
$C_0 = \left(1.19 U_{\text{sig}}\right)^2$	4	0.052
$C_0 = U_{\text{sig}} - 0.647 \,(1000 d_{50})$	5	0.048
$C_0 = \left(\dfrac{0.235 U_{\text{sig}}}{(1000 d_{50})}\right)^2$	7	0.048
$C_0 = \left(\dfrac{0.328 U_{\text{sig}}}{0.0688 + (1000 d_{50})}\right)^2$	9	0.045
$C_0 = \left(1.27 \sqrt{U_{\text{sig}}} - 1.21 \,(1000 d_{50})\right)^2$	12	0.045
$C_0 = \dfrac{0.179 U_{\text{sig}}^2 - 0.00538}{d_0 (1000 d_{50})} + \dfrac{0.0185 + 0.179 U_{\text{sig}}^2 d_0 - 0.179 U_{\text{sig}}^2 - 0.0319 U_{\text{sig}}^4}{(1000 d_{50})}$	41	0.043

depth of 7 m. Data from three experiments (Green et al., 2004; Trembanis et al., 2004) were collected from separate locations in a field of sorted bedforms (669, 126, and 554 measurements). A single instrument frame was located in a domain composed of coarse sand (22 m depth) and two instrument frames were located in fine sand domains (15 and 22 m depth). The fifth experiment was deployed off of a headland in 25 m of water depth (56 measurements; Vincent and Green, 1999). The final experiment in the database collected 241 measurements in a microtidal estuary in a mean water depth of 1.7 m (Green and MacDonald, 2001). All data were gathered in burst mode, with burst durations ranging from 4.267 to 17.06 min. In addition to the multiple settings and significant amount of data, this data set is ideal for application in the sorted bedform model because three of the six experiments in the composite data set are derived from a sorted bedform field (Green et al., 2004; Trembanis et al., 2004).

2.2 Selection of training, validation, and testing data sets

The database is split into three subsets to be used as training, validation, and testing. The training data set is used to develop candidate solutions. The validation data set is used to evaluate the generality of a predictor, the fitness of GP-derived solutions against more data, and ultimately to determine which predictors persist. The testing data set is unused and unseen by the GP algorithm; it is reserved as an independent test of the final predictors (and other published predictors). Because our database does not cover the entirety of the forcing space with equal density (Fig. 3), the selection and partitioning of data into these three categories is crucial for developing a well-performing predictor applicable to a range of environments (e.g., Bowden et al., 2002). The C_0 data set is sparse in areas because of a lack of collected data, while dense in other regions of phase space as a result of similar field settings, forcing conditions, and the number of data points collected in a given experiment. If the data are randomly divided, there is a potential that the

training data exclude data from sparse regions in the data set (i.e., coarse-grained and/or strong hydrodynamic data). However, in the genetic programming literature we could find no proven "best practice" for selection of the data subsets or an optimal percentage of training, validation, and testing data (Kuschu, 2002; Panait and Luke, 2003; Gagné et al., 2006); we therefore use a technique that was successful in a previous study (Goldstein et al., 2013).

Informed data selection has been shown to produce better results with ML predictors than "blind" or random data selection (e.g., Bowden et al., 2002; May et al., 2010). In this study we select training data through the use of a maximum dissimilarity algorithm (MDA; Camus et al., 2011). This algorithm is not a clustering routine (where centroids denote a representative value of the data in the cluster), but is instead a selection routine (where a centroid represents the most dissimilar data point from the previous centroids; Camus et al., 2011). This selection routine allows the use of a minimum of training data that is able to capture the variance present in the entire data set while leaving the majority of the data to be utilized as validation and testing.

The maximum dissimilarity algorithm is described in Camus et al. (2011) and we review the method. Selection starts with the linear normalization of the independent variables to a value between 0 (minimum value of a given variable) and 1 (maximum value of a given variable). A single data point, a "seed", is selected as the first centroid. The algorithm then selects the additional centroids (the number determined by the user) through an iterative process: each data point is a four-dimensional vector (normalized T_{mean}, U_{sig}, d_0, d_{50} space) and is associated with a distance to the nearest centroid. The single data point with the maximum distance between itself and the nearest centroid is selected as the next centroid (Camus et al., 2011). The MDA routine continues until the user-defined number of centroids is reached and the data are then denormalized.

There remains significant ambiguity in determining the appropriate number of centroids needed to accurately represent

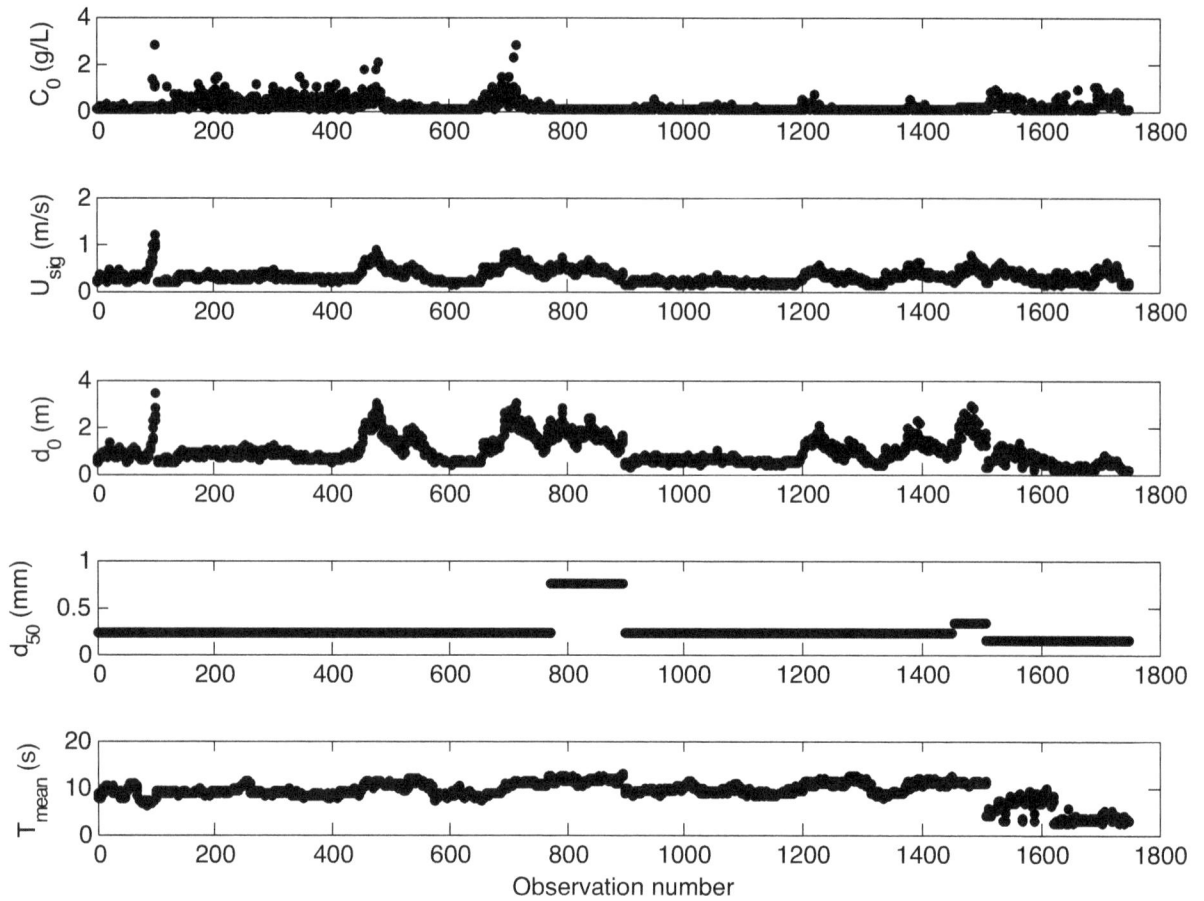

Figure 2. Observations of suspended sediment reference concentration data set C_0 and concomitant measurements of significant wave velocity at the bed (U_{sig}), wave orbital excursion at the bed (d_0), mean grain size of bed material (d_{50}), and mean spectral wave period at the bed (T_{mean}). Note that mean grain size of bed material is shown here in millimeters. A similar figure appears in Oehler et al. (2012).

a data set, especially continuous data (e.g., May et al., 2010; Goldstein et al., 2013). Selecting too many centroids can rob the validation and testing data sets of poorly represented data (e.g., large T_{mean}, U_{sig}, d_0, d_{50}) and may tend to cause the GP to produce overly complex predictors (e.g., Gonçalves and Silva, 2013; Oates and Jensen, 1997, 1998). The selection of too few centroids can leave the testing data with too few data points to capture the variability in the data set (Goldstein et al., 2013). We use 40 centroids for the prediction of C_0 (centroid locations can be seen in Fig. 3), the same as Goldstein et al. (2013). Data selected as the centroid locations are used for the training data, while the remaining data are used for validation and testing data. The data set is split between validation and testing randomly, without using a selection routine. The final breakdown for the data sets is ~ 2 % training, ~ 49 % validation, and ~ 49 % testing.

2.3 Genetic programming

We operate on this data set using the ML technique of genetic programming (GP; Koza, 1992; Poli et al., 2008), where can-

didate solutions (i.e., randomly generated initial equations) are evaluated and subsequently modified by adjusting the independent variables as well as the mathematical relationships between variables (i.e., the mathematical form). Independent variables used in this study to predict C_0 are T_{mean}, U_{sig}, d_0, and d_{50}. We use T_{mean}, U_{sig}, and d_0 as separate independent variables for input to the GP (though they are related) in an attempt to introduce no additional information about which of these parameters is most relevant. Mathematical operators used in this study are + (addition), − (subtraction), × (multiplication), ÷ (division), and $\sqrt{\ }$ (square root), as well as integer powers (e.g., x^2, x^3, etc.). We omit logical functions in this analysis (e.g., if-then-else) because we aim to develop a smooth final solution.

Candidate solutions are evaluated based on a "fitness function", a user-defined error metric that determines how well a given candidate fits the validation data. Mean squared error (MSE) is used as the fitness function:

$$\text{MSE} = \frac{\sum_{i=1}^{n}(p_i - b_i)^2}{n}, \tag{1}$$

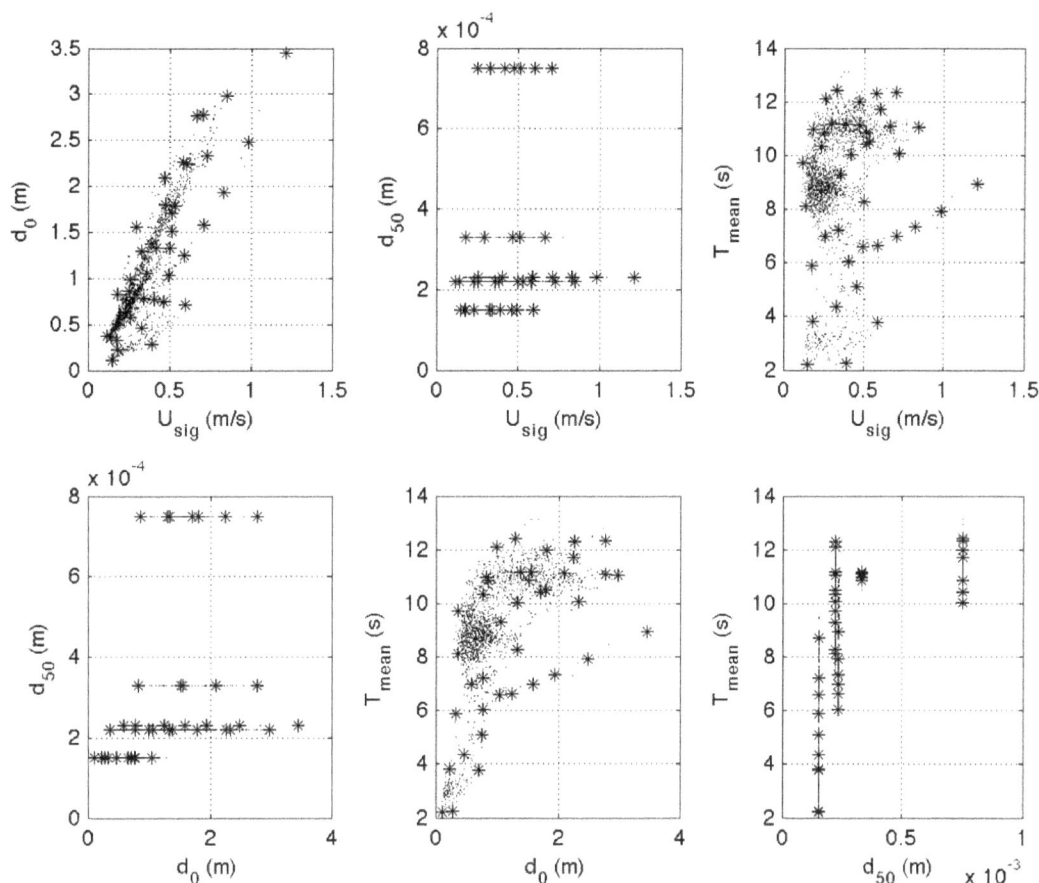

Figure 3. Visualization of the range of conditions in the C_0 data set. Each plot represents a two-dimensional projection of the entire data set onto the set of axes shown. For instance, the first panel with data projected onto the $d_0 - U_{sig}$ plane shows no information about d_{50} or T_{mean}. Stars denote centroid locations (training data), while points denote unselected data (validation and testing). Note that centroids are distributed throughout the data set.

where n is the sample size, p are the predicted values, and b are the observed values. Candidate solutions that minimize mean squared error are retained and poor performing solutions are discarded. Retained solutions are rearranged, combined, and manipulated in a probabilistic manner according to combinatorial processes: solutions "crossover" by combining elements of other solutions to develop a new solution and "mutations" develop new mathematical expression to substitute or tack on to a previous solution. Candidate solutions are commonly encoded in GP software as graphs or "trees". The evolutionary processes that modify candidate solutions (change of variables and/or mathematical expression) is accomplished by adjusting tree "limbs" (Fig. 4). Predictors range from simple (small trees) to complex (large trees) as they are recombined in a variety of ways. The range of candidate solutions enables the searching of a large solution space, and the search process continues until a solution with zero error is found or the routine is halted.

In this study we use a proven software package developed by Schmidt and Lipson (2009, 2013). This software package,

"Eureqa", outputs a suite of solutions with increasing mathematical "complexity", where complexity is a count of the numbers of operators and variables are used in the candidate solution. Each solution of a given complexity represents the equation with the least error compared to identically "complex" candidate solutions. Additionally, solutions must have less error compared to all previous less complex solutions. The line that traces the suite of solutions in complexity–fitness space is the "Pareto front", and is a graphical representation of increasing fitness with increasing complexity. Many predictors along the Pareto front, from simple to complex, are retained in the solution set, requiring the user to pick a single solution as the final predictor of choice.

In the results presented here there is no single zero-error solution found; therefore we cease the search after roughly 10^{10} formulas have been evaluated – continued search shows only marginal increases in predictive power (and this increase occurs only on more complex, likely overfitted, predictors). Several methods exist for eliminating overfitted solutions (e.g., Gonçalves et al., 2012). We use several

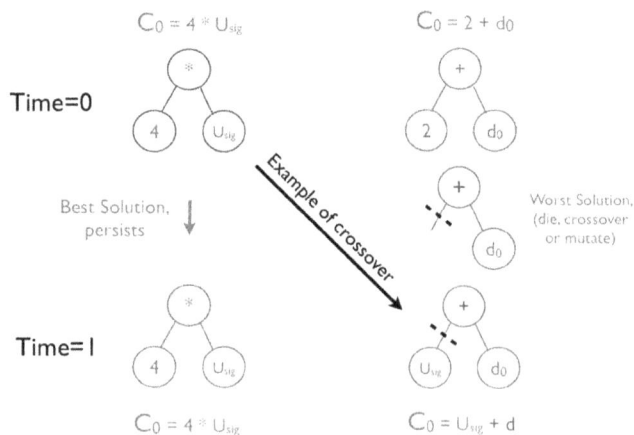

Figure 4. Example of the genetic programming process. Potential solutions are encoded as a population of trees. Here a hypothetical population of two solutions is shown. The first solution has a low MSE and therefore persists to the next iteration. The second solution has a high MSE and therefore is subject to removal, mutation, or crossover. An example of "crossover" is shown here, whereby the old solution is combined with parts of other, better performing solutions to create a new potential solution in the next iteration.

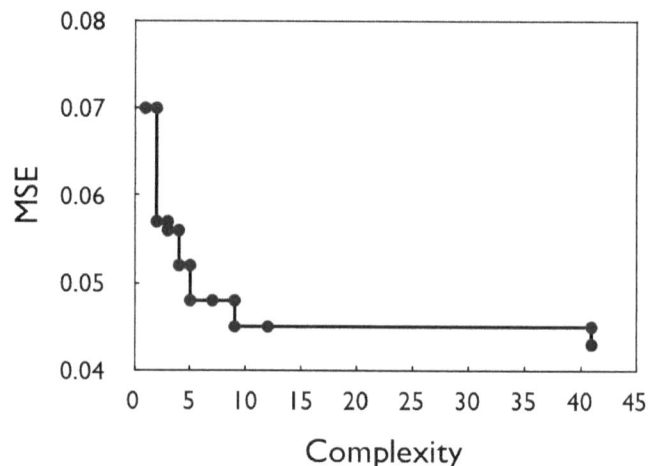

Figure 5. Reference concentration Pareto front; MSE is mean squared error of candidate solution versus the validation data set. Complexity is a quantification of the candidate solution length (both mathematical operators and variables).

techniques in parallel to determine a single appropriate solution: (1) bias toward shorter, physically reasonable solutions, (2) examining "cliffs" in the Pareto front, and (3) examination of solution fit.

Compact, simple solutions tend to offer more generalization power and are likely less overfitted (the minimum description length principle; e.g., O'Neill et al., 2010). Additionally, shorter solutions reappear with repeat initialization of the genetic programming algorithm, suggesting that these reappearing candidates represent the globally optimum solutions for a given function size. Longer solutions do not tend to reappear, a result of a large search space that is not repeated during repeat initializations or the presence of multiple, equally optimal solutions in the large phase space (i.e., local minima). The inherent reproducibility of simple, weakly nonlinear solutions suggests their use as predictors until further data can be used to justify the use of highly nonlinear predictors.

Areas along the Pareto front where large gains in prediction are obtained with small gains in solution complexity, "cliffs", are a natural place to observe potential solutions (Fig. 5). Schmidt and Lipson (2009) observed many physically relevant solutions at the bottom of the last cliff of a given Pareto front, and therefore we focus our search for a final solution at the cliffs. Additionally, as candidate solutions are evaluated by minimizing error functions, solutions occasionally minimize mean squared error but are unphysical (e.g., functions that have poor extrapolation ability beyond the domain of the training data). These solutions must be manually disregarded, as there is as yet no means of excluding them.

Once a single predictor is selected, it is evaluated using the independent testing data (data that the ML algorithm has not seen) with the normalized root-mean-squared error (NRMSE):

$$\text{NRMSE} = \frac{\sqrt{\text{MSE}}}{\bar{b}}, \qquad (2)$$

where $\bar{b}$ is the mean of the observed values. Additionally we report the correlation coefficient (Pearson's r) for each predictor evaluated against the independent testing data. The NRMSE and correlation coefficient are also reported for the reference concentration predictor of Nielsen (1986) and Lee et al. (2004) evaluated against the independent testing data.

3 GP results

The GP algorithm output is shown in Table 2 (note that numerical coefficients listed in the table are dimensional). This experiment evaluated 10^{10} formulas to develop the Pareto front shown in Fig. 5. Cliffs occur along the Pareto front at complexities of 2, 4, 5, 9, and 41 (Fig. 5). Predictors generally show nonlinear dependence on U_{sig}/d_{50}, qualitatively similar to the predictors developed by Nielsen (1986) and Lee et al. (2004), which both show dependence on the modified Shields parameter. We focus our analysis on the last cliff before the proliferation of very complex, nonlinear terms (solution 9):

$$C_0 = \left(\frac{0.328 U_{\text{sig}}}{0.0688 + (1000 d_{50})} \right)^2. \qquad (3)$$

Note that the coefficients of Eq. (3) are dimensional. Reserved testing data are used as an independent data set to

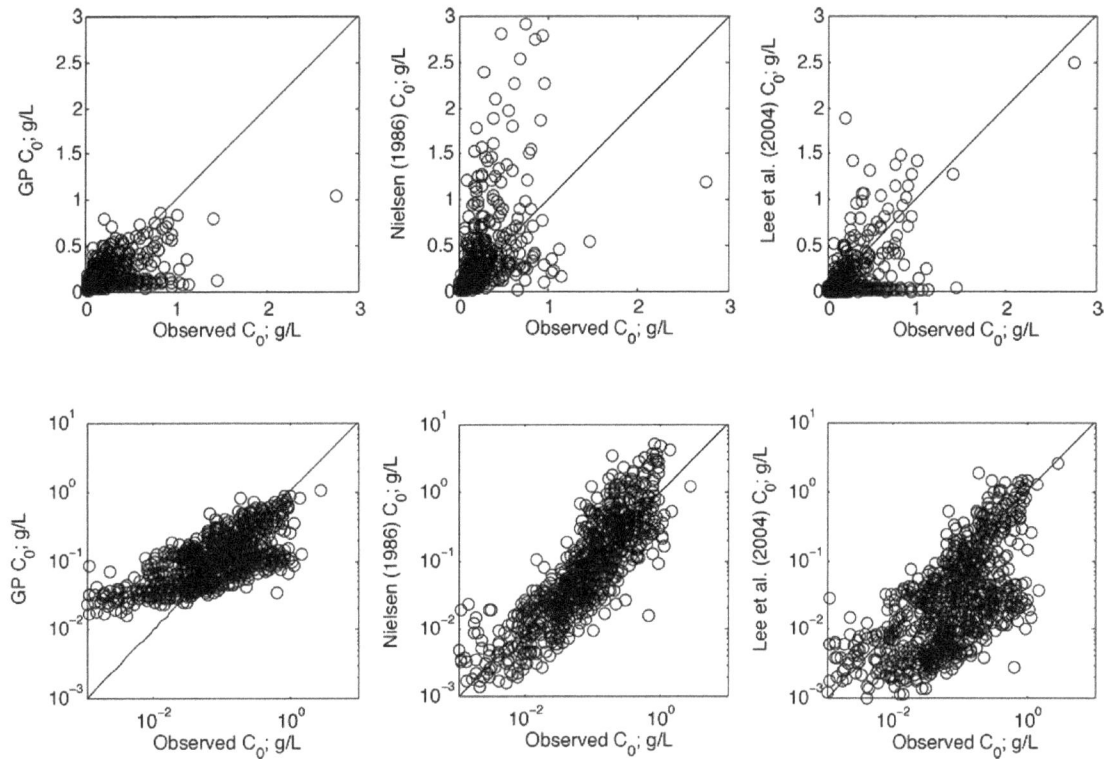

Figure 6. GP predictor of C_0, Nielsen (1986) and Lee et al. (2004) predictor evaluated using only the independent testing data set. Top row shows the predictors in linear space; bottom row shows log–log space.

compare the GP predictor as well as those developed by Nielsen (1986) and Lee et al. (2004): the NRMSE for each predictor is 1.1, 2.6, and 1.3, respectively, and the correlation coefficient is 0.58, 0.58, and 0.57, respectively. Results are shown in Fig. 6. The GP-derived predictor outperforms other predictors based on the NRMSE and is roughly identical to the other predictors based on correlation coefficient. However, we note that at very low concentrations the performance of Eq. (3) deteriorates.

4 Hybrid sorted bedform model overview

We now incorporate this new C_0 predictor into a previously described model of inner-shelf sorted bedforms developed by Coco et al. (2007a) that is based on the initial work of Murray and Thieler (2004). We briefly review the model below; a detailed treatment of the sediment transport relations, hydrodynamic equations and their computational implementation are presented in Coco et al. (2007a). A three-dimensional model domain with periodic horizontal boundary conditions is used to represent a seabed composed of two grain sizes ($d_{coarse} = 0.0005$ m and $d_{fine} = 0.0002$ m; fall velocity $w_{coarse} = 0.07$ m s^{-1} and $w_{fine} = 0.02$ m s^{-1}). An initially flat bed (with slight bathymetric perturbation below 0.01 m) has a bulk composition of 70 % fine sediment and

30 % coarse sediment with individual cells that deviate from this ratio no more than 10 %. The model domain has a plan view size of 500 m × 500 m, a vertical resolution of 0.05 m and a horizontal resolution of 5 m. Small-scale sorted bedforms are modeled in the interest of computational efficiency (observed sorted bedforms range from the scale modeled to kilometers in plan view). In the experiments presented the initial water depth is 9 m, the wave period is 10 s, wave height is 2 m, the mean current is 0.2 m s^{-1}, and the current is unidirectional. Sediment transport, computed independently for each size fraction, occurs only as suspended load and results in the change of bed elevation.

Suspended sediment transport is based on a simplified advection–diffusion framework, neglecting horizontal diffusion and assuming steady-state suspended sediment concentration profiles (Murray and Thieler, 2004; Coco et al., 2007a). The flux of suspended sediment ($q_{susp,s}$), evaluated separately for each size fraction s, is the vertically integrated product of the current velocity profile ($V(z)$) and the suspended sediment concentration profile ($C_s(z)$, where z is the vertical coordinate) combined with a "morphodynamic diffusion" term to incorporate the role of bed slope (∇z) on sediment transport:

$$q_{susp,s} = \int C_s V dz - \gamma_s \frac{1}{5w_s} U_w^5 \nabla z, \tag{4}$$

$$\gamma_s = \gamma_c \frac{16E\rho}{3\pi w_s} C_d, \qquad (5)$$

where U_w is the maximum wave orbital speed at the bed (m s^{-1}; evaluated with linear wave theory), γ_c is the morphodynamic diffusion coefficient, ρ is the density of water, C_d is the drag coefficient, and E is an efficiency factor (set to 0.035). The integration of suspended sediment flux begins at the height where reference concentration is defined. The second term in Eq. (4) represents a "morphodynamic diffusion" term derived from energetics arguments (Bowen, 1980; Bailard, 1981). The calibration parameter in this framework is γ_c and is adjusted to maintain an order of magnitude difference between the two terms on the right-hand side of Eq. (4), similar to the methodology of Calvete et al. (2001). For all experiments in this contribution, $\gamma_c = 0.07$. The role of this parameter is addressed further in the discussion section.

Previous work by Coco et al. (2007a) demonstrates negligible sensitivity to different vertical current profile parameterizations (i.e., descriptions that include current–wave interactions). In these experiments we use a logarithmic vertical current profile:

$$V(z) = \frac{1}{\kappa} U^* \log \frac{z}{z_0}. \qquad (6)$$

where U^* is the shear velocity and κ is the von Kármán constant. The current profile begins at the roughness height z_0, which is related to wave-generated ripples (van Rijn, 1993):

$$z_0 = \frac{1}{30} (2d_{50} + 28\eta\vartheta), \qquad (7)$$

where η is ripple height and ϑ is ripple steepness.

The wave-period-averaged vertical suspended sediment profile above wave-generated ripples (C_s) is calculated based on Nielsen (1992):

$$C_s(z) = C_{0,s} e^{-\frac{w_s z}{\varepsilon_s}} \qquad (8)$$

where $C_{0,s}$ is the near-bed reference concentration for grain size s and ε_s is the vertical sediment diffusivity. Coco et al. (2007a) relied on the formulation developed by Nielsen (1986) to determine the near-bed reference concentration. We use the new GP-derived formulation developed in the previous section. To make the GP-derived C_0 predictor compatible with this model formulation, we assume $U_{sig} = U_w$ and $d_{50} = d_s$, and therefore Eq. (3) becomes

$$C_0 = \left(\frac{0.328 U_w}{0.0688 + (1000 d_s)} \right)^2. \qquad (9)$$

The reference concentration is applied at the height of the ripple crest, as in Coco et al. (2007a). In contrast to the work of Coco et al. (2007a) in this work we evaluate the sediment diffusion coefficient based on the work of Nielsen (1992):

$$\varepsilon_s = \Omega k_s U_w, \qquad (10)$$

$$k_s = 25\eta\vartheta, \qquad (11)$$

where k_s is the equivalent roughness and Ω is a scaling coefficient. Thorne et al. (2009) demonstrated that this parameterization underpredicts vertical sediment diffusivity by a factor of ~ 2 when using the original value of $\Omega = 0.016$ suggested by Nielsen (1992). We therefore set $\Omega = 0.032$. Ripple prediction is performed using a new equilibrium scheme developed using GP by Goldstein et al. (2013):

$$\eta = \frac{0.313 d_0 (1000 d_{50})}{1.12 + 2.18 (1000 d_{50})}, \qquad (12)$$

$$\vartheta = \frac{3.42}{22 + \left(\frac{d_0}{1.12(1000 d_{50}) + 2.18(1000 d_{50})^2} \right)^2}. \qquad (13)$$

We evaluate the mean grain size at each model cell i ($d_{50,i}$) at each time step as

$$d_{50,i} = (1 - B_{\text{coarse},i}) d_{\text{fine}} + B_{\text{coarse},i} d_{\text{coarse}}, \qquad (14)$$

where $B_{\text{coarse},i}$ is the percentage of coarse sediment in the active layer at location i, and d_{fine} and d_{coarse} are the diameter of the fine and coarse fraction, respectively. An active layer vertically restricts sediment–flow interactions. All experiments presented here have a constant active layer thickness of 0.15 m. Sensitivity analyses performed by Coco et al. (2007a) demonstrate that the nature of the sorting feedback is not changed by modification of the active layer thickness.

5 Hybrid sorted bedform model results

The initially flat, well-mixed conditions can be seen in Fig. 7. This configuration is unstable, and sorted bedforms emerge within 50 model days to form the rhythmic segregated pattern shown in Fig. 7. This self-organization is a consequence of the sorting feedback. Compared to previous modeling, bedforms develop more slowly in the hybrid model. The flux of suspended sediment is smaller for the hybrid model because of the change in reference concentration predictor. Bedforms show an abundance of pattern defects (bifurcations, terminations, and "eyes"), and after initial development the pattern continues to develop through time as a result of bedform interactions: a process of coarsening and pattern maturation occurs as defects move through the system and coarse domains merge to form combined features. This leads to fewer pattern elements (coarse domains) seen through time in Fig. 7. Under unidirectional forcing the sorted bedforms migrate slowly in the direction of the current and profile views show that coarse sediment domains are located along the updrift flank. Fine material is advected downdrift and deposited on the lee side of the coarse domains. Coarse sediment is also transported downdrift, but its mobility is limited on upslope surfaces and in fine domains (where wave-generated bedforms

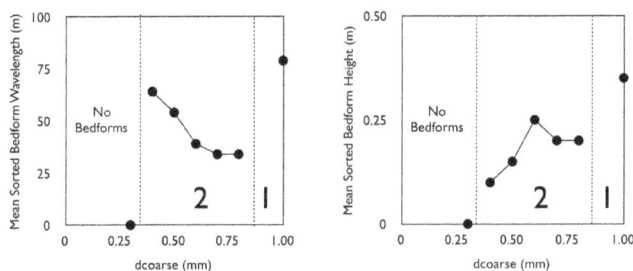

Figure 8. Variations in sorted bedform characteristics (wavelength and height) after 100 days when coarse grain size is held constant. No bedforms appear when the coarse material is too fine. Mode 1 bedforms (long wavelength, larger relief, coarse domains in trough) appear when coarse grain size is large and relatively immobile. Mode 2 bedforms (short wavelength, low relief, coarse domains on updrift flank) appear when coarse grain is between these two limits. No clear pattern was observed after 100 days when $d_{coarse} = 0.9$ mm.

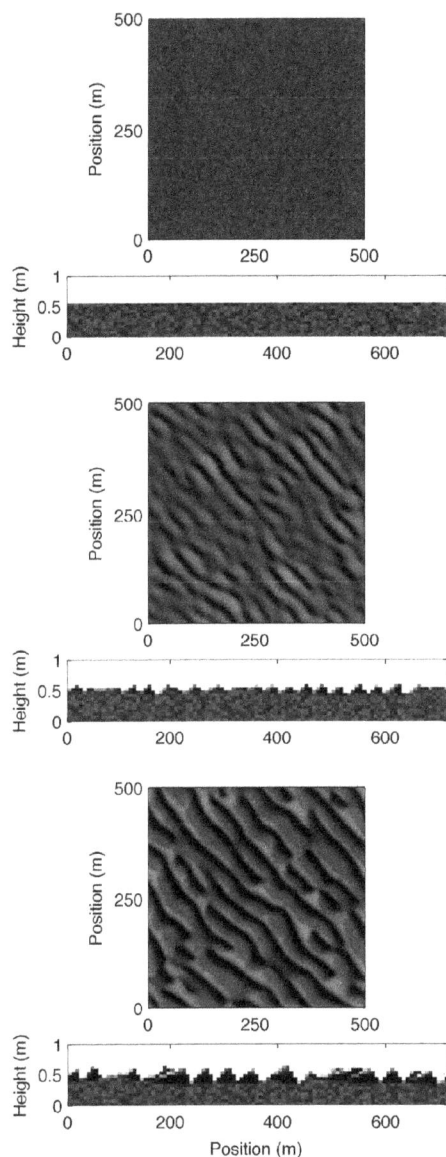

Figure 7. Plan view and profile view of sorted bedform model output (note the vertical exaggeration of profile view). Black and white pixels indicate fine ($d_{fine} = 0.0002$ m) and coarse ($d_{coarse} = 0.0005$ m) sediment, respectively. Current direction is from lower left to upper right and the profile is taken along this axis. The well-mixed and flat initial condition is shown in the top panels. Sorted bedforms appear within 50 days (middle panels) and are well developed by model day 100 (bottom panels). These are mode 2 bedforms; note that coarse domains appear on the updrift flank of the bedforms and wavelength and height are relatively small

are smaller), and therefore it tends to occupy the updrift flank of the bedform only.

Previous work by Coco et al. (2007a) showed the effect of variations in the size of the fine fraction while the coarse fraction size was held constant. In these experiments we evaluate the reverse: fine fraction diameter is held constant ($d_{fine} =$

0.0002 m; $w_{fine} = 0.02$ m s^{-1}), while the coarse fraction diameter is varied between 0.0003 and 0.001 m ($w_{coarse} = 0.04$–0.12 m s^{-1}). This range of sizes for the coarse fraction is similar to the values found in sorted bedform fields worldwide (Coco et al., 2007b).

Results from this analysis can be seen in Fig. 8 (sorted bedform wavelength and height are evaluated after 100 model days). Similar to Coco et al. (2007a) sorted bedforms do not appear when the grain size contrast between size fractions is too small ($d_{fine}/d_{coarse} < 0.5$). When coarse grains range from 0.004 to 0.008 m in diameter, larger coarse sediment tends to cause sorted bedforms to appear faster, decrease in wavelength, and increase in height. Within this range of grain sizes the coarse domain is located along the updrift flank and bedforms migrate in the current direction.

When coarse sediment diameter is larger than 0.008 m, bedforms are strikingly different: bedforms develop faster, wavelengths and height increase significantly, coarse sediment is only present in the trough of the bedform (not along the updrift flank), and bedforms migrate upstream (Fig. 9). This behavior is autogenic in the hybrid sorted bedform model. This pattern configuration is not observed under steady wave climates in the Coco et al. (2007a) model and only appears as the result of specific changes in forcing (Coco et al., 2007b). Bedforms migrate rapidly upcurrent as a result of the decreased mobility of coarse sediment: coarse material is mobile but is not transported significantly up the flank of the bedform and instead remains predominantly in the trough. This is a result of low coarse sediment mobility relative to the downslope transport term in Eq. (4). As fine sediment is advected past the coarse domain in the bedform trough, it can be deposited on the updrift side of the bedform (there is no coarse sediment to prevent its deposition). Along the downdrift side of the bedform the downstream increases in downslope gradient (convex-upward curvature) tends to cause the erosion of bed material and its suspension. This

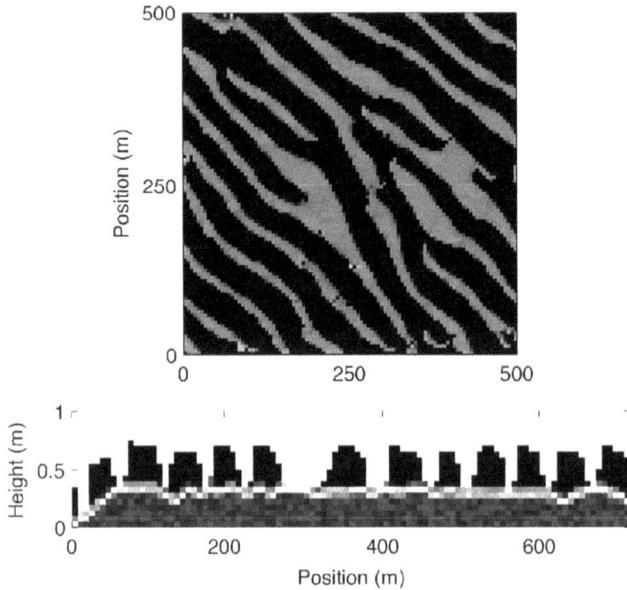

Figure 9. Plan view and profile view of mode 1 sorted bedforms after 50 days. Conditions are identical to Fig. 7 except $d_{coarse} = 0.001$ m. From identical initial conditions sorted bedforms appear much faster and are prominent features by 50 model days. Note that coarse domains appear solely in the bathymetric trough of the bedforms and wavelength and height are relatively large.

suspended material is advected over the coarse domain (the bedform trough) and subsequently deposited on the updrift side of the following (downdrift) bedform.

In profile view a contiguous layer of coarse sediment exists directly below the sorted bedform field (Fig. 9). This coarse layer occurs at the interface between the well-mixed sediment below (the undisturbed model initial conditions) and the reworked sediment above, a consequence of limited coarse sediment mobility and bedform migration (Goldstein et al., 2011). As bedforms migrate, the position of the sorted bedform trough changes. Fine sediment under the bedform trough, once too deep to experience fluid–sediment interactions, is excavated and suspended. Winnowing of fine sediment and coarsening locally in the bedform trough, repeated as the bedforms migrate, results in the development of a horizontal layer of buried coarse sediment, a "sorting lag".

In all results presented here, bedforms migrate and bedform wavelength continues to grow through the model run and wavelength does not saturate. This perpetual coarsening of wavelength under conditions of unidirectional currents is identical to the behavior of the Coco et al. (2007b) and Murray and Thieler (2004) model under unidirectional current forcing. (In the previous results, wavelength coarsening also occurs under the more realistic conditions of an asymmetrically reversing current, although coarsening is more gradual than under a unidirectional current.)

6 Discussion

6.1 GP-derived C_0 predictor

The newly developed C_0 predictor has a nonlinear dependence on d_{50} and U_{sig}, similar to other previous empirical predictors (Nielsen, 1986; Lee et al., 2004). This dependence is not imposed, but instead a result of the data sets used in the GP algorithm.

The GP reference concentration predictor relies on U_{sig}, while the sorted bedform model uses U_w. In the hybrid model we assume $U_{sig} = U_w$, where U_w is calculated from linear wave theory. We direct the reader to other methods available to estimate U_{sig} from surface wave parameters (e.g., Wiberg and Sherwood, 2008). We force the sorted bedform model with a constant monochromatic wave field (height and period) to eliminate the chance that changes in wave characteristics influence the simulated seabed evolution. Therefore the assumption of $U_{sig} = U_w$ does not impact model results shown here.

Ripple geometry was not used as an independent variable in the construction of the C_0 predictor. Dolphin and Vincent (2009) recently suggested that ripple geometry may not aid in the prediction of C_0, contrary to Nielsen (1986) and Green and Black (1999). Though we do not have data to either support or refute this claim, we can offer our results as an example of a well-performing prediction of reference concentration without the explicit inclusion of ripple geometry. However, the nonlinear nature of the reference concentration prediction and the constants embedded within Eq. (3) suggest that ripple configuration may be encoded within the predictor, either as a cause of the nonlinearity or a determinant of the constants.

The C_0 predictor does not explicitly account for near-bed currents that may be important mechanisms for enhancing suspension in sorted bedform fields (e.g., Gutierrez et al., 2005). The C_0 predictor developed in this study is an equilibrium predictor; therefore the role of time variance of C_0 is not addressed (e.g., Vincent and Hanes, 2002). However, the data were collected in burst mode, a technique that involves time averaging. Burst measurements may reduce the effect of some time-dependent processes (e.g., advected clouds of sediment, wave groups, etc.). The GP predictor is constructed solely with regard to the measurement data and is not based on "first principles". Using the independent testing data, the new GP predictor has a lower NRMSE and identical correlation coefficient than the Nielsen (1986) and Lee et al. (2004) predictors; however the GP predictor does not perform well at low concentrations (Fig. 6). The poor performance may be the result of nonlinearities in sediment transport that are not captured by the prediction scheme, noise in the experimental signal at low concentrations, or other as yet unknown reasons. Notably, more energetic conditions are required to move sediment using the GP predictor than compared to the Nielsen (1986) prediction scheme previously used in the sorted bedform model. This result is similar to previous work that suggests the Nielsen (1986) predictor may overestimate reference concentration (Bolaños et al., 2012; Thorne et al., 2002).

6.2 Hybrid sorted bedform model

The hybrid version of the sorted bedform model is able to reproduce the sorting feedback using new parameterizations built from data. The sorting feedback hypothesized by Murray and Thieler (2004) is robust to changes in the mathematical description of the processes in sediment transport and hydrodynamics on the continental shelf, and hybrid model results are comparable to previous modeling efforts (Murray and Thieler, 2004; Murray et al., 2005; Coco et al., 2007a). The behavior of the hybrid model and the Coco et al. (2007a) model under identical hydrodynamic forcing is different because there are quantitative differences between the mathematical description of sediment transport processes. For instance, using the baseline conditions of the Coco et al. (2007a) model the hybrid model produces no sorted bedforms. This is a direct result of changing the C_0 predictor from the Nielsen (1986) formula (which overpredicts sediment transport; Fig. 6) to the new GP-derived C_0 predictor. Changes to the sediment transport formulas prohibit us from directly comparing the three models under identical forcing conditions. Instead we offer this hybrid model as a refined version of the Coco et al. (2007a) model. The hybrid model has additional advantages beyond being more tightly coupled to observational data, most notably in favorable comparison to previous observational work.

Results shown in this contribution use two new prediction schemes based on GP (i.e., ripple morphology and reference concentration). We believe the new ripple prediction scheme of Goldstein et al. (2013) is an improvement over the previous method used in the Coco et al. (2007a) model; however ripples in this model only significantly impact the vertical sediment diffusivity (ε_s) and the roughness height (z_0). The reference concentration, since it sets the magnitude of suspended sediment, is more strongly related to the new behaviors in the model, and as a result we focus our analysis on the reference concentration.

Observational work has previously detected several distinct varieties of sorted bedforms – those with coarse sediment in the trough and those where coarse sediment appears either in the trough and bedforms where coarse sediment is located on the flank (both the updrift and/or downdrift; e.g., Goff et al., 2005; Ferrini and Flood, 2005). Van Oyen et al. (2010, 2011) found that these two pattern configurations appear in linear stability analysis as a result of two separate feedback mechanisms. Mode 1 bedforms (flow–topography feedback), where coarse domains are located in the bedform trough, have a faster growth rate when waves and currents are weaker and result in bedforms with longer wavelength, larger amplitude, and faster migration rates. Mode 2 bedforms (sorting or "roughness" feedback), where coarse grains appear along the updrift and downdrift flank of the bedform, have a faster growth rate when waves and currents are stronger and result in bedforms with smaller wavelengths, smaller heights, and slower migration rates. Yet results from linear stability analysis are applicable only at the scale of an infinitesimal perturbation.

Results from the finite-amplitude hybrid model also show that coarse domains can occur either on the updrift flank of the sorted bedform or collocated with the bedform trough, matching some aspects of previous observation work. However instead of relying on two separate feedback mechanisms, the hybrid model is able to reproduce these two pattern configurations solely via the sorting mechanism. The presence of two distinct pattern modes occurs while current and wave conditions remain unchanged but coarse grain size is varied. When coarse grains are smaller (essentially identical to increasing wave conditions in terms of increasing coarse sediment mobility) bedforms conform to the description of the mode 2 features of Van Oyen et al. (2010, 2011) with smaller features, slower migration rates, and coarse sediment along the updrift flank of bedforms. When coarse grains are larger (essentially identical to decreasing wave conditions in terms of decreasing coarse sediment mobility) bedforms show characteristics of the mode 1 features of Van Oyen et al. (2010, 2011) with larger bedforms, faster migration rates, and coarse sediment in the bedform trough. We again note this behavior occurs solely from a sorting feedback. Bedform wavelength continues to grow in all model results shown here as a result of unidirectional current. However, results in this contribution show that, for any given instant in model time, modeled sorted bedform patterns display relatively homogenous wavelength and height (similar to Coco et al. (2007a) and Murray and Thieler (2004)). Observational work shows sorted bedform fields have a well-defined pattern scale (i.e., a similar height and wavelength throughout the entire bedform field; see the compilation of observed bedform features in Coco et al. (2007b) for more details). It remains unknown whether the well-defined pattern scale of observed sorted bedforms reflects a saturated (steady state) wavelength or the uniformity of bedform wavelength and height at a given moment of pattern evolution.

Several features of mode 1 bedforms in the hybrid model warrant additional attention. Linear stability analysis (Van Oyen et al., 2010, 2011) suggests infinitesimal mode 1 bedforms should migrate in the current direction. The large-scale mode 1 bedforms formed in the finite-amplitude hybrid model show upcurrent migration, which has not previously been observed in field examples of sorted bedforms. Furthermore, mode 1 bedforms develop in the linear stability analysis as a result of a flow–bathymetry feedback (Van Oyen et al., 2010, 2011). The finite-amplitude hybrid model presented here does not parameterize hydrodynamics at small enough scales to permit the development of bedforms as a result of a flow–bathymetry feedback. In contrast to the linear stability analysis, mode 1 bedforms in the hybrid model develop as result of the sorting feedback operating at finite amplitude. Future work with more detailed hydrodynamic parameterizations could shed light on the interplay between flow–bathymetry interactions and the sorting feedback in the mode 1 regime at finite amplitudes. However, these results do suggest that the finite-amplitude hybrid model may be able to capture the dynamics observed in the field. The presence of two distinct pattern modes in the hybrid model is a direct result of incorporating new data-driven parameterizations of the sediment transport process. In this contribution we explore only one specific mechanism that results in mode 1 sorted bedforms, increasing the diameter of the coarse grain size fraction. There are likely other mechanism by which mode 1 bedforms may develop instead of mode 2 bedforms, notably by increasing water depth, decreasing wave forcing, or decreasing current velocity.

There are additional pattern-scale consequences to adjusting the sediment transport formulations. The new C_0 predictor requires energetic conditions to move coarse sediment. This matches the observations and interpretations of Green et al. (2004), Trembanis et al. (2004), and Trembanis and Hume (2011), who suggest that energetic conditions are the only time when the coarse sediment of sorted bedforms is mobile. However lower coarse sediment mobility results in the creation of more pattern defects, a common feature of field examples of sorted bedforms (e.g., Fig. 1). Furthermore, after the work of Werner and Kocurek (1997, 1999), defects have been recognized as a fundamental variable in pattern-scale dynamics of bedforms (Huntley et al., 2008; Maier and Hay, 2009; Goldstein et al., 2011; Skarke and Trembanis, 2011). The presence of additional defects in the hybrid model may exert fundamental controls on pattern evolution.

The hybrid model is able to reproduce sorting feedback and two pattern modes when successfully calibrated. Calibration is accomplished by adjusting the variable γ_c in the morphodynamic diffusion term, Eqs. (4) and (5). The results shown in this contribution have $\gamma_c = 0.07$. The sorting feedback and the development of two sorted bedform pattern modes occur in the range of $\gamma_c = 0.05$–0.08. This range contrasts with the work of Coco et al. (2007a, b), where the γ_c term could be adjusted at least one order of magnitude. This more limited calibration is the result of using multiple nonlinear elements in the construction of the model. Specifically the morphodynamic diffusion term (that γ_c modifies) is highly nonlinear (i.e., $\propto U_w^5$) and is built from energy-based theory (Bowen, 1980; Bailard, 1981). Coco et al. (2007a) relied on a parameterization of C_0 that scaled with U_w^6, effectively scaling the two terms of Eq. (4) in a similar manner. In contrast our new C_0 predictor scales with U_w^2, and therefore does not scale in a similar manner to the morphodynamic term (U_w^5). We suggest that this mismatch, coupled with the strong forcing condition that is required to move sediment in the model (i.e., large U_w), has lead to a smaller permissible parameter space where the morphodynamic term and the

new GP derived predictor are interoperable. We define the permissible parameter space by the scaling argument made previously by Calvete et al. (2001): γ_c should be set to a value that maintains the ratio between the two terms on the right side of Eq. (4) to ~ 1 order of magnitude. If γ_c is set too high, the slope-dependent term is too strong and no bathymetric perturbations develop. If γ_c is set too low, nonphysically steep bathymetric perturbations develop. These results highlight the need to test the Bailard (1981) term in a range of conditions to see whether this description (or others) is valid. Though this morphodynamic diffusion term is often used in morphodynamic models, we could find no instance where this term has been tested in a wide range of conditions.

Finally, the promising results of data-driven parameterizations as components in the sorted bedform model suggests that this approach could be extended to other morphodynamic models and other parameterizations. A specific example from this work is the parameterization of vertical sediment diffusivity (or, more generally, the shape function that described the vertical suspended sediment concentration profile). Recent work has begun to investigate the fast scale dynamics of vertical sediment diffusion over ripples (e.g., Davies and Thorne, 2005; van der Werf et al., 2007; O'Hara Murray et al., 2011) and how best to parameterize this process in large-scale coastal models (Amoudry and Souza, 2011; Amoudry et al., 2013). Traditional equilibrium parameterizations have also been evaluated with newly collected data (e.g., Thorne et al., 2002, 2009; Bolaños et al., 2012). More data, collected in a range of conditions, would enable a data-driven approach to the parameterization of the vertical suspended sediment profile shape.

7 Conclusion

A new predictor for near-bed reference concentration developed using genetic programming performs as well or better than previous empirical parameterizations. However the GP predictor shows poor performance at low concentrations. This predictor is incorporated, along with previously developed predictors for ripple morphology (developed by GP), into a new "hybrid" model of sorted bedforms. This modeling strategy is a viable option when large data sets can be used to construct data-driven subcomponents of a morphodynamic model. The sorting feedback is relatively invariant to changes in hydrodynamic and sediment transport parameterizations. However, the new hybrid model is able to generate novel autogenic behavior in the sorted bedform model: sorted bedform morphology changes when the size of the coarse fraction is modified. This model behavior more closely resembles field observations showing sorted bedform coarse domains that occur in multiple positions along the bedform (however downdrift coarse domains still do not appear in this model)

Acknowledgements. We thank Paula Camus for sharing her clustering algorithm and to Terry Hume (NIWA, NZ) for providing the image of Tairua Beach, NZ, bedforms seen in Fig. 1. E. B. Goldstein thanks "IH Cantabria" for funding during his stay, where this work was started. G. Coco acknowledges funding from the Cantabria Campus Internacional (Augusto Gonzalez Linares program). We thank an anonymous reviewer and Tomas Van Oyen for stimulating comments that improved the manuscript.

Edited by: F. Metivier

References

Amoudry, L. O. and Souza, A. J.: Deterministic coastal morphological and sediment transport modeling: A review and discussion, Rev. Geophys., 49, RG2002, doi:10.1029/2010RG000341, 2011.

Amoudry, L. O., Bell, P. S., Thorne, P. D., and Souza, A. J.: Toward representing wave-induced sediment suspension over sand ripples in RANS models, J. Geophys. Res.-Oceans, 118, 1–15, 2013.

Bailard, J. A.: An energetics total load sediment transport model for a plane sloping beach, J. Geophys. Res., 86, 10938–10954, 1981.

Baptist, M. J., Babovic, V., Uthurburu, J. R., Keijzer, M., Uittenbogaard, R. E., Mynett, A., and Verwey, A.: On inducing equations for vegetation resistance, J. Hydraul. Res., 45, 435–450, 2007.

Bolaños, R., Thorne, P. D., and Wolf, J.: Comparison of measurements and models of bed stress, bedforms and suspended sediments under combined currents and waves, Coast. Eng., 62, 19–30, 2012.

Bowden, G. J., Maier, H. R., and Dandy, G. C.: Optimal division of data for neural network models in water resources applications, Water Resour. Res., 38(2), 2-1–2-11, 2002.

Bowen, A. J.: Simple models of nearshore sedimentation: Beach profiles and long-shore bars, in: The Coastline of Canada, edited by: McCann, S. B., 111 pp., Geol. Surv. of Can., Ottawa, 1980.

Cacchione, D. A., Thorne, P. D., Agrawal, Y., and Nidzieko, N. J.: Time-averaged near-bed suspended sediment concentrations under waves and currents: comparison of measured and model estimates, Cont. Shelf Res., 28, 470–484, 2008.

Calvete, D., Falqués, A., de Swart, H. E., and Walgreen, M.: Modelling the formation of shoreface-connected sand ridges on storm-dominated inner shelves, J. Fluid Mech., 441, 169–193, 2001.

Camus, P., Mendez, F. J., Medina, R., and Cofiño, A. S.: Analysis of clustering and selection algorithms for the study of multivariate wave climate, Coast. Eng., 58, 453–462, 2011.

Coco, G., Murray, A. B., and Green, M. O.: Sorted bedforms as self-organized patterns: 1. Model development, J. Geophys. Res., 112, F03015, doi:10.1029/2006JF000665, 2007a.

Coco, G., Murray, A. B., Green, M. O., Thieler, E. R., and Hume, T. M.: Sorted bedforms as self-organized patterns: 2. Complex forcing scenarios, J. Geophys. Res., 112, F03016, doi:10.1029/2006JF000666, 2007b.

Corzo, G. A., Solomatine, D. P., Hidayat, de Wit, M., Werner, M., Uhlenbrook, S., and Price, R. K.: Combining semi-distributed process-based and data-driven models in flow simulation: a case study of the Meuse river basin, Hydrology and Earth System Sciences, 13, 9, 1619–1634, 2009.

Cummings, D. I., Dumas, S., and Dalrymple, R. W.: Fine-grained versus coarse-grained wave ripples generated experimentally under large-scale oscillatory flow, J. Sediment. Res., 79, 83–93, 2009.

Davies, A. G. and Thorne, P. D.: Modeling and measurement of sediment transport by waves in the vortex ripple regime, J. Geophys. Res., 110, C05017, doi:10.1029/2004JC002468, 2005.

Dolphin, T. and Vincent, C. E.: The influence of bed forms on reference concentration and suspension under waves and currents, Cont. Shelf Res., 29, 424–432, 2009.

Ferrini, V. L. and Flood, R. D.: A comparison of rippled scour depressions identified with multibeam sonar: Evidence of sediment transport in inner shelf environments, Cont. Shelf Res., 25, 1979–1995, 2005.

Gagné, C., Schoenauer, M., Parizeau, M., and Tomassini, M.: Genetic programming, validation sets, and parsimony pressure, in: Genetic Programming, 9th European Conference, EuroGP2006, Lecture Notes in Computer Science, LNCS 3905, edited by: Collet, P., Tomassini, M., Ebner, M., Gustafson, S., and Ekárt, A., Springer, Berlin, Heidelberg, New York, 2006, 109–120, 2006.

Goff, J. A., Mayer, L. A., Traykovski, P., Buynevich, I., Wilkens, R., Raymond, R., Glang, G., Evans, R. L., Olson, H., and Jenkins, C.: Detailed investigation of sorted bedforms, or "rippled scour depressions", within the Martha's Vineyard Coastal Observatory, Massachusetts, Cont. Shelf Res., 25, 461–484, 2005.

Goldstein, E. B., Murray, A. B., and Coco, G.: Sorted bedform pattern evolution: Persistence, destruction and self-organized intermittency, Geophys. Res. Lett., 38, L24402, doi:10.1029/2011GL049732, 2011.

Goldstein, E. B., Coco, G., and Murray, A. B.: Prediction of wave ripple characteristics using genetic programming, Cont. Shelf Res., 71, 1–15, 2013.

Gonçalves, I. and Silva, S.: Balancing Learning and Overfitting in Genetic Programming with Interleaved Sampling of Training Data, in: EuroGP 2013. LNCS, edited by: Krawiec, K., Moraglio, A., Hu, T., Etaner-Uyar, A., and Hu, B., Springer, Heidelberg, 7831, 73–84, 2013.

Gonçalves, I., Silva, S., Melo, J., and Carreiras, J.: Random Sampling Technique for Overfitting Control in Genetic Programming, in: EuroGP 2012. LNCS, edited by: Moraglio, A., Silva, S., Krawiec, K., Machado, P., and Cotta, C., Springer, Heidelberg, 7244, 218–229, 2012.

Green, M.: Introducing ALICE, Water and Atmosphere, 4, 8–10, 1996.

Green, M. O.: Test of sediment initial-motion theories using irregular-wave field data, Sedimentology, 46, 427–441, 1999.

Green, M. O. and Black, K. P.: Suspended-sediment reference concentration under waves: field observations and critical analysis of two predictive models, Coast. Eng., 38, 115–141, 1999.

Green, M. O. and MacDonald, I. T.: Processes driving estuary infilling by marine sands on an embayed coast, Mar. Geol., 178, 11–37, 2001.

Green, M. O., Vincent, C. E., and Trembanis, A. C.: Suspension of coarse and fine sand on a wave-dominated shoreface, with implications for the development of rippled scour depressions, Cont. Shelf Res., 24, 317–335, 2004.

Gutierrez, B. T., Voulgaris, G., and Thieler, E. R.: Exploring the persistence of sorted bedforms on the inner-shelf of Wrightsville Beach, North Carolina, Cont. Shelf Res., 25, 65–90, 2005.

Huntley, D. A., Coco, G., Bryan, K. R., and Murray, A. B.: Influence of "defects" on sorted bedform dynamics, Geophys. Res. Lett., 35, L02601, doi:10.1029/2007GL030512, 2008.

Karunarathna, H. and Reeve, D. E.: A hybrid approach to model shoreline change at multiple timescales, Cont. Shelf Res., 66, 29–35, 2013.

Kitsikoudis, V., Sidiropoulos, E., and Hrissanthou, V.: Derivation of Sediment Transport Models for Sand Bed Rivers from Data-Driven Techniques, in: Sediment Transport Processes and Their Modelling Applications, edited by: Manning, A., ISBN: 978-953-51-1039-2, InTech, doi:10.5772/53432, 2013.

Knaapen, M. A. F. and Hulscher, S. J. M. H.: Regeneration of sand waves after dredging, Coast. Eng., 46, 277–289, 2002.

Knaapen, M. A. F. and Hulscher, S. J. M. H.: Use of a genetic algorithm to improve predictions of alternate bar dynamics, Water Resour. Res., 39, 1231, doi:10.1029/2002WR001793, 2003.

Koza, J. R.: Genetic Programming, On the Programming of Computers by Means of Natural Selection, MIT Press, Cambridge, MA, USA, 1992.

Krasnopolsky, V. M. and Fox-Rabinovitz, M. S.: A new synergetic paradigm in environmental numerical modeling: Hybrid models combining deterministic and machine learning components, Ecol. Model., 191, 5–18, 2006.

Kushchu, I.: An evaluation of evolutionary generalisation in genetic programming, Artif. Intell. Rev., 18, 3–14, 2002.

Lee, G., Dade, W. B., Friedrichs, C. T., and Vincent, C. E.: Examination of reference concentration under waves and currents on the inner shelf, J. Geophys. Res., 109, C02021, doi:10.1029/2002JC001707, 2004.

Maier, I. and Hay, A. E.: Occurrence and orientation of anorbital ripples in near-shore sands, J. Geophys. Res., 114, F04022, doi:10.1029/2008JF001126, 2009.

May, R. J., Maier, H. R., and Dandy, G. C.: Data splitting for artificial neural networks using SOM-based stratified sampling, Neural Networks, 23, 283–294, 2010.

Murray, A. B. and Thieler, E. R.: A new hypothesis and exploratory model for the formation of large-scale inner-shelf sediment sorting and "rippled scour depressions", Cont. Shelf Res., 24, 295–315, 2004.

Murray, A. B., Coco, G., Green, M. O., Hume, T., and Thieler, E. R.: Different approaches to modeling inner shelf sorted bedforms, in: Proceedings of the Conference "River, Coastal and Estuarine Morphodynamics", edited by: Parker, G. and Garcia, M., Taylor and Francis, London, 1009–1015, 2005.

Nielsen, P.: Suspended sediment concentrations under waves, Coast. Eng., 10, 23–31, 1986.

Nielsen, P.: Coastal Bottom Boundary Layers and Sediment Transport, World Sci., Singapore, 1992.

Oates, T. and Jensen, D.: The effects of training set size on decision tree complexity, Proceedings of the Fourteenth International Conference on Machine Learning, Madison, WI, Morgan Kaufmann, 254–262, 1997.

Oates, T. and Jensen, D.: Large datasets lead to overly complex models: An explanation and a solution, Proceedings of the Fourth International Conference on Knowledge Discovery and Data Mining, New York, NY, AAAI Press, 294–298, 1998.

Oehler, F., Coco, G., Green, M. O., and Bryan, K. R.: A data driven approach to predict suspended-sediment reference concentration under non-breaking waves, Cont. Shelf Res., 46, 96–106, 2012.

O'Hara Murray, R. B., Thorne, P. D., and Hodgson, D. M.: Intrawave observations of sediment entrainment processes above sand ripples under irregular waves, J. Geophys. Res., 116, C01001, doi:10.1029/2010JC006216, 2011

O'Neill, M., Vanneschi, L., Gustafson, S., and Banzhaf, W.: Open issues in genetic programming, Genet. Program. Evol. M., 11, 339–363, 2010.

Panait, L. and Luke, S.: Methods for Evolving Robust Programs, in: Genetic and Evolutionary Computation-GECCO 2003, Vol. 2724, no. 66, Berlin, Heidelberg, Springer Berlin Heidelberg, 1740–175, 2003.

Pape, L., Ruessink, B. G., Wiering, M. A., and Turner, I. L.: Recurrent neural network modeling of nearshore sandbar behavior, Neural Networks, 20, 509–518, 2007.

Pape, L., Kuriyama, Y., and Ruessink, B. G.: Models and scales for cross-shore sandbar migration, J. Geophys. Res., 115, F03043, doi:10.1029/2009JF001644, 2010.

Poli, R., Langdon, W. B., and McPhee, N. F.: A field guide to genetic programming, Lulu Enterprises Uk Limited, 2008

Ruessink, B. G.: Calibration of nearshore process models: Application of a hybrid genetic algorithm, J. Hydroinformatics, 7, 135–149, 2005.

Schmidt, M. and Lipson, H.: Distilling free-form natural laws from experimental data, Science, 324, 81–85, 2009.

Schmidt, M. and Lipson, H.: Eureqa (Version 0.98 beta) [Software], available at: http://www.eureqa.com/, 2013.

Skarke, A. and Trembanis, A. C.: Parameterization of bedform morphology and defect density with fingerprint analysis techniques, Cont. Shelf Res., 31, 1688–1700, 2011.

Thieler, E. R. Foster, D. S., Himmelstoss, E. A., and Mallinson, D. J.: Geologic framework of the northern North Carolina, USA inner continental shelf and its influence on coastal evolution, Mar. Geo., 348, 113–130, 2014.

Thorne, P. D., William, J. J., and Davies, A. G.: Suspended sediments under waves measured in a large scale flume facility, J. Geophys. Res., 107, 4.1–4.16, 2002.

Thorne, P. D., Davies, A. G., and Bell, P. S.: Observations and analysis of sediment diffusivity profiles over sandy rippled beds under waves, J. Geophys. Res., 114, C02023, doi:10.1029/2008JC004944, 2009.

Trembanis, A. C. and Hume, T. M.: Sorted bedforms on the inner shelf off northeastern New Zealand: spatiotemporal relationships and potential paleo-environmental implications, Geo-Mar Letters, 31, 203–214, 2011.

Trembanis, A. C., Wright, L. D., Friedrichs, C. T., Green, M. O., and Hume, T.: The effects of spatially complex inner shelf roughness on boundary layer turbulence and current and wave friction: Tairua embayment, New Zealand, Cont. Shelf Res., 24, 1549–1571, 2004.

van der Werf, J. J., Doucette, J. S., Donoghue, T., and Ribberink, J. S.: Detailed measurements of velocities and suspended sand concentrations over full?scale ripples in regular oscillatory flow, J. Geophys. Res., 112, F02012, doi:10.1029/2006JF000614, 2007.

van Maanen, B., Coco, G., Bryan, K. R., and Ruessink, B. G.: The use of artificial neural networks to analyze and predict alongshore sediment transport, Nonlin. Processes Geophys., 17, 395–404, 2010, http://www.nonlin-processes-geophys.net/17/395/2010/.

Van Oyen, T., de Swart, H. E., and Blondeaux, P.: Bottom topography and roughness variations as triggering mechanisms to the formation of sorted bedforms, Geophys. Res. Lett., 37, L18401, doi:10.1029/2010GL043793, 2010.

Van Oyen, T., de Swart, H. E., and Blondeaux, P.: Formation of rhythmic sorted bed forms on the continental shelf: an idealised model, J. Fluid Mech., 684, 475–508, 2011.

van Rijn, L. C.: Principles of Sediment Transport in Rivers, Estuaries and Coastal Seas, Aqua, Amsterdam, 1993.

Vincent, C. E. and Green, M. O.: The control of resuspension over megaripples on the continental shelf, in: Proceedings of Coastal Sediments 99, New York, USA, ASCE, 269–280, 1999.

Vincent, C. E. and Hanes, D. M.: The accumulation and decay of near-bed suspended sand concentration due to waves and wave groups, Cont. Shelf Res., 22, 1987–2000, 2002.

Werner, B. T. and Kocurek, G.: Bedform dynamics: Does the tail wag the dog?, Geology, 25, 771–774, 1997.

Werner, B. T. and Kocurek, G.: Bedform spacing from defect dynamics, Geology, 27, 727–730, 1999.

Wiberg, P. L. and Sherwood, C. R.: Calculating wave-generated bottom orbital velocities from surface-wave parameters, Comput. Geosci., 34, 1243–1262, 2008.

Yoon, H.-D., Cox, D. T., and Kim, M.: Prediction of time-dependent sediment suspension in the surf zone using artificial neural network, Coast. Eng., 71, 78–86, 2013.

9

The role of velocity, pressure, and bed stress fluctuations in bed load transport over bed forms: numerical simulation downstream of a backward-facing step

M. W. Schmeeckle

School of Geographical Sciences and Urban Planning, Arizona State University,
Tempe, Arizona, USA

Correspondence to: M. W. Schmeeckle (schmeeckle@asu.edu)

Abstract. Bed load transport over ripples and dunes in rivers exhibits strong spatial and temporal variability due to the complex turbulence field caused by flow separation at bedform crests. A turbulence-resolving flow model downstream of a backward-facing step, coupled with a model integrating the equations of motion of individual sand grains, is used to investigate the physical interaction between bed load motion and turbulence downstream of separated flow. Large bed load transport events are found to correspond to low-frequency positive pressure fluctuations. Episodic penetration of fluid into the bed increases the bed stress and moves grains. Fluid penetration events are larger in magnitude near the point of reattachment than farther downstream. Models of bed load transport over ripples and dunes must incorporate the effects of these penetration events of high stress and sediment flux.

1 Introduction

The details of turbulent flow over dunes and ripples in rivers and oceans have been described by field and laboratory experiments (see Best, 2005, for an extensive review), as well as high-resolution, turbulence-resolving numerical simulations (Shimizu et al., 1999, 2001; Nelson et al., 2006; Zedler and Street, 2001; Omidyeganeh and Piomelli, 2011; Grigoriadis et al., 2009; Stoesser et al., 2008; Chang and Constantinescu, 2013). However attempts to couple turbulence to the transport of sediment over bedforms have usually relied on empirical formulas, wherein the sediment flux is either a direct function of boundary shear stress or indirectly through entrainment rate and deposition rate formulas (Niemann et al., 2011; Nguyen and Wells, 2009; Giri and Shimizu, 2006; Chou and Fringer, 2010; Paarlberg et al., 2009; Kraft et al., 2011) (although see Nabi et al., 2013, and Penko et al., 2013). Unlike suspended sediment fields, experiments detailing the spatiotemporal pattern of bed load transport over ripples and dunes have not been reported.

Grass and Ayoub (1982) hypothesized that the mean bed load sediment flux could be calculated as the integral of the probability density of bed stress due to turbulence times the sediment flux as a function of stress. They further hypothesized that the bed stress distribution could be determined from the distribution of near-bed downstream velocity. The experiments of Nelson et al. (1995) simultaneously measuring sediment flux and near-bed fluid velocity over a flat bed and downstream of a backward-facing step showed that the relationship between near-bed fluid Reynolds stress and bed load transport was not simple, and the spatially varying distribution of velocity fluctuations relative to the shear velocity must be considered in formulating transport relationships over ripples and dunes. They also found that there was not a simple monotonic relationship between instantaneous, downstream, near-bed velocity and sediment flux. Specifically, longer duration positive fluctuations of near-bed velocity were found to transport more sediment per unit time than shorter duration events. As such, the hypothesis of Grass and Ayoub (1982) needs significant modification to be useful downstream of separated flows.

Experimental measurement of turbulence is often limited to a time series of fluid velocity components at a single point or, more rarely, several points. In such instances, the detection of spatially and temporally evolving turbulence structures is difficult. Quadrant analysis has been used to detect certain types of turbulence structures (Lu and Willmarth, 1973; Bogard and Tiederman, 1986). Quadrant analysis involves joint examination of the fluctuating components of fluid velocity in the downstream, x, and bed-perpendicular, z, directions. u' and w' are the downstream and bed-perpendicular fluctuating components of fluid velocity. With u' and w' measurements plotted on a two-dimensional graph, the first quadrant (Q1) is a point with $u' > 0$ and $w' > 0$; this is also known as an outward interaction. Quadrant 2 events (Q2, $u' < 0$ and $w' > 0$) are termed bursts or ejections, quadrant 3 events (Q3, $u' < 0$ and $w' < 0$) are called inward interactions, and quadrant 4 events (Q4, $u' > 0$ and $w' < 0$) are known as sweeps. Q2 and Q4 events transport downstream momentum toward the bed, and are thus positive contributions to the Reynolds stress component, $-\rho\overline{u'w'}$, whereas Q1 and Q3 events are negative contributions.

Schmeeckle (2014) used a coupled turbulence-resolving numerical model of flow and a particle model of sediment motion to simulate the interaction between turbulence and sediment movement over a flat bed. Vortical structures embedded within broader sweep structures were found to bring fluid into and out of the bed, and were sites of sediment entrainment and transport. In this article I extend the model of Schmeeckle (2014) and Furbish and Schmeeckle (2013) to the case of bed load transport downstream of a backward-facing step, largely matching the experiments of Nelson et al. (1995). Quadrant analysis is extended to include sediment flux, bed stress, and fluid pressure. Flow over a backward-facing step, like that over bed forms, causes flow separation, but does not have the complicating effect of flow acceleration by an upstream sloping bed.

2 Methodology

The fluid is modeled by the large eddy simulation (LES) technique in which the spatially filtered Navier–Stokes equations are integrated using the finite-volume method. The equations of motion of each sediment grain are integrated over time using the distinct element method (DEM). The sand grains are assumed to be spheres, and forces between particles are calculated when grain boundaries overlap. The LES and DEM models are coupled in momentum. The flow field is interpolated to the particle centers and used to derive fluid forces on the particles. In turn, each fluid force acting on the particles is given as a resistance term to the fluid momentum equations at the fluid cell containing the center of the particle. Only drag, pressure gradient, and buoyancy forces are included as fluid–particle forces. The bed of particles is about three to four grain diameters thick above the lower fluid boundary, and non-moving particles of the particle bed rapidly damp the fluid velocity. The details of the numerical model reported here are the same as reported in Schmeeckle (2014).

The flow magnitude, step height (x_{step}), particle diameter (D) and density (ρ_s), and flow depth of the numerical simulations are specified to nearly match the experiments of Nelson et al. (1995). The computational domain extends 0.2 m upstream of the 0.04 m backward-facing step and 1.2 m (30 step heights) downstream of the step, and 0.1 m (2.5) step heights across stream. The fluid domain in the vertical dimension extends 0.16 m above the step, and downstream of the step the vertical domain is 0.2025 m. At rest, the topmost particles are roughly 0.0025 m above the lower wall. Thus, the flow height downstream of the step is about 0.2 m from the bed of particles to the top of the numerical domain. The grid used in this study is a structured mesh of 4 655 000 hexagonal cells. The grid is evenly spaced in the downstream and cross-stream directions. The downstream and cross-stream grid lengths are 0.002 and 0.00143 m, respectively. The vertical grid spacing is nonuniform, with smaller grid cells containing the particles and near-bed flow. The vertical grid dimension is also significantly reduced in a zone containing the separation bubble shear layer. The vertical dimension of cells containing particles at the bottom of the numerical domain is 0.00025 m. The downstream and cross-stream grid dimensions are slightly larger than the diameter of the particles, but the particle diameters are about 3.6 times larger than the vertical grid dimension. As such there is rarely more than one particle in a grid cell.

If fluid and particles are coupled in mass and momentum, the fluid solver becomes unstable when the particles are of the same size as the fluid grid cells. Each particle is treated as a source of resistance in the cell where the particle(s) center is located. It is possible to smooth the effect of a particle over a broader number of cells to achieve a stable algorithm, but this smooths the sharp interface between the bed and the overlying relatively sediment-free fluid. Smoothing of the bed interface was deemed not appropriate, because most of the fluid momentum is damped within a grain diameter below the bed interface. Here, fluid and particles are coupled in momentum but not in mass. The flow around each particle is not directly modeled; only the damping of flow by the integrated force of each particle is modeled. Thus, the flow separation and turbulence generated by flow around particles and in the interstices of particles is not modeled. Further, the calculated pressure field within the bed may be different than reality, because the fluid continuity equation does not account for the reduced volume of fluid within the bed. It is difficult to predict how these assumptions degrade the fidelity of the simulations. However, the integrated momentum effect of each particle on each fluid cell (and vice versa) is calculated in the model and should lead to relatively accurate fluid simulations at the scale of the grid.

The domain of the DEM model begins at the step in the x direction but otherwise coincides with the fluid domain boundaries. The particle domain is periodic in the downstream and cross-stream directions. The diameters of the 415 000 particles in this simulation are randomly drawn from a normal distribution with a median of 0.9 mm and a standard deviation of 0.1 mm. The diameters are varied to avoid close packing arrangement of the bed during the simulation. The particle parameters for the DEM model are the same as in Schmeeckle (2014) except that the Young's modulus is increased to 5×10^6 Pa.

Boundary conditions for fluid velocity are no shear at the upper boundary, periodic conditions in the cross-stream direction, and zero gradient at the outlet. The no-slip condition is applied at the lower boundary, but the fluid velocity becomes negligible before reaching this boundary because of the presence of a bed of particles above it. The inlet boundary condition is specified as the velocity 0.15 m downstream of the inlet. This is similar to a periodic boundary condition wherein the inlet and outlet are the same velocity, but the recycled velocity is taken before the backward step, thus ensuring fully developed boundary layer turbulence upstream of the backward-facing step.

3 Results

Prior to recording simulation results, the flow reached dynamic equilibrium after about 30 s of simulated time. Results reported here are for 20 s of simulated time. This length of time provided adequate statistics, but it was not so long that the bed of particles developed bedforms and areas of the bed without sufficient numbers of particles. However, bed elevation changes of one to two particle diameters were apparent in response to the passage of individual turbulent events. The position, velocity, and fluid force of each particle of known diameter is recorded at 40 Hz. Similarly, the fluid velocity and pressure are saved simultaneously at 40 Hz along two near-bed horizontal slices at $z = 1$ mm and $z = 5$ mm and along a slice perpendicular to the cross-stream at the center of the numerical domain, $y = 0.05$ m. The lower boundary of the fluid and particle domain is at $z = -0.0025$ m and the topmost particles of the bed at rest are at approximately $z = 0$. A local depth-integrated downstream sediment flux, q_{sx}, is calculated by summing the product of each particle volume and velocity that is found in a 0.01×0.01 m horizontal area of the bed and then dividing by the local bed area (i.e., division by 0.0001 m^2). Downstream bed shear stress, τ_{bx} is calculated by summing the downstream component of fluid force acting on all particles with centers contained in the same 0.01×0.01 m horizontal grid, and then dividing by the local bed area. The saved fluid velocity and pressure at $z = 1$ mm are extracted at points in the center of the grid of local bed areas used to calculate boundary stress and sediment flux. In this manner, the data examined in this article

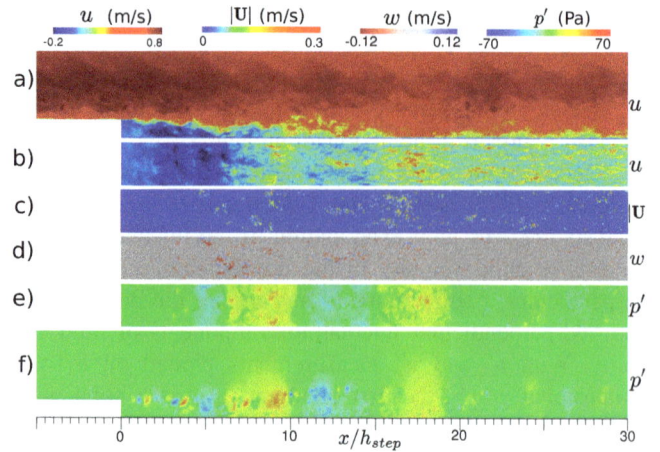

Figure 1. Visualization of the downstream fluid velocity, u; vertical velocity, w; particle velocity magnitude, $|\mathbf{U}|$; and fluid pressure fluctuation, p'; at an instant in time. (a) Downstream velocity on a vertical slice at the middle of the cross-stream domain. (b) Downstream velocity on a horizontal slice at $z = 1$ mm. (c) Particle velocity magnitude. (d) Vertical velocity on a horizontal slice at $z = 1$ mm. (e) Fluid pressure fluctuation on a horizontal slice at $z = 1$ mm. (f) Fluid pressure at the middle of the cross-stream domain. An animation associated with this figure can be found in the Supplement.

are of 20 s of simulated time at a rate of 40 Hz at 12 000 (1200 downstream by 10 cross-stream) points of fluid velocity and pressure, boundary shear stress, and depth-integrated downstream sediment flux.

It should be noted that the bed stress, as defined here, is the sum particle force per bed area. It is not the near-bed Reynolds stress component, $-\rho \overline{u'w'}$. Averaged over sufficient time, these two quantities are equivalent by a balance of forces. However, at a particular instant in time, there are fluid accelerations, and $\tau_{bx} \neq -\rho u'w'$, except by coincidence.

Simultaneous visualization of the fluid pressure fluctuation, $p' = p - \overline{p}$; downstream fluid velocity, u; and particle velocity magnitude, $|\mathbf{U}|$, reveals a positive covariance of particle motion with near-bed, downstream fluid velocity and fluid pressure (Fig. 1). Low-frequency, cross-channel variations in pressure are apparent in Fig. 1, which were also noted in the direct numerical simulations of Le et al. (1997). Figure 1 shows that positive pressure fluctuations are associated with both large near-bed downstream fluid and particle velocity magnitude. There are small areas of the bed with large negative vertical velocity (red areas of Fig. 1d). Fluid that penetrates the bed leads to neighboring areas where fluid exits the bed (blue areas of Fig. 1d). These large fluctuations in vertical velocity are associated with significant sediment motion (Fig. 1c).

Figure 2 shows the temporal statistics (mean, 10th, and 90th percentile) vs. downstream distance for u, w, p, τ_{bx}, and q_{sx}. The position of the point of reattachment is plotted at a vertical dotted line in the five plots of Fig. 2. Figure 2a

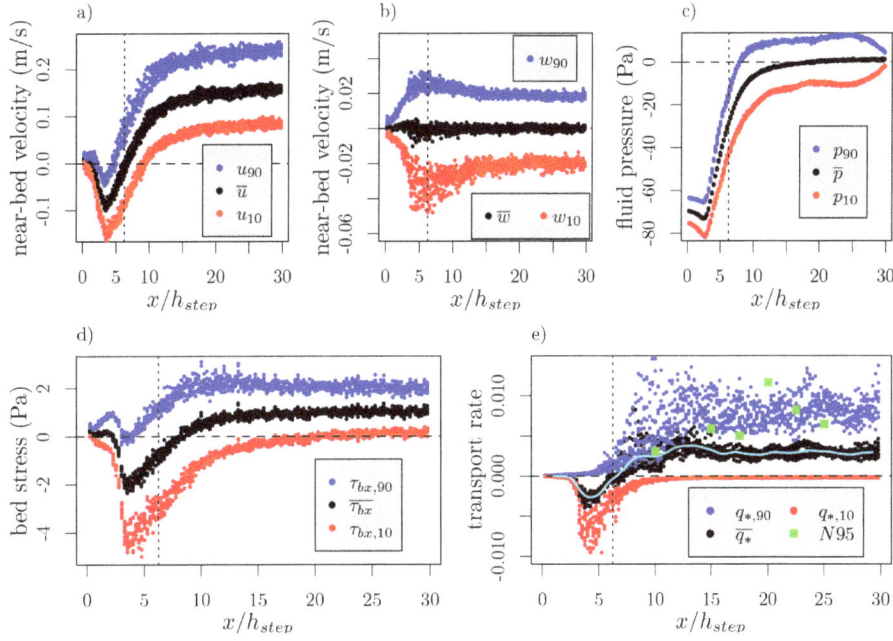

Figure 2. Temporal mean and the 10th and 90th percentile flow and transport parameters are plotted against distance downstream relative to step height, x/h_{step}. **(a)** Downstream fluid velocity at $z = 1$ mm, **(b)** vertical fluid velocity at $z = 1$ mm, **(c)** fluid pressure at $z = 1$ mm, **(d)** bed shear stress, and **(e)** depth-integrated downstream sediment flux. N95 is the measured sediment flux of Nelson et al. (1995). A smoothed line of a moving average of $\overline{q_*}$ of all points within 0.025 m upstream and downstream is also shown.

shows that the mean near-bed velocity (at $z = 1$ mm) increases rapidly near the point of reattachment, and it increases, albeit much less rapidly, all the way to the downstream outlet. Interestingly, the difference between the 90th and 10th percentile of velocity is larger near the outlet than in the reattachment zone. However, the difference between the 90th and 10th percentile of bed stress (Fig. 2d) is smaller near the outlet. The 10th percentile transport rates are essentially zero (or slightly negative) downstream of flow reattachment (Fig. 2e). The mean sediment transport rate (Fig. 2e) increases rapidly downstream of reattachment and does not show a peak in transport at 20 step heights as do the results of Nelson et al. (1995). However, they suggested that the peak could be the result of sampling error.

The fluid pressure (Fig. 2c) rises rapidly from the recirculation region through the zone of reattachment (from about $x/h_{step} = 3$ to 7). The largest magnitude of this upstream-directed pressure gradient is about 400 Pa m^{-1}, which leads to a stress of about -0.5 Pa at $x/h_{step} = 5$. This "pressure gradient stress" is about one-third to one-quarter of the negative bed stress in the recirculation and reattachment zone. However, this stress is distributed throughout the bed of particles, and the resulting pressure force on individual grains is more than an order of magnitude smaller than is required to entrain the topmost grains that are able to move.

While Fig. 1 qualitatively shows the spatial covariance of some of the fluid and particle variables, Fig. 3 shows some of the significant temporal correlation pairs of variables u, $|w'|$,

τ_{bx}, and p. The absolute magnitude of the vertical velocity fluctuation, $|w'|$ is used rather than w because transport was found to peak when the fluctuations of vertical velocity were high. It is perhaps unsurprising that u' is positively correlated with τ_{bx} and q_{sx}, but Fig. 3 also shows the positive correlation with fluid pressure, p.

4 Discussion

4.1 Permeable splat events

Given that the force on bed grains results primarily from fluid drag, it is somewhat paradoxical that the temporal variance in the bed stress is much larger near reattachment, despite the smaller variance in downstream velocity at reattachment, relative to farther downstream. This apparent paradox is due to the fluctuations in vertical velocity being much larger near the point of reattachment than farther downstream (Fig. 2b). A large negative vertical velocity brings high downstream fluid velocity into the bed, thus creating peak bed stresses. Consider the plots in Fig. 2 between about $x/h_{step} = 8$ and $x/h_{step} = 12$. Figure 2a shows that the 90th percentile of the near bed velocity continues to increase downstream. However, Fig. 2d shows that the 90th percentile in boundary shear stress is as high or higher than farther downstream. Figure 2e shows that the transport, similarly, is as large as farther downstream, despite having a lower near-bed downstream velocity. Figure 2b shows that the magnitude of the 10th percentile

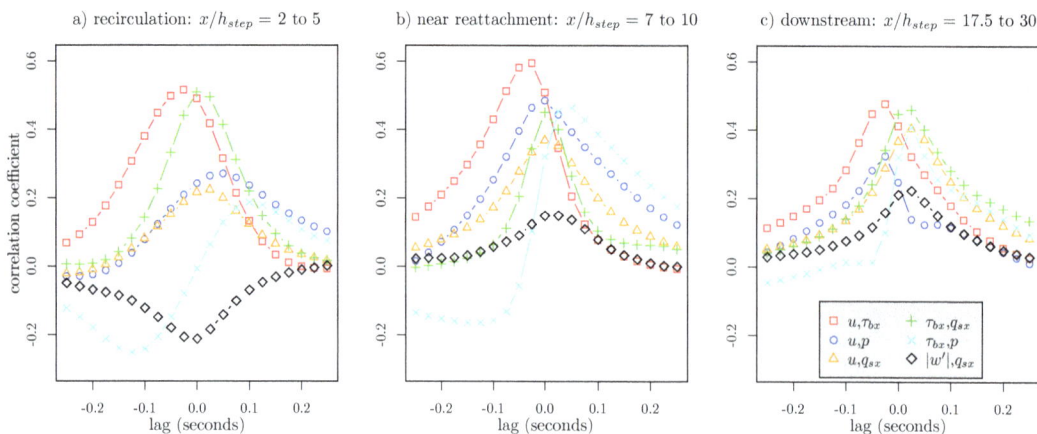

Figure 3. Correlation coefficients vs. time lag for various pairs of flow and transport variables as indicated in the legend in (**c**). The lag is of the second variable relative to the first variable in the legend shown in (**c**).

of vertical velocity is larger in this zone than farther downstream. These large negative vertical velocities bring high-momentum fluid into the bed, increasing the force on grains and causing transport.

When a localized volume of fluid approaches and impinges on an impermeable boundary, the boundary-normal velocity must stagnate, and the fluid gets redirected to move parallel to the wall. Perot and Moin (1995) refer to these wall impingements as "splat events", and Stoesser et al. (2008) note the occurrence of splats near flow reattachment in their simulations of turbulence over dunes. In the simulations reported here, the bed is a permeable boundary, and splats can penetrate the bed. To satisfy fluid continuity, infiltration of the bed by a splat must be accompanied by exfiltration of the bed surrounding the splat. Permeable splat events are apparent near reattachment in Fig. 1d (areas of intense red and blue) and the dynamics of the splat events are apparent in the Supplement animation of Fig. 1. Schmeeckle (2014) remarked that significant entrainment of bed load grains occurs on the boundaries between areas of bed infiltration and exfiltration.

The very large negative stresses in the recirculation region which peak at about $x/h_{\text{step}} = 4$ in Fig. 2d are due to a negative mean vertical velocity, $\overline{w}$, at the particle bed (Fig. 2b) and the large negative vertical velocity fluctuations, w_{10}. There is a mean penetration of fluid into the bed, and there are permeable splat events. The downstream fluid velocity is also negative in the bed of particles due to the adverse pressure gradient. However, once again, the drag forces produced on the grains in this region are more broadly distributed through the bed, in contrast to the bed well downstream of flow reattachment, where the boundary shear stress is concentrated on only the topmost particles. This set of conditions also explains why the mean transport rate and near-bed downstream velocity are negligible even though Fig. 2d

shows that the mean boundary shear stress is negative at the point of reattachment.

4.2 Quadrant analysis

Recall that the simulation data were collected for u, w, q_{sx}, p, and τ_{bx} simultaneously at a horizontal grid of points. In Fig. 4 all of the data were aggregated from all points downstream of $x/h_{\text{step}} = 12.5$. Figure 4a shows the frequency of u'–w' paired bins, and the predominance of burst and sweep events is apparent. In Fig. 4b, q_{sx} is summed for each u'–w' bin. The bins are then normalized by dividing all bins by the bin with the maximum sum of q_{sx}. Figure 4b shows that most of the transport (about 80 %) takes place during sweeps and outward interactions. This result is consistent with Nelson et al. (1995). In Fig. 4c, τ'_{bx} is summed for each u'–w' bin, and each bin is normalized by the largest magnitude bin. Percentages for each quadrant in Fig. 4c are given by the sum $\Sigma \tau'_{\text{bx}}$ and divided by the total deviation, $\Sigma |\tau'_{\text{bx}}|$. Sweeps are associated with high bed stress, and bursts are associated with low bed stress. The pressure deviation is summed in each u'–w' bin in Fig. 4d and normalized by the magnitude of the bin with the largest magnitude. Percentages for each quadrant are given by the sum $\Sigma p'$ and divided by the total deviation, $\Sigma |p'|$. Sweeps and outward interactions are associated with high-pressure events and bursts and inward interactions are associated with low-pressure events.

The spatial correlation between sweeps and outward interactions and between bursts and inward interactions is apparent in Fig. 5. Areas of the bed occupied predominantly by sweeps and outward interactions are also areas with high fluid pressure, large particle forces, and large sediment fluxes. Conversely, bursts and inward interactions are associated low pressure, small particle forces, and small sediment fluxes. Sweeps and inward interactions occur together when a broad volume of fluid moves toward the bed, bringing with it high downstream velocity. Such a situation is apparent in

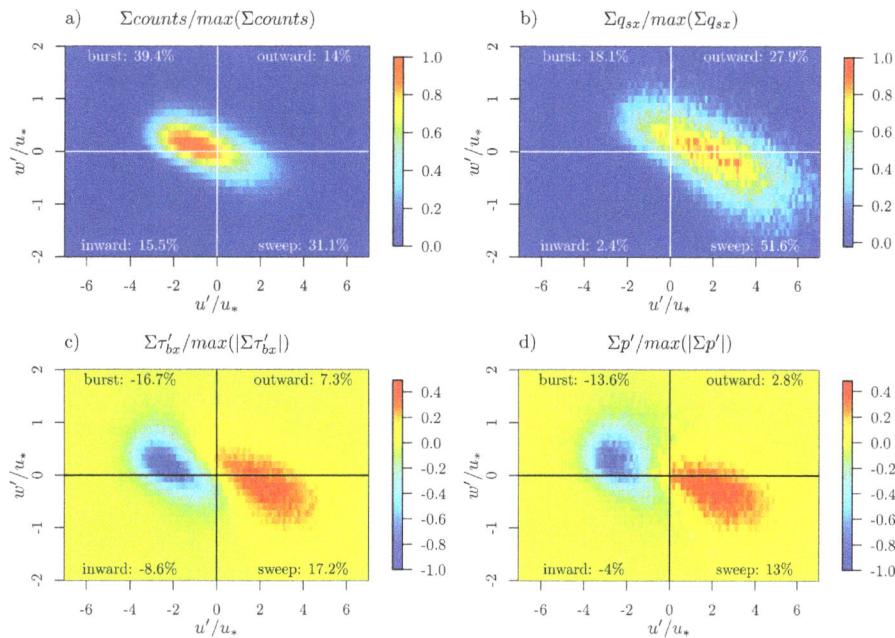

Figure 4. Flow and transport data binned by downstream and vertical velocity fluctuation pairs, $u'-w'$. Data are aggregated for all points downstream of $x/h_{\text{step}} = 12.5$. (**a**) Bin counts normalized by the largest bin. The total percentage of counts for each quadrant are shown. (**b**) The sum of downstream sediment transport in each bin, Σq_{sx}, normalized by the bin with the largest transport sum. The percentage of transport for each quadrant is shown. (**c**) The sum of the downstream bed stress fluctuation for each bin, normalized by the magnitude of the bin with the largest magnitude. Percentages shown in each quadrant are for the sum of the stress fluctuation, $\Sigma \tau'_{\text{bx}}$, divided by the total absolute deviation, $\Sigma |\tau'_{\text{bx}}|$. (**d**) The sum of the fluid pressure fluctuation for each bin, $\Sigma p'$, normalized by the magnitude of the bin with the largest magnitude. Percentages shown in each graph are for the sum of the pressure fluctuation divided by the total absolute pressure deviation, $\Sigma |p'|$.

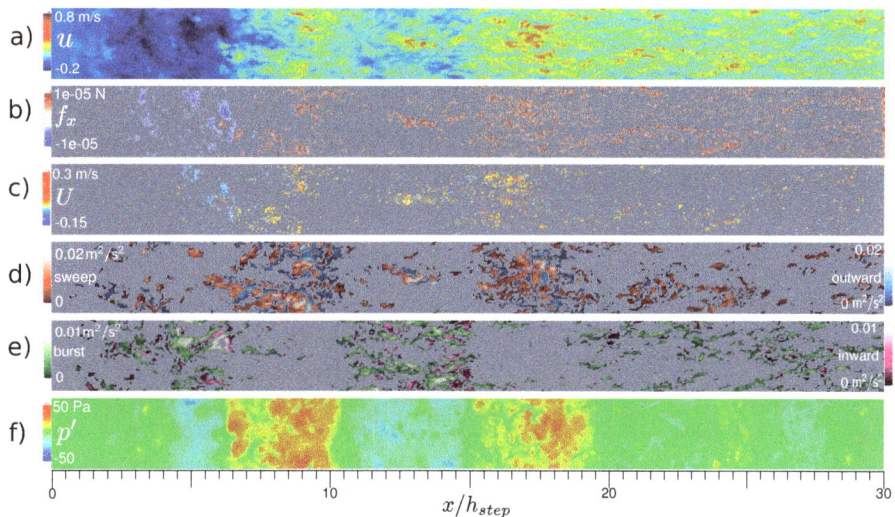

Figure 5. Flow, velocity quadrant, and transport variables at a time instant. (**a**) Downstream velocity at $z = 1\,\text{mm}$ as shown in Fig. 1. (**b**) Downstream force, f_x, on particles. Particles with $|f_x| < 2 \times 10^{-6}\,\text{N}$ are not shown. (**c**) Downstream particle velocity. Particles with $|U| < 0.007\,\text{m s}^{-1}$ are not shown. (**d**) Sweeps and outward interactions at $z = 5\,\text{mm}$. Areas with $|u'w'| < 0.0004\,\text{m}^2\,\text{s}^{-2}$ are not shown. (**e**) Bursts and inward interactions at $z = 5\,\text{mm}$. Areas with $|u'w'| < 0.0004\,\text{m}^2\,\text{s}^{-2}$ are not shown. (**f**) Near-bed pressure fluctuation at $z = 1\,\text{mm}$ as shown in Fig. 1. An animation associated with this figure can be found in the Supplement.

Figs. 1 and 5 at $x/h_{\text{step}} \approx 17$. When a broad sweep impinges on the permeable bed, there is infiltration and exfiltration at spatial scales smaller than the broader sweep structure. Areas of exfiltration are apparent as outward interactions.

Downstream-elongated structures of high- and low-speed fluid begin to emerge downstream of flow reattachment (Fig. 5a) (as also noted by Le et al., 1997). These emerging streaks also produce streaks of high particle forces (Fig. 4b) and particle motion (Fig. 4c).

5 Conclusions

Temporally averaged bed stress is not sufficient to specify the rate of bed load transport downstream of separated flow (compare Fig. 2d and e). Most of the transport takes place at high-stress events that are associated with both high downstream velocity and high-magnitude vertical velocity events (Fig. 4b). The temporal distribution of bed stress is broader near flow reattachment than farther downstream (Fig. 4d), even though the temporal distribution of near-bed downstream velocity is less broad near flow reattachment than downstream. "Near-bed" and "in the bed" fluid velocities are different. In this study near-bed was specified at $z = 1$ mm, which is about one sand grain diameter above the top of the bed. Negative vertical velocity events (splats) bring high downstream momentum fluid into the bed, and those bed penetration events are stronger near flow reattachment (Fig. 2b). Consequently, the 90th percentile of stress and the mean sediment flux reaches a peak in a relatively short distance downstream of reattachment. This provides a probable explanation of the findings of Nelson et al. (1995) that instantaneous, near-bed downstream velocity was not sufficient to specify the instantaneous sediment flux; the actual force on bed particles is also dependent on the penetration of turbulence structures into the bed. The upstream inclination of the stoss of bed forms, relative to the flat bed considered here, is expected to increase the intensity of fluid penetration events near flow reattachment.

Acknowledgements. This material is based upon work supported by the National Science Foundation under grant no. 1226288. All simulations were performed at the Arizona State University Advanced Computing Center (A2C2). Data produced in making this article can be obtained upon email request to the author.

Edited by: D. Parsons

References

Best, J.: The fluid dynamics of river dunes: a review and some future research directions, J. Geophys. Res.-Earth, 110, F04502, doi:10.1029/2004JF000218, 2005.

Bogard, D. and Tiederman, W.: Burst detection with single-point velocity measurements, J. Fluid Mech., 162, 389–413, 1986.

Chang, K. and Constantinescu, G.: Coherent structures in flow over two-dimensional dunes, Water Resour. Res., 49, 2446–2460, doi:10.1002/wrcr.20239, 2013.

Chou, Y.-J. and Fringer, O. B.: A model for the simulation of coupled flow-bed form evolution in turbulent flows, J. Geophys. Res.-Oceans, 115, C10041, doi:10.1029/2010JC006103, 2010.

Furbish, D. J. and Schmeeckle, M. W.: A probabilistic derivation of the exponential-like distribution of bed load particle velocities, Water Resour. Res., 49, 1537–1551, doi:10.1002/wrcr.20074, 2013.

Giri, S. and Shimizu, Y.: Numerical computation of sand dune migration with free surface flow, Water Resour. Res., 42, W10422, doi:10.1029/2005WR004588, 2006.

Grass, A. and Ayoub, R.: Bed load transport of fine sand by laminar andturbulent flow, Coastal Eng. Proc., 1, 1589–1599, 1982.

Grigoriadis, D. G. E., Balaras, E., and Dimas, A. A.: Large-eddy simulations of unidirectional water flow over dunes, J. Geophys. Res.-Earth, 114, F02022, doi:10.1029/2008JF001014, 2009.

Kraft, S., Wang, Y., and Oberlack, M.: Large eddy simulation of sediment deformation in a turbulent flow by means of level-set method, J. Hydraul. Eng., 137, 1394–1405, doi:10.1061/(ASCE)HY.1943-7900.0000439, 2011.

Le, H., Moin, P., and Kim, J.: Direct numerical simulation of turbulent flow over a backward-facing step, J. Fluid Mech., 330, 349–374, 1997.

Lu, S. and Willmarth, W.: Measurements of the structure of the Reynolds stress in a turbulent boundary layer, J. Fluid Mech., 60, 481–511, 1973.

Nabi, M., de Vriend, H. J., Mosselman, E., Sloff, C. J., and Shimizu, Y.: Detailed simulation of morphodynamics: 3. Ripples and dunes, Water Resour. Res., 49, 5930–5943, doi:10.1002/wrcr.20457, 2013.

Nelson, J. M., Shreve, R. L., McLean, S. R., and Drake, T. G.: Role of near-bed turbulence structure in bed load transport and bed form mechanics, Water Resour. Res., 31, 2071–2086, doi:10.1029/95WR00976, 1995.

Nelson, J. M., Burman, A. R., Shimizu, Y., McLean, S. R., Shreve, R. L., and Schmeeckle, M.: Computing flow and sediment transport over bedforms, in: Proceedings of the 4th IAHR Symposium on River, Coastal and Estuarine Morphodynamics, 4–7 October 2005, Urbana, Illinois, USA, vol. 2, edited by: Parker, G. and Garcia, M., Taylor & Francis Group, London, UK, 861–872, 2006.

Nguyen, Q. and Wells, J. C.: A numerical model to study bedform development in hydraulically smooth turbulent flows, J. Hydraul. Eng., 53, 157–162, 2009.

Niemann, S., Fredsøe, J., and Jacobsen, N.: Sand dunes in steady flow at low froude numbers: dune height evolution and flow resistance, J. Hydraul. Eng., 137, 5–14, doi:10.1061/(ASCE)HY.1943-7900.0000255, 2011.

Omidyeganeh, M. and Piomelli, U.: Large-eddy simulation of two-dimensional dunes in a steady, unidirectional flow, J. Turbul., 12, 1–31, 2011.

Paarlberg, A. J., Dohmen-Janssen, C. M., Hulscher, S. J. M. H., and Termes, P.: Modeling river dune evolution using a parameterization of flow separation, J. Geophys. Res.-Earth, 114, F01014, doi:10.1029/2007JF000910, 2009.

Penko, A., Calantoni, J., Rodriguez-Abudo, S., Foster, D., and Slinn, D.: Three-dimensional mixture simulations of flow over dynamic rippled beds, J. Geophys. Res.-Oceans, 118, 1543–1555, doi:10.1002/jgrc.20120, 2013.

Perot, B. and Moin, P.: Shear-free turbulent boundary layers, Part 1. Physical insights into near-wall turbulence, J. Fluid Mech., 295, 199–227, 1995.

Schmeeckle, M. W.: Numerical simulation of turbulence and sediment transport of medium sand, J. Geophys. Res.-Earth, 119, 1240–1262, doi:10.1002/2013JF002911, 2014.

Shimizu, Y., Schmeeckle, M. W., Hoshi, K., and Tateya, K.: Numerical simulation of turbulence over two-dimensional dunes, river, coastal and estuarine morphodynamics, in: Proceedings International Association for Hydraulic Research Symposium, Genova, Italy, 251–260, 1999.

Shimizu, Y., Schmeeckle, M. W., and Nelson, J. M.: Direct numerical simulation of turbulence over two-dimensional dunes using CIP method, J. Hydrosci. Hydraul. Eng., 19, 85–92, 2001.

Stoesser, T., Braun, C., García-Villalba, M., and Rodi, W.: Turbulence structures in flow over two-dimensional dunes, J. Hydraul. Eng., 134, 42–55, doi:10.1061/(ASCE)0733-9429(2008)134:1(42), 2008.

Zedler, E. A. and Street, R. L.: Large-eddy simulation of sediment transport: currents over ripples, J. Hydraul. Eng., 127, 444–452, doi:10.1061/(ASCE)0733-9429(2001)127:6(444), 2001.

10

Analysis of the drainage density of experimental and modelled tidal networks

Z. Zhou[1], L. Stefanon[2], M. Olabarrieta[3], A. D'Alpaos[4], L. Carniello[2], and G. Coco[1]

[1]Environmental Hydraulics Institute, "IH Cantabria", University of Cantabria, Santander, Spain
[2]Department of Civil, Environmental and Architectural Engineering, University of Padova, Padova, Italy
[3]Department of Civil and Coastal Engineering, University of Florida, Florida, USA
[4]Department of Geosciences, University of Padova, Padova, Italy

Correspondence to: Z. Zhou (zeng.zhou@unican.es)

Abstract. Based on controlled laboratory experiments, we numerically simulate the initiation and long-term evolution of back-barrier tidal networks in micro-tidal and meso-tidal conditions. The simulated pattern formation is comparable to the morphological growth observed in the laboratory, which is characterised by relatively rapid initiation and slower adjustment towards an equilibrium state. The simulated velocity field is in agreement with natural reference systems such as the micro-tidal Venice Lagoon and the meso-tidal Wadden Sea. Special attention is given to the concept of drainage density, which is measured on the basis of the exceedance probability distribution of the unchannelled flow lengths. Model results indicate that the exceedance probability distribution is characterised by an approximately exponential trend, similar to the results of laboratory experiments and observations in natural systems. The drainage density increases greatly during the initial phase of tidal network development, while it slows down when the system approaches equilibrium. Due to the larger tidal prism, the tidal basin has a larger drainage density for the meso-tidal condition (after the same amount of time) than the micro-tidal case. In both micro-tidal and meso-tidal simulations, it is found that there is an initial rapid increase of the tidal prism which soon reaches a relatively steady value (after approximately 40 yr), while the drainage density adjusts more slowly. In agreement with the laboratory experiments, the initial bottom perturbations play an important role in determining the morphological development and hence the exceedance probability distribution of the unchannelled flow lengths. Overall, our study indicates an agreement of the geometric characteristics between the numerical and experimental tidal networks.

1 Introduction

Channel networks (e.g. mountainous, fluvial or tidal) exhibit a variety of morphologies and their shape and functioning has attracted considerable attention in the research sphere (Montgomery et al., 1997; Rinaldo et al., 1998, 1999a; D'Alpaos et al., 2005). A comprehensive understanding of channel networks' morphologies as well as their long-term evolutions is fundamental to address their response to environmental variations under increasing climate change and human pressure (French, 2006; D'Alpaos, 2011; Coco et al., 2013). Branching tidal networks are typical landscapes in back-barrier tidal basins such as the Wadden Sea (Wang et al., 2012) and the Venice Lagoon (D'Alpaos et al., 2007a). These systems consist of various geomorphological elements (Fig. 1) that are shaped by a variety of processes. Ebb deltas and barrier islands act as natural shelters for flood deltas and tidal flats (and associated salt marshes). Tidal flats are often dissected by dendritic channel networks that serve as effective nutrient pathways for flora and fauna, and provide valuable accommodating space for coastal ecosystems (Allen, 2000; Townend et al., 2011).

Different methodologies (e.g. Montgomery and Dietrich 1988; Fagherazzi et al., 1999; Passalacqua et al., 2010; Jiménez et al., 2013a) have been developed in order to extract the structure of channel networks and gain insight into their geomorphic characteristics. Due to the high variability in morphologies of channel networks, it is in fact convenient to study these systems using a statistical approach (Marani et al., 2003; Passalacqua et al., 2013). Remote sensing, associated with these specifically developed techniques to analyse digital terrain maps (DTM), has enabled a more accurate measure and understanding of the geomorphic characteristics of tidal networks (Fagherazzi et al., 1999; Rinaldo et al., 1999a, b; Mason et al., 2006; Vandenbruwaene et al, 2012).

Analyses of these field data indicate that scale invariance, as widely observed in river networks, does not necessarily hold for tidal networks (Rinaldo et al., 1999a) because of the more complicated interplay between various competing landscape-forming processes operating in estuarine systems (e.g. bi-directional tidal flow and the presence of locally generated wind waves). Recently, Passalacqua et al. (2013) analysed the channel network of the Ganges–Brahmaputra–Jamuna delta using a statistical framework based on several descriptors (island area, shape factor, aspect ratio and nearest-edge distance), and their results indicated that the statistical behaviour of these descriptors could distinguish the coastal region and the area along the main rivers; hence geomorphic signatures of processes and vegetation types responsible for delta formation and evolution could be identified.

Drainage density, a measure of channelisation in watersheds, has been found to be of fundamental importance for the understanding of channel network systems since it reflects a balance between climate, geomorphology and hydrology (Moglen et al., 1998). This quantity was first defined by Horton (1932, 1945) as the ratio between the total channelised stream length and the watershed area: $D = \sum L_c / A$. Although widely adopted, the "laws of drainage networks" developed by Horton (1945) are still under debate. Analysing unbiased samples of networks generated with a Monte Carlo method, Kirchner (1993) found that almost all possible networks follow Horton's laws, indicating the distinctiveness of an individual channel network actually cannot be revealed by Horton's theory (see also Rinaldo et al., 1998). This was also confirmed by Marani et al. (2003) through analysis of both natural networks in the Venice Lagoon and highly schematic test settings. In order to overcome the drawback embedded in Horton's laws, Marani et al. (2003) proposed another measure based on the statistical distribution of the unchannelled flow length (l), i.e. the mean distance from a point on the lagoon platform to the nearest channel. This methodology, based on a Poisson-type hydrodynamic model (Rinaldo et al., 1999a), has been used in many studies afterwards (D'Alpaos et al., 2005, 2007a, b; Feola et al., 2005), proving to be a useful geomorphic tool for analysing channel network systems.

Field observations, together with different geometric analyses (e.g. Rinaldo et al., 1999a; Marani et al., 2003; Novakowski et al., 2004; Vandenbruwaene et al., 2012a, b), have confirmed that tidal networks show distinctive landscapes and geometric characteristics, because of the highly

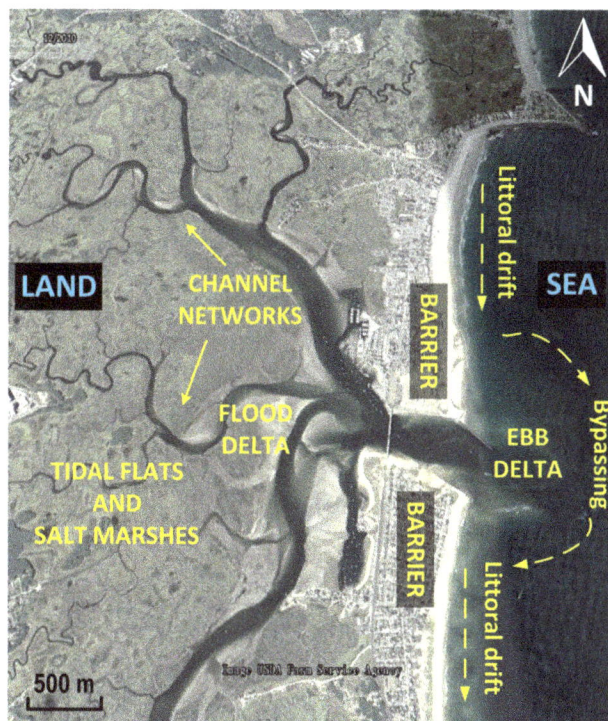

Figure 1. Demonstration of a typical barrier tidal basin system, showing various geomorphological features including flood and ebb deltas, channel networks, barrier beaches, tidal flats and salt marshes. The system is Hampton Harbour inlet, located at 42°53′48″ N and 70°49′06″ W on the east coast of the US. The image was taken from Google Earth (dated 1 January 2010), accessed on 11 October 2013. ©Image USDA, Farm Service Agency, Google Earth 2013.

complicated interactions between various processes such as waves, tides, biological activities and sea level rise. By limiting the number of active processes, controlled laboratory experiments provide an important alternative to gain insight (Stefanon et al., 2010, 2012; Vlaswinkel and Cantelli, 2011; Kleinhans et al., 2012; Iwasaki et al., 2013). Moreover, the data obtained from these experiments (or after proper scaling analyses) can be utilised to benchmark numerical models (Tambroni et al., 2010).

Stefanon et al. (2010, 2012) successfully reproduced the major features of the evolution of tidal networks in the laboratory, which are comparable to natural systems in terms of both the morphological development and the geomorphic characteristics. Zhou et al. (2014) performed a comparative study between these laboratory experiments and numerical simulations at both laboratory and natural scales, exploring the effects of different parameterisations on the long-term morphological evolution of tidal networks. They found that eddy viscosity, sediment transport formulation and bed slope terms played an important role in the resulting morphologies of tidal networks. In this study, emphasis is given to the comparison of the general geomorphic characteristics of tidal

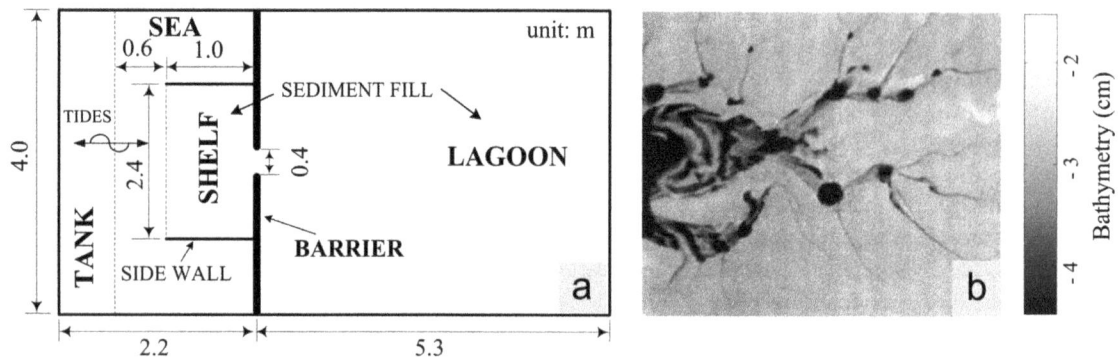

Figure 2. Laboratory experimental setting and results of Stefanon et al. (2010). (**a**) Schematic overview of the experimental apparatus; (**b**) final stable channel network obtained in the lagoon of the experiment Run 4.

Table 1. Experimental setting and numerical model parameters after scaling analysis of Stefanon et al. (2010). W, L and D represent width, length and depth scales of the basins; U denotes the typical velocity in tidal channels.

Parameters	a (m)	T (s)	ρ_s (kg m^{-3})	D_{50} (μm)	$W \times L$ (m^2)	D (m)	U (m s^{-1})
Experiment	0.005	480	1041	800	4×5.2	0.05	0.05
Micro-tidal case	0.5	43 200	2650	234	3600×4700	5.0	0.5
Meso-tidal case	1.0	43 200	2650	234	5000×6700	10.0	0.7

network systems using model and laboratory experiments. Specifically, we will explore whether the numerically modelled channel networks are comparable with the experimental (Stefanon et al., 2010, 2012) or natural ones from the perspective of drainage density. The variation of drainage density during the morphological development will also be examined in order to shed light on tidal network ontogeny. The remaining of the paper is organised as follows: Sect. 2 describes the experimental and numerical models, Sects. 3 and 4 present the results and associated discussions, and concluding remarks are made in Sect. 5.

2 Methods

2.1 Physical model and scaling analysis

Stefanon et al. (2010, 2012) carried out experiments under a range of different settings. One of these experiments (indicated by "Run 4" in Stefanon et al., 2010) was conducted considering a constant mean water level. Run 4 is the focus of this study. The experimental domain included an initially flat lagoon, a gently sloping shelf and a deep sea (Fig. 2a). The lagoon was connected with the shelf and the sea via an inlet which was bounded by non-erodible barriers. Initially, both the lagoon and shelf were filled with coarse, low-density non-cohesive grains (medium size $D_{50} = 0.8$ mm and density $\rho_s = 1041$ kg m^{-3}), providing an erodible sediment layer thick enough to prevent the erosion down to the non-erodible concrete bottom. A harmonic tide, with a tidal amplitude

$a = 0.005$ m and period $T = 480$ s, was simulated at the sea by a vertically oscillating weir.

The final stable network of Run 4, observed in the laboratory after approximately 8000 tidal cycles, is shown in Fig. 2b. Through scaling analysis (Stefanon et al., 2010), it was possible to "transfer" this experiment to typical natural systems where tidal networks are often observed, i.e. micro-tidal and meso-tidal basins like the Venice Lagoon and the Wadden Sea (see Table 1 for the results of the scaling analysis). The reader is referred to Stefanon et al. (2010) for a detailed description of the experiments and the scaling analysis.

2.2 Numerical model and setup

The numerical model Delft3D is used to simulate the "morphodynamic loop", which in this study consists of the depth-averaged shallow water equations, sediment transport (van Rijn, 1993) and bed level updating equations. At each hydrodynamic time step, the morphological change is calculated feeding back on the hydrodynamics (i.e. "online approach"). In order to reduce the computational cost, the model introduces a so-called "morphological accelerating factor" (*MorFac*) which linearly scales up the bed level change; see Lesser et al. (2004) and van der Wegen and Roelvink (2008) for more details.

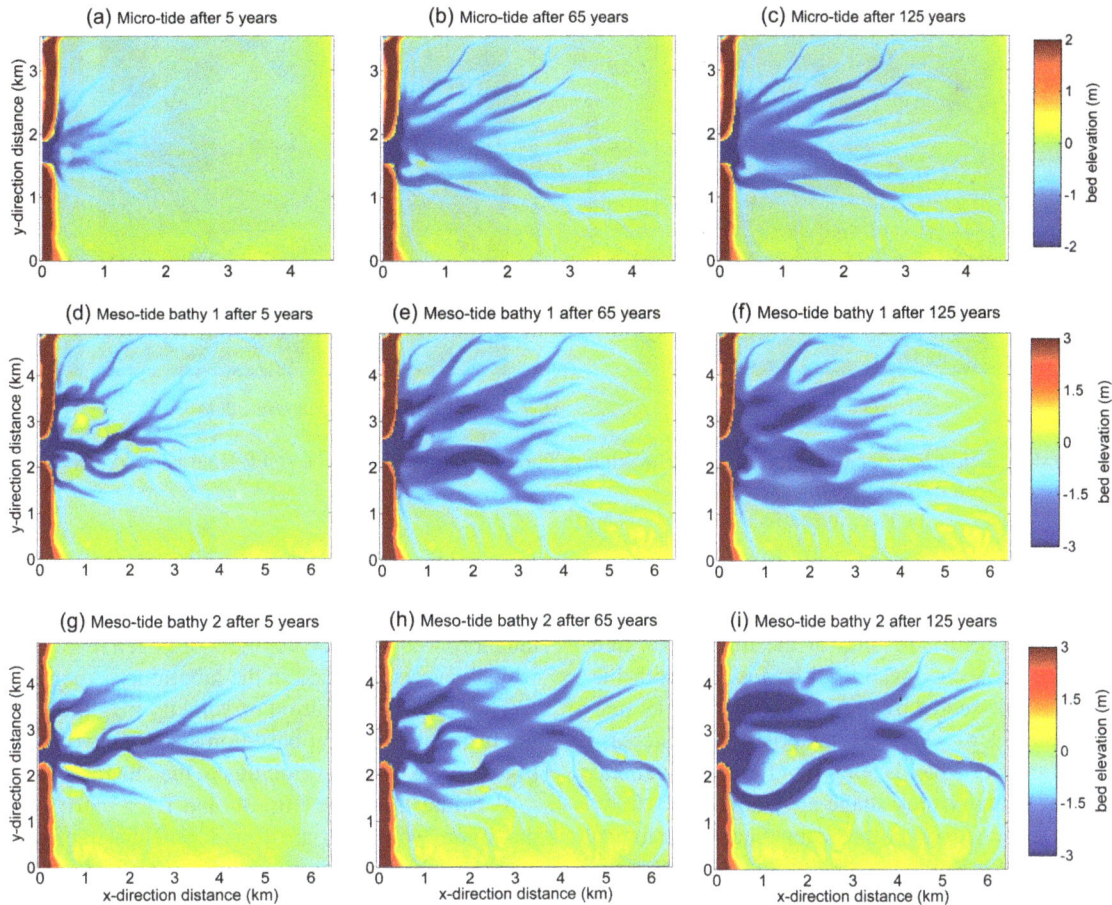

Figure 3. Morphological pattern formations after 5, 65 and 125 yr simulated by the numerical model. **(a–c)** Micro-tidal condition, **(d–f)** meso-tidal condition with the first bathymetry, and **(g–i)** meso-tidal condition with the second bathymetry.

Both the scaled-up micro- and meso-tidal cases are simulated in this study according to the scaling laws given in Table 1. A grid mesh of 48 300 rectangular cells is adopted for both cases with different cell sizes (micro-tidal case: 22.5 m × 22.5 m; meso-tidal case: 31 m × 31 m) and their input bathymetries are deduced by scaling up the bottom perturbations of the initial measured experimental bathymetry with proper multiplier factors. The values of the multiplier factors are chosen on the basis of the scaling analysis presented by Stefanon et al. (2010). In particular, the scaling ratios were derived based on the similarities of hydrodynamics (Froude similitude and the similitude of local inertia and advection) and sediment transport (similitude of Shields parameter and particle Reynolds number) between the prototype and physical model. In order to maintain those similarities, the vertical depth scaling factor for micro-tidal and meso-tidal cases is 100 and 200, respectively (Table 1). The bottom perturbation of the initial experimental bathymetry of Run 4 is approximately within ±0.004 m; hence the perturbations of the micro-tidal and meso-tidal initial bathymetries are set within ±0.4 and ±0.8 m, respectively. As for the

meso-tidal case, a second initial bathymetry with different bottom perturbations is considered in order to investigate the role of different random bed perturbations on the morphological evolution of tidal networks. The magnitude of the bottom perturbations for the second bathymetry is the same as the first one, whereas the spatial distributions which were generated randomly are different. Some other key model parameters are chosen through sensitivity tests (Zhou et al., 2014), including time step ($\Delta t = 30$ s), morphological accelerating factor (*MorFac* $= 20$), transversal bed slope term ($\alpha_{\mathrm{bn}} = 50$) and Chézy friction coefficient ($C = 65\,\mathrm{m}^{1/2}\,\mathrm{s}^{-1}$).

3 Results

3.1 Pattern formation and velocity field

The morphological configurations of the micro-tidal (tidal range: 1 m) and meso-tidal (tidal range: 2 m) basins after 5, 65 and 125 yr are shown in Fig. 3. Much like the laboratory observations, the initial phase of the channel development proved to be rapid. By scouring the inlet, a central channel was formed within 5 yr. The channel subsequently branched

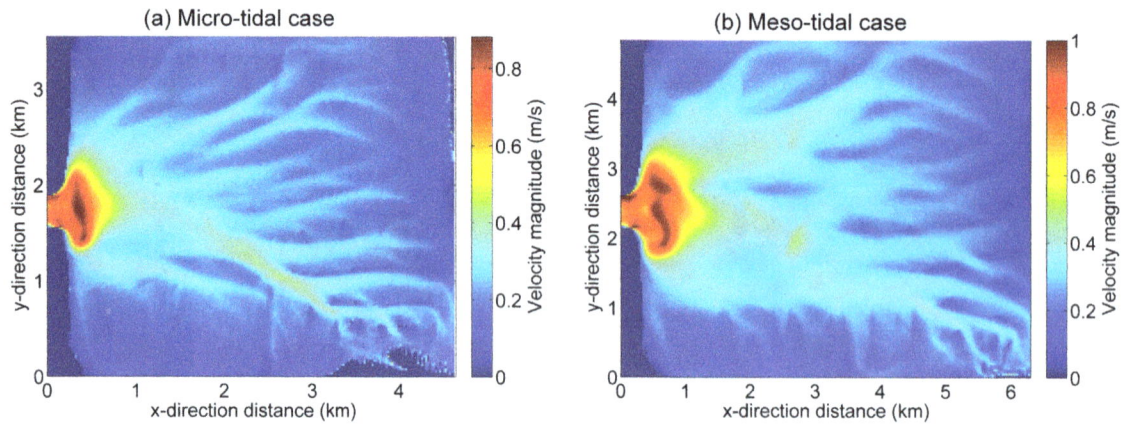

Figure 4. Spatial distribution of velocity magnitude at the maximum flood condition based on a stable tidal network: **(a)** micro-tidal case and **(b)** meso-tidal case with the first bathymetry.

into smaller tributaries which grew wider and deeper and continued branching, meandering and elongating. After 5 yr, approximately one-third of the area of the micro-tidal and over half of the area of the meso-tidal basins (for both the bathymetries) were dissected by the expanding channel network (Fig. 3a, d and g). As the tidal prism kept increasing, the network continued to develop rapidly until a general quasi-stable tidal network took shape after around 50–60 yr (Fig. 3b, e and h). The growth of the channel network slowed down and adjusted to a stable planar configuration after approximately 120 yr (Fig. 3c, f and i). After that, only the inlet area was still subject to small erosion (leading to further inlet deepening) until the morphological equilibrium state was reached after around 200 yr.

The average depth of the inlet area of the stable tidal network observed in the laboratory is around 0.05–0.06 m (Fig. 2b), and the corresponding depths of the numerically modelled micro-tidal and meso-tidal systems are 5–6 m and 10–11 m, respectively. This agrees well with the scaling analysis of Stefanon et al. (2010) (see Table 1). Overall, the average channel depths of the micro-tidal and meso-tidal simulated networks are approximately 2–3 m and 3–4 m, respectively, with the average tidal flat depths of about 0.3–0.4 m and 0.5–0.6 m, respectively. These morphological features generally resemble natural examples such as micro-tidal Venice Lagoon and meso-tidal Dutch Wadden Sea (Stefanon et al., 2010). Although sharing a similar trend in channel network development, the detailed morphological evolution of the micro-tidal and meso-tidal simulations differs in primarily two aspects. First, these two networks display different final morphologies: the channels formed in the meso-tidal case are more meandering and occupy a larger portion of the basin area than in the micro-tidal case. Second, the channel network developed more rapidly in the meso-tidal case and the channels were wider and deeper after the same amount of simulation time. This was also found by van Maanen et al. (2013), who showed that tidal networks grow more

rapidly for larger tidal range. Finally, our simulations show that the initial bottom perturbations play an important role in the morphological development of tidal networks (Fig. 3d–i), which was also noticed in the laboratory experiments.

The spatial distribution of velocity magnitude at the maximum flood condition based on stable networks is shown in Fig. 4. As expected, velocities in the channels are larger than in the tidal flats (see Fig. 3c and f for morphologies). The maximum velocities are found near the inlet area, with an average value of approximately 0.6–0.7 m s^{-1} for the micro-tidal case (Fig. 4a) and 0.7–0.8 m s^{-1} for the meso-tidal case (Fig. 4b). These values fall within a reasonable range when compared to natural lagoon systems and are also consistent with the scaling analysis (Table 1) of Stefanon et al. (2010).

3.2 Drainage density

Following the method proposed by Marani et al. (2003) in the framework of tidal systems, the drainage density of tidal networks is evaluated on the basis of the exceedance probability distribution of the unchannelled flow lengths, determined using the Poisson-type hydrodynamic model developed by Rinaldo et al. (1999a). Previous studies on natural tidal networks indicate that the probability distribution of unchannelled flow lengths is exponential (Marani et al., 2003; Feola et al., 2005; D'Alpaos et al., 2007; Vandenbruwaene et al., 2012a). The probability density function shows a linear trend when it is plotted on a semilogarithmic scale, and the inverse of the slope represents the mean unchannelled flow length. The drainage density is larger when the slope is steeper, suggesting that the channel networks are more efficient in draining tidal flats and marshes (Marani et al., 2003).

Figure 5 shows the exceedance probability distribution of tidal networks after 5, 35, 65 and 125 yr under micro-tidal conditions. After 5 yr, the slope of the semilogarithmic distribution is relatively gentle when less than one-third of the tidal flats is drained. The probability distributions after 35,

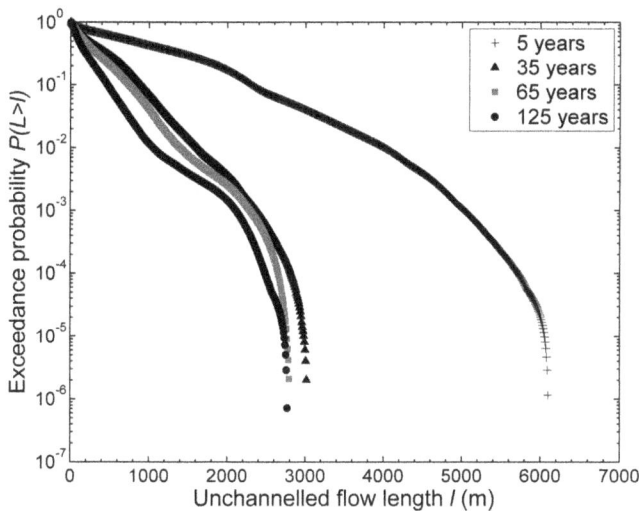

Figure 5. Semilog plot of the exceedance probability $P(L > l)$ of the unchannelled flow length l, computed for the configurations of channel networks under micro-tidal conditions after 5, 35, 65 and 125 yr; see also Fig. 3a–c for the morphologies of corresponding networks.

65 and 125 yr are close since the basic structure of the tidal network has taken shape after 35 yr and only some minor morphological adjustments occur afterwards. The slope of the semilogarithmic distribution increases with time, indicating a decreasing mean unchannelled flow length and an increasing drainage density (Fig. 6). The mean unchannelled flow length decreases sharply in the first 35 yr, indicating a rapid expansion of tidal networks. The decrease of the later phase (when the system is evolving towards equilibrium) is slower (Fig. 6a). Generally, the characteristic of the probability distribution (Fig. 5), as well as the evolution of the mean unchannelled flow length and drainage density (Fig. 6), is in accordance with the trend of the morphological evolution: the initial phase of development is rapid and slows down with time, especially when the morphological equilibrium approaches. Similar results were also presented by Vandenbruwaene et al. (2012a, b), who conducted a 5 yr field study of spontaneous formation and evolution of a tidal network in the Scheldt Estuary, Belgium.

The morphological evolution of the micro-tidal and meso-tidal cases differs pronouncedly (Fig. 3), which is also reflected by the exceedance probability distribution. After the same amount of time (i.e. 5 and 125 yr), the unchannelled flow length of the low exceedance probability is larger under meso-tidal conditions (Fig. 7a) than micro-tidal ones (Fig. 5), which occurs because the dimension of the meso-tidal basin is larger. In order to compare the laboratory-scale experiment Run 4 with the micro-tidal and meso-tidal simulations, the unchannelled flow lengths are scaled with the corresponding basin widths (see Fig. 7b). The basin is drained more efficiently under the meso-tidal condition than

micro-tidal conditions after the same time (5 and 125 yr; see also Fig. 8b) since the dimensionless exceedance probability curve of the meso-tidal case displays a larger slope. This is consistent with the smaller mean unchannelled flow length shown by Fig. 8a. The drainage densities of both the micro-tidal and meso-tidal simulated stable networks are larger than those found in the laboratory experiment (represented by the marker "+" in Fig. 7b). Overall, different initial conditions can result in wide differences in either the experimental (see Fig. 13 of Stefanon et al., 2010) or the numerically simulated drainage systems (Figs. 7 and 9).

Stefanon et al. (2010) performed two experiments using different initial bottom perturbations under the same tidal forcing and found that the final patterns obtained in the laboratory differed pronouncedly. Their finding was also reflected by the distinctive exceedance probability distributions of the unchannelled flow lengths. Our numerical model results are consistent with the laboratory observations. The evolved tidal network after 5 yr, obtained using the first bathymetry as initial condition, showed less drained area than the one obtained when considering the second bathymetry (characterised by different randomly generated bed perturbations with respect to the first bathymetry) as an initial condition, which is in agreement with the larger slope of the exceedance probability distribution (in a semilog plot) of the unchannelled flow lengths obtained starting from the second bathymetry (Fig. 9). However, the first bathymetry leads to a larger final drainage density than the second when the morphological equilibrium configuration is reached. This indicates that the lagoon of the second bathymetry can be better drained near the inlet area, while the inner lagoon is drained less efficiently when compared to the lagoon of the first bathymetry. This is simply the result of different spatially distributed random perturbations of the bottom and the morphodynamic feedbacks they generated. Analogous to the micro-tidal case shown in Fig. 6a, the mean unchannelled flow length decreases rapidly in the beginning phase and tends to be stable when morphological equilibrium is approached (blue lines in Fig. 8a).

The evolution of the mean unchannelled flow length and drainage density display asymptotic behaviour (Fig. 8), as also found by Vandenbruwaene et al. (2012a). As a primary land-forming force shaping channel networks, the tidal prism of both micro-tidal and meso-tidal simulations also shows a similar evolution (Fig. 10). The tidal prism increases sharply in the first 35 yr when the channel network expands, quickly dissecting the tidal flats, and slows down afterwards. Although sharing a similar trend, the tidal prism of the meso-tidal case is larger than the micro-tidal one due to the different tidal ranges and basin sizes (Fig. 10). Initially, the tidal prism adapts to the morphological evolutions more rapidly than the drainage density does and then the two quantities vary with similar temporal scales, as also found by Stefanon et al. (2012).

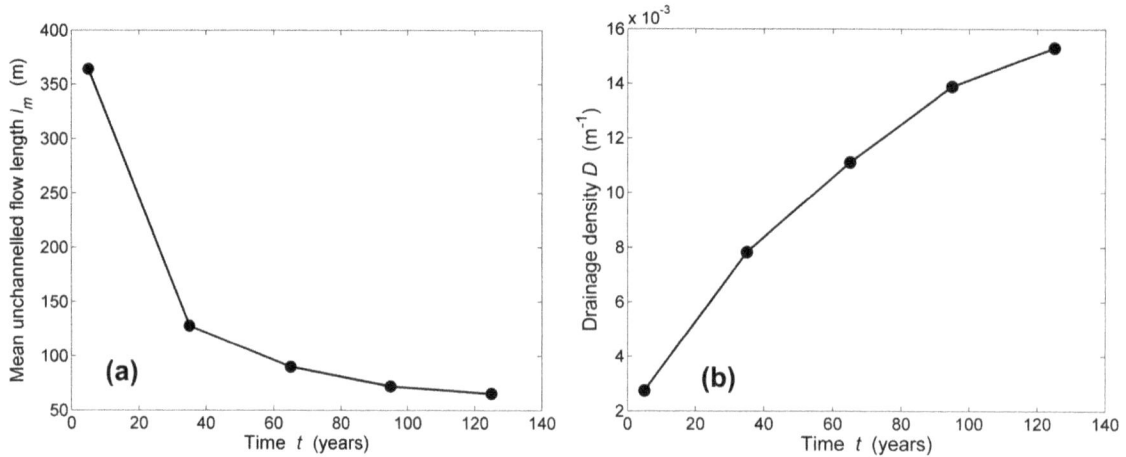

Figure 6. (a) Evolution of the mean unchannelled flow length l_{m} with time, and (b) evolution of the drainage density D of the developing tidal networks under micro-tidal conditions.

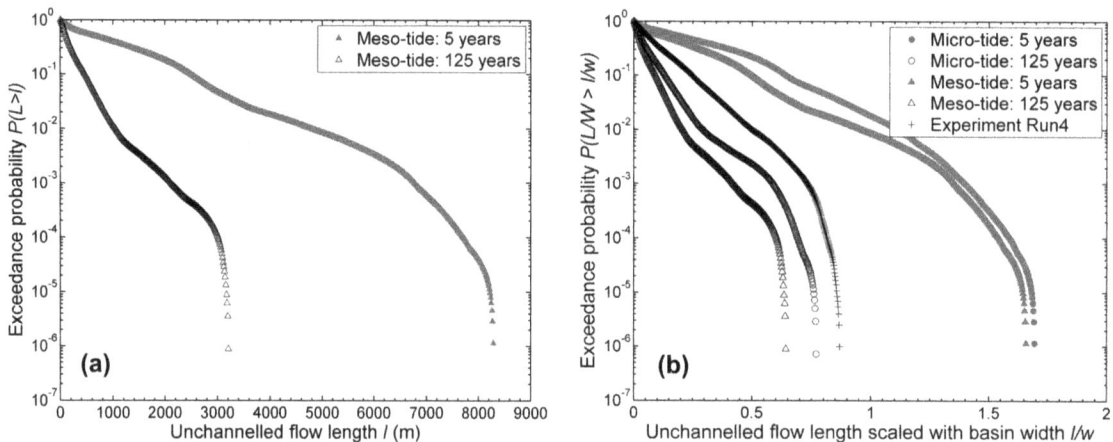

Figure 7. (a) Semilog plot of the exceedance probability $P(L > l)$ distribution of the unchannelled flow length l, computed for the configurations of numerical tidal networks under meso-tidal conditions after 5 and 125 yr, and (b) semilog plot of the exceedance probability $P(L/W > l/w)$ of dimensionless unchannelled flow length l/w (scaled with basin width w), computed for the configurations of stable experimental tidal networks (Run 4) and numerical tidal networks under both micro-tidal and meso-tidal conditions after 5 and 125 yr; see also Fig. 3 for the morphologies of corresponding networks.

Both the evolution of drainage density and the tidal prism displays asymptotic behaviour (Fig. 10). Therefore, numerical simulations predict that the drainage density of a stable channel network system should correspond to a certain tidal prism. In fact, using varying water levels to simulate different sea level conditions (hence different morphological equilibrium systems), the laboratory experiments conducted by Stefanon et al. (2012) indicate an approximately linear relationship between the land-forming tidal prism and drainage density, which is also in qualitative agreement with the results of Marani et al. (2003) for natural salt-marsh basins.

4   Discussion

The existence of feedback mechanisms between various landscape-forming processes (e.g. tides, waves, biological effects, sea level rise and human intervention) limits our understanding of the ontogeny of tidal networks (Rinaldo et al., 1999a; Coco et al., 2013). By reducing the number of processes affecting the evolution of tidal networks, controlled laboratory experiments provide a good base for gaining insight into these systems. While tidal flows are the external driver of bathymetric changes, the slowly evolving estuarine bathymetry can also constrain the tidal flow and determine flow patterns. Therefore, estuarine landscapes are the result of the mutual adaptation between flow and morphology (i.e. morphodynamic feedback).

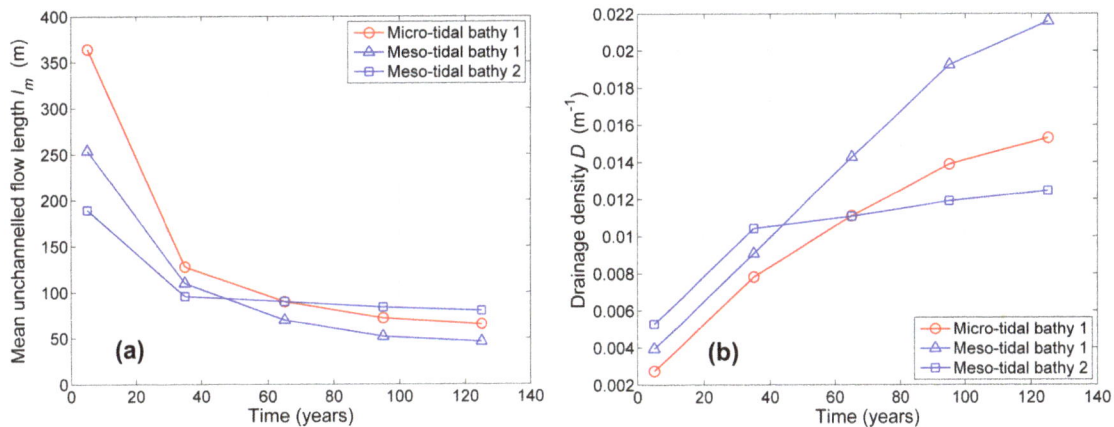

Figure 8. Comparison of the evolution of micro-tidal first, meso-tidal first and meso-tidal second bathymetry with regard to **(a)** the mean unchannelled flow length and **(b)** the drainage density; see also Fig. 3.

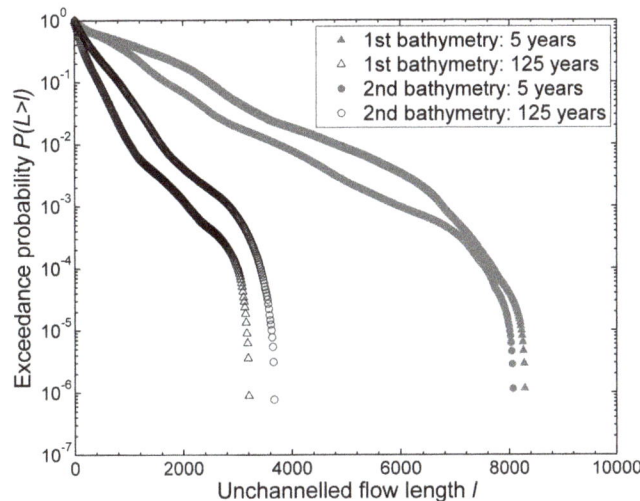

Figure 9. Semilog plot of the exceedance probability distribution of the unchannelled flow length, computed for the configurations of channel networks with different initial bottom perturbations under meso-tidal conditions after 5 and 125 yr.

The laboratory-observed tidal networks of Stefanon et al. (2010, 2012) are morphologically comparable to natural ones. However, their quantitative analysis almost certainly suffers from inevitable scaling effects, which can be difficult to evaluate, especially when translated to long-term numerical modelling. Based on scaling arguments (Stefanon et al., 2010), we set up numerical simulations in an attempt to understand the evolution of drainage density of tidal networks from a numerical modelling point of view and also examine to which degree numerical model results agree with the scaling laws. Quantitatively, the numerically simulated tidal networks resemble those obtained in the laboratory experiments (Stefanon et al., 2010, 2012; Kleinhans et al., 2012) and some natural back-barrier tide-dominated systems (e.g. the Venice Lagoon and the Dutch Wadden Sea). The flow velocities fall in a reasonable range both in the tidal channels and the flats

and are in good agreement with the scaling arguments. Overall, the numerical model appears to be capable of reproducing the major features of tidal network ontogeny, i.e. from initiation to equilibrium. In this context, it is important to specify that the equilibrium (or stable) state considered in this study is not a "frozen" or "static" equilibrium state (strictly null sediment fluxes and no bed level change), which is highly unfeasible in reality (especially if one considers that external forcing is constantly changing). Instead, a different definition, sometimes indicated as "dynamical equilibrium" and characterised by null gradients in sediment fluxes (and no bed level change) over a specific timescale, is more appropriate to describe our numerical simulations. In fact our results indicate that the gradients in sediment fluxes, and for continuity bed level changes, decreased over time and approached zero after approximately 200 yr, even though sediment fluxes

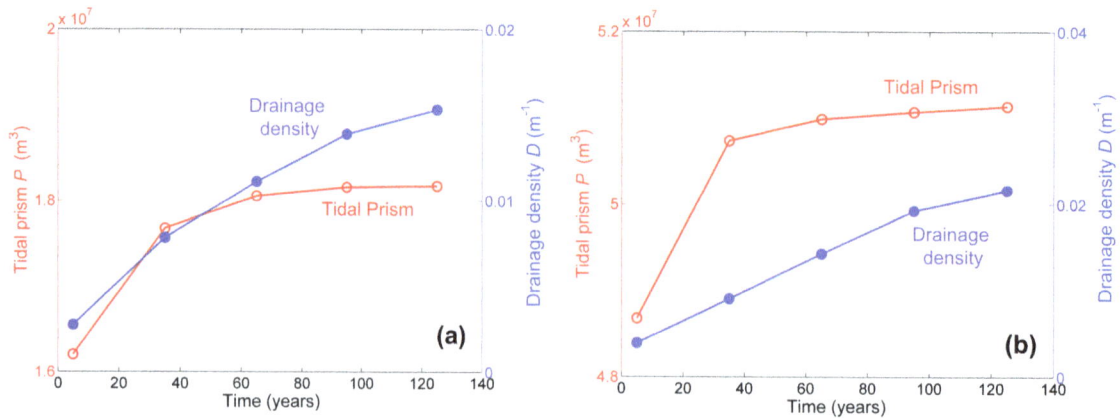

Figure 10. Evolution of tidal prism and drainage density for (**a**) the micro-tidal case and (**b**) the meso-tidal case.

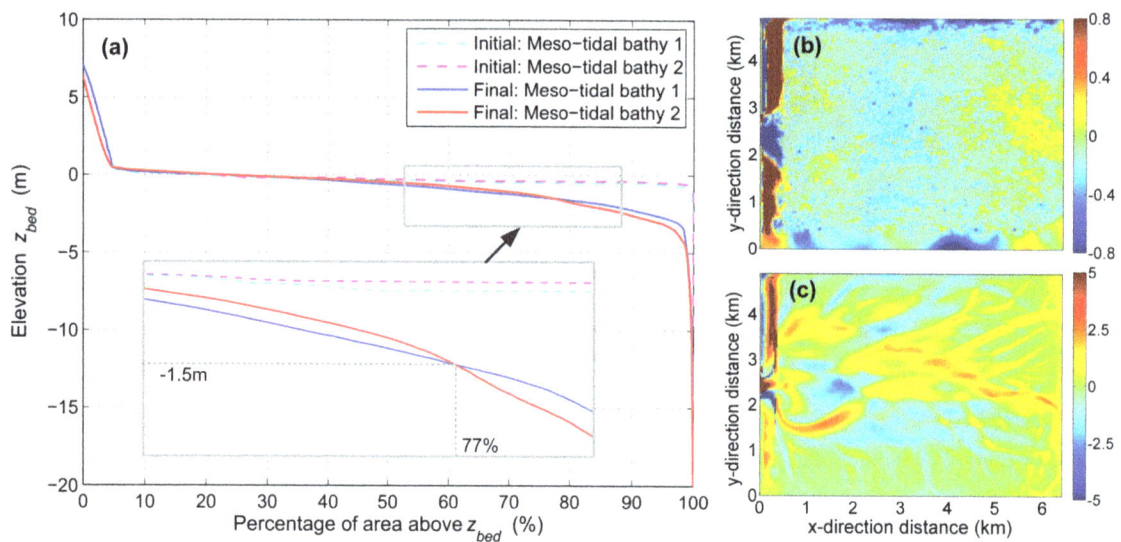

Figure 11. (**a**) Initial and final hypsometries of the first and second meso-tidal bathymetries; (**b**) and (**c**) are the initial and final differences of bed elevation between bathymetry 1 and 2, respectively (note that the colour bar scales in panel (**b**) and (**c**) are different). For bed elevation, the upward direction is positive.

were never strictly null. Our definition of dynamic equilibrium is also different from the usual definition of statistical equilibrium, which implies that small fluctuations in the system do not necessarily cease even if the overall property of the network remains unchanged.

Consistent with the field observation of the natural systems in the Scheldt Estuary, Belgium (Vandenbruwaene et al., 2012a), the drainage density of the numerically simulated tidal networks is asymptotic towards an equilibrium configuration. The exceedance probability distribution of the unchannelled flow lengths of the simulated tidal networks displays an exponential trend, in analogy with the laboratory experiments and natural tidal networks in the Venice Lagoon (Marani et al., 2003; D'Alpaos et al., 2007a). The exceedance probability distributions of the unchannelled flow lengths for the micro-tidal and meso-tidal scale are different after the

same amount of simulation time, suggesting, from this perspective, the absence of scale-invariant processes. As previously indicated, uncertainties in scaling of laboratory observations could potentially cascade on the scaling adopted in the numerical simulations. Model results indicate that the drainage density of the meso-tidal basin is larger due to the larger tidal prism. The tidal prism increases rapidly during the beginning phase of tidal network development and soon reaches a relatively steady value, while the drainage density adjusts more slowly, which is consistent with Stefanon et al. (2012).

Model results also show that the random bottom perturbations play a considerable role in determining the long-term morphological development of tidal networks, and hence the overall evolution of the drainage density. Consistent with laboratory observations (Stefanon et al., 2010), the different exceedance probability distributions of the unchannelled flow length (Fig. 9) reflect the pronounced difference of final configurations of tidal networks. This indicates that small differences in the initial characteristics of tidal lagoons affect the later evolution and ultimately result in pronouncedly different landscapes. Another metric to quantify the general differences between the resulting bathymetries is the hypsometric curves shown in Fig. 11a. The cyan and magenta dashed lines represent the meso-tidal starting hypsometries for bathymetry 1 and 2, respectively. The blue and red solid lines indicate the corresponding final hypsometries at equilibrium. These two numerical simulations started with input bathymetries characterised by different spatial random perturbations (the magnitude of the perturbations was the same) and nearly 95 % of the bed elevation was around the mean sea level. The existence of higher areas near the barrier islands (the red areas in Fig. 3) approximately accounts for the remaining 5 % of the lagoon (notice that in the numerical simulations these higher areas remained unchanged since most of that area was dry even at high tide). Due to the statistical similarity in the seabed perturbations, the two initial bathymetries (dashed lines) show similar hypsometries (the elevation of the first bathymetry is slightly lower). The resulting final hypsometries also share a similar morphological pattern, but some noticeable differences are present. The two hypsometric curves intersect so that the second hypsometry (red line) is higher than the first one (blue line) for areas lower than 77 %, while the opposite occurs for areas larger than 77 %. Overall, the shallower areas are less channelised, while the deep areas are more incised (second case), consistent with the final simulated morphologies (Fig. 3d–i), which show that less area is drained and the main channels are deeper in the second case. This can also be noted from Fig. 11b and c, which are the initial and final differences of bed elevation between the first and second bathymetries, respectively. The initial differences are generally within the range of −0.8 to 0.8 m, while the final differences can reach 5 m with different spatial distribution of channels and tidal flats. Nonetheless, hypsometric curves could only capture a general difference between the two final morphologies due to the aggregated nature of the method. Instead, the probability distributions of unchannelled flow lengths (Fig. 9) provided a better way to investigate the differences between the two cases. Some other metrics (e.g. evolution of lagoon bed level change, channel width-to-depth ratios and the relationship between tidal prism and cross-sectional area) have been discussed in detail elsewhere (Zhou et al., 2014), and overall, they indicate close similarity between experimental and numerical tidal networks.

Nevertheless, there are limitations in the current study that need to be addressed. Firstly, both the experiment and numerical models do not include the effects of waves, biological activities, mud (or sand–mud mixture) and possible river inflow, all of which are commonly present in estuarine environments. Therefore, more processes should be added in

both the physical and numerical models so that a more realistic condition can be considered in the future. Secondly, the Poisson-type hydrodynamic model (Rinaldo et al., 1999a), as adopted to calculate the exceedance probability distribution of the unchannelled flow length (and hence the drainage density), assumes friction terms dominate over inertial and advective terms in the momentum equations. This assumption holds well for shallow basins in which the water depth is normally smaller than 5 m (e.g. the Venice Lagoon), while it may not be valid for the deep area in the model domain. As for the current study, the effect of this limitation is minor for the micro-tidal case (water depth is generally below 5 m), whereas the effect is more pronounced for the meso-tidal case since channels are more incised with a larger tidal prism, especially near the inlet area. This limitation can be overcome by using a more sophisticated hydrodynamic model (e.g. Delft3D); however, this may be computationally demanding. The reader is referred to Jiménez et al. (2013b) for some ongoing research on this topic.

5 Conclusions

We adopt a state-of-the-art numerical model (Delft3D) to investigate the morphological development of tidal networks and compare it to a controlled laboratory experiment. The numerical model is able to reproduce the general behaviour of tidal network ontogeny and the modelled velocity field agrees with the scaling analysis and different natural reference systems.

The comparison primarily focuses on the analysis of the drainage density of tidal networks. Instead of using the classic Horton's method, we analyse the drainage density based on the exceedance probability distribution of the unchannelled flow lengths (the mean of this exponential probability distribution, i.e. the mean unchannelled flow length, is the drainage density). Much like laboratory experiments and natural systems, analysis of the numerically simulated tidal networks shows that the probability distribution of unchannelled lengths displays an exponential trend. In accordance with the observed morphological evolution, the simulated drainage density increases rapidly during the initial phase of tidal network development and slows down when equilibrium is approached. Because of the larger tidal prism, the meso-tidal basin is more drained than the micro-tidal case when stable channel networks are attained. The initial increase of the tidal prism is more rapid than that of the drainage density, while later the increase is slower, indicating that the tidal prism adjusts faster to a stable condition. The initial bottom perturbations are found to play an important role in determining the morphological development, which is also captured in the exceedance probability distribution of the unchannelled flow length.

Using the unchannelled flow length as a descriptor for statistical analysis, we have shown that the numerically modelled tidal networks are comparable to the experimental and natural ones. We have also shown that tidal networks display distinctive morphological features that can be influenced by various factors (e.g. tidal range, initial bottom perturbations). Finally, this study highlights the benefit (and hence need) of complementary studies between physical and numerical modelling in gaining insight into the long-term morphological evolution of estuarine systems.

Acknowledgements. This work is supported by the Augusto González Linares programme (AGL) from the University of Cantabria. Z. Zhou would like to acknowledge Mirian Jiménez for the interesting and fruitful discussions during the course of this study. We are grateful to the reviewers for their insightful comments and suggestions, which have certainly improved the quality of the paper.

Edited by: F. Metivier

References

Allen, J. R. L.: Morphodynamics of Holocene salt marshes: A review sketch from the Atlantic and Southern North Sea coasts of Europe, Quat. Sci. Rev., 19, 1155–1231, doi:10.1016/S0277-3791(99)00034-7, 2000.

Coco, G., Zhou, Z., van Maanen, B., Olabarrieta, M., and Tinoco, R.: Morphodynamics of tidal networks: advances and challenges, Mar. Geol., 346, 1–16, doi:10.1016/j.margeo.2013.08.005, 2013.

D'Alpaos, A.: The mutual influence of biotic and abiotic components on the long-term ecomorphodynamic evolution of salt-marsh ecosystems. Geomorphology, 126, 269–278, doi:10.1016/j.geomorph.2010.04.027, 2011.

D'Alpaos, A., Lanzoni, S., Marani, M., Fagherazzi, S., and Rinaldo, A.: Tidal network ontogeny: Channel initiation and early development, J. Geophys. Res., 110, F02001, doi:10.1029/2004JF000182, 2005.

D'Alpaos, A., Lanzoni, S., Marani, M., Bonometto, A., Cecconi, G., and Rinaldo, A.: Spontaneous tidal network formation within a constructed salt marsh: Observations and morphodynamic modelling, Geomorphology, 91, 186–197, doi:10.1016/j.geomorph.2007.04.013, 2007a.

D'Alpaos, A., Lanzoni, S., Marani, M., and Rinaldo, A.: Landscape evolution in tidal embayments: Modeling the interplay of erosion, sedimentation, and vegetation dynamics, J. Geophys. Res., 112, F01008, doi:10.1029/2006JF000537, 2007b.

Feola, A., Belluco, E., D'Alpaos, A., Lanzoni, S., Marani, M., and Rinaldo, A.: A geomorphic study of lagoonal landforms, Water Resour. Res., 41, W06019, doi:10.1029/2004WR003811, 2005.

French, J.: Tidal marsh sedimentation and resilience to environmental change: exploratory modelling of tidal, sea-level and sediment supply forcing in predominantly allochthonous systems, Mar. Geol., 235, 119–136, doi:10.1016/j.margeo.2006.10.009, 2006.

Horton, R. E.: Drainage-basin characteristics, Transactions, American geophysical union, 13 , 350–361, 1932.

Horton, R. E.: Erosional development of streams and their drainage basins; hydrophysical approach to quantitative morphology, Geol. Soc. Am. Bull., 56, 275–370, 1945.

Iwasaki, T., Shimizu, Y., and Kimura, I.: Modelling of the initiation and development of tidal creek networks. Proceedings of the ICE – Mar. Engin., 166, 76–88, doi:10.1680/maen.2012.12, 2013.

Jiménez, M., Zhou, Z., Castanedo, S., Coco, G., Medina, R., Rodriguez-Iturbe, I.: Tidal channel networks: hydrodynamic controls of their topological structure, The 8th Symposium on River, Coastal and Estuarine Morphodynamics (RCEM), Santander, Spain, 2013a.

Jiménez, M., Castanedo, S., Zhou, Z., Coco, G., Medina, R., Rodriguez-Iturbe, I.: On the sensitivity of tidal network characterization to power law estimation, Adv. Geosci., 1, 1–5, doi:10.5194/adgeo-1-1-2014, 2013b.

Kirchner, J. W.: Statistical inevitability of Horton's laws and the apparent randomness of stream channel networks, Geology, 21, 591–594, 1993.

Kleinhans, M. G., van der Vegt, M., van Scheltinga R. T., Baar, A. W., and Markies, H.: Turning the tide: experimental creation of tidal channel networks and ebb deltas, Geol. Mijnb./Neth. J. Geosci., 91, 311–323, 2012.

Lesser, G. R., Roelvink, J. A., van Kester J. A. T. M., and Stelling, G. S.: Development and validation of a three-dimensional morphological model, Coast. Engin., 51, 883–915, doi:10.1016/j.coastaleng.2004.07.014, 2004.

Marani, M., Belluco, E., D'Alpaos, A., Defina, A., Lanzoni, S., and Rinaldo, A.: On the drainage density of tidal networks, Water Resour. Res., 39, 1040, doi:10.1029/2001WR001051, 2003.

Mason, D. C., Scott, T. R., and Wang, H.-J.: Extraction of tidal channel networks from airborne scanning laser altimetry, ISPRS J. Photogramm. Remote Sens., 61, 67–83, doi:10.1016/j.isprsjprs.2006.08.003, 2006.

Moglen, G. E., Eltahir, E. A. B., and Bras, R. L.: On the sensitivity of drainage density to climate change, Water Resour. Res., 34, 855–862, doi:10.1029/97WR02709, 1998.

Montgomery, D. and Dietrich, W.: Where do channels begin? Nature, 336, 232–234, 1988.

Montgomery, D. R. and Buffington, J. M.: Channel-reach morphology in mountain drainage basins. Geological Society of America Bulletin, Geol. Soc. Am., 109, 596–611, 1997.

Novakowski, K. I., Torres, R., Gardner, L. R., and Voulgaris, G.: Geomorphic analysis of tidal creek networks, Water Resour. Res., 40, W05401, doi:10.1029/2003WR002722, 2004.

Passalacqua, P., Do Trung, T., Foufoula-Georgiou, E., Sapiro, G., and Dietrich, W. E. A.: Geometric framework for channel network extraction from lidar: Nonlinear diffusion and geodesic paths, J. of Geophys. Res. Earth Surf., 115, F01002, doi:10.1029/2009JF001254, 2010.

Passalacqua, P., Lanzoni, S., Paola, C., and Rinaldo, A.: Geomorphic signatures of deltaic processes and vegetation: The Ganges-Brahmaputra-Jamuna case study, J. Geophys. Res. Earth Surf., 118, 1838–1849, doi:10.1002/jgrf.20128, 2013.

Rinaldo, A., Rodriguez-Iturbe, I., and Rigon, R.: Channel networks. Annu. Rev. Earth Planet. Sci., 26, 289–327, doi:10.1146/annurev.earth.26.1.289, 1998.

Rinaldo, A., Fagherazzi, S., Lanzoni, S., Marani, M., and Dietrich, W. E.: Tidal networks: 2. Watershed delineation and comparative network morphology, Water Resour. Res., 35, 3905–3917, doi:10.1029/1999WR900237, 1999a.

Rinaldo, A., Fagherazzi, S., Stefano, L., Marani, M., and Dietrich, W. E.: Tidal networks: 3. Landscape-forming discharges and studies in empirical geomorphic relationships. Water Resour. Res., 35, 3919–3929, doi:10.1029/1999WR900238, 1999b.

Rodriguez-Iturbe, I. and Rinaldo, A.: Fractal River Basins: Chance and Self-Organization, Cambridge Univ. Press, New York, 1997.

Stefanon, L., Carniello, L., D'Alpaos, A., and Lanzoni, S.: Experimental analysis of tidal network growth and development, Cont. Shelf Res., 30, 950–962, doi:10.1016/j.csr.2009.08.018, 2010.

Stefanon, L., Carniello, L., D'Alpaos, A., and Rinaldo, A.: Signatures of sea level changes on tidal geomorphology: Experiments on network incision and retreat, Geophys. Res. Lett., 39, L12402, doi:10.1029/2012GL051953, 2012.

Tambroni, N., Ferrarin, C., and Canestrelli, A.: Benchmark on the numerical simulations of the hydrodynamic and morphodynamic evolution of tidal channels and tidal inlets, Cont. Shelf Res., 30, 963–983, doi:10.1016/j.csr.2009.12.005, 2010.

Townend, I., Fletcher, C., Knappen, M., and Rossington, K.: A review of salt marsh dynamics, Water Environ. J., 25, 477–488, doi:10.1111/j.1747-6593.2010.00243.x, 2011.

Van der Wegen, M. and Roelvink, J. A.: Long-term morphodynamic evolution of a tidal embayment using a two-dimensional, process-based model, J. Geophys. Res., 113, C03016, doi:10.1029/2006JC003983, 2008.

Van Maanen, B., Coco, G., and Bryan, K. R.: Modelling the effects of tidal range and initial bathymetry on the morphological evolution of tidal embayments, Geomorphology, 191, 23–34, doi:10.1016/j.geomorph.2013.02.023, 2013.

Van Rijn, L. C.: Principle of Sediment Transport in Rivers, Estuaries and Coastal Seas, Aqua Publications, the Netherlands, 1993.

Vandenbruwaene, W, Meire, P., and Temmerman, S.: Formation and evolution of a tidal channel network within a constructed tidal marsh, Geomorphology, 151/152, 114–125, doi:10.1016/j.geomorph.2012.01.022, 2012a.

Vandenbruwaene, Wouter, Bouma, T. J., Meire, P., and Temmerman, S.: Bio-geomorphic effects on tidal channel evolution: impact of vegetation establishment and tidal prism change, Earth Surf. Process. Landforms, 38, 122–132, doi:10.1002/esp.3265, 2012b.

Vlaswinkel, B. M. and Cantelli, A.: Geometric characteristics and evolution of a tidal channel network in experimental setting, Earth Surf. Process. Landforms, 36, 739–752, doi:10.1002/esp.2099, 2011.

Wang, Z. B., Hoekstra, P., Burchard, H., Ridderinkhof, H., De Swart, H. E., and Stive, M. J. F.: Morphodynamics of the Wadden Sea and its barrier island system, Ocean and Coastal Management, 68, 39–57, doi:10.1016/j.ocecoaman.2011.12.022, 2012.

Zhou, Z., Olabarrieta, M., Stefanon, L., D'Alpaos, A., Carniello, L., and Coco, G.: A comparative study of physical and numerical modelling of tidal network ontogeny, J. Geophys. Res., in review, 2014.

11

Numerical modelling of glacial lake outburst floods using physically based dam-breach models

M. J. Westoby[1], J. Brasington[2], N. F. Glasser[3], M. J. Hambrey[3], J. M. Reynolds[4], M. A. A. M. Hassan[5], and A. Lowe[6]

[1]Geography, Engineering and Environment, Northumbria University, Newcastle upon Tyne, UK
[2]School of Geography, Queen Mary, University of London, London, UK
[3]Department of Geography and Earth Sciences, Aberystwyth University, Aberystwyth, Wales, UK
[4]Reynolds International Ltd, Suite 2, Broncoed House, Broncoed Business Park, Mold, UK
[5]HR Wallingford Ltd, Howbery Park, Wallingford, Oxfordshire, UK
[6]CH2M HILL, 304 Bridgewater Place, Warrington, Cheshire, UK

Correspondence to: M. J. Westoby (mjwestoby@gmail.com)

Abstract. The instability of moraine-dammed proglacial lakes creates the potential for catastrophic glacial lake outburst floods (GLOFs) in high-mountain regions. In this research, we use a unique combination of numerical dam-breach and two-dimensional hydrodynamic modelling, employed within a generalised likelihood uncertainty estimation (GLUE) framework, to quantify predictive uncertainty in model outputs associated with a reconstruction of the Dig Tsho failure in Nepal. Monte Carlo analysis was used to sample the model parameter space, and morphological descriptors of the moraine breach were used to evaluate model performance. Multiple breach scenarios were produced by differing parameter ensembles associated with a range of breach initiation mechanisms, including overtopping waves and mechanical failure of the dam face. The material roughness coefficient was found to exert a dominant influence over model performance. The downstream routing of scenario-specific breach hydrographs revealed significant differences in the timing and extent of inundation. A GLUE-based methodology for constructing probabilistic maps of inundation extent, flow depth, and hazard is presented and provides a useful tool for communicating uncertainty in GLOF hazard assessment.

1 Introduction

1.1 Moraine-dammed glacial lakes

Glacier recession is occurring globally as a result of recent climatic change (Oerlemans, 1994; Kaser et al., 2006; Zemp et al., 2009; Bolch et al., 2011, 2012). The exposure of terminal and lateral moraine complexes is becoming increasingly commonplace as a result of glacier recession, particularly in high-mountain regions (Hambrey et al., 2008). Moraines reflect the historical maximum extent of a given glacier, and are typically composed of poorly consolidated glacial material. A latero-terminal moraine can present a physical barrier to drainage of glacial meltwater and, in such cases, result in the formation of a moraine-dammed lake (Costa and Schuster, 1988) and create the potential for a glacial lake outburst flood (GLOF) hazard (e.g. Clague and Evans, 2000; Benn et al., 2012; Westoby et al., 2014a; Worni et al., 2014).

Moraine-dammed lakes form through one of two mechanisms: recession of the glacier terminus and ponding of water in the proglacial moraine basin (Frey et al., 2010; Westoby et al., 2014b) or via the coalescence and expansion of supraglacial ponds on heavily debris-covered glaciers (Reynolds, 1998, 2000; Richardson and Reynolds, 2000; Benn et al., 2001, 2012; Thompson et al., 2012). Following expansion, such lakes are capable of impounding volumes of water in excess of 10^6–10^7 m^3 (Lliboutry, 1977; Vuichard and Zimmerman, 1987; Watanabe et al., 1995; Sakai et al., 2000; Janský et al., 2009; Somos-Valenzuela et al., 2014).

Breaching of a moraine dam can result in the generation of a GLOF (Lliboutry 1977; Vuichard and Zimmerman, 1987; Clague and Evans, 2000; Kershaw et al., 2005; Harrison et al., 2006; Osti and Egashira, 2009; Worni et al., 2012, 2014; Westoby et al., 2014a, b). These sudden-onset floods represent high-magnitude, low-frequency catastrophic phenomena that have enormous potential for geomorphological reworking of channel and floodplain environments (Cenderelli and Wohl, 2003; Worni et al., 2012; Westoby et al., 2014b). These floods can pose significant hazards, resulting in the destruction of in-channel and riparian assets, including hydroelectric power facilities, trekking routes, and impact on settlements with ensuing loss of life (Vuichard and Zimmerman, 1987; Lliboutry et al., 1977; Watanabe and Rothacher, 1996). Of the various glacial hazards, GLOFs have far-reaching, distal impacts, with destruction often reported tens or hundreds of kilometres downstream of their source (Vuichard and Zimmerman, 1987; Clague and Evans, 2000; Richardson and Reynolds, 2000).

1.2 Modelling glacial lake outburst floods

A number of approaches to model GLOFs have been presented; these are summarised by Westoby et al. (2014a). At the outset, it is worth highlighting that it is currently not expedient to employ our modelling workflow for GLOF analysis at catchment or basin scales but rather as a logical next step based on the results from a broader hazard assessment screening exercise. Such exercises should, in the first instance, identify and quantify the risk posed by individual glacial lakes in a region through the use of multi-criteria hazard analysis techniques (e.g. RGSL, 2003; Reynolds, 2014) and may also involve the application of DEM-based flood-routing models to provide rapid, first-pass assessments of likely patterns of inundation (e.g. Huggel et al., 2004) that form the basis for subsequent detailed GLOF analysis.

Most GLOF modelling approaches typically employ an empirical or numerical dam-breach model to derive a breach outflow hydrograph, and couple this to a numerical hydrodynamic model to simulate the downstream propagation of the flood wave. Such loose model coupling, or "process chain" simulation, is now a well-established practice and has been applied for reconstructive (e.g. Osti and Egashira, 2009; Klimeš et al., 2014; Westoby et al., 2014b; Worni et al., 2014) and predictive GLOF modelling studies (e.g. Xin et al., 2008; Worni et al., 2013).

Dam-breach modelling typically relies on an empirical or analytical formulation (e.g. Walder and O'Connor, 1997). Empirical models return a single breach hydrograph descriptor, usually peak discharge or flood time to peak, to which a hydrograph is fitted. Such models are derived from historical case study data and use simple geometric descriptors of the moraine or lake, such as dam height and lake surface area, as the sole inputs. Their robustness may be questionable because of their explicit reliance on the case-study data

and their representativeness. Analytical models are similar to their empirical counterparts, but require that the user know the final breach dimensions and formation time, to which an analytical solution for the rate of breach growth is fitted. This approach suffers equally where limited data are available to describe the breach and its rate of formation (i.e. in particular where dams have yet to fail and the breach form and dimensions must be estimated a priori).

Physically based numerical models represent the current state of the art in dam-breach modelling, and combine numerical erosion and sediment transport models in combination with a 1-D or 2-D flow hydraulics solver to simulate breach expansion and channel flow (Worni et al., 2012; Westoby et al., 2014a, b). Despite their advantages, these models are not yet in widespread use by the glacial hazard communities, with recent studies still preferring to adopt established empirical approaches (e.g. Byers et al., 2013). In large part, this slow rate of adoption may reflect the high data requirements of physically based models. For example, these physics-based simulation require knowledge of the mean particle size (or a full particle-size distribution), internal angle of friction, cohesion, and porosity, amongst other physical attributes of the material. Such data are scale-dependent and at best require detailed field or laboratory investigation, which may be logistically challenging in remote, high-altitude environments and may be impossible to obtain for some reconstruction studies.

1.3 Motivation for this study and modelling approach

Numerous sources of predictive uncertainty exist in the parameterisation of contemporary numerical dam-breach and hydrodynamic models when used in either reconstructive or predictive GLOF simulation (Westoby et al., 2014a). Reconstructions of historic events where no field-based or published data exist to describe the geometric and material characteristics of the dam represent the most poorly constrained case. However, the characteristic compositional heterogeneity of moraines implies that even data-rich scenarios are likely to undersample the actual system complexity. Moreover, reconciling the often disparate spatial scales of processes, field observations, and effective model parameters presents a persistent ambiguity even under optimal conditions. For such ill-conditioned problems, an increasingly popular approach is to accept uncertainty in the a priori model parameters, and rather than seek the optimal parameter set through calibration, numerical methods are used to quantify the associated predictive uncertainty and present simulation output probabilistically (Beven, 2005).

To date, few studies have sought to explore the predictive uncertainty of numerical dam-breach models and link this to the downstream consequences in terms of flood wave propagation and inundation (e.g. Wang et al., 2012; Worni et al., 2012; Westoby et al., 2014b). In this paper we address this research gap through the development and demonstration of a unifying framework for cascading uncertainty in a GLOF model chain, illustrated for a reconstruction of a major historical outburst flood in the Nepalese Himalaya. We demonstrate how numerical models representing components of the GLOF process can be coupled to provide probabilistic predictions of the breaching process and downstream flood propagation using generalised likelihood uncertainty estimation (GLUE; Beven and Binley, 1992).

2 Uncertainty and equifinality in numerical modelling

2.1 Types, sources, and the potential significance of uncertainty

Pervasive uncertainty surrounds almost all aspects of environmental modelling, and may be broadly classified as *aleatory* or *epistemic* (Beven et al., 2011). Aleatory uncertainties relate to fundamental random variability in the modelling process. Epistemic uncertainties are associated with a lack of knowledge of system characteristics or architecture. Specific, epistemic uncertainties relate to boundary conditions, initial conditions, parameter values, and model structures (Beven et al., 2014) – in essence, elements that are common to almost all environmental modelling problems.

Similar sources of uncertainty exist in the dam-breach and flood-routing components of the GLOF model chain (Westoby et al., 2014a). In the case of the former, appreciable, and predominantly epistemic, sources of uncertainty surround the establishment of initial conditions (e.g. dam geometry, reservoir hypsometry), dam material parameterisation (e.g. grain-size distribution curves, material porosity, density, cohesion, roughness coefficients, internal angle of friction) and the establishment of computational constraints (e.g. model time step and grid discretisation).

The construction of high-resolution digital terrain models (DTMs) using novel geomatics technologies such as terrestrial laser scanning and low-cost, structure-from-motion photogrammetry have the potential to help constrain these uncertainties (e.g. Westoby et al., 2012, 2014b). For example, these data can be used to extract accurate models of the cross-sectional geometry of a moraine dam and the bathymetry of a (drained) lake basin. However, the accurate quantification of the material characteristics of the dam structure is more difficult to sample effectively and requires logistically challenging fieldwork (e.g. Hanson and Cook, 2004; Osti et al., 2011; Worni et al., 2012). Furthermore, the heterogeneity of moraine sediments is so high that field observations inevitably undersample this variability, so that even the most detailed field measurements must result in significant spatial averaging

The predictive performance of linked hydrodynamic models is strongly influenced by model dimensionality (e.g. Alho and Aaltonen, 2008; Bohorquez and Darby, 2008), grid discretisation strategies (e.g. Sanders, 2007; Huggel et al.,

2004), DEM quality or the frequency of cross-section data (e.g. Castellarin et al., 2009), and the parameterisation of in-channel and floodplain roughness coefficients (Wohl, 1998; Hall et al., 2005; Pappenberger et al., 2005, 2006). Recent studies have undertaken various forms of sensitivity analyses and uncertainty estimation to quantify the effects of uncertainty on numerical dam-breach model output (e.g. IMPACT, 2004; Xin et al., 2008; Dewals et al., 2011; Zhong et al., 2011; Worni et al., 2012).

The uncertainties surrounding model parameterisation are likely to translate into varying characteristics of the breach-outflow hydrograph, specifically peak discharge, time to peak, and the hydrograph form. When these data are then used subsequently as the upstream boundary condition for hydrodynamic modelling, their effects can be manifest in significantly different simulations of downstream flood propagation. This can give rise to strong variations in the celerity of the flood wave, time-varying flow depths and velocities, and variations in inundation extent. Importantly, it is these data that form the basis for flood hazard assessment. Consequently, it is essential not just to better quantify the predictive uncertainty associated with the component models but also to extend this to consider how uncertainty propagates through the model chain and ultimately might influence the development of effective flood mitigation and management strategies (Beven et al., 2014).

2.2 Equifinality in numerical modelling

One consequence of the parametric and structural complexity of many environmental models is the possibility to obtain similar or even identical model outputs through different combinations of input parameters, initial conditions, and model structures (Beven and Binley, 1992; Beven and Freer, 2001; Beven, 2005). Beven and Binley (1992) termed this behaviour model "equifinality". The concept is linked to the more generic form of landscape equifinality used in geomorphology, in which interactions between different processes can give rise to similar landscapes or landform assemblages through differing but equally plausible genetic mechanisms (e.g. Nicholas and Quine, 2010; Stokes et al., 2011). Both forms of equifinality have their origins in systems theory (von Bertalanffy, 1968) and has been identified and quantified in a range of geoscience settings (e.g. Beven and Binley, 1992; Kuczera and Parent, 1998; Romanowicz and Beven, 1998; Beven and Freer, 2001; Blazkova and Beven, 2004; Hunter et al., 2005; Hassan et al., 2008; Vasquez et al., 2009; Vrugt et al., 2009; Franz and Hogue, 2011).

For event-specific reconstructions, calibration of input parameters is often undertaken to identify optimal model parameter sets (e.g. Kidson et al., 2006; Cao and Carling, 2002). Such calibration typically involves a search of the model parameter space to identify optimal model performance with respect to an observed set of observations (e.g. Refsgaard, 1997; Beven and Freer, 2001; Hunter et al.,

2005; Westerberg et al., 2011). However, a problem emerges when the observed calibration data contain insufficient information to uniquely identify a single optimal parameter set, and may highlight the presence of multiple optima within the parameter space that may correspond to different mechanistic solutions to achieve the same end results obtained in the calibration data. This "equifinality" is particularly problematic in the case of complex, spatially distributed models where the number of model parameters vastly outweighs the information content of calibration data, which are often spatially averaged model responses, such as flood wave celerity of peak discharge.

Multiple strategies can be adopted to minimise these effects, and range from reducing model complexity to increasing the variety of calibration criteria (for example involving multiple data streams). However, where an imbalance in between model parameterisation and calibration is persistent, Beven and Binley (1992) argued that any combination of model input that reproduces the observed outcome, within acceptable limits, must be considered equally likely as a simulator of the system under investigation (Beven and Binley, 1992). This principle has been used to advocate a quantitative method to assess model performance, and present results in a probabilistic framework, based on a method termed GLUE (Beven and Binley, 1992; Beven, 2005). In this article, we adopt this numerical approach to quantify the predictive uncertainty associated with a physically based dam-breach model for the reconstruction of a historical outburst flood in the Nepalese Himalaya. In light of the considerable uncertainty that surrounds dam-breach model parameterisation, the application of the GLUE method enables the quantification of the influence which this uncertainty exerts over derivation dam-breach outflow hydrographs. Our approach to probabilistic dam-breach modelling is described below.

3 Numerical dam-breach modelling

Simulations of incipient breach development were based on HR BREACH, a physically based, numerical dam-breach model. This simulation tool predicts the progressive growth of a dam breach initiated by either the overtopping or piping of non-cohesive and cohesive embankment materials (Morris et al., 2008; Westoby, 2013; Westoby et al., 2014b). The model employs physically based hydraulic, sediment-erosion and discrete embankment stability modules to calculate the evolving breach geometry and associated drainage outflow hydrograph (Morris et al., 2008). This approach offers a significant advance over simplified, empirically derived breaching models or analytical methods that fit the pattern of breach expansion to user-defined final breach dimensions and formation time.

Breach enlargement is simulated through the interaction of two mechanisms: (i) continuous hydraulic erosion based on either equilibrium sediment-transport equations or erosion-depth equations and (ii) discrete mass failures as a consequence of side-slope instability (Mohamed et al., 2002). In this study, erosional processes were modelled using the erosion-depth equation for non-cohesive embankments, after Chen and Anderson (1986):

$$\varepsilon_r = K_d (\tau_e - \tau_c)^a, \tag{1}$$

where ε_r is the detachment rate per unit area (e.g. $m^3 s^{-1} m^{-2}$), τ_e is the flow shear stress (kN) at the breach boundary, τ_c is the critical shear stress required to initiate particle detachment (kN), and K_d and a are dimensionless coefficients based on the sediment properties. Bending (tensional) failure is represented by a moment of force, M_o:

$$M_o = We + W_s e_s + W_u e_u + H_2 e_2 - H_1 e_1, \tag{2}$$

where W is the weight of a dry block of soil (kN); W_s is the weight of a saturated block of soil (kN); W_u is the weight of a submerged block of soil (kN); H_1 is the hydrostatic pressure force in the breach channel (kN); H_2 is the hydrostatic pressure force inside the dam structure (kN), e, e_s, and e_u are weight-force eccentricities (m); and e_1 and e_2 are hydrostatic pressure-force eccentricities (m) (Hassan and Morris, 2012). The occurrence of shear-type failure is evaluated through the analysis of factor-of-safety (FoS) values using the following equation:

$$FoS = \frac{cL + H_1 \tan \Phi}{W + W_s + W_u + H_2 \tan \Phi}, \tag{3}$$

where c is soil cohesion (kN m^{-2}), L is the length of the failure plane (m), and Φ is the soil angle of friction ($^\circ$).

Subaerial flows across and through the evolving breach are represented using a steady-state one-dimensional flow model (Mohamed et al., 2002; Morris et al., 2008). This model combines a weir discharge equation and a simplified version of the Saint-Venant equations (Chanson, 2011) to simulate breach flow and flow descending the distal face of the dam, respectively. A variable weir discharge coefficient (C_d) is used to reflect the changing profile of the dam crest and is incorporated into calculations of breach discharge (Q_b) as follows (Hassan and Morris, 2012):

$$Q_b = C_d L H, \tag{4}$$

where L and H are the length and height of the dam crest, respectively. While a significant simplification, this approach accounts for some of the complexity that arises during breach development. Primary model output is time-step-specific breach discharge, which reflects water-only breach flow.

Application of the model requires definition of the following key boundary and initial conditions: (i) geometric descriptors of the dam, specifically proximal and distal dam face slope, dam length, crest width, spillway dimensions (width and height), and downstream valley slope; (ii) lake or reservoir hypsometry based on either stage–area or stage–volume curves. In addition, parameter range estimates and their likelihood distributions (either uniform, linear, or triangular), and computational settings (model time step, total run time, and minimum wetted depth threshold) must be defined. The version of the model used in this study incorporates a Monte Carlo parameter sampling routine to automatically search the model parameter set and generate multiple realisations of the potential drainage hydrographs and predicted breach geometry at specific time steps. The input scenarios modelled are described in Sect. 5.3.

4 Study site

The Dig Tsho moraine-dammed lake complex is located at the head of the Langmoche Valley in the western sector of Sagarmatha (Mt. Everest) National Park in the Khumbu Himal, Nepal (Fig. 1; 27°52'23.89" N, 86°35'14.91" E). A relict moraine-dammed lake, oriented west–east, is impounded by a latero-terminal moraine complex to the north and east. The basin is bounded by a near-vertical bedrock face to the south. The parent Langmoche Glacier has receded approximately 2 km since its maximum Neoglacial terminal position and is now confined to the upper north-east face of Tangi Ragi Tau (6940 m).

The moraine possesses steep (25–30°) ice-distal and ice-proximal faces along the northern lateral moraine. In contrast, ice-proximal faces of the terminal moraine are generally shallower, but they are more topographically complex than their distal counterparts. The moraine is composed of sandy boulder gravel, and includes rare large (> 10 m diameter) boulders. A 200 m wide, 400 m long, 60 m deep breach dissects the northern sector of the terminal moraine (Fig. 1; Vuichard and Zimmerman, 1987; Westoby et al., 2012). Breach dimensions conform to those for other documented Nepalese moraine-dam failures (Mool et al., 2001).

On 4 August 1985 an ice avalanche from the receding Langmoche Glacier traversed the steep (~30°) avalanche and debris cone into the proglacial lake and triggered a major displacement wave. This initial wave is believed to have overtopped the moraine dam and initiated its failure by scour of the moraine crest. Although no directly documented observations are available, anecdotal accounts suggest the breach formation occurred within 4–6 h (Vuichard and Zimmerman, 1987). The geomorphological effects of the Dig Tsho GLOF have been well documented elsewhere (Vuichard and Zimmerman, 1986, 1987; Cenderelli and Wohl, 2001, 2003). Downstream, the flood created a major hazard, destroying a newly completed hydroelectric power installation at Thamo

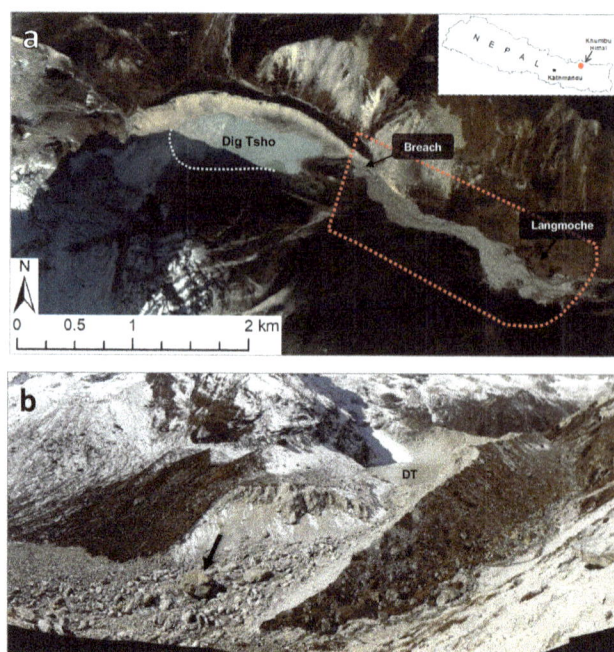

Figure 1. The Dig Tsho moraine dam and upper reaches of the Langmoche Khola. (**a**) True-colour GeoEye imagery of the study site. Extensive reworking of the valley floor by escaping floodwaters is evident immediately downstream of the moraine breach. Red dashes indicate two-dimensional model domain. (**b**) The moraine-dam breach, formed by the 1985 GLOF. For scale, the boulder indicated by the black arrow is > 10 m in height and diameter. The breach is ~ 200 m at its widest point, and has a maximum depth of 60 m. The relict Dig Tsho glacial lake is visible in the background (DT). The progressive concentration of such boulders in the breach thalweg during moraine down-cutting and side-wall failure may have served to modify the final breach geometry.

(~ 11.5 km from source) and all trekking bridges for a distance of > 30 km. Up to five people were reported to have been killed (Galay, 1985).

Existing observational data for the site include first-hand descriptions of the condition of the glacial lake prior to, and immediately following, moraine dam failure in 1985, as well as a description of the pre-GLOF nature of the valley-floor sedimentology immediately downstream (Vuichard and Zimmerman, 1987). From these observations, an estimate of the total mass of material removed from the breach was calculated as 9×10^5 m^3, and that of the drained volume of the moraine basin as 5×10^6 m^3. The vast majority of the dam material eroded during breach development was deposited within 2 km of the moraine. More recent reconstructions of the moraine and lake basin involving high-precision 3-D photogrammetric reconstructions by Westoby et al. (2012) have challenged these initial estimates and suggest the volume of the contemporary breach (surveyed in 2010) to be approximately 5.8×10^5 m^3, and the volume of water released by the 1985 GLOF to be ~ 7.3×10^6 m^3. Sedimentological (D_{50})

sampling of matrix material exposed in the breach sidewalls was undertaken during field investigation, and was used to aid dam-breach model parameterisation.

5 Method

In order to quantify the predictive uncertainty associated with the coupled modelling system used here, a sequential workflow was developed that involved the following steps:

i. Topographic data were derived from digital terrain modelling using terrestrial structure-from-motion (SfM) photogrammetry to reconstruct pre-existing moraine and floodplain topography and extract glacial lake bathymetry and dam geometry data.

ii. A priori parameter values for the numerical dam-breach model were estimated using material properties obtained from field investigation and from the published literature.

iii. A series of potential breach initiation scenarios were hypothesised to account for uncertainty in the principal driving factors behind failure (overtopping and mechanical failure).

iv. For each scenario, the multiple model outputs were weighted using a multi-criterion likelihood statistic based on a comparison of the modelled and observed final breach geometries. This provided a means to weight the ensemble of predicted discharges at each time step and derive probabilistic hydrograph forecasts for each scenario.

v. Two-dimensional hydrodynamic modelling was used to simulate the downstream propagation of the probabilistic outflow hydrographs, using upper and lower 95 % confidence intervals to describe the range of potential flows. These simulations provide a range of estimates for the potential inundation extent, flow velocity, and flow depth data for use as input to probabilistic GLOF hazard mapping.

The following sections describe each method in detail.

5.1 Topographic data acquisition

Three topographic models are required for this analysis: (i) a reconstruction of the pre-failure moraine dam and lake basin; (ii) detailed models of the breach topography to determine the position, width, length, and slope of the breach and reconstruct the drainage volume; and (iii) the downstream floodplain topography. These models were derived using terrestrial photogrammetry, based on novel structure-from-motion and multi-view stereo (SfM-MVS). A full description of the image processing is beyond the scope of this paper and the reader is referred to more comprehensive accounts elsewhere

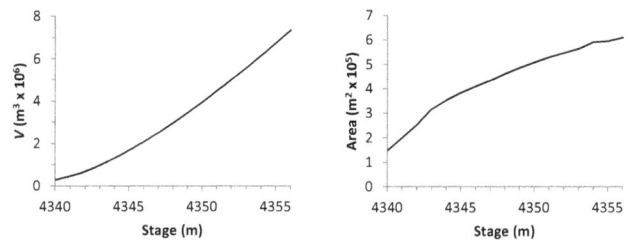

Figure 2. Bathymetric data describing (**a**) volume–stage and (**b**) area–stage relationships, extracted from a fine-resolution, structure-from-motion-derived digital terrain model of the Dig Tsho moraine-dammed lake basin (Westoby et al., 2012).

(James and Robson, 2012; Westoby et al., 2012; Fonstad et al., 2013; Javernick et al., 2014). Briefly, SfM-MVS involves the reconstruction of 3-D point cloud data from highly redundant photographic data sets that have a high degree of image overlap. Unlike traditional photogrammetry, SfM uses feature tracking between images and bundle adjustment methods to reconstruct the camera alignment and reconstruct the 3-D scene geometry (Lowe, 2004; Snavely, 2008; Snavely et al., 2008).

In this study, we use published high-resolution reconstructions of the pre- and post-flood moraine topography derived by Westoby et al. (2012). These terrain models were used to quantify the moraine geometry and lake basin hypsometry (Fig. 2) for the establishment of initial conditions for numerical dam-breach modelling.

A topographic model of the downstream floodplain was obtained using similar methods. The analysis was based on two photosets, incorporating 226 and 303 photographs obtained from south- and north-facing transects along the valley flanks. Photographs were acquired with a Panasonic DMC-G10 (12 MP) digital camera. Automatic focusing and exposure settings were enabled during photograph acquisition. The freely available software bundle SFMToolkit3 (Astre, 2010) was used to process the input images using feature extraction (Lowe, 2004) and sparse and dense SfM-MVS reconstruction algorithms. Following manual outlier removal and editing, final dense point clouds numbered 4.8×10^6 and 1.3×10^6 points for north- and south-facing photosets, respectively. Data transformation was achieved through the identification of clearly identifiable boulders from kite aerial photography and an existing SfM-DTM of the easternmost sector of the floodplain domain, as well as boulders visible in the georeferenced DTM of the moraine dam. Averaged georegistration residual errors for both dense point clouds were 1.37, 0.30, and 0.06 m for x, y, and z dimensions, respectively.

Georeferenced point cloud data were merged and decimated to improve data handling whilst preserving sub-grid statistics using the C++/Python-based Topographic Point Cloud Analysis Toolkit point cloud decimation toolkit (Brasington et al., 2012; Rychkov et al., 2012). A bare-earth DTM at 8 m spatial resolution was extracted for hydrodynamic GLOF routing. Sensitivity analyses have previously revealed this particular grid discretisation to produce time-varying inundation extents and wetting-front travel times of comparable (relative) accuracies to those produced using finer grid resolutions, and of a higher accuracy than coarser grids (Westoby et al., 2014a).

5.2 Dam-breach model setup

5.2.1 Model parameterisation

The first step in the GLUE workflow is to establish the parameters for inclusion and their respective ranges. In the absence of any prior information regarding parameter and range choice, all available input parameters and their entire simulation range should be included. In practice, complete uncertainty regarding parameter and range choice is unlikely, since a combination of initial sensitivity analyses, modelling guidelines, and basic intuition and reasoning can typically be used to assist in constraining their choice (Beven and Binley, 1992). The parameters and their ranges used in this study (including computational settings) are displayed in Table 1.

Information regarding initial, or a priori, parameter distributions is also required, reflecting the modeller's prior knowledge of the parameter space (Beven and Binley, 1992). In the absence of any information pertaining to a priori parameter probability distributions at Dig Tsho, uniform distributions were used for stochastic sampling of all available parameter spaces.

5.2.2 Model perturbations

In many instances, the specific mechanisms of moraine dam failure are poorly documented, often as a result of a lack of direct observation. Post-event glaciological and geomorphological analysis of a moraine, its parent glacier, and their surroundings may provide clues as to the triggering event that causes moraine failure or a stand-alone GLOF event (e.g. Hubbard et al., 2005; Osti et al., 2011). However, a degree of uncertainty may still surround the establishment of the precise trigger(s) and the nature of dam failure.

Our approach to dam-breach modelling comprises the simulation of three modes of breach initiation, namely (i) a control scenario where lake waters have risen gradually to a point where the overtopping discharge is large enough to trigger sustained down-cutting and breach development; (ii) breach initiation by a series of overtopping waves, such as those resulting from the rapid input of a mass of rock or ice, that traverse the lake and overtop the moraine, thereby initiating its failure; and (iii) instantaneous mechanical failure of the dam face. All three modes are documented triggers for moraine-dam failure (Vuichard and Zimmerman, 1986; Richardson and Reynolds, 2000; Worni et al., 2012). The trigger mechanism for the failure of the Dig Tsho moraine is generally accepted to be repeated overtopping and down-cutting of the

Table 1. Parameter ranges and geometric characteristics of the Dig Tsho moraine dam used for input to the HR BREACH numerical dam-breach model. Input parameter ranges were established from a combination of initial experimentation and parameter sensitivity analysis (E), in situ field observation (F) and data from similar sites stated in the literature (L). All dam geometry data were extracted from the SfM-DTMs of the moraine and floodplain (SfM-DTM), with the exception of downstream Manning's n, which reflects a sedimentological valley-floor characterisation of gravels, cobbles, and rare large boulders (after Chow, 1959).

Input parameter/ geometric characteristic	Range/ value	Source (E/F/L)
Parameter		
Sediment flow factor	0.8–1.2	E
Erosion width : depth ratio	0.5–2	E
D_{50} (mm)	28–200	F, L (Worni et al., 2012; Xin et al., 2008)
Porosity (% voids)	0.01–0.3	E
Density (KN m^{-3})	19–24	E
Manning's n (m$^{-1/3}$ s)	0.02–0.05	F, L
Internal angle of friction (°)	25–42	E, L (Lebourg et al., 2004)
Cohesion (KN m^{-2})	0–100	E
Dam geometry		
Crest level (m OD)	4356	
Foundation level (m OD)	4316	
Crest length (m)	250	
Crest width (m)	30	SfM-DTM
Distal face slope (1 : x)	3.1	
Proximal face slope (1 : x)	15.2	
Downstream valley slope (1 : x)	0.19	
Downstream Manning's n	0.04	L (Chow, 1959)

moraine by waves generated from an ice avalanche entering the lake (Vuichard and Zimmerman, 1986). The purpose of including two additional breach initiation scenarios was to explore whether equifinal final breach morphologies would be produced from a variety of breach initiation scenarios. Scenario design is described in the following sections.

Control scenario

A control scenario was formulated in which breach formation was initiated through down-cutting of a predefined spillway. Inclusion of an existing spillway conforms to pre-GLOF observations of the Dig Tsho moraine by Vuichard and Zimmerman (1986), and we note that many extant moraine-dammed lakes are drained in this manner (e.g. Hambrey et al., 2008). This down-cutting is a result of flow produced by the pressure head associated with our specified initial lake level (which mirrored the reconstructed dam crest elevation of 4356 m a.s.l.). The modelled spillway measured 0.5 m wide and 0.5 deep and extended from the upstream end of the moraine crest to the dam toe. We note that, in the absence of detailed observations of the spillway prior to dam failure (Vuichard and Zimmerman, 1987), these dimensions are hypothetical and do not necessarily replicate the precise spillway conditions prior to dam failure in 1985.

In addition to a control scenario (DT_control), two styles of system perturbation scenario were introduced to the dam-breach models to explore and quantify the impact of system-scale perturbations on model output. These perturbations were (i) the introduction of overtopping waves of varying magnitude and (ii) the instantaneous removal of material from the downstream face of the dam immediately prior to breach development

Overtopping waves

The failure of Dig Tsho has been attributed to the overtopping of the terminal moraine by waves produced by an ice avalanche from the receding Langmoche Glacier (Vuichard and Zimmerman, 1987). Presently, most numerical dam-breach models, including HR BREACH, are unable to explicitly simulate the dynamic effects of avalanche–lake interactions. This necessitated an inventive yet relatively simple approach to reconstructing overtopping behaviour in HR BREACH. Instead of simulating the passage of a series of gradually attenuating displacement or seiche waves and dynamic interaction with the dam structure, a solution was devised which involved the rapid increase and subsequent decrease of the lake water surface elevation. These artificial water level variations prompted short-lived overtopping of the dam structure at predefined intervals. Temporary increases in lake level were achieved through systematic variation of reservoir inflow to introduce maximum overtopping discharges of 4659, 3171, and 1809 $m^3 s^{-1}$, representing initial volumes of the triggering avalanche of 2.0×10^5, 1.5×10^5, and $1.0 \times 10^5 m^3$, respectively. This overtopping discharge range reflects the estimated volumetric range provided by Vuichard and Zimmerman (1987). Each initial wave is followed by successive waves of exponentially decaying magnitude. Overtopping wave scenarios were named DT_overtop_min, DT_overtop_mid, and DT_overtop_max, respectively.

Instantaneous mass removal

Richardson and Reynolds (2000) suggested that the initial failure of a moraine dam may be "explosive" in nature. This explosive force is reflected by the mass of individual transported clasts, which can exceed $> 100 t$ (Richardson and Reynolds, 2000). Osti et al. (2011) documented the destabilisation and landsliding of morainic material from the Tam Pokhari moraine dam, Nepal. Seismic activity and excessive rainfall caused oversaturation and subsequent partial failure of a section of the proximal dam structure, which contributed to its eventual failure and the generation of a GLOF.

HR BREACH is currently unable to simulate "explosive" or rapid, large-scale rotational mass failures. In order to simulate the instantaneous removal of material from the dam structure, three mass-removal scenarios were developed. HR BREACH requires that the user specify an initial breach spillway in the dam crest and downstream face of the dam. We use these spillway dimensions to represent the mass of morainic material "instantaneously" removed from the crest and distal face of the dam. However, this is far from an ideal numerical realisation of the precise mechanics of catastrophic, near-instantaneous dam failure, and represents a highly simplified approximation.

The default spillway dimensions used in the control and overtopping experiments were 0.5 m wide and 0.5 m deep. Perturbed spillways cross-sectional dimensions were 1, 3, and 5 m^2, representing total removal volumes of 159, 1435, and 3987 m^3 of material and named DT_instant_1, DT_instant_3, and DT_instant_5, respectively,

5.3 Stochastic parameter space sampling

Although a range of stochastic sampling methods are available for use in the GLUE workflow (e.g. Metropolis, 1987; Iman and Helton, 2006), perhaps the most widely used approach is Monte Carlo analysis (e.g. Beven and Binley, 1992; Kuczera and Parent, 1998; Aronica et al., 1998, 2002; Blazkova and Beven, 2004). With the Monte Carlo method, values from individual parameter spaces are sampled in a truly random manner, thereby eliminating any subjectivity which might be introduced at this stage. The method is fully capable of accounting for predefined probability distribution data, making it a simple yet effective tool for rapidly generating the random parameter ensembles required for model input. However, with a poorly defined prior distribution, or a small number of model simulations, point clustering in specific regions of the parameter space may occur. Clustering can be easily avoided by undertaking a straightforward investigation into patterns of histogram convergence of an output variable (or variables), whereby the minimum number of simulations required for the production of an acceptable level of convergence is established. Equally, alternative stochastic sampling methods, including Latin hypercube sampling, may be employed (e.g. Hall et al., 2005; Iman and Helton, 2006). When faced with multi-dimensional problems, the Latin hypercube method can be used to partition the probability distribution of each input parameter, before proceeding to sample from each partition. This method thereby avoids the clustering that is associated with an insufficient number of Monte Carlo samples and ensures that the final parameter ensembles are representative of the full sampling space of each parameter.

In this study, sustained histogram convergence became minimal after the execution of 1000 model runs. This number of simulations was therefore deemed acceptable for stochastic sampling. However, we note that this number of simulations might not be directly applicable to other sites, and we highly recommend that modellers undertake their own pre-

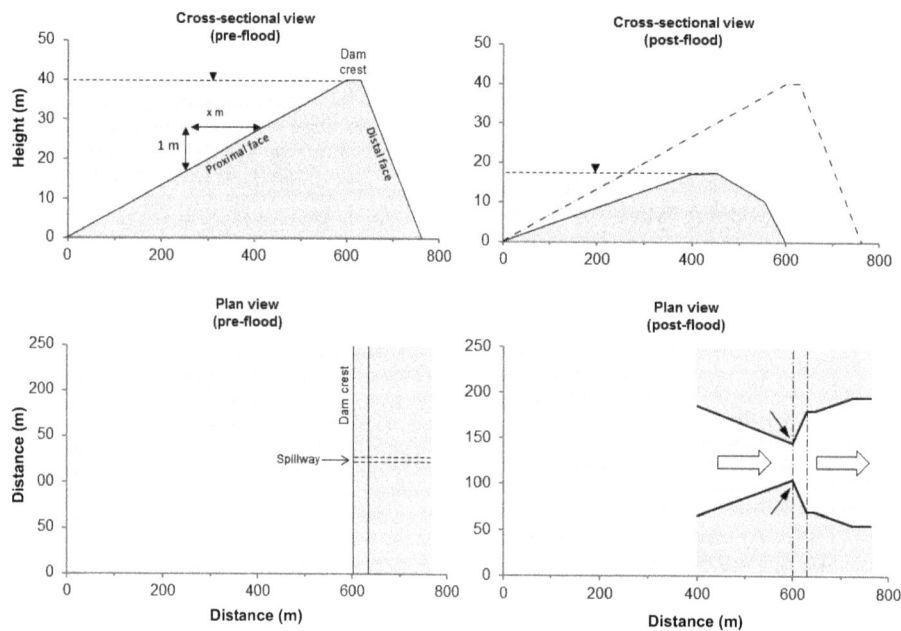

Figure 3. Schematic diagrams of initial (pre-flood) and an example of post-flood geometry for the Dig Tsho moraine dam (left and right columns, respectively), as modelled by HR BREACH. Initial dam face slope angle is represented as a ratio of the form $1:x$. White arrows indicate flow direction. Solid black arrows (post-flood, plan view panel) highlight location of the critical flow constriction and used as a likelihood measure for model evaluation. Note vertical exaggeration of dam elevation.

liminary sensitivity analyses before establishing a sufficient number of simulations to undertake.

5.4 Model evaluation

5.4.1 Likelihood measures and functions

Model evaluation is achieved through quantification of how well a parameter ensemble performs at reproducing a series of observable system-state variables, or "likelihood measures". Parameter ensembles that are unable to do so are deemed to be non-behavioural and are assigned a likelihood score of zero. In contrast, ensembles that reproduce these variables within acceptable limits are deemed to be behavioural and accepted for further analysis. It is not uncommon for all ensembles to be rejected (e.g. Parkin et al., 1996; Freer et al., 2002), thereby suggesting that it is the model structure that is incapable of reproducing the observed data, instead of individual parameter combinations (Beven and Binley, 1992). Such a situation may be overcome by widening the limits of acceptability, but at the potential cost of decreasing confidence in any newly accepted ensembles (Beven and Binley, 1992).

Behavioural ensembles were assigned positive likelihood values in the range 0–1, where 1 represents an ensemble that is capable of perfectly replicating the observed data. Likelihood functions are specific to each likelihood measure used for model evaluation. Three likelihood measures were used to evaluate model performance: (i) final upstream

breach depth (LH1); (ii) the residual sum of squared errors of the final longitudinal elevation profile of the breach (LH2); and (iii) the location of the critical flow constriction (LH3, Fig. 3). These morphological variables are directly quantifiable by comparing final modelled breach geometry with that extracted from the SfM-DTM of the breached moraine dam. Dam breaching is a fully three-dimensional problem that involves progressive backwasting, mass slumping, and downcutting of morainic material by escaping lake waters. The likelihood measures described above are two-dimensional approximations of the breach geometry, and were deemed appropriate descriptors of the breaching process since in combination they describe the breach end states in the vertical and both horizontal dimensions and relate to the modelled and observed lateral, longitudinal, and vertical expansion of the breach.

An observed final upstream breach depth of 16 ± 2 m was used as the first likelihood measure. We note that this value essentially reflects the maximum depth of lake lowering and, by definition, the volume of released lake water. This depth, or lowering, was quantified using an assumed dam freeboard of zero prior to breach initiation, and differs from the previously described maximum breach depth (~ 60 m) owing to the slope gradient of the breach channel. The error range was designed to account for observed SfM model georegistration errors, which may have resulted in elevation of the lake exit being under- or overestimated. This range also accounts for any post-GLOF lake lowering or seasonal variations in lake level, which remain unquantified. A triangular likelihood function was used, with an observed breach depth of 16 m and upper and lower limits defined as 14 and 18 m, respectively.

The second likelihood measure is a direct comparison between observed (post-GLOF) and modelled elevation profiles of the breach thalweg. This measure represents a distributed method of quantifying the performance of HR BREACH in producing post-GLOF thalweg elevation profiles that replicate that observed in the field. Thalweg elevation data were directly extracted from the SfM-DTM (Fig. 12a; Westoby et al., 2012). The residual sum of squares (RSS) method was used to quantify the deviation between the observed and modelled data:

$$RSS = \sum_{i=1}^{n} \left(Z_{obs.} - Z_{pred.} \right)^2, \qquad (5)$$

where $Z_{obs.}$ is the observed elevation (m) of the breach thalweg and $Z_{pred.}$ is the elevation of the modelled thalweg. For the attribution of a scaled likelihood value, an RSS of 0 corresponded to a likelihood score of 1, whilst the lowest RSS for all scenarios (169 221, dimensionless) was used to represent the lower, non-behavioural limit (i.e. a likelihood of zero). A simple linear likelihood function was applied between these upper and lower values.

Choice of an observed value for the location of the critical flow constriction was complicated by the asymmetry of the observed breach planform, whereby flow constrictions on either side of the breach are offset by a distance of approximately 40 m. This asymmetry is most likely a function of complex flow hydraulics and patterns of erosion of the moraine during development and expansion of the breach. Specifically, the wide grain-size distribution of the moraine, which comprises material ranging from silts and sands to boulders with intermediate diameters greater than 5 m, combined with its unconsolidated nature causes the breach enlargement process to differ markedly from a systematic, largely uniform style of expansion typically modelled by numerical breach models. Specifically, side-wall detachment and emplacement of large boulders in the breach thalweg may serve to impede or divert breach flow, such that the breach planform adjusts in response to locally altered flow directions and magnitudes. This behaviour may be one explanation for the observed breach asymmetry. Whilst our numerical model accounts for undercutting and mechanical failure of the breach walls, it is unable to resolve flow obstructions caused by individual large clasts.

Use of a trapezoidal likelihood function would have been possible, whereby any modelled values that fell within the observed constriction "offset zone" would be assigned a likelihood of 1. However, such an approach was deemed inappropriate, because it would render a significant proportion of parameter ensembles as absolutely behavioural. Instead, a triangular likelihood function was used. The mid-point of the observed offset was used as the central, observed value,

and a range of ± 40 m was applied to encompass observed asymmetry in flow constriction location.

5.4.2 Likelihood updating

Where multiple likelihood measures are used, it is necessary to arrive at a final, global likelihood value for each behavioural parameter ensemble. Bayesian updating represents a statistically robust method for combining multiple likelihood values. It is able to account for the influence of prior likelihood values on the generation of updated values as more data become available. In the context of this study, the initial prior likelihood value relates to final breach depth. Likelihood values from the second (breach-elevation profile) and third (location of the critical flow constriction) are subsequently combined with this initial likelihood through implementation of Bayesian updating, which is summarised as (modified from Lamb et al., 1998)

$$L\left(\Theta_i | Y_{1,2}\right) = \frac{L\left(\Theta_i | Y_1\right) L\left(\Theta_i | Y_2\right)}{C}, \qquad (6)$$

with

$$C = \left(L\left(\Theta_i | Y_1\right) L\left(\Theta_i | Y_2\right)\right) + \left(L\left(\Theta_i | Y_1^i\right) L\left(\Theta_i | Y_2^i\right)\right), \quad (7)$$

where $L\left(\Theta_i | Y_{1,2}\right)$ is the conditional posterior likelihood for a parameter ensemble, Θ_i, conditioned on two sets of observations, Y_1 and Y_2; $L\left(\Theta_i | Y_1\right)$ is the likelihood of Θ_i conditioned on an initial set of observations (Y_1); $L\left(\Theta_i | Y_2\right)$ becomes the likelihood conditioned on a second, or additional, set of observations (Y_2); and C is a conditional operator, or scaling constant, to ensure the cumulative posterior likelihood equals unity (Lamb et al., 1998; Hunter et al., 2005). The "i" superscript refers to the inverse of the relevant conditional likelihood values.

5.4.3 Cumulative distribution function generation

Final likelihood values associated with each behavioural parameter ensemble reflect the confidence of the modeller in the ability of each ensemble to reproduce an observed data set. Considering the cumulative distribution of these global likelihood values as a probabilistic function facilitates an assessment of the degree of uncertainty associated with the behavioural predictions (Beven and Binley, 1992). These data are referred to as cumulative distribution functions (CDFs).

Measure-specific likelihood values for each behavioural ensemble were re-scaled to sum to unity. The final measure can be treated as a surrogate for true probability, but cannot be used for subsequent statistical inference (Hunter et al., 2005). Weighted and re-scaled ensembles are ranked and plotted as a CDF curve (Fig. 4), from which cumulative prediction limit data can be extracted (Beven and Binley, 1992). The generation of weighted CDFs is unique to the GLUE approach and represents a multivariate, additive method that accounts for ensemble performance.

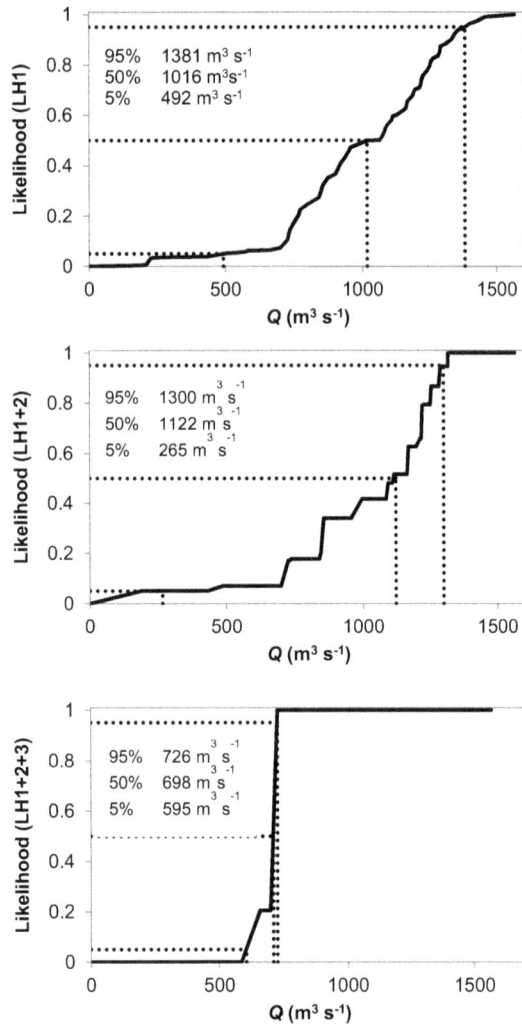

Figure 4. DT_control cumulative density function (CDF) data for $T = 150$ min (solid line). Changes to likelihood values resulting from Bayesian updating are ultimately translated to changes in the form of the CDF curve, from which percentile Q data are extracted at 5, 50, and 95 % confidence intervals (dashes). LH1 = final breach depth; LH1 + 2 = final breach depth + centreline elevation profile; LH1 + 2 + 3 = final breach depth + centreline elevation profile + flow constriction location.

5.5   Two-dimensional hydrodynamic modelling

5.5.1   ISIS 2-D

The shallow-water flow model ISIS 2-D (Halcrow, 2012) was used to simulate GLOF propagation. The model includes alternating direction implicit (ADI) and MacCormack total variation diminishing (TVD) two-dimensional solvers for hydrodynamic simulation. The ADI scheme solves the shallow-water equations (SWEs) over a regular grid of square cells. Water depth is calculated at cell centres, and flow discharges at cell boundaries. The SWEs are solved by subdividing the computation into x and y dimensions. Water depth (m) and discharge (q_x, per unit width) are solved in the x dimension in the first half of the time step. Water depth and discharge (q_y) in the y dimension are solved in the latter half of the time step. Other variables are represented explicitly for each stage. Since water depths are calculated at the cell centre, depths at cell boundaries are interpolated. Different interpolation methods are used, depending on water depth (Liang et al., 2006). Treatment of the friction term is also depth-dependent, such that, below a user-defined threshold, a semi-implicit scheme is used to improve model stability with decreasing flow depth (Halcrow, 2012). The ADI scheme provides accurate solutions for flows where spatial variations are smooth. Sudden changes in water elevation and flow velocity may give rise to numerical oscillations, making it largely unsuitable for the simulation of transcritical flow regimes (Liang et al., 2006).

In contrast, the MacCormack-TVD scheme uses predictor and corrector steps to compute depth and discharge for successive time steps. A TVD term, Var(h), is added at the corrector step to suppress numerical oscillations near sharp gradients, making it highly suited for the simulations of rapidly evolving, transcritical, and supercritical flows (Liang et al., 2006), including sudden-onset floods arising from dam breach (Liang et al., 2007; Liao et al., 2007). Var(h) is calculated as

$$\text{Var}(h) = \int \left| \frac{\delta h}{\delta x} \right| dx. \tag{8}$$

A far smaller time step is required to achieve numerical stability, resulting in extended model run times (Halcrow, 2012).

5.5.2   Hydrodynamic solver comparison

A comparison was carried out to assess which of the two-dimensional solvers would be more appropriate for GLOF simulation. The comparison comprised the simulation of a single dam-breach hydrograph across a reconstructed digital terrain model of the Langmoche Khola (Fig. 5). A $4\,\text{m}^2$ grid discretisation was used, in conjunction with a 0.1 and 0.04 s time step for the ADI and TVD schemes, respectively (following guidance in the model documentation). A global Manning's n roughness coefficient of 0.05 was used and reflects a channel bed and margins composed predominantly of pebbles, cobbles, and large boulders, which is characteristic of floodplains in alpine settings, and also reflects our field-based sedimentological observations. The spatial resolution of the valley-floor grid was finer than that used for subsequent probabilistic GLOF mapping, and was intended to represent the finest amount of topographic complexity on the floodplain. An ASTER GDEM-derived digital elevation model was appended to the downstream boundary of the SfM-derived DTM of the Langmoche Khola to allow water to exit the domain of interest unimpeded and eliminate any artificial upstream tailwater effects.

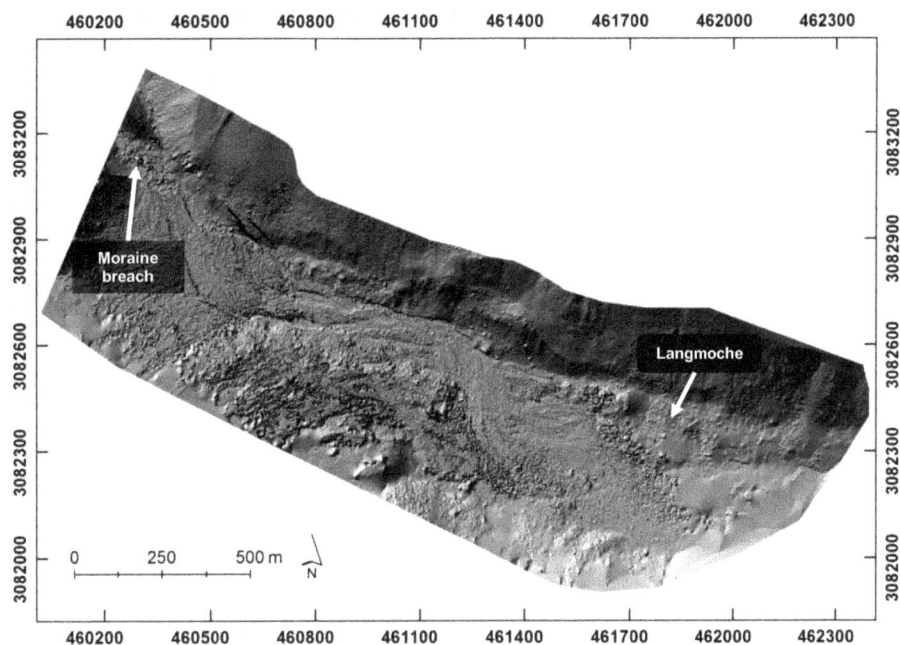

Figure 5. Hillshaded DTM of the Langmoche Valley floor (0–2.2 km from breach), produced using terrestrial photography from the valley flanks in combination with structure-from-motion photogrammetric processing techniques. Elevation data were extracted at 1 m^2 grid resolution (from detrended mean cell elevation) from the geo-referenced data (UTM Zone 45 N). DTMs of the breached moraine-dam complex are displayed in Westoby et al. (2012). Individual boulders up to ~5 m in diameter are resolved. Areas of weak reconstruction to the south-east of the domain are attributed to poor photographic density and topographic obscurement.

Routing of a breach hydrograph of 5 h duration took approximately 2.5 and 0.5 h of simulation time for the TVD and ADI solvers, respectively. The computational burden of the ISIS 2-D TVD solver far exceeds that of the ADI scheme for identical model setups, owing to the finer temporal resolution required by the TVD solver. Depth-based inundation maps for the results of each solver were created in ISIS Mapper® and exported for display in ArcGIS (Fig. 6). A difference image of inundation was also created. Floodwaters follow the channel thalweg for both solvers during the 1 h. Thereafter, increasing flow stage results in the inundation of a wide reach between 1.1 and 1.7 km. Total inundation of this particular reach is achieved by 01:15 h for the ADI solver, and ~02:00 h for the TVD solver. In the early stages of the GLOF (0–45 min), the travel distance of the flood wave front is slightly greater for the TVD solver. After 1 h, areas of floodplain inundated exclusively by the TVD code are confined to a zone immediately south of the moraine breach. This difference becomes increasingly evident following the onset of floodwater recession (~3 h onwards).

Inundated area is greater for the ADI solver for all time steps (Fig. 6). Maximum difference in inundation extent (122 816 m^2) coincides with the upstream flood peak at 2.5 h. Significant dynamic differences also arise between the two solvers. The most prominent of these is the onset of severe oscillatory behaviour from an early stage (~0.5 h) in the ADI data (Fig. 6). These oscillatory behaviours, which are

manifested as discrete, arcuate "waves" of excessive flow depth (often > 10 m), are oriented perpendicular to flow, span the entire width of the main channel in places, and occur at regularly spaced intervals. These oscillatory features in fact signify dramatic numerical oscillations in the ADI code and may be the result of numerical "shocks" being aligned with the topographical grid (e.g. Meselhe and Holly, 1997; Venutelli et al., 2002). In contrast, the TVD data reproduce flow channelisation far more clearly, apparently free of any significant oscillatory behaviour.

The results support the use of the two-dimensional TVD solver for GLOF simulation. The lack of any significant numerical instability, otherwise prevalent in the ADI results, is the predominant advantage of the TVD solver. Although processing times are considerably lengthier for the TVD solver, these were deemed acceptable. The solver used in this study simulates only clear-water flows, with no consideration of sediment entrainment, transfer, and depositional dynamics, including any impact on flow rheology. By default, our breach model does not output time-step-specific sediment outflow discharges. It is possible to undertake a crude calculation of time-varying sediment production through the interpolation and differencing of successive breach cross-sectional geometries (see Fig. 7 in Westoby et al., 2014a). These data would in theory be suitable for input to multi-phase hydrodynamic modelling.

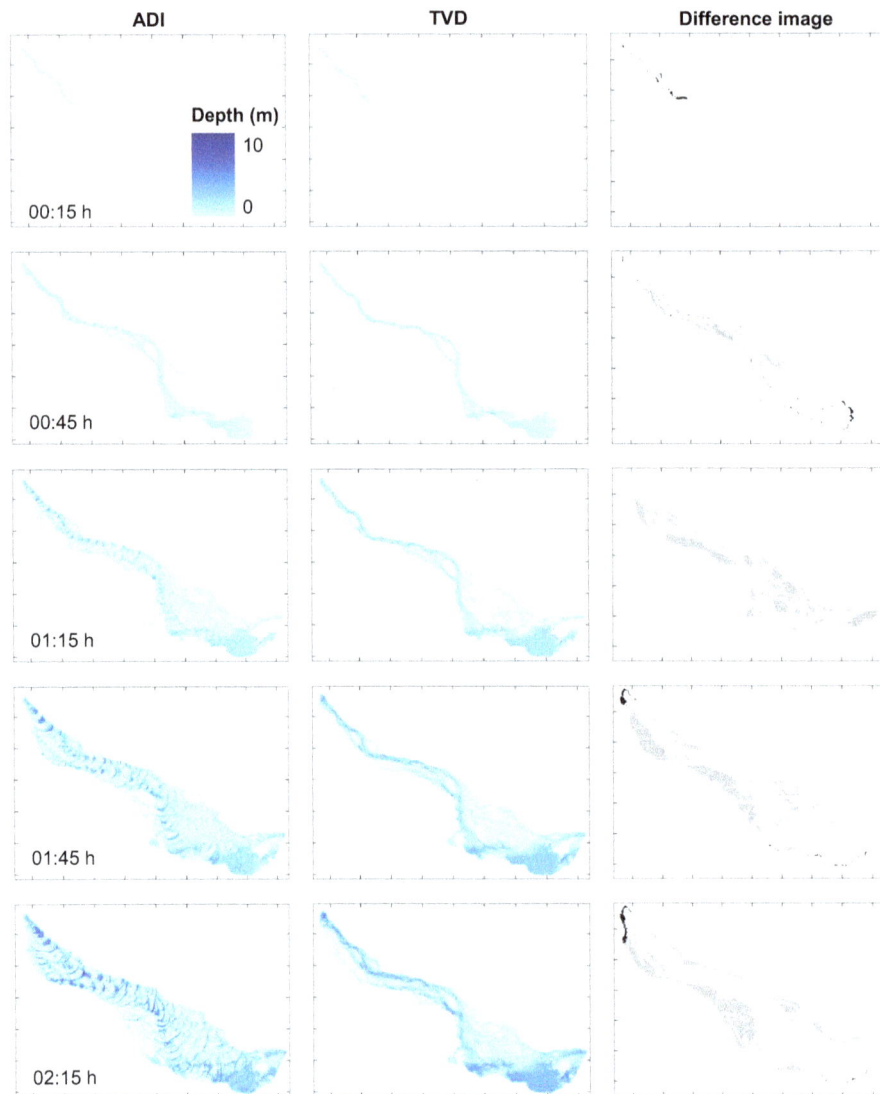

Figure 6. Comparison of GLOF inundation in the upper reaches of the Langmoche Khola using the ISIS 2-D alternating implicit direction (ADI) and total variation diminishing (TVD) solvers at selected time steps. Also shown is a difference image of inundation, where black and grey shading corresponds to areas inundated exclusively by the TVD and ADI solvers, respectively. See Fig. 1a for location. Inset tick marks spaced at 200 m intervals.

In addition, the DTM that was used to represent the floodplain domain immediately downstream of the moraine dam reflects post-GLOF valley-floor topography. As such, derived maps of inundation and flow depth should not be taken as indicative of the passage of the 1985 GLOF.

6   Results

The following sections present the results of our probabilistic, coupled dam-breach–GLOF simulation experiments. The performance of the dam-breach model at reproducing observed breach geometry is first evaluated (Sect. 6.1.1) before attention turns to issues associated with the extraction

of useful, probabilistic breach hydrograph data for use as input to GLOF modelling (Sects. 6.1.2 and 6.1.3). A variety of approaches for translating the impact of dam-breach model parameter uncertainty into probabilistic maps of GLOF inundation and hazard are presented in Sect. 6.2.

6.1   Dam-breach modelling

6.1.1   Model evaluation

Simulations that were deemed non-behavioural were assigned a likelihood value of 0 and not considered for further analysis (Table 2). Analysis of parameter-specific likelihood data reveal that weak correlations exist for all input param-

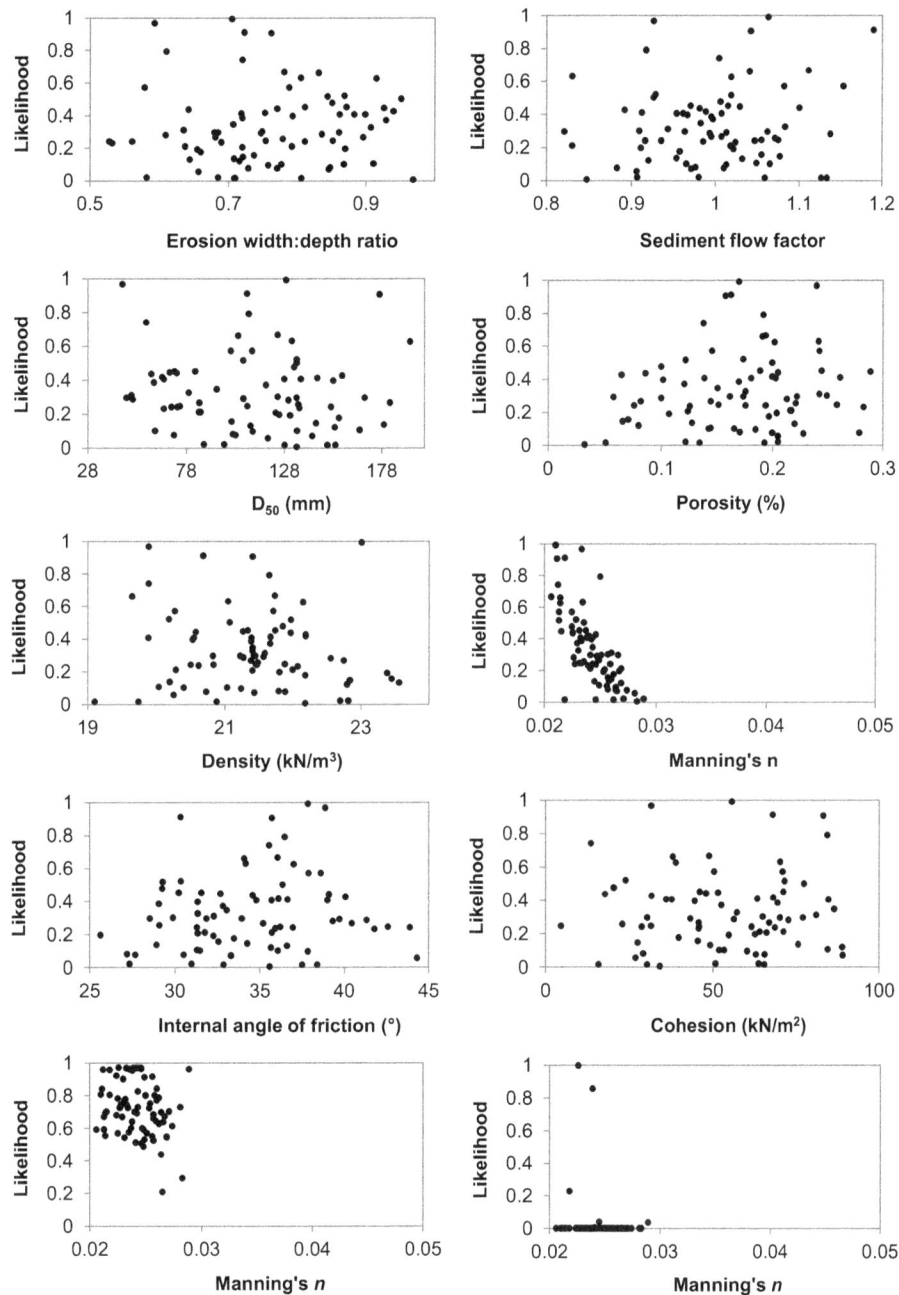

Figure 7. Scatter, or "dotty", plots of likelihood values for HR BREACH input parameters for behavioural Monte Carlo simulations of DT_control, conditioned on final breach depth. The bottom two panels represent the Manning's n likelihood surface for results conditioned on final breach depth and residual slope error (bottom left panel); and final breach depth, residual thalweg elevation error, and final flow constriction location (bottom right panel).

eters except Manning's n (Fig. 7), whereby increasing the roughness coefficient is associated with decreasing peak discharges. However, this inverse relationship is evident only for the first and second likelihood measures. Near-identical relationships were identified for all additional, perturbation-focused scenarios. Behavioural parameter ensembles possessed a low range of peak discharge (Q_p) values relative to the entire range of unconditioned data; maximum non-behavioural Q_p values exceeded $18\,000\ \mathrm{m^3\ s^{-1}}$ for most scenarios, whereas maximum behavioural Q_p was found typically to be approximately $\sim 2000\ \mathrm{m^3\ s^{-1}}$ (Fig. 10) and is in line with previous empirical estimates for the Dig Tsho failure (Cenderelli and Wohl, 2001), but smaller than other documented Himalayan GLOFs (e.g. Osti and Egashira, 2009).

Whilst the number of simulations retained for the overtopping scenarios was broadly similar, the rather more extended range possessed by the instantaneous mass-failure scenarios reflects an inverse correlation between the initial volume of material removed and number of simulations retained as behavioural. Full hydrographs were obtained for each of the retained simulations. The range of behavioural peak discharges of similar magnitude to previous estimates provided for the Dig Tsho GLOF (Vuichard and Zimmerman, 1987; Cenderelli and Wohl, 2001, 2003), and of equivalent or lower magnitude to palaeo-GLOFs reported from other regions (e.g. Clague and Evans, 2000; Huggel et al., 2004; Kershaw et al., 2005; Worni et al., 2012).

The maximum and minimum range of behavioural Q_p values for all scenarios are strikingly similar (Table 3). These similarities imply that scenarios with considerably different modes of breach initiation are capable of producing a broadly similar range of behavioural peak discharges. This finding suggests that additional factors, including the constraints on breach development imposed by the initial dam geometry and the various parameter sampling ranges, exert an overriding influence on this aspect of breach outflow.

Maximum and minimum behavioural likelihood scores after the data had been conditioned using the RMSE of modelled elevation profile of the breach thalweg varied from 0.97 to 0.014. Within this range, the distribution of likelihood scores was comparable for the control scenario and all overtopping scenarios (0.970, 0.969, 0.969, and 0.970 for DT_control, min, mid, and max overtopping scenarios, respectively). Whilst the maximum likelihood score for DT_instant_1 was almost equally as high (0.821), this value decreases with increasing volume of mass removed (0.819 and 0.744 for DT_instant_3 and DT_instant_5). All of these scenarios possessed minimum likelihood scores that were appreciably lower than the control and overtopping scenarios. Within these ranges, likelihoods were distributed relatively evenly. Accordingly, the instantaneous mass-removal scenarios, particularly DT_instant_3 and DT_instant_5 exhibited the poorest performance against this likelihood function.

Behavioural elevation profiles are displayed in Fig. 8. The well-defined break in slope which is identifiable in the observed data at ~550 m is reproduced by the output from HR BREACH. However, HR BREACH almost consistently underestimates the distance along the breach at which this break occurs. The majority of modelled profiles in the behavioural DT_instant_3 and DT_instant_5 simulations are located ~50–100 m further upstream than the equivalent SfM-derived, observed profile (Fig. 8). It is unclear why this systematic underestimation occurs, although an important consideration is that the observed centreline profile, viewed from above, is not linear. Instead, the breach thalweg meanders along its length, having exploited localised weaknesses in the degrading dam structure and breach-flow dynamics as it developed. Consequently, the observed and modelled profiles are not truly comparable, introducing a fundamental source of error into the resulting likelihood scores (and also

Table 2. Scenario-specific behavioural simulation count for individual likelihood measures. Note: all simulations deemed to be behavioural after application of LH1 (final breach depth) were retained for conditioning on LH2 (breach thalweg elevation profile).

	LH1, LH2	LH3
DT_control	76	7
DT_overtop_min	98	33
DT_overtop_mid	93	33
DT_overtop_max	91	29
DT_instant_1	90	49
DT_instant_3	55	17
DT_instant_5	20	2

accounting for the discrete zone of variance between 300 and 370 m on all plots (Fig. 8).

Modelled planforms are broadly similar in form both between and within each scenario (Fig. 9). Observed and modelled planforms gradually taper towards a flow constriction, beyond which the breach width expands to form a bell-shaped exit. Flow-constriction location varies considerably, with a substantial number of parameter ensembles deemed non-behavioural following conditioning using this likelihood measure (Table 2). The majority of non-behavioural simulations located the flow constriction upstream of the behavioural limit (Fig. 9). However, no discernible relationship between input parameters and flow constriction location was identified in the parameter-specific likelihood data. Only seven DT_control parameter ensembles were retained (0.7 % of the original simulation pool), following conditioning on this likelihood measure, and only two (0.2 %) of the DT_instant_5 simulations remained. Further reductions in the number of retained simulations were imposed for all scenarios (Table 2).

6.1.2 Percentile hydrograph extraction

CDF curves were extracted from behavioural, scenario-specific likelihood data (Fig. 4). Using these data, time-step-specific percentile discharges were extracted and combined to construct probabilistic breach outflow hydrographs for each scenario (Fig. 10). Similarities between the percentile hydrographs for each scenario are striking, particularly between data conditioned on modelled final upstream breach depth (LH1) and modelled upstream breach depth and breach centreline elevation profile data (LH1 + 2). All 95th percentile hydrographs possess steep rising limbs and comparatively shallower falling limbs, and they generally trace the upper boundary of the behavioural hydrograph envelope. Exceptions to this rule are DT_overtop_min, DT_overtop_max, and DT_instant_1, whose hydrographs dip below this upper bound by a noticeable margin. DT_instant_3 and DT_instant_5 possess shorter-duration hydrographs – a direct consequence of the form of the behavioural envelope

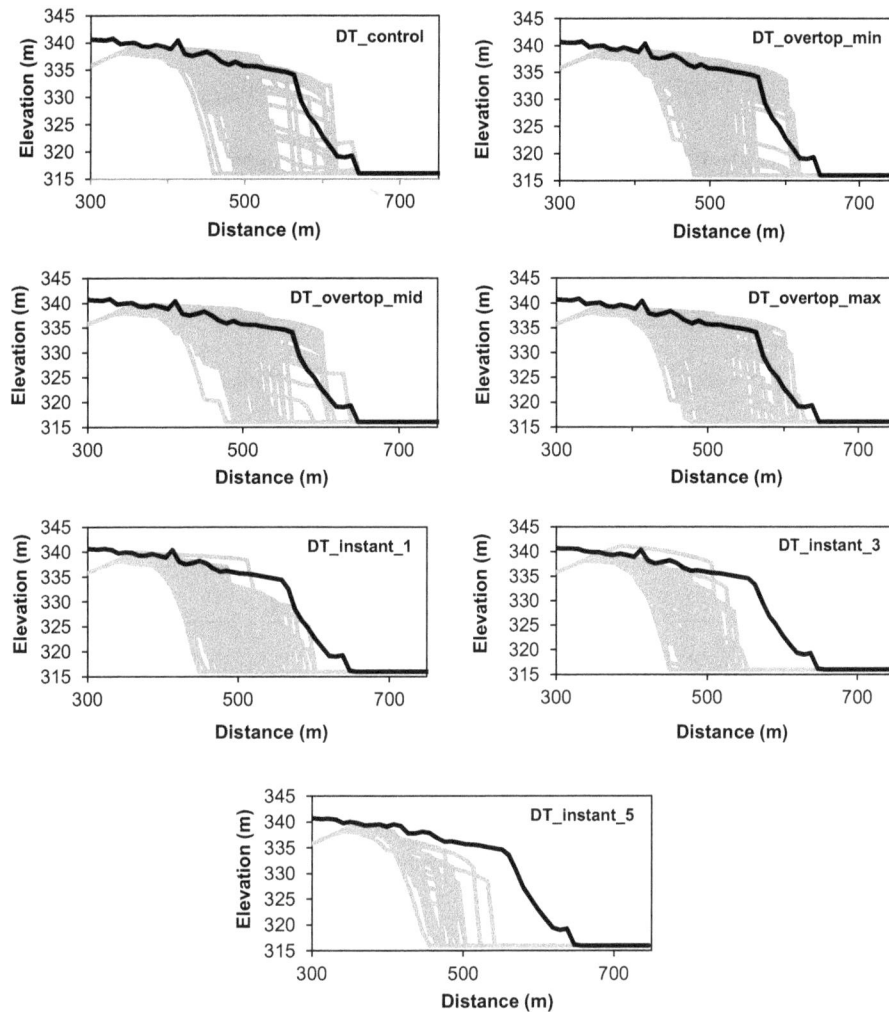

Figure 8. Final centreline breach elevation profiles for the Dig Tsho simulations. Black: observed profile; grey: modelled (behavioural) profiles.

from which the CDFs were derived. Shortening of the hydrograph duration results in greater concentration of all percentile hydrographs for these scenarios, with the result being reduced variation in percentile-specific Q_p and Q_{vol} (Table 4).

Individual overtopping waves are preserved for each percentile in the relevant scenarios (Fig. 10). Median (50th) percentile hydrographs exhibit slightly more variation, both between scenarios and following conditioning using the second likelihood measure. This conditioning step results in a decrease in median percentile Q_p for control and overtopping scenarios, but it has a negligible effect on the mass-removal simulations (Fig. 10). For all scenarios, 5th percentile hydrographs generally trace the lower boundary of the behavioural hydrograph envelope, and appear to be largely unaffected by additional conditioning using the elevation profile of the breach centreline.

The most noticeable impact on percentile hydrograph form is caused by the additional conditioning of the data on the final likelihood measure. Mass-removal scenarios appear to be affected to a far lesser degree than the control and overtopping scenarios. The exception is DT_instant_5, where discharges for 50th and 5th percentile hydrographs are increased in the first ~ 80 min, after which 95th percentile discharges decrease, resulting in substantial concentration of the percentile hydrographs (Fig. 10). Conditioning on this final likelihood measure causes increased Q_p for DT_overtop_min and DT_overtop_mid median percentile hydrographs, whilst the hydrograph for DT_overtop_max is perturbed between 140 and 200 min. Notably, DT_control exhibits increased discharges for the 5th and 50th percentiles, coupled with decreased Q_p and steepening of the falling limb of the 95th percentile hydrograph.

Crucially, percentile hydrograph form is dictated by the time-step-specific CDF data. In turn, CDF form is deter-

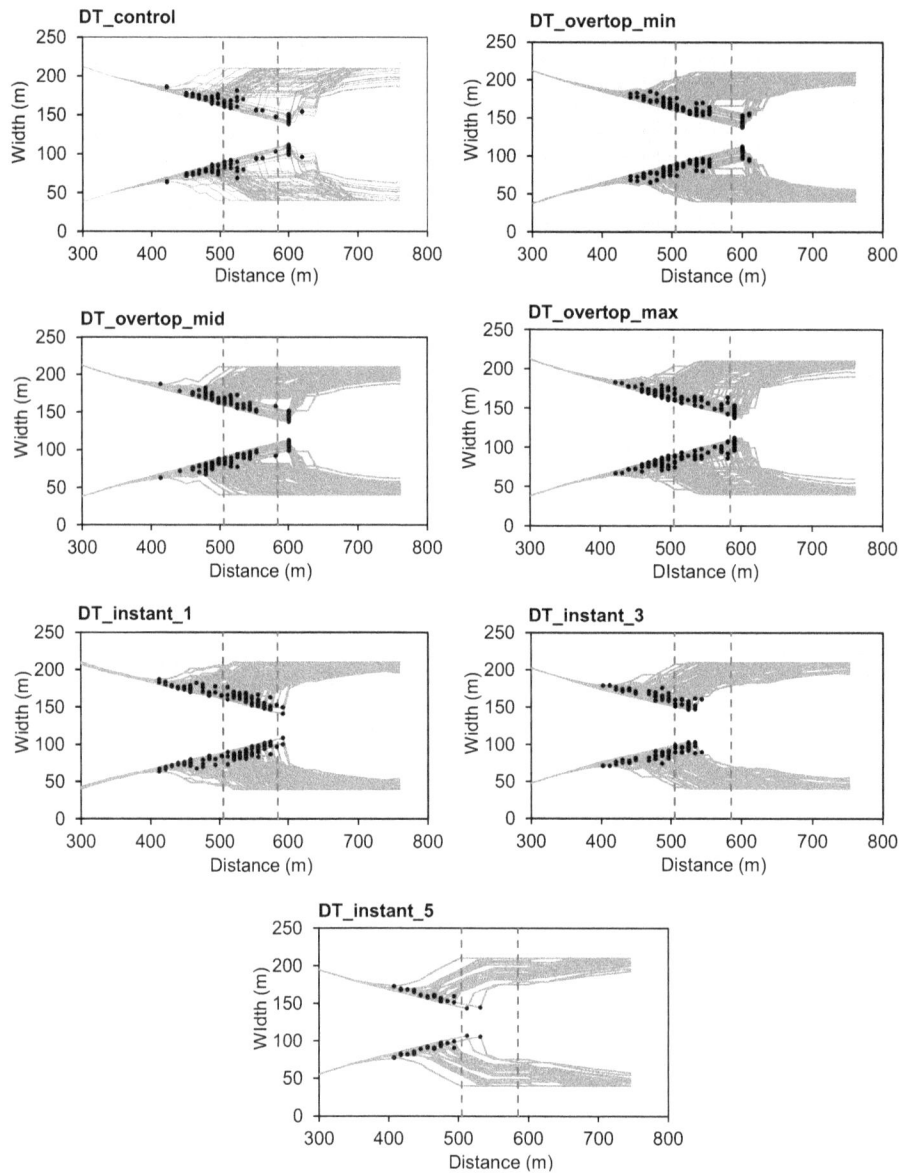

Figure 9. Final breach planforms (grey lines) for all modelled scenarios. Precise locations of critical flow constriction are highlighted by black dots. Simulations possessing flow constriction locations that fell between the red dashed lines (distance 505–585 m) were deemed to be behavioural and assigned positive likelihood values.

mined by variations in the likelihood of individual behavioural hydrographs, and the cumulative distribution of their associated discharges (for each time step). The vast number of simulations which, following conditioning on flow constriction location, were subsequently deemed to be non-behavioural (Table 1) serves to alter the form of scenario-specific CDFs. This effect is particularly dramatic for DT_control, where the number of behavioural simulations reduces from 76 to 7 following conditioning on final breach depth, breach centreline elevation profile, and flow constriction location (LH1 + 2 + 3). Similarly, CDF data for DT_instant_5 are derived from just two behavioural hydro-

graphs, which explains both the smaller hydrograph envelope and the high concentration of the percentile hydrographs (Fig. 10).

6.1.3 Data clustering

Issues of mass conservation arose with the extraction of behavioural, percentile-derived breach hydrographs. Both 5th and 50th percentile hydrographs for all scenarios consistently under-predicted the volume of lake water released during breach development ($\sim 7.3 \times 10^6 \, \mathrm{m}^3$), whereas volumetric integration of 95th percentile hydrograph data produced substantially increased values of Q_{vol} (Table 4). The median

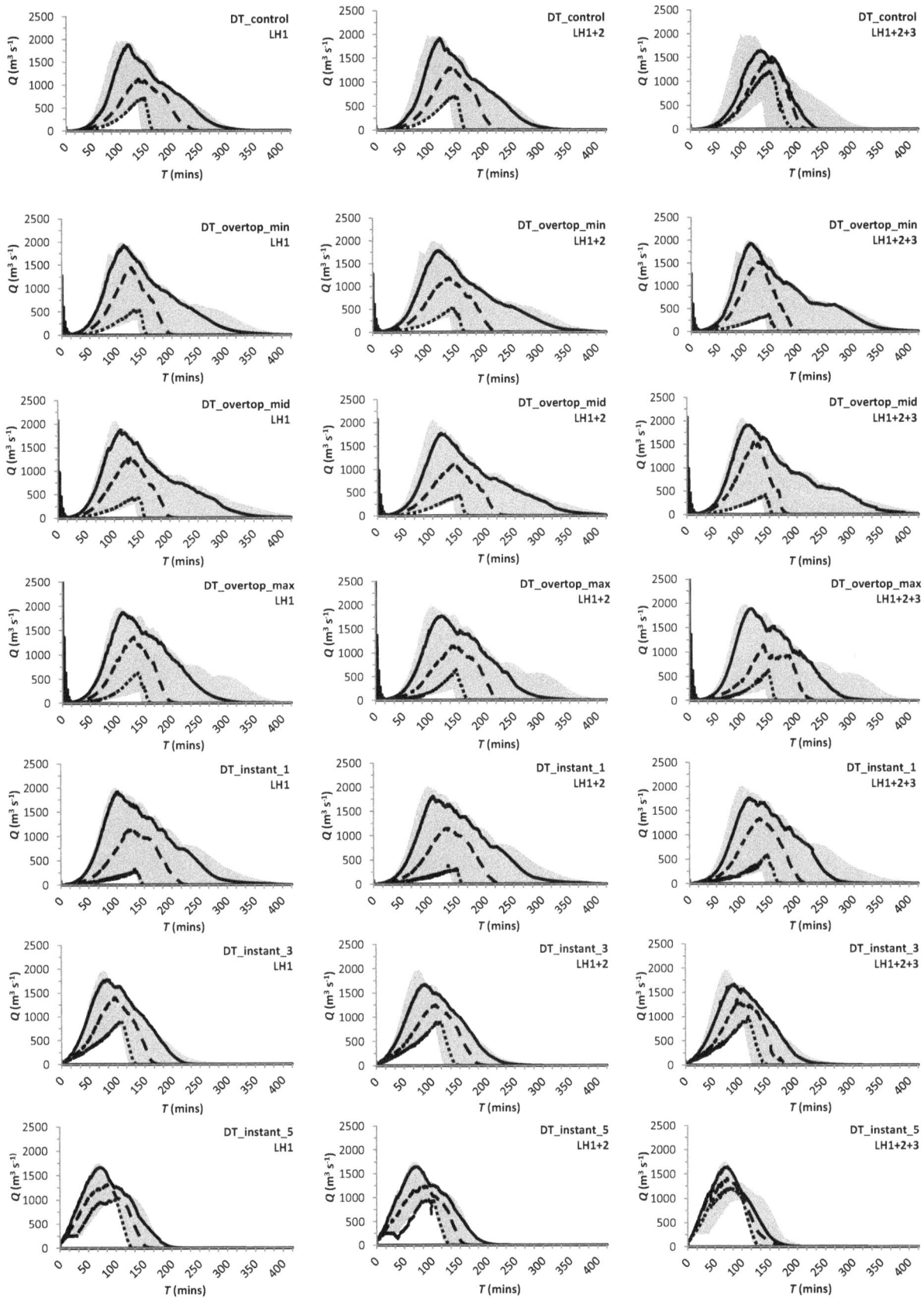

Figure 10. Percentile hydrographs derived from behavioural Dig Tsho simulations, for successive likelihood updating steps ("LH1", "LH1 + 2", "LH1 + 2 + 3"). Small outdashes: 5th percentile; long dashes: 50th (median) percentile; solid black: 95th percentile. Behavioural hydrograph envelope is displayed in grey.

Table 3. Scenario-specific maximum ($Q_{p\,max}$), minimum ($Q_{p\,min}$), median ($Q_{p\,med}$), and optimal ($Q_{p\,opt}$) peak discharges.

($m^3\,s^{-1}$)	DT_control	DT_overtop_min	DT_overtop_mid	DT_overtop_max	DT_instant_1	DT_instant_3	DT_instant_5
$Q_{p\,max}$	2005	2009	2082	2006	2029	1975	1757
$Q_{p\,min}$	703	593	542	598	706	904	977
$Q_{p\,med}$	1484	1489	1375	1495	1605	1468	1352
$Q_{p\,opt}$	882	1441	1325	1403	1320	1363	1505

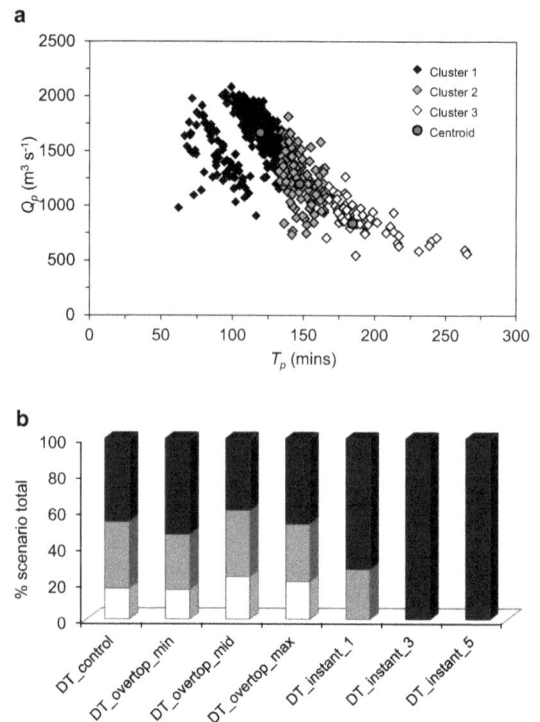

Figure 11. Data clustering. **(a)** Cluster membership following application of a subtractive clustering algorithm in MATLAB®; **(b)** cluster membership as a function of total scenario-specific behavioural simulations (%). For key see **(a)**.

percentile performs best at replicating total flood volume, under-predicting Q_{vol} by between 0.2 and $0.7 \times 10^6\,m^3$, 0.4 and $0.7 \times 10^6\,m^3$, and 0.1 and $1.3 \times 10^6\,m^3$ for LH1, LH1 + 2, and LH1 + 2 + 3, respectively (Table 4).

From a hydrodynamic modelling perspective, an additional and equally important observation is the form of the percentile hydrographs, which generally do not mirror the form of any of the behavioural hydrographs used as input. When combined, issues of mass conservation and the unrepresentative form of the percentile breach hydrographs render them largely unsuitable for use as an upstream boundary condition for subsequent hydrodynamic modelling.

In an effort to further refine the behavioural simulations and improve the representativeness of the derived percentile hydrographs, all behavioural data were clustered, regardless of their inclusion of any modelled system perturbations (Fig. 11). Data clustering was undertaken in an effort to characterise "styles" of breaching, such as those characterised by low peak discharge magnitude and lengthy time to peak or high peak discharge and short, sharp rise to peak. Clustering used Q_p and T_p data as the sole input, and was independent of the various breaching scenarios that were modelled. An automated, subtractive clustering algorithm was applied to the unified Q_p and time-to-peak (T_p) data. This method identifies natural cluster centroids in raw point data sets, and quantifies the density of points relative to one another. Each point is assumed to represent a potential cluster centroid, and a measure of likelihood is calculated based on the density of surrounding points. Following initial cluster identification, the density function is revised and subsequent cluster centres identified in the same manner until a sufficient number of natural clusters are deemed to have been obtained (Math-Works, 2012). Point-cluster membership was determined by calculation of the minimum Euclidean distance between each data point and cluster centre. The subtractive method eliminates any subjectivity associated with manual cluster identification.

Clusters are broadly defined by T_p range. Cluster 1 contains all simulations with T_p of approximately 60–130 min, cluster 2 is defined by T_p of $\sim$ 130–170 min, and cluster 3

possesses T_p values in the range $\sim$ 170–270 min. Ranges of Q_p overlap substantially between cluster 1 and cluster 2 ($\sim$ 900–2100 and $\sim$ 700–1800 $m^3\,s^{-1}$, respectively), and to a lesser degree between cluster 2 and cluster 3. These clusters may be taken as approximately representing a number of "types" of breach hydrograph, namely (relatively) high-magnitude, short-duration (cluster 1), moderate peak magnitude and mid-range duration (cluster 2) and low-magnitude, extended duration (cluster 3) GLOF hydrographs.

Cluster membership is not as clear-cut as might be anticipated (Fig. 11). The clustering results appear to imply that pigeonholing different breaching scenarios by hydrograph type, or style, is virtually impossible. However, the exceptions to this rule are the instantaneous mass-removal scenarios, which almost exclusively produce high-magnitude, short-duration hydrographs. This finding would appear to imply that alternative factors are required to explain the similarity in the range of hydrograph forms that are produced by each scenario.

Percentile hydrographs were also extracted from the clustered data. Deviations between observed and modelled median percentile Q_{vol} data are in the range −41 to −6 %, demonstrating a minimal improvement over unclustered and scenario-specific values of modelled Q_{vol} (Table 4). Clus-

Table 4. Total percentile hydrograph outflow volumes for individual scenarios and following clustering.

| | Total percentile hydrograph volume ($\times 10^6$ m^3) | | | | | | | | |
| | LH1 | | | LH1 + 2 | | | LH1 + 2 + 3 | | |
Scenario/cluster	V_5	V_{50}	V_{95}	V_5	V_{50}	V_{95}	V_5	V_{50}	V_{95}
DT_control	2.2	6.6	13.1	2.2	6.6	13.3	4.8	7.0	9.5
DT_overtop_min	1.8	6.5	14.4	1.7	6.5	14.2	1.2	6.6	15.2
DT_overtop_mid	1.5	6.4	15.0	1.5	6.2	14.6	1.3	6.3	16.0
DT_overtop_max	1.9	6.5	14.2	1.8	6.5	13.8	1.7	6.0	14.1
DT_instant_1	1.0	6.4	15.2	1.1	6.5	14.6	1.8	6.9	13.4
DT_instant_3	3.2	6.9	11.8	3.4	6.8	11.3	3.6	6.8	11.1
DT_instant_5	4.4	7.1	10.2	3.6	6.9	10.2	5.7	7.3	8.8
Cluster 1	2.1	6.6	13.8	2.4	6.8	12.9	1.6	4.4	8.8
Cluster 2	3.2	6.8	11.5	3.2	6.8	11.6	2.5	6.9	12.1
Cluster 3	3.2	6.8	11.2	3.1	6.8	11.5	3.0	6.7	11.6

tered 5th and 95th percentile Q_{vol} data vastly under- and overestimate observed Q_{vol}.

Clustering was largely unsuccessful at improving the utility of percentile-based breach hydrographs for use as hydrodynamic input. This result necessitated the exploration of alternative methods for cascading likelihood-weighted estimates of dam-breach parameter ensemble performance through to the simulation and mapping of GLOF inundation and hazard.

6.2 Hydrodynamic modelling

6.2.1 The optimal hydrograph

Deterministic approaches to flood reconstruction require the identification of the optimal model, and its subsequent use for predictive flood-forecasting. To illustrate the variability between scenario-specific optimal hydrograph routing patterns, the optimal hydrographs for DT_control, DT_overtop_max and DT_instant_5 were used as upstream input for simulation in ISIS 2-D. Maps of inundation extent and flow depth (Fig. 12) reveal prominent inter-scenario differences in the spatial extent of inundation as the respective hydrographs and GLOF floodwaters progress downstream. Variations include the initial downstream transmission of the DT_overtop_max overtopping wave, which triggers rapid inundation of the entire reach. However, initially high-flow stages are not maintained; these only rise once again with increasing breach discharge associated with breach expansion. Use of the DT_instant_5 hydrograph produces spatial and temporal patterns of inundation and wetting front travel time similar to that of DT_overtop_max (Fig. 12).

6.2.2 GLUE-based GLOF reconstruction

Probabilistic maps of inundation extent and flow depth were constructed through the retention and evaluation of scenario-

specific and likelihood-weighted breach hydrographs. In the example presented herein, we simulated the propagation of 76 individual moraine-breach hydrographs using the ISIS 2-D TVD solver (with an 8 m topographic grid discretisation and a 0.04 s time step), representing the behavioural DT_control parameter ensembles after conditioning on final breach depth. However, the method is equally applicable to the use of several, hundreds, or thousands of individual simulations. For each time step, per-cell CDF curves of flow depth were assembled, from which percentile flow depths were extracted and plotted (Fig. 13). Given the inherent uncertainty surrounding the precise mode of moraine-dam failure and outflow hydrograph form, these data effectively convey the resulting variability in likelihood-weighted predictions of reconstructed inundation extent, whilst preserving time-step-specific percentile flow depths. Because of the nature of their construction, these data do not relate to a specific event or hydrograph but instead provide an indication as to the potential uncertainty in GLOF inundation extents and flow depths associated with a range of behavioural breach hydrographs.

6.2.3 Probabilistic hazard mapping

The final output of a GLOF hazard assessment comprises the production of maps of flood hazards, conditioned by one or more directly quantifiable flood-intensity indicators (e.g. Aronica et al., 2012). Whilst inundation depth is arguably the most significant flood-intensity indicator for predicting monetary losses associated with individual flood events (Merz and Thieken, 2004; Vorogushyn et al., 2010, 2011), its combination with flow velocity is regarded as an improved indicator of hazard to human life (Aronica et al., 2012).

A global hazard index proposed by Aronica et al. (2012) was used to construct maps of GLOF hazard (Fig. 14). Taking probabilistic flow depth and velocity data as input, probabilistic GLOF hazard maps were produced for the

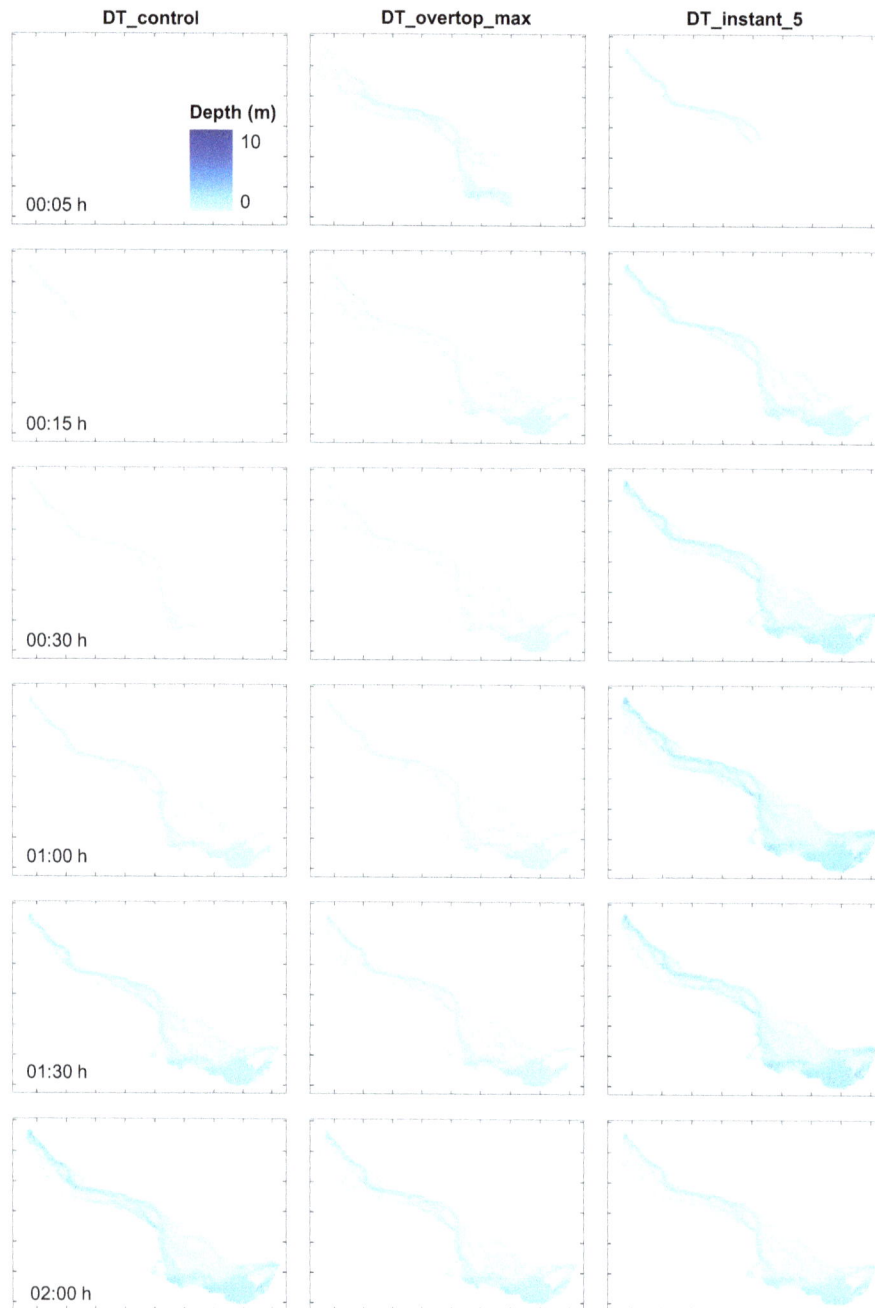

Figure 12. Inundation extent and flow depth distribution for selected time steps for the DT_control, DT_overtop_max, and DT_instant_5 optimal flood even. See Fig. 1a for location. Inset tick marks spaced at 200 m intervals.

DT_control scenario. Four hazard classes are defined and shaded for distinction (H_1 to H_4, in order of increasing hazard). Hazard distribution is characterised by extensive zones of H_1 for the 5th and 50th percentiles in the first 0.5 h, with higher hazard classes becoming increasingly prevalent in the 95th percentile data (Fig. 14). As breach outflow increases and the maximum inundation extent is approached, the zone immediately adjacent to the channel thalweg is classed as "very high" (H_4) hazard, and is associated with flow depths in excess of 1.5 m and regardless of velocity. Classes H_1, H_3, and H_4 dominate for all time steps, whilst H_2 is noticeably under-represented. Inundated channel-marginal areas are generally classed as being of low hazard, whilst the classification of the remaining inundated area as H_3 or H_4 indicates the presence of flow velocities that either exceed the prescribed product of depth and velocity ($0.7 \, \mathrm{m^2 \, s^{-1}}$) or depths in excess of 1.5 m. Rather than representing the observed flood hazard zonation observed with the Dig Tsho

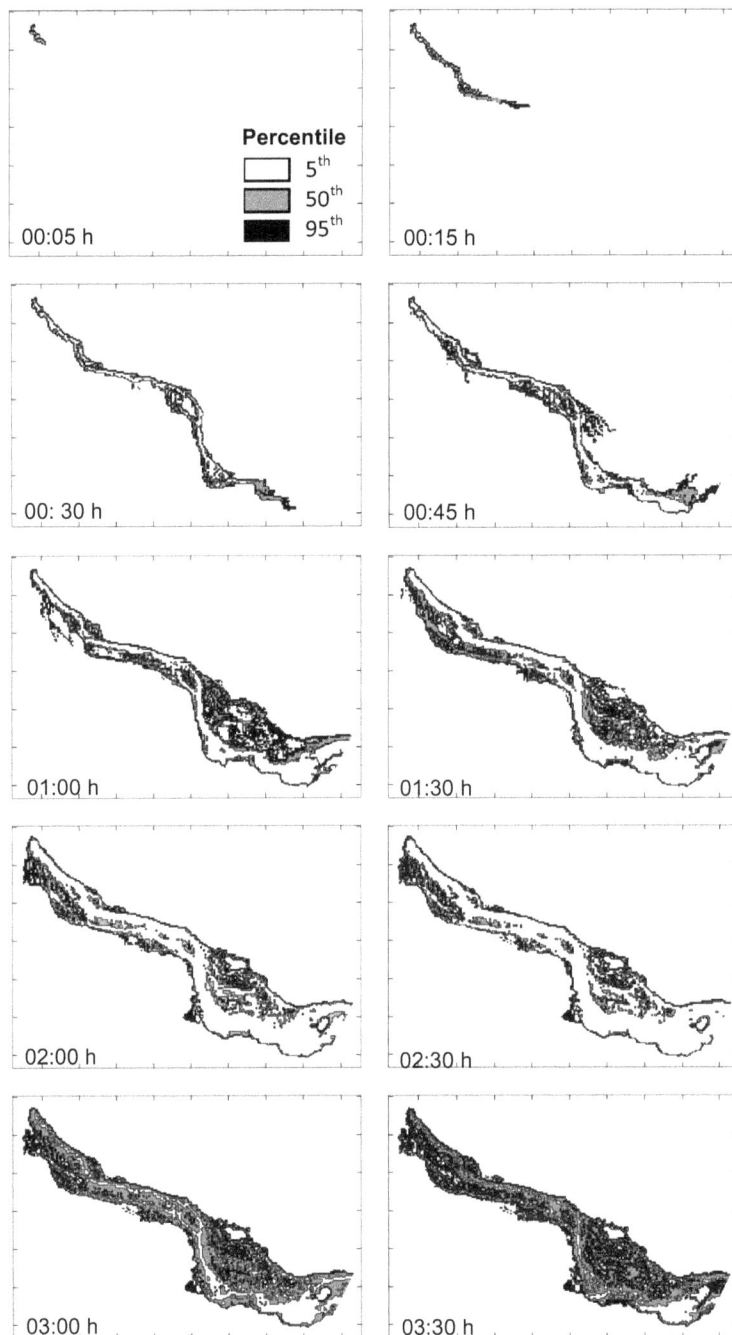

Figure 13. GLUE-based percentile maps of inundation for DT_control. 5th, 50th, and 95th percentiles maps of inundation represent the water depth that would be exceeded with 95, 50, and 5 % probability, respectively. See Fig. 1a for location. Inset tick marks spaced at 200 m intervals.

GLOF, the purpose of including these data is to highlight the impact of breach model parametric uncertainty on one of the final products in a typical GLOF hazard assessment exercise. The representivity of these data is associated with a number of caveats, namely the routing of the flood across a post-GLOF digital elevation model and the application of water-only hydrodynamic modelling. In reality, the flow may have exhibited transient debris-flow characteristics given the availability of sediment for entrainment and transport from the eroding breach and unconsolidated valley floor (Huggel et al., 2004; Westoby et al., 2014a).

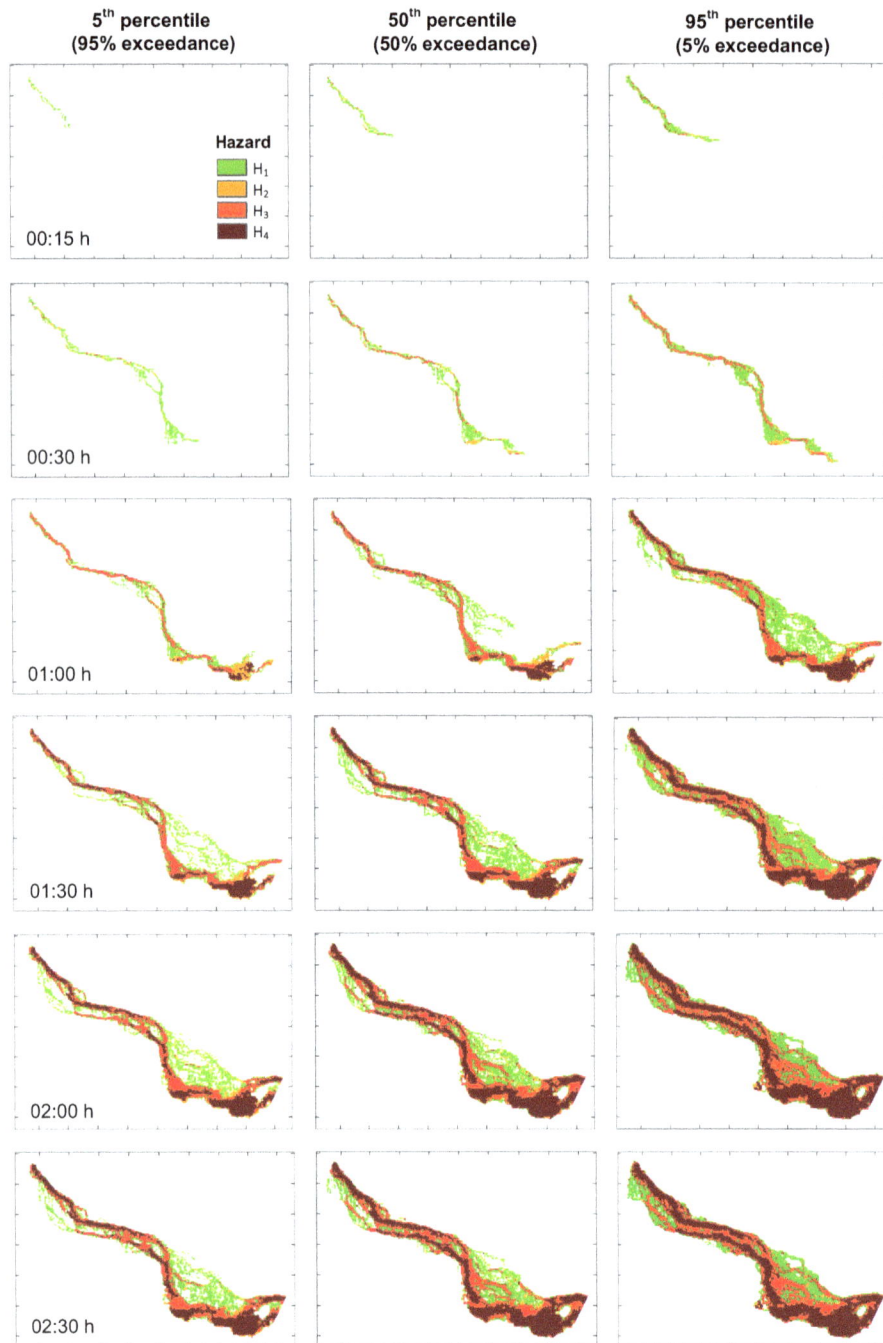

Figure 14. Percentile flood-hazard maps, based on a global hazard index forwarded by Aronica et al. (2012). Such data facilitate a probabilistic evaluation of the evolution of GLOF flood hazard. See Fig. 1a (this paper) for location, and Fig. 1b in Aronica et al. (2012) for description of the hazard index.

7   Discussion

7.1   Key controls on breach development

We have demonstrated that the propagation, or cascading, of the parametric uncertainty and equifinality through the dam-breach and hydrodynamic modelling components of the GLOF model chain is not only possible but may also be of considerable value to flood-risk practitioners. A key contribution of the research is the demonstration that the predictive limits of numerical models, in this instance applied to the reconstruction of historical moraine breaching, can be quantified through the use of a weighted, probabilistic modelling framework (Fig. 15). Our approach illustrates how para-

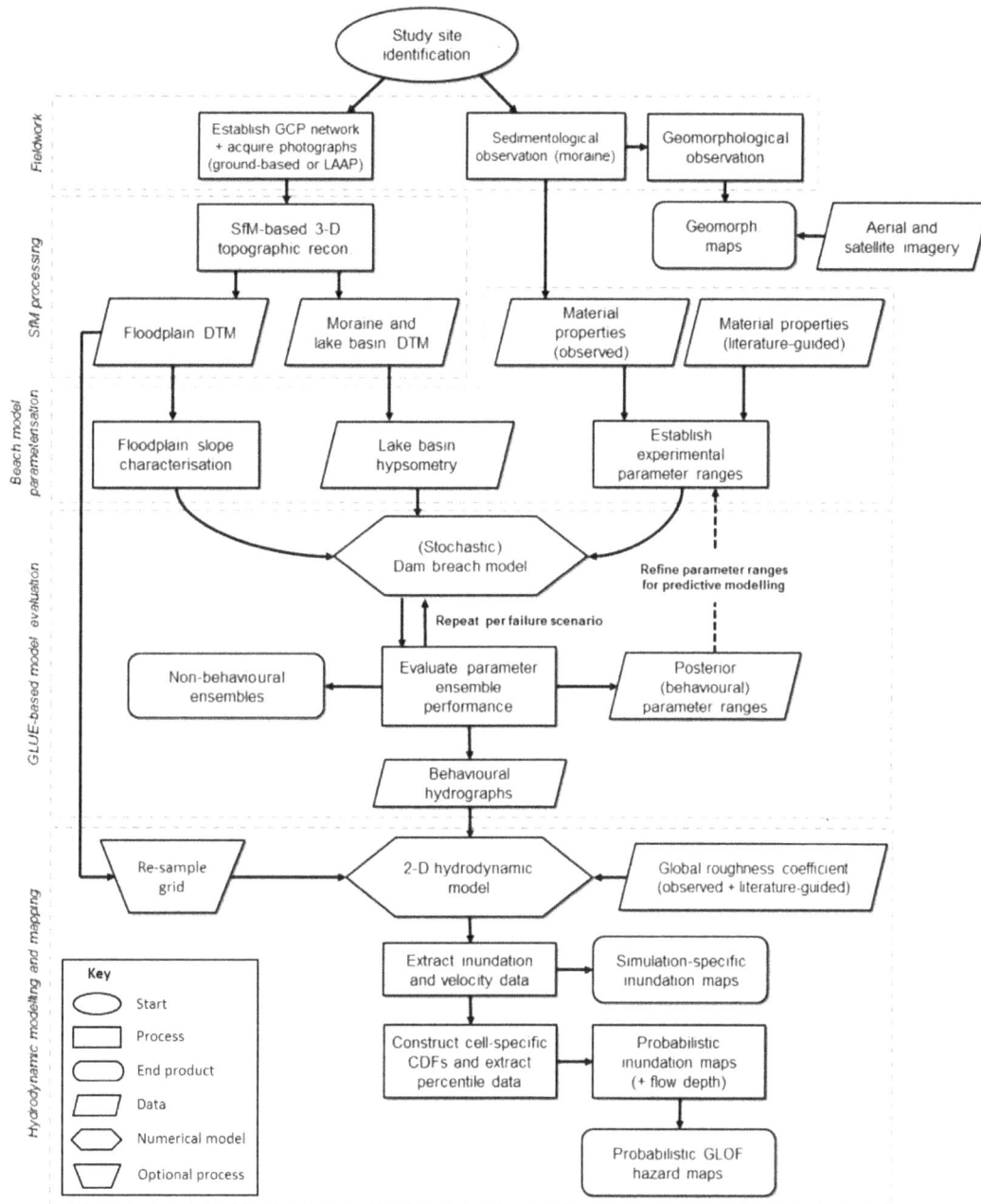

Figure 15. A unifying, "end-to-end" framework for probabilistic GLOF reconstruction incorporating high-resolution photogrammetry and probabilistic, GLUE-based numerical dam-breach and hydrodynamic modelling approaches.

metric uncertainties may be propagated through the GLOF model chain when the output from one model is used as the input to another. The approach can therefore be used to isolate the most sensitive components of a predictive system – e.g. Manning's n – and thus indicate the most important areas for future model development or data collection to support informed model parameterisation and testing.

Our results highlight the primary influence of the material roughness of moraine material indicating HR BREACH parameter ensemble, and therefore breach-hydrograph, per-

formance. Specifically, behavioural Manning's n coefficients were found to be in the range 0.020–0.029 $m^{-1/3}$ s (Fig. 7), representing a significant refinement of the a priori parameter range (0.02–0.05 $m^{-1/3}$ s). In contrast, no significant refinement of the remaining material characteristic parameters was made following model evaluation. However, posterior Manning's n values appear to be unrealistic; the Dig Tsho moraine is composed predominantly of pebble-, cobble-, and boulder-sized material, which is more likely to be associated with larger roughness-coefficient values (Chow, 1959).

If all additional resistance effects, including form-, spill-, and curvature, are fully accounted for in the model solution, then the resistance term will reduce to account for the particle, or skin, resistance alone. This would account for the retention of such "low" Manning values. However, since 2-D fluid flow is modelled in HR BREACH at comparatively coarse spatial resolutions relative to the scale of the particles that moraine is composed of, it is reasonable to expect that the effective values of n will incorporate the aforementioned macroscopic properties. In such a case we would expect values of n to be larger. For example, in a reconstruction of the Chukhung moraine breach, also in the Khumbu Himal, Westoby et al. (2014b) found behavioural roughness values for overtopping-type failures to be in the range 0.042 to 0.049. Our results suggest that a reduction in n is more likely to reflect parameter compensation or interaction effects elsewhere in the model chain. Specifically, a reduction in n will increase shear stress, which will in turn offset other parameters that serve to reduce bed erodibility. Given the complexity of the model parameter space, it is these effects that account for the low values of n, rather than uniquely the identification of appropriately scaled particle-roughness values.

The observation that broadly similar behavioural peak discharges are associated with different modes of breach initiation is a particularly significant one (Table 3). This finding appears to suggest that additional factors, such as the sampling ranges for the various input parameter ranges and model boundary conditions (dam geometry and lake bathymetry), exert an overriding influence on this breach hydrograph characteristic. However, the significance of breach initiation mode is a primary control over downstream GLOF wetting front travel times and inundation extent. This is illustrated in Fig. 12, and reinforces the need for moraine-dam failure modelling exercises to consider, wherever possible, the complete range of potential breach-initiation scenarios at the model design stage.

The numerical physicality of the dam-breach model used in this research represents a notable improvement over empirical or analytical models. However, many of the geometric and material characteristics of the moraine and lake complex remain highly simplified. This simplification is a necessary compromise related to our use of a dam-breach model, the primary intended application of which is the investigation of (relatively) simple artificial earthen or concrete dam constructions. Alternative dam-breach models, including unstructured mesh-based variants (e.g. Worni et al., 2012), have been demonstrated to perform well at reproducing historical moraine-dam failure dynamics. In combination with stochastic parameter sampling functionality, they would represent a powerful combination that could be easily incorporated into the framework we present herein (Fig. 15).

7.2 Wider use of probabilistic GLOF modelling

The model chain presented in this article is parametrically complex, and whilst there remains an imbalance between the parametric degrees of freedom in the model setup and the objective descriptors of post-GLOF geometry used to evaluate model performance, this imbalance is one common to many areas of numerical modelling in the Earth surface sciences. The value of the GLUE method in this respect is its suitability for actively exploring and quantifying the extent of uncertainty and equifinality in dam-breach model output that results from poorly constrained conditioning of the model input parameters. Similarly, this problem is common to virtually all dam-breach modelling efforts, not only moraine dam breaching. This research has provided specific tools, in the form of probability-based maps of GLOF inundation and hazard, to communicate the effects of model uncertainty to potential end users in a manner that is both open and objective.

Whilst the extraction of percentile maps of inundation extent and flow depth is not necessarily an entirely new concept (e.g. McMillan and Brasington, 2007, 2008; Vorogusyhn et al., 2010, 2011), the application to GLOF reconstruction presented here is an original and novel one. This approach represents a significant improvement in the effective communication of the likelihood associated with a range of moraine-dam failure scenarios. Significantly, our probability-based flood hazard maps (Fig. 14) can be visualised in a GIS environment, and provide a clear picture of patterns of flood hazard zonation, at varying levels of confidence and at successive model time steps, that would prove useful to disaster risk managers.

The likelihood of multiple GLOFs occurring from an individual moraine-dammed lake complex is low. In most cases, the breached moraine dam can comfortably accommodate relict lake discharges. Therefore, the identification and use of posterior parameter distribution data for predictive GLOF forecasting is of limited utility if these ranges prove to be highly site-specific. The identification of a suite of universal or region-specific material characteristics and their probabilistic distributions would facilitate their use in predictive GLOF simulation efforts. We believe that the Dig Tsho failure is regionally representative of comparable styles of glacial lake systems, and as such our results are of value for extension of the technique to similar glaciated regions, with the caveat that a first-pass assessment of GLOF hazard should be first undertaken to identify "priority" glacial lakes (Huggel et al., 2004). In identifying the importance of constraining specific parameters for dam-breach model parameterisation (e.g. Manning's n), we have highlighted the importance of identifying model-critical data, which will aid the design of future field campaigns.

Probabilistic approaches have clear advantages over the deterministic approaches traditionally used for GLOF reconstruction such as the use of palaeohydraulic techniques for at-a-point or reach-scale peak discharge estimation (e.g. Cenderelli and Wohl, 2003; Kershaw et al., 2005; Bohorquez and Darby, 2008). Probabilistic methods embrace and attempt to convey the influence of uncertainty and equifinality in model input on subsequent output. It might be argued that their value outweighs the additional processing time required for their implementation, which may involve the execution of hundreds or thousands of individual simulations.

In considering the source of uncertainty in the GLOF modelling process, we have focused on its influence over the moraine dam-failure process. However, numerous additional sources of uncertainty are present at various stages in the workflow, such as the reconstruction of lake-basin bathymetry, and merit further investigation (Westoby et al., 2014a). The logistical impracticalities of identifying and addressing all sources of uncertainty in the GLOF model chain are currently a significant hindrance to applied modelling efforts. However, simple sensitivity analyses remain of value to quantify the impacts of individual sources of uncertainty on numerical dam-breach and hydrodynamic simulation, and might be incorporated straightforwardly into our modelling framework (Fig. 15).

8 Conclusions

This paper has outlined and presented results from a workflow for cascading uncertainty and equifinality through the glacial lake outburst flood (GLOF) model chain using a combination of advanced, physically based numerical dam-breach and hydrodynamic models. Dam material roughness is the dominant influence on outflow hydrograph form. Morphological characteristics of a GLOF breach are appropriate measures for assessing the performance of individual simulations, or parameter ensembles, at reproducing observed breach morphology. Breach morphology is reproducible by parameter ensembles associated with differing breach-initiation scenarios, lending support for the adoption of probabilistic, as opposed to deterministic, methods for dam-breach outburst-flood reconstruction. We also demonstrate an effective approach for cascading dam-breach simulation likelihood data through to the construction of probability-based maps of GLOF inundation extent and flow depth, and the subsequent derivation of event-specific maps of flood hazard.

Acknowledgements. M. J. Westoby was funded by a NERC Open CASE award (NE/G011443/1) in partnership with Reynolds International Ltd. Additional funding to support field activities from the Department of Geography and Earth Sciences Postgraduate Discretionary Fund (Aberystwyth University) is duly acknowledged. J. Balfour, P. Cowley, S. Doyle, H. Sevestre, C. Souness, R. Taylor, and guides and porters from Summit Trekking, Kathmandu, assisted with data collection in the Khumbu Himal. HR Wallingford Ltd and the Halcrow Group Ltd are thanked for the provision of academic licences for the use of HR BREACH and ISIS 2-D, respectively. GeoEye imagery (Fig. 1b) was provided free of charge by the GeoEye Foundation. ASTER GDEM data were downloaded free of charge from http://asterweb.jpl.nasa.gov/gdem.asp. Subtractive clustering of breach hydrograph data was undertaken in MathWorks' MATLAB®. We thank the associate editor, Katharine Huntington, for handling the manuscript, and the reviewers for providing constructive comments on initial and re-submitted iterations.

Edited by: K. W. Huntington

References

Alho, P. and Aaltonen, J.: Comparing a 1D hydraulic model with a 2D hydraulic model for the simulation of extreme glacial outburst floods, Hydrol. Process., 22, 1537–1547, 2008.

Aronica, G., Hankin, B., and Beven, K.: Uncertainty and equifinality in calibrating distributed roughness coefficients in a flood propagation model with limited data, Adv. Water Resour., 22, 349–365, 1998.

Aronica, G., Bates, P. D., and Horritt, M. S.: Assessing the uncertainty in distributed model predictions using observed binary pattern information within GLUE, Hydrol. Process., 16, 2001–2016, 2002.

Aronica G., Candela, A., Fabio, P., and Santoro, M.: Estimation of flood inundation probabilities using global hazard indexes based on hydrodynamic variables, Phys. Chem. Earth, 42–44, 119–129, 2012.

Astre, H.: SFMToolkit3, availabl at: http://www.visual-experiments.com/demos/sfmtoolkit (last access: 15 February 2015), 2010.

Benn, D. I., Wiseman, S., and Hands, K. A.: Growth and drainage of supraglacial lakes on debris-mantled Ngozumpa Glacier, Khumbu Himal, Nepal, J. Glaciol., 47, 626–638, 2001.

Benn, D. I., Bolch, T., Hands, K., Gulley, J., Luckman, A., Nicholson, L.I., Quincey, D.J., Thompson, S., Toumi, R., and Wiseman, S.: Response of debris-covered glaciers in the Mount Everest region to recent warming and implications for outburst flood hazards, Earth-Sci. Rev., 114, 156–174, 2012.

Beven, K.: A manifesto for the equifinality thesis, J. Hydrol., 320, 18–36, 2005.

Beven, K. and Binley, A.: The future of distribution models: model calibration and uncertainty prediction, Hydrol. Process., 6, 279–298, 1992.

Beven, K. and Freer, J.: Equifinality, data assimilation, and uncertainty estimation in mechanistic modelling of complex environmental systems using the GLUE methodology, J. Hydrol., 249, 11–29, 2001.

Beven, K. Smith, P.J., and Wood, A.: On the colour and spin of epistemic error (and what we might do about it), Hydrology and Earth System Sciences, 15, 3123-3133, 2011.

Beven, K., Lamb, R., Leedal, D., and Hunter, N.: Communicating uncertainty in flood inundation mapping: a case study, Int. J. River Basin Manage., doi:10.1080/15715124.2014.917318, in press, 2014.

Blazkova, S. and Beven, K.: Flood frequency estimation by continuous simulation of subcatchment rainfalls and discharges with the aim of improving dam safety assessment in a large basin in the Czech Republic, J. Hydrol., 292, 153–172, 2004.

Bohorquez, P. and Darby, S. E.: The use of one- and two-dimensional hydraulic modelling to reconstruct a glacial outburst flood in a steep Alpine valley, J. Hydrol., 361, 240–261, 2008.

Bolch, T., Kulkarni, A., Kääb, A., Huggel, C., Paul, F., Cogley, J.G., Frey, H., Kargel, J.S., Fujita, K., Scheel, M., Bajracharya, S., and Stoffel, M.: The state and fate of Himalayan glaciers, Science, 336, 310–314, 2012.

Bolch, T., Pieczonka, T., and Benn, D. I.: Multi-decadal mass loss of glaciers in the Everest area (Nepal Himalaya) derived from stereo imagery, The Cryosphere, 5, 349–358, doi:10.5194/tc-5-349-2011, 2011.

Brasington, J., Vericat, D., and Rychkov, I.: Modelling river bed morphology, roughness and surface sedimentology using high resolution Terrestrial Laser Scanning, Water Resour. Res., 48, W11519, doi:10.1029/2012WR012223, 2012.

Byers, A. C., McKinney, D. C., Somos-Valenzuela, M., Watanabe, T., and Lamsal, D.: Glacial lakes of the Hinku and Hongu valleys, Makalu Barun National Park and Buffer Zone, Nepal, Nat. Hazards, 69, 115–139, 2013.

Cao, Z. and Carling, P.: Mathematical modelling of alluvial rivers: reality and myth, Part 1: general overview, Mar. Eng., 154, 207–219, 2002.

Castellarin, A., Merz, R., and Bloschl, G.: Probabilistic envelope curves for extreme rainfall events, J. Hydrol., 378, 263–271, 2009.

Cenderelli, D. A. and Wohl, E. E.: Peak discharge estimates of glacial-lake outburst floods and "normal" climatic floods in the Mount Everest region, Nepal, Geomorphology, 40, 57–90, 2001.

Cenderelli, D. A. and Wohl, E. E.: Flow hydraulics and geomorphic effects of glacial-lake outburst floods in the Mount Everest region, Nepal, Earth Surf. Proc. Land., 28, 385–407, 2003.

Chanson, H.: The Hydraulics of Open Channel Flow, 2nd Edn., Elsevier Butterworth-Heinemann, Oxford, 650 pp., 2011.

Chen, Y. H. and Anderson, B. A.: Development of a methodology for estimating embankment damage due to flood overtopping, US Federal Highway Administration Report No. FHWA/RD-86/126, US Federal Highway Administration, Washington, D.C., 1986.

Chow, V. T.: Open-channel hydraulics, McGraw-Hill, New York, 677 pp., 1959.

Clague, J. J. and Evans, S. G.: A review of catastrophic drainage of moraine-dammed lakes in British Columbia, Quaternary Sci. Rev., 19, 1763–1783, 2000.

Costa, J. E. and Schuster, R. L.: The formation and failure of natural dams, Geol. Soc. Am. Bull., 100, 1054–1068, 1988.

Dewals, B., Erpicum, S., Detrembleur, S., Archambeau, P., and Pirotton, M.: Failure of dams arranged in series or in complex, Nat. Hazards, 56, 917–939, 2011.

Fonstad, M. A., Dietrich, J. T., Courville, B. C., Jensen, J. L., and Carbonneau, P. E.: Topographic structure from motion: a new development in photogrammetric development, Earth Surf. Proc. Land., 38, 421–430, 2013.

Franz, K. J. and Hogue, T. S.: Evaluating uncertainty estimates in hydrologic models: borrowing measures from the forecast verification community, Hydrol. Earth Syst. Sci., 15, 3367–3382, doi:10.5194/hess-15-3367-2011, 2011.

Freer, J., Beven, K. J., and Peters, N. E.: Multivariate seasonal period model rejection within the generalised likelihood uncertainty estimation procedure, in: Calibration of Watershed Models, edited by: Duan, Q., Gupta, H., Sorooshian, S., Rousseau, A. N., and Turcotte, R., AGU Books, Washington, 69–87, 2002.

Frey, H., Haeberli, W., Linsbauer, A., Huggel, C., and Paul, F.: A multi-level strategy for anticipating future glacier lake formation and associated hazard potentials, Nat. Hazards Earth Syst. Sci., 10, 339–352, doi:10.5194/nhess-10-339-2010, 2010.

Galay, V.: Glacier Lake Outburst Flood (Jökulhlaup) on the Bothe/Dudh Kosi, August 4, 1985, Internal Report, Water and Energy Committee, Ministry of Water Resources, His Majesty's Government of Nepal, Kathmandu, 1985.

Halcrow: ISIS 2D by Halcrow, available at: http://www.halcrow.com/isis/ (last access: 3 March 2013), 2012.

Hall, J., Tarantola, S., Bates, P. D., and Horritt, M.: Distributed sensitivity analysis of flood inundation model calibration, J. Hydraul. Eng., 131, 117–126, 2005.

Hambrey, M. J., Quincey, D. J., Glasser, N. F., Reynolds, J. M., Richardson, S. J., and Clemmens, S.: Sedimentological, geomorphological and dynamic context of debris-mantled glaciers, Mount Everest (Sagarmatha) region, Nepal, Quaternary Sci. Rev., 28, 2361–2389, 2008.

Hanson, G. J. and Cook, K. R.: Apparatus, test procedures, and analytical methods to measure soil erodibility in situ, Appl. Eng. Agricult., 20, 455–462, 2004.

Harrison, S., Glasser, N. F., Winchester, V., Haresign, E., Warren, C., and Jansson, K.: A glacial lake outburst flood associated with recent mountain glacier retreat, Patagonian Andes, Holocene, 16, 611–620, 2006.

Hassan, A. E., Bekhit, H., and Chapman, J. B.: Uncertainty assessment of a stochastic groundwater flow model using GLUE analysis, J. Hydrol., 362, 89–109, 2008.

Hassan, M. and Morris, M.: HR-BREACH Model Documentation, HR Wallingford Ltd, Wallingford, Oxfordshire, 2012.

Hubbard, B., Heald, A., Reynolds, J. M., Quincey, D., Richardson, S. D., Zapata Luyo, M., Santillan Portilla, N., and Hambrey, M. J.: Impact of a rock avalanche on a moraine-dammed proglacial lake: Laguna Safuna Alta, Cordillera Blanca, Peru, Earth Surf. Proc. Land., 30, 1251–1264, 2005.

Huggel, C., Haeberli, W., Kääb, A., Bieri, D., and Richardson, S.: An assessment procedure for glacial hazards in the Swiss Alps, Can. Geotech. J., 41, 1068–1083, 2004.

Hunter, N. M., Bates, P. D., Horritt, M. S., De Roo, A. P. J., and Werner, M. G. F.: Utility of different data types for calibrating flood inundation models within a GLUE framework, Hydrol. Earth Syst. Sci., 9, 412–430, doi:10.5194/hess-9-412-2005, 2005.

Iman, R. L. and Helton, J. C.: An investigation of uncertainty and sensitivity analysis techniques for computer models, Risk Analysis, 8, 71–90, 2006.

IMPACT: Investigation of Extreme Flood Processes and Uncertainty: Risk and Uncertainty (WP5) – Technical Report, 44 pp., available at: http://www.impact-project.net (last access: 3 March 2014), 2004.

James, M. R. and Robson, S.: Straightforward reconstruction of 3D surfaces and topography with a camera: Accuracy and geoscience applications, J. Geophys. Res., 117, F03017, doi:10.1029/2011JF002289, 2012.

Janský, B., Engel, Z., Šobr, M., Beneš, V., Špaček, K., and Yerokhin, S.: The evolution of Petrov lake and moraine dam rupture risk (Tien-Shan, Kyrgyzstan), Nat. Hazards, 50, 83–96, 2009.

Javernick, L., Brasington, J., and Caruso, B.: Modelling the topography of shallow braided rivers using Structure-from-Motion photogrammetry, Geomorphology, 213, 166–182, 2014.

Kaser, G., Cogley, J. G., Dyurgerov, B., Meier, M. F., and Ohmura, A.: Mass balance of glaciers and ice caps: Consensus estimates for 1961–2004, Geophys. Res. Lett., 33, 1–5, 2006.

Kershaw, J. A., Clague, J. J., and Evans, S. G.: Geomorphic and sedimentological signature of a two-phase outburst flood from moraine-dammed Queen Bess Lake, British Columbia, Canada, Earth Surf. Proc. Land., 30, 1–25, 2005.

Kidson, R. L., Richards, K. S., and Carling, P. A.: Hydraulic model calibration for extreme floods in bedrock-confined channels: case study from northern Thailand, Hydrol. Process., 20, 329–344, 2006.

Klimeš, J., Benešová, M., Vilímek, V., Bouška, P., and Rapre, A. C.: The reconstruction of a glacial lake outburst flood using HEC-RAS and its significance for future hazard assessments: an example from Lake 513 in the Cordillera Blanca, Peru, Natural Hazards, 71, 1617–1638, 2014.

Kuczera, G. and Parent, E.: Monte Carlo assessment of parameter uncertainty in conceptual catchment models: the Metropolis algorithm, J. Hydrol., 211, 69–85, 1998.

Lamb, R., Beven, K., and Myrabo, S.: Use of spatially distributed water table observations to constrain uncertainty in a rainfall-runoff model, Adv. Water Resour., 22, 305–317, 1998.

Lebourg, T., Riss, J., and Pirard, E.: Influence of morphological characteristics of heterogeneous moraine formations on their mechanical behaviour using image and statistical analysis, Eng. Geol., 73, 37–50, 2004.

Liang, D., Falconer, R. A., and Lin, B.: Comparison between TVD-MacCormack and ADI-type solvers of the shallow water equations, Adv. Water Resour., 29, 1833–1845, 2006.

Liang, D., Lin, B., and Falconer, R. A.: Simulation of rapidly varying flow using an efficient TVD-MacCormack scheme, Int. J. Numer. Meth. Fluids, 53, 811–826, 2007.

Liao, C. B., Wu, M. S., and Liang, S. J.: Numerical simulation of a dam break for an actual river terrain environment, Hydrol. Process., 21, 447–460, 2007.

Lliboutry, L., Morales, B., Pautre, A., and Schneider, B.: Glaciological problems set by the control of dangerous lakes in Cordillera Blanca, Peru, I: Historical failure of morainic dams, their causes and prevention, J. Glaciol., 18, 239–254, 1977.

Lowe, D. G.: Distinctive image features from scale-invariant keypoints, Int. J. Comput. Vis., 60, 91–110, 2004.

MathWorks®: MATLAB (version 7.6), available at: http://www.mathworks.co.uk (last access: 15 December 2012), 2012.

McMillan, H. K. and Brasington, J.: Reduced complexity strategies for modelling urban floodplain inundation, Geomorphology, 90, 226–243, 2007.

McMillan, H. K. and Brasington, J.: End-to-end flood risk assessment: A coupled model cascade with uncertainty estimation, Water Resour. Res., 44, W03419, doi:10.1029/2007WR005995, 2008.

Merz, B. and Thieken, A.: Flood risk analysis: Concepts and challenge, Oester. Wasser Abfallwirt., 56, 27–34, 2004.

Meselhe, E. A. and Holly Jr., F. M.: Invalidity of Preissmann scheme for transcritical flow, J. Hydraul. Eng., 123, 652–655, 1997.

Metropolis, N.: The beginning of the Monte Carlo method, Los Alamos Science, 15, 125–130, 1987.

Mohamed, M. A. A., Samuels, P. G., Morris, M., and Ghataora, G. S.: Improving the accuracy of prediction of breach formation through embankment dams and flood embankments, in: River Flow 2002, edited by: Bousmar, D. and Zech, Y., 1st International Conference on Fluvial Hydraulics, 3–6 September 2002, Louvain-la-Neuve, Belgium, 10 pp., 2002.

Mool, P. K., Bajracharya, S. R., and Joshi, S. P.: Inventory of glaciers, glacial lakes and glacial lake outburst floods, Nepal, International Centre for Integrated Mountain Development, Kathmandu, Nepal, 2001.

Morris, M., Hanson, G., and Hassan, M. A. A.: Improving the accuracy of breach modelling: why are we not progressing faster?, J. Flood Risk Manage., 1, 150–161, 2008.

Nicholas, A. P. and Quine, T. A.: Quantitative assessment of landform equifinality and palaeoenvironmental reconstruction using geomorphic models, Geomorphology, 121, 167–183, 2010.

Oerlemans, J.: Quantifying global warming from the retreat of glacier, Science, 264, 243–245, 1994.

Osti, R. and Egashira, S.: Hydrodynamic characteristics of the Tam Pokhari Glacial Lake outburst flood in the Mt. Everest region, Nepal, Hydrol. Process., 23, 2943–2955, 2009.

Osti, R., Bhattarai, T. N., and Miyake, K.: Causes of catastrophic failure of Tam Pokhari moraine dam in the Mt. Everest region, Nat. Hazards, 58, 1209–1223, 2011.

Pappenberger, F., Beven, K. J., Hunter, N. M., Bates, P. D., Gouweleeuw, B. T., Thielen, J., and de Roo, A. P. J.: Cascading model uncertainty from medium range weather forecasts (10 days) through a rainfall-runoff model to flood inundation predictions within the European Flood Forecasting System (EFFS), Hydrol. Earth Syst. Sci., 9, 381–393, doi:10.5194/hess-9-381-2005, 2005.

Pappenberger, F., Matgen, P., Beven, K. J., Henry, J. B., Pfister, L., and de Fraipont, P.: Influence of uncertain boundary conditions and model structure on flood inundation predictions, Adv. Water Resour., 29, 1430–1449, 2006.

Parkin, G., O'Donnell, G., Ewen, J., Bathurst, J. C., O'Connell, P. E., and Lavabre, J.: Validation of catchment models for predicting land-use and climate change impacts: 2. Case study for a Mediterranean catchment, J. Hydrol., 175, 595–613, 1996.

Refsgaard, J. C.: Parameterisation, calibration and validation of distributed hydrological models, J. Hydrol., 198, 69–97, 1997.

Reynolds, J. M.: High-altitude glacial lake hazard assessment and mitigation: a Himalayan perspective, in: Geohazards in Engineering, edited by: Maund, J. G. and Eddleston, M., Engineering Geology Special Publications, Geological Society, London, 15, 25–34, 1998. Reynolds, J. M.: On the formation of supraglacial lakes on debris-covered glaciers, in: Debris-Covered Glaciers, edited by: Nakawo, M., Raymond, C. F., and Fountain, A., Proceedings of the Seattle Workshop, September 2000, USA, IAHS Publ., 264, 153–161, 2000.

Reynolds, J. M.: Assessing glacial hazards for hydro development in the Himalayas, Hindu Kush and Karakoram, Hydropower Dams, 2, 60–65, 2014.

RGSL – Reynolds Geo-Sciences Ltd: Development of glacial hazard and risk minimisation protocols in rural environments: Guidelines for the management of glacial hazards and risks. Reynolds Geo-Sciences Ltd Report R7816.142, available at: http://www.reynolds-international.co.uk/dfid (last access: 14 February 2015), 2003.

Richardson, S. D. and Reynolds, J. M.: An overview of glacial hazards in the Himalayas, Quatern. Int., 65/66, 31–47, 2000.

Romanowicz, R. and Beven, K.: Dynamic real-time prediction of flood inundation probabilities, Hydrolog. Sci. J., 43, 181–196, 1998.

Rychkov, I., Brasington, J., and Vericat, D.: Computational and methodological aspects of terrestrial surface analysis based on point clouds, Comput. Geosci., 42, 64–70, 2012.

Sakai, A., Chikita, K., and Yamada, T.: Expansion of a moraine-dammed glacial lake, Tsho Rolpa, in Rolwaling Himal, Nepal Himalaya, Limnol. Oceanogr., 45, 1401–1408, 2000.

Sanders, B. F.: Evaluation of on-line DEMs for flood inundation modelling, Adv. Water Resour., 30, 1831–1843, 2007.

Snavely, N.: Scene Reconstruction and Visualization from Internet Photo Collections, unpublished PhD thesis, University of Washington, USA, 2008.

Snavely, N., Seitz, S. N., and Szeliski, R.: Modelling the world from internet photo collections, Int. J. Comput. Vis., 80, 189–210, 2008.

Somos-Valenzuela, M. A., McKinney, D. C., Rounce, D. R., and Byers, A. C.: Changes in Imja Tsho in the Mount Everest region of Nepal, The Cryosphere, 8, 1661–1671, doi:10.5194/tc-8-1661-2014, 2014.

Stokes, C. R., Spagnolo, M., and Clark, C. D.: The composition and internal structure of drumlins: complexity, commonality, and implications for a unifying theory of their formation, Earth-Sci. Rev., 107, 398–422, 2011.

Thompson, S. S., Benn, D. I., Dennis, K., and Luckman, A.: A rapidly-growing moraine-dammed glacial lake on Ngozumpa Glacier, Nepal, Geomorphology, 145–146, 1–11, 2012.

Vasquez, R. F., Beven, K., and Feyen, J.: GLUE based assessment on the overall predictions of a MIKE SHE application, Water Resour. Manage., 23, 1325–1349, 2009.

Venutelli, M.: Stability and accuracy of weighted four-point implicit finite difference schemes for open channel flow, J. Hydraul. Eng., 128, 281–288, 2002.

von Bertalanffy, L.: General System Theory – Foundations, Development, Applications, 2nd Edn., George Braziller, New York, 295 pp., 1968.

Vorogushyn, S., Merz, B., Lindenschmidt, K. E., and Apel, H.: A new methodology for flood hazard assessment considering dike breaches, Water Resour. Res., 46, W08541, doi:10.1029/2009WR008475, 2010.

Vorogushyn, S., Apel, H., and Merz, B.: The impact of the uncertainty of dike breach development time on flood hazard, Phys. Chem. Earth, 36, 319–323, 2011.

Vrugt, J. A., ter Braak, C. J. F., Gupta, H. V., and Robinson, B. A.: Equifinality of formal (DREAM) and informal (GLUE) Bayesian approaches in hydrologic modelling?, Stoch. Environ. Res. Risk A., 23, 1011–1026, 2009.

Vuichard, D. and Zimmerman, M.: The Langmoche flash-flood, Khumbu Himal, Nepal, Mount. Res. Develop., 6, 90–94, 1986.

Vuichard, D. and Zimmerman, M.: The 1985 catastrophic drainage of a moraine-dammed lake, Khumbu Himal, Nepal: Cause and consequences, Mount. Res. Develop., 7, 91–110, 1987.

Walder, J. S. and O'Connor, J. E.: Methods for predicting peak discharge of flood caused by failure of natural and constructed earth dams, Water Resour. Res., 33, 2337–2348, 1997.

Wang, W., Yang, X., and Yao, T.: Evaluation of ASTER GDEM and SRTM and their suitability in hydraulic modelling of a glacial lake outburst flood in southeast Tibet, Hydrol. Process., 26, 213–225, 2012.

Watanabe, T. and Rothacher, D.: The 1994 Lugge Tsho Glacial Lake Outburst Flood, Bhutan Himalaya, Mount. Res. Develop., 16, 77–81, 1996.

Watanabe, T., Kameyama, S., and Sato, T.: Imja Glacier dead-ice melt rates and changes in a supraglacial lake, 1989–1994, Khumbu Himal, Nepal: danger of lake drainage, Mount. Res. Develop., 15, 293–300, 1995.

Westerberg, I. K., Guerrero, J.-L., Younger, P. M., Beven, K. J., Seibert, J., Halldin, S., Freer, J. E., and Xu, C.-Y.: Calibration of hydrological models using flow-duration curves, Hydrol. Earth Syst. Sci., 15, 2205–2227, doi:10.5194/hess-15-2205-2011, 2011.

Westoby, M. J.: The development of a unified framework for low-cost Glacial Lake Outburst Flood hazard assessment, Unpublished PhD thesis, Department of Geography, History and Politics, Aberystwyth University, Aberystwyth, 456 pp., 2013.

Westoby, M. J., Brasington, J., Glasser, N. F., Hambrey, M. J., and Reynolds, J. M.: 'Structure-from-Motion' photogrammetry: A low-cost, effective tool for geoscience applications, Geomorphology, 179, 300–314, 2012.

Westoby, M. J., Glasser, N. F., Brasington, J., Hambrey, M. J., Quincey, D. J., and Reynolds, J. M.: Modelling outburst floods from moraine-dammed glacial lakes, Earth-Sci. Rev., 134, 137–159, doi:10.1016/j.earscirev.2014.03.009, 2014a.

Westoby, M. J., Glasser, N. F., Hambrey, M. J., Brasington, J., and Mohamed, M. A. A. M.: Reconstructing historic Glacial Lake Outburst Floods through numerical modelling and geomorphological assessment: Extreme events in the Himalaya, Earth Surf. Proc. Land., 39, 1675–1692, 2014b.

Wohl, E. E.: Uncertainty in flood estimates associated with roughness coefficient, J. Hydraul. Eng., 124, 219–223, 1998.

Worni, R., Stoffel, M., Huggel, C., Volz, C., Casteller, A., and Luckman, B.: Analysis and dynamic modelling of a moraine failure and glacier lake outburst flood at Ventisquero Negro, Patagonian Andes (Argentina), J. Hydrol., 444–445, 134–145, 2012.

Worni, R., Huggel, C., and Stoffel, M.: Glacial lakes in the Indian Himalayas – From an area-wide glacial lake inventory to on-site and modeling based risk assessment of critical glacial lakes, Sci. Total Environ., 468–469, S71–S84, 2013.

Worni, R., Huggel, C., Clague, J. J., Schaub, Y., and Stoffel, M.: Coupling glacial lake impact, dam breach, and flood processes: A modeling perspective, Geomorphology, 224, 161–176, 2014.

Xin, W., Shlyln, L., Wanqin, G., and Junll, X.: Assessment and simulation of Glacier Lake Outburst Floods for Longbasaba and Pida Lakes, China, Mount. Res. Develop., 28, 310–317, 2008.

Zemp, M., Hoelzle, M., and Haeberli, W.: Six decades of glacier mass balance observations – a review of the worldwide monitoring network, Ann. Glaciol., 50, 101–111, 2009.

Zhong, D., Sun, Y., and Li, M.: Dam break threshold value and risk probability assessment for an earth dam, Nat. Hazards, 59, 129–147, 2011.

12

Constraining the stream power law: a novel approach combining a landscape evolution model and an inversion method

T. Croissant[1,*] **and J. Braun**[1]

[1]ISTerre, Université Grenoble-Alpes and CNRS, BP 53, 38041 Grenoble Cedex 9, France
[*]Now at Geosciences Rennes, Université de Rennes 1 and CNRS, Rennes, France

Correspondence to: T. Croissant (thomas.croissant@univ-rennes1.fr)

Abstract. In the past few decades, many studies have been dedicated to the understanding of the interactions between tectonics and erosion, in many instances through the use of numerical models of landscape evolution. Among the numerous parameterizations that have been developed to predict river channel evolution, the stream power law, which links erosion rate to drainage area and slope, remains the most widely used. Despite its simple formulation, its power lies in its capacity to reproduce many of the characteristic features of natural systems (the concavity of river profile, the propagation of knickpoints, etc.). However, the three main coefficients that are needed to relate erosion rate to slope and drainage area in the stream power law remain poorly constrained. In this study, we present a novel approach to constrain the stream power law coefficients under the detachment-limited mode by combining a highly efficient landscape evolution model, FastScape, which solves the stream power law under arbitrary geometries and boundary conditions and an inversion algorithm, the neighborhood algorithm. A misfit function is built by comparing topographic data of a reference landscape supposedly at steady state and the same landscape subject to both uplift and erosion over one time step. By applying the method to a synthetic landscape, we show that different landscape characteristics can be retrieved, such as the concavity of river profiles and the steepness index. When applied on a real catchment (in the Whataroa region of the South Island in New Zealand), this approach provides well-resolved constraints on the concavity of river profiles and the distribution of uplift as a function of distance to the Alpine Fault, the main active structure in the area.

1 Introduction

Because their geometry is very sensitive to external forcing such as climate or tectonics, rivers are ideal natural laboratories for studying the interactions of the various processes at play during orogenesis over geological timescales (Kirby and Whipple, 2001; Montgomery and Brandon, 2002; Duvall et al., 2004; Whittaker et al., 2007; Kirby and Whipple, 2012). For this purpose many parameterizations of fluvial incision have been developed (Kooi and Beaumont, 1994; Sklar and Dietrich, 1998; Whipple and Tucker, 1999). The most widely used, the so-called stream power law (SPL) (Howard, 1994; Whipple and Tucker, 1999), relates incision rate $\dot{\epsilon}$ to both drainage area A, a proxy for local discharge, and local slope S in the following manner:

$$\dot{\epsilon} = KA^m S^n; \qquad (1)$$

K is a proportionality coefficient called the "erosion efficiency" or "erodibility" that mostly depends on lithology and climate, while m and n are positive exponents that mostly depend on catchment hydrology and the exact nature of the dominant erosional mechanism such as plucking, abrasion, dissolution or weathering.

Although the SPL is widely used in the community and has been implemented in various landscape evolution models (LEMs) (Crave and Davy, 2001; Tucker et al., 2001)

(Braun and Willett, 2013), the values of K, m and n remain poorly constrained. These parameters depend on numerous factors and cannot easily be measured from direct field observations. At best, one is conventionally required to fix the value of one or two of them in order to deduce the value of the other parameters from observational constraints (Stock and Montgomery, 1999; Kirby and Whipple, 2001). A more commonly approach is to compare the long-term predictions of an LEM with observational constraints on the rate of change of a given landform to infer the value of the SPL parameters (van Der Beek and Bishop, 2003; Tomkin et al., 2003). However, most LEMs require a fine spatial and temporal discretization and are commonly limited by their computational cost (Tucker and Hancock, 2010). The use of inversion or optimization methods that require a thorough search through parameter space has been limited by these computational limitations. An alternative is to limit the computation and the comparison with observations to 1-D river profiles, as was done by Roberts and White (2010) in Africa and Roberts et al. (2012) in the Colorado Plateau to deduce information about the geometry and timing of uplift.

In the past year, major advances have been made in improving the efficiency of the surface process models (SPM) solving the SPL, and an algorithm has been developed (Braun and Willett, 2013) that is implicit in time and $O(n_p)$; in other words, computational time increases linearly with n_p, the number of points used to discretize the landscape. This new algorithm, called FastScape, is sufficiently efficient to be used inside an inversion procedure that requires tens of thousands of runs to search through parameter space while still using a very high spatial discretization (10^8 nodes).

We present here a novel approach that we have developed to constrain the parameters of the SPL in environments that have reached geomorphic steady state, i.e., a local equilibrium between uplift and erosion. The objective is to determine the best combination of the K, n and m parameters that will maintain a given landform in its starting geometry after applying a known or arbitrary uplift and eroding it according to the SPL. To achieve this we used the LEM FastScape combined with the neighborhood algorithm (NA) inversion method. We first applied our approach to a synthetic landscape for which the value of the SPL parameters are known to test its validity and usefulness. We then applied it to a digital elevation model (DEM) from the Whataroa Valley in New Zealand, a region that has very likely reached geomorphic steady state (Adams, 1980; Herman et al., 2010).

2 The erosion law

Using the SPL, we can predict river channel evolution in detachment-limited systems (bedrock rivers) undergoing constant and uniform uplift by using the following mass balance equation:

$$\frac{\partial h}{\partial t} = U - \dot{\epsilon} = U - KA^m S^n, \tag{2}$$

where h is the elevation of the channel, t is time and U is rock uplift rate relative to a fixed or known base level (Whipple and Tucker, 1999). As explained above, constraining the exact value of K, m and n from natural landscapes is relatively complex. The value of these parameters is still debated and is likely to depend on the geomorphological, climatic and tectonic context but the following ranges are commonly admitted:

- $0 < m < 2$;

- $0 < n < 4$;

- K varies by several orders of magnitude as it depends not only on many factors such as lithology, climate, sedimentary flux or river channel width but also on the value of the other two parameters m and n.

Assuming steady state, an expression for equilibrium channel gradient or slope, S_e, can be easily obtained from Eq. (2):

$$S_e = \left(\frac{U}{K}\right)^{1/n} A^{-m/n} = k_s A^{-\theta}, \tag{3}$$

which shows that, in situations where an equilibrium between uplift and incision has been reached, one can obtain information about the ratio $\theta = m/n$ by simply computing the relationship that must exist between drainage area and local slope. The results of such studies are numerous and yield values in the range $\theta = 0.35 - 0.6$ (Whipple and Tucker, 1999; Whipple, 2004; Kirby and Whipple, 2012). This ratio is called the concavity as its value is mostly constrained by the concavity of river profiles. The other parameter, k_s, relating equilibrium slope to drainage area is called the "steepness index". Its use is a direct consequence of our realization that erodibility can only be constrained where uplift rate is known or, more exactly, that we should focus on constraining the relative response of a river to tectonic uplift, not its intrinsic erosional efficiency.

3 Combining FastScape and NA

As stated earlier, we used the FastScape algorithm (Braun and Willett, 2013) to solve the SPL and predict landscape evolution in a given tectonic and geomorphic setting. The efficiency and stability of this algorithm make it well suited to be used inside an inversion scheme that requires a large number of model runs. As seen previously, the number of parameters used in the SPL (K, m and n) is relatively small, but these combined with the unknown uplift rate U and the fact that each parameter varies over a relatively wide

Table 1. Parameterization of the different runs

RUN	m	n	K (m^{1-2m} yr^{-1})	U (m yr^{-1})	N_{total}	N_{init}	N_{it}
Synthetic cases:							
Reference model	0.4	1	10^{-5}	0.0005			
nm	0.1/2	0.2/4	fixed	fixed	22 500	5	10 000
nmk	0.1/2	0.2/4	$10^{-6}/10^{-4}$	fixed	90 000	10	30 000
unmk	0.1/2	0.2/4	$10^{-6}/10^{-4}$	0.0003/0.0008	90 000	10	30 000
Whataroa case:							
nmk	0.1/2	0.2/4	$10^{-13}/10^{-4}$	fixed	90 000	10	30 000
αnmk	0.1/2	0.2/4	$10^{-13}/10^{-4}$	$\alpha:0/1$	90 000	10	30 000

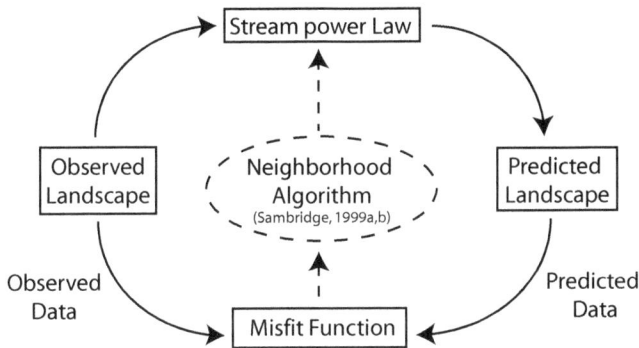

Figure 1. Scheme for the inversion. Observed landscape is extracted from a DEM or obtained by running the SPM to steady state. The predicted landscape is obtained by running the SPL with known parameter (U, K, m and n) values selected by the NA in order to minimize the misfit function obtained by comparing the observed and predicted landscapes.

range of values makes an exhaustive search through parameter space a rather tedious exercise that would require a large number of forward runs. Consequently we attempted to minimize the computational cost by using the neighborhood algorithm (NA) (Sambridge, 1999a, b), an inversion method that is well adapted to solving nonlinear problems. This optimization method is based on two separate stages. The first one, called the "sampling" stage, consists in finding an ensemble of best fitting models (combinations between U, K, m and n) that reproduce well the observed data or, in our case, that maintain the landscape at steady state. The second one, called the "appraisal" stage, consists in deriving quantitative and statistically meaningful estimates of each parameter from the ensemble of models generated in the first stage.

In order to compare the observed or reference landscape with the predicted one, one needs to construct a misfit function. In our case, we consider that both landscapes must have reached steady state ($\partial h/\partial t = 0$). We use the observed or reference landscape as the starting condition for the LEM to which we apply a uniform uplift increment; we then compute the resulting erosion over a time step of length Δt. We

define the misfit, ϕ, as the square root of the $L2$ norm of the change in height between observed h_{obs} and predicted h_{pred} topographies over the time step, Δt:

$$\phi = \sqrt{\sum_{i=1}^{N} \frac{(h_{i,pred} - h_{i,obs})^2}{\Delta t^2\, U_{\text{obs}}^2}}, \tag{4}$$

where N is the number of pixels in the landscape. The scheme is illustrated in Fig. 1. Because Eq. (2) only applies to river profile evolution, the optimization method is applied only on river pixels and the summation in Eq. (4) is limited to the nodes that have a drainage area larger than a specified minimum. To normalize the misfit function, we decided not to use the error on the observed topography (as is usually the case) because this error is very small in comparison to other potential sources of error inherent to our assumptions of steady state and, more importantly, to the assumption that the SPL controls the evolution of stream profiles; in its place, we use the imposed or known uplift rate, U, such that the misfit becomes a measure of the proportion of the imposed uplift rate that can be eroded back using the SPL.

To provide robust estimations of the parameter values during the appraisal stage of the inversion, the posterior probability density functions (PDFs) are based on the likelihood function L, defined as

$$L = \exp\left(-\frac{1}{2}\phi^2\right). \tag{5}$$

4 Application to synthetic landscapes

In order to demonstrate the validity of our approach, we first perform some tests by using synthetic landscapes created by FastScape as our starting condition (or in place of a natural steady-state landscape). We thus run FastScape with a set of known parameter values until steady state is reached. Our objective is then to retrieve these parameter values through the inversion procedure described above. The value of the parameters for the reference model are given by Table 1, as well as the range of parameter values tested during the inversion

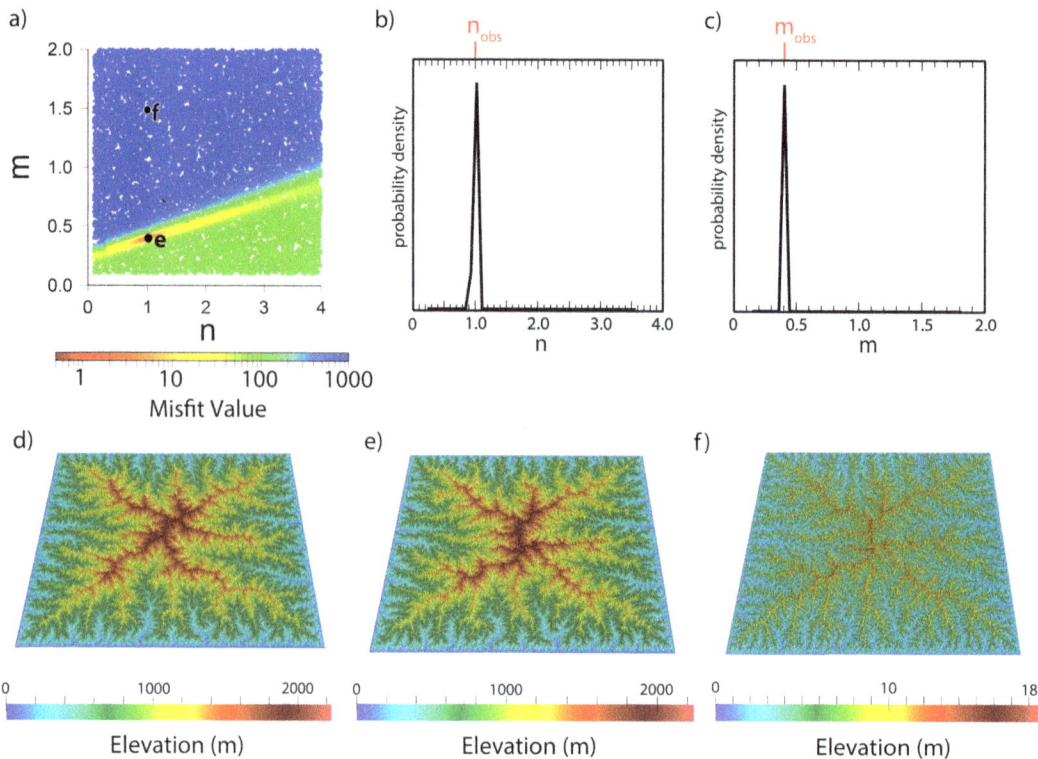

Figure 2. Results from inversion for the free parameters m and n. (**a**) Scatter plot showing the results from NA sampling stage. (**b**), (**c**) PDFs of the two parameters resulting from the NA appraisal stage . (**d**) Reference model topography. (**e**) Topography of the best fit model. (**f**) Topography of a high misfit value model.

and the number of model runs. NA has a few free parameters: these are N_{init}, the number of model runs in the first iteration (i.e., for which random values of the model parameters are used), N_{runs} the number of model runs that are resampled at each subsequent iteration (they correspond to the model runs that have given the smallest misfit value during the preceding iteration) and N_{it} the number of iterations. The total number of model runs is given by $N_{tot} = N_{init} + N_{it} \times N_{runs}$. The value of each of these NA parameters is also given in Table 1.

We tested various possible combinations of free parameters (i.e., those that are tested by the inversion scheme) among n, m, K and U to see whether we could retrieve them independently of each other or whether some combinations would be better constrained than others. We present the results of the following combinations:

– n and m are free, here because their ratio is supposed to control the concavity of a river profile at equilibrium;

– n, m and K are free, here because in many circumstances we may have independent evidence on the value of U, for example by interpreting cooling ages of rocks obtained by thermochronology;

– n, m, K and U are free; this would correspond to the most common situation in natural systems.

4.1 n and m are free

In this inversion, we take for U and K the values used to create the steady-state, reference landscape and let m and n vary freely in their preset ranges. The results of the sampling stage are shown in Fig. 2a, where each colored circle in parameter space corresponds to a model run and thus to a combination of model parameters. The color of each circle is a function of the misfit value, with the smaller misfit values corresponding to the warmer (red) colors. NA is designed to find the minimum value of the misfit function and converges towards $m = 0.4$ and $n = 1$ (as shown also by Supplement Fig. 1), which are the values used to create the initial reference landscape. The PDFs of each of the two parameters (Fig. 2b,c) show a narrow peak around these values, confirming that, in this configuration, we can constrain the exact value of the two parameters and a fortiori their ratio. The same result is obtained for a reference topography generated with a non-linear erosion law, i.e., $n \neq 1$ (see Supplement Fig. 2) This is an interesting result that shows that if a natural landscape is at equilibrium and has been created by processes obeying the SPL exactly, and if we know the value of U and K, then we should be able to retrieve not only the value of the ratio $\theta = m/n$, which has already been demonstrated to be controlled not only by the concavity of river profiles but also the exact value of each of the two exponents, m and n.

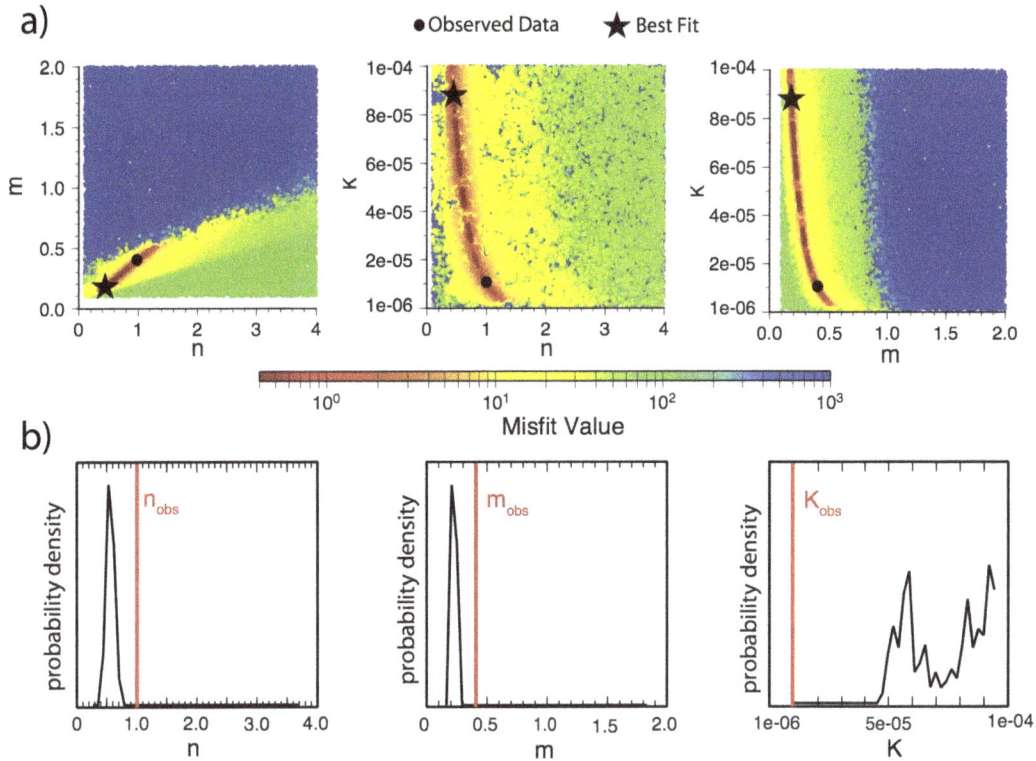

Figure 3. Results from inversion for the free parameters m, n and K. (**a**) Scatter plots showing the results from NA sampling stage. (**b**) PDFs of the three parameters resulting form the appraisal stage of NA.

4.2 n, m and K are free

In this case we assume that only the uplift rate, U, is known, and we fix it to the value used to construct the reference model by driving FastScape to steady state. The results from the sampling stage show several important points (Fig. 3a). First, the reduction of the misfit function leads to a trade-off between the m and n. All models runs that have a common m/n ratio of 0.4 are characterized by the smallest misfit values; they appear in the first panel of Fig. 3a as a red line. Note that this ratio between m and n is the same as the one used to compute the reference landscape ($m/n = 0.4/1.0$). Second, the absolute minimum is not located at these values for m and n, as shown by the PDFs shown in Fig. 3b. Thus, if we do not know the erodibility, K, we cannot retrieve the values of the m and n exponents, only their ratios as demonstrated in previous slope-area studies by Lague et al. (2000) and Snyder et al. (2000). Third, the other two scatter plots of Fig. 3a show that there is also a trade-off between $m - n$ and K. The same low value of the misfit function can be achieved with high values for K and small values of both m and n, or, conversely, with small values of K and large values of m and n in their permissible ranges. Note that in Fig. 3, K varies logarithmically along the vertical axis. This is easily explained by the asymptotic behavior of the SPL in the

vicinity of $m = n = 0$:

$$\frac{\partial h}{\partial t} = U - KA^0 S^0 = U - K, \tag{6}$$

which shows that steady state can only be reached when $K = U$, where $U = 5 \times 10^{-4}$ is the uplift rate imposed to compute the steady-state reference landscape. This also explains why the optimum values of the parameters m and n obtained from the inversion are smaller than they should be (Fig. 3b and Supplement Fig. 3) – because the misfit function contains an intrinsic minimum as m and n tend toward 0. This can be illustrated by computing the difference map between the reference target topography and the topography computed with various combinations of K, m and n during the inversion procedure (Supplement Fig. 4), which clearly show that the difference is zero when $m = n = 0$ and $K = U$.

The abrupt termination of the alignment (or line) of red circles between high misfit values domain (in blue) and the rest in Fig. 3a is an artifact of the range imposed on K. To produce small misfit landscapes for values of m and n larger than 0.6 and 1.5, respectively, would require values of K that are smaller than the smallest value permitted, i.e., 10^{-6}.

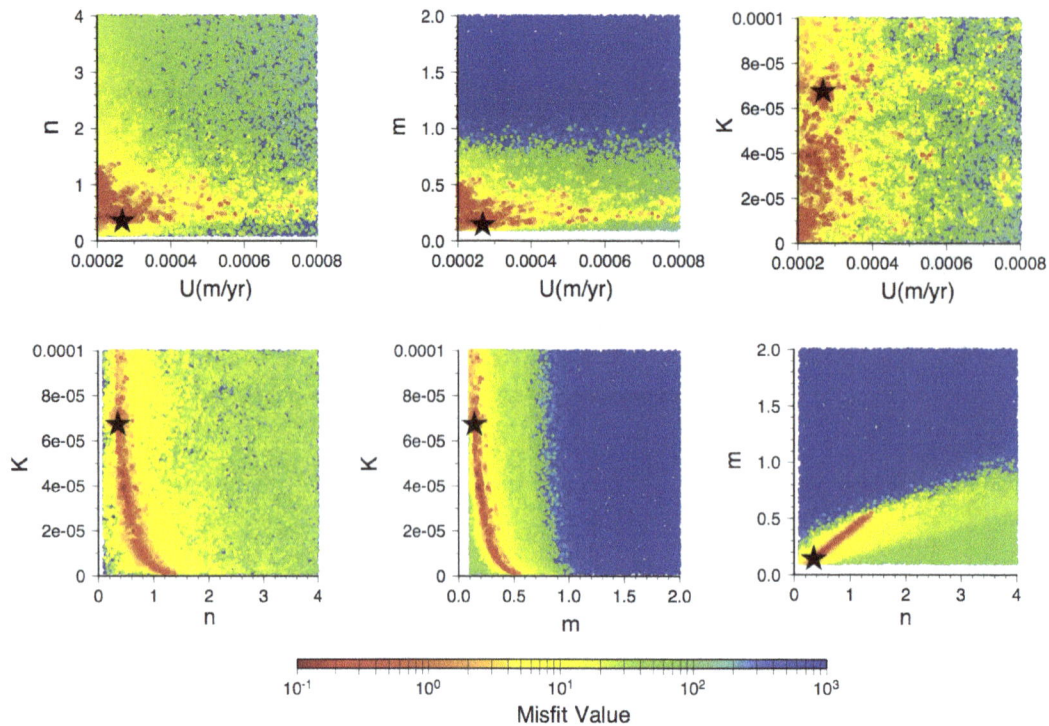

Figure 4. Results from inversion for the free parameters U, K, m and n. Scatter plots showing the results from NA sampling stage.

4.3 n, m, K and U are free

In this case, all parameters m, n, K and U are left free during the inversion. The scatter plots illustrating the behavior of the misfit function as a function of m, n and K are shown in Fig. 4 and are very similar to those of Fig. 3a. The ratio between m and n is properly retrieved and converges towards the imposed value of 0.4. The other scatter plots indicate that the uplift rate is poorly constrained and that no clear relationship can be evidenced between the parameter U and the other parameters. This result clearly demonstrates that the ratio $\theta = m/n$ can indeed be constrained from a steady-state landscape but that neither K nor U nor independent values for m and n can be constrained because of the presence of a spurious solution to the problem corresponding to $m = n = 0$ and $K = U$. Moreover, if neither U nor K is constrained, their ratio is itself unconstrained, as is the steepness index, $k_s = (U/K)^{1/n}$.

We realize, however, that this result may depend on the way we have constructed the misfit function to compare the reference and predicted landscape. Other definitions of the misfit function could prove to be more constraining, especially concerning the steepness index. However, if we assume that the shape of the reference or observed landform is the only information we possess, the definition of the misfit function we have used here makes full use of the information content of the observables, and it is difficult to see how it could be improved.

5 Application to the Whataroa catchment

5.1 Tectonic, climatic and geomorphic context

The Southern Alps, New Zealand, are the surface expression of the ongoing oblique collision between the Australian and Pacific Plate (DeMets et al., 1990; Norris et al., 1990). This zone is characterized by very high uplift rates of up to $10\,\mathrm{km\,yr^{-1}}$ on the west side of the orogen (Wellman, 1979; Tippett and Kamp, 1993; Batt et al., 2000). This results in part from the high rate of convergence between the two plates (8–$12\,\mathrm{mm\,yr^{-1}}$) and in part from the strong orographic control on precipitations that results in precipitation rate of the order of 10–$13\,\mathrm{m\,yr^{-1}}$ on the west coast of the island, in comparison to the much drier climate of the east coast ($1\,\mathrm{m\,yr^{-1}}$) (Griffiths and McSaveney, 1983).

We focus our study on the Whataroa catchment (Fig. 5), which is located in the central Southern Alps along the West Coast and presents ones of the highest uplift rates in the region (≈ 6–$8\,\mathrm{mm\,yr^{-1}}$) (Tippett and Kamp, 1993; Herman et al., 2010). The profile of the Whataroa River can be divided into three distinct zones that correspond to different erosional and depositional environments or mechanisms (Herman and Braun, 2006). At high elevations, the dominant mechanism is glacial erosion. Between elevations of 200 and 1200 m, the valley cross section is markedly V-shaped, which indicates that the river is incising or in a "detachment-limited" state. Below 200 m elevation, the river dynamics change drastically with the formation of large meanders and

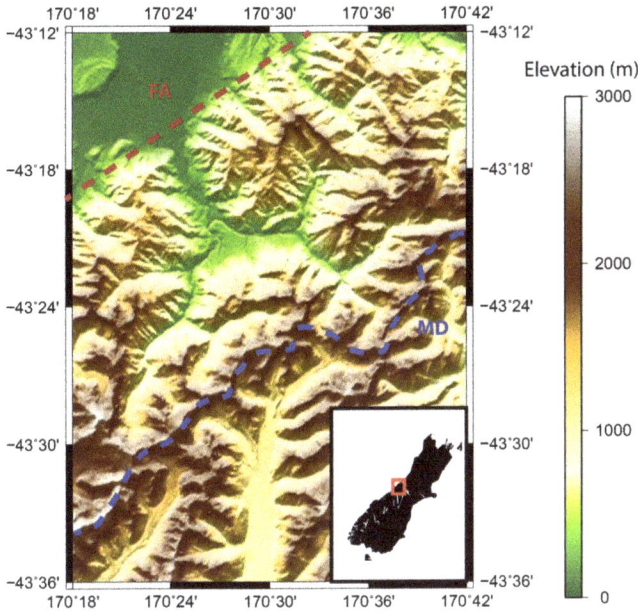

Figure 5. DEM of the Whataroa catchment in the central Southern Alps, New Zealand. AF: Alpine Fault; MD: Main Divide.

an array of braided channels during low flow periods, indicating that the river is in a "transport-limited" state, i.e., transporting sediments eroded in the upstream part of its catchment towards base level.

We have used the present-day topography of the Whataroa catchment area as initial and reference landscapes in our inversion scheme. For that, we used a DEM obtained from the SRTM3 mission (resolution of 3 arcsec). We corrected it for the presence of iso-elevation areas by using the geometry of the current drainage system. Using ESRI ArcGIS9.3, we modified the value of each of the pixels of the DEM by an infinitesimal amount that is inversely proportional to the discharge computed by using the "real" and thus known drainage geometry. In doing so, we ensure that the discharge computed by FastScape is done in accordance with the real drainage network. We only applied the inversion procedure to the pixels that have a computed drainage area superior to a critical area of $10 \, \text{km}^2$ (Supplement Fig. 5). In doing so, we impose that solely the elevation within the main river trunk and its main tributaries are used to compute the misfit function. This prevents the potential bias that might be introduced by the parts of the landscape that is still glaciated and/or controlled by hillslope processes only. For consistency, we did not include the parts of the landscape that are below 200 m in elevation where the evolution of the landscape is likely to be transport-limited and where the SPL is unlikely to apply.

5.2 Inversion results

5.2.1 Constraint of the SPL

We applied the inversion scheme to the Whataroa catchment, by letting three parameters be free – m, n and K – and imposing, for each of them, a range of values that is commonly admitted in the literature (see Table 1). The Whataroa catchment is potentially ideally suited for this exercise, as the lithology is relatively spatially invariant with surface rocks consisting mostly of the mildly metamorphosed Otago schists. As the catchment is relatively small (15 km in length), it is characterized by a relatively uniform and high precipitation rate. This implies that K should be spatially uniform ($\partial K / \partial x = 0$) if we neglect the effect of fracturing. Its value is, however, poorly constrained, and we will assume that it can potentially vary by up to 9 orders of magnitude. We will assume that the uplift rate is well constrained by a broad range of thermochronological data to a mean value of $6 \, \text{km} \, \text{Myr}^{-1}$ (Tippett and Kamp, 1993; Batt et al., 2000; Herman et al., 2010) and is spatially uniform. The misfit function is identical to what we used for the synthetic cases and given by Eq. (4). The values of the NA parameters N_{init}, N_{it} and N_{runs} are given in Table 1.

The results of the inversion (Fig. 6) show that the misfit function displays a minimum for a constant ratio between the parameters m and n in a very similar way to the results of the synthetic runs shown previously. The relationship between m and n that minimizes the misfit function is, however, not so well defined. Two peaks characterize the PDFs (Fig. 6b) for m and n, but these values must be considered with great care as the inversion is similarly attracted by the spurious solution corresponding to $m = n = 0$ as shown in the two panels of Fig. 6a that illustrate the behavior of the misfit function with K. It is interesting to note, however, that the optimum values for m and $n - 0.3$ and 0.5, respectively – are somewhat different from the extremum values allowed by the imposed ranges ($0.2 < m < 1.5$ and $0.4 < n < 2.5$), which could suggest that these values are potentially meaningful. Regardless of these considerations, it is the ratio of the two exponents, i.e., the concavity, that is best constrained at a value of 0.6. Similar to the synthetic cases, no constraint can be obtained for the value of K, as illustrated by the PDF of this parameter in the last panel of Fig. 6b.

As in the synthetic cases, the optimization method proves to be efficient in constraining the ratio m/n. For the Whataroa catchment and under the assumption that it is in geomorphic steady state, the optimum value is found to be 0.6, which sits within the range of acceptable values but towards their upper bound.

5.2.2 Constraint on the uplift geometry

The distribution of uplift (and thus exhumation, under the assumption of geomorphic steady state) in the vicinity of the Alpine Fault is a matter of debate (Braun et al., 2010).

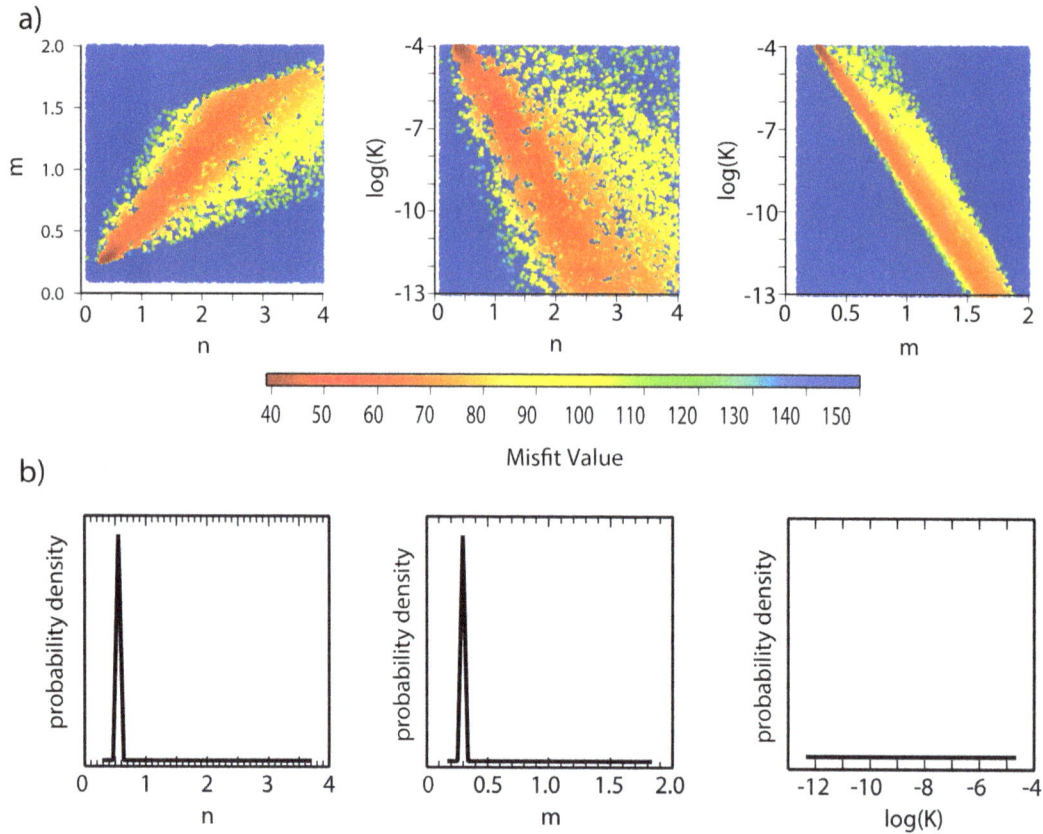

Figure 6. Results from inversion for the free parameters K, m and n. (**a**) Scatter plots showing the results from NA sampling stage. (**b**) PDFs of the three parameters resulting form the appraisal stage of NA.

Low-temperature thermochronology data from rocks exposed along the western side of the Southern Alps are commonly interpreted as evidence for an increase in exhumation rate towards the Alpine Fault (Tippett and Kamp, 1993). Higher temperature thermochronological data as well as first-order structural evidence that all of the structures east of and including the Alpine Fault are east-dipping reverse faults imply that the uplift and exhumation rates should be maximum near the present-day divide and thus decrease towards the Alpine Fault (Braun et al., 2010).

In order to test these two hypotheses, we allowed the uplift rate to vary linearly between the base of the Whataroa catchment near the Alpine Fault and the position of the main divide at the top of the catchment in such a way that either of the two scenarios can be reproduced by varying a single coefficient, α, introduced in the definition of the uplift rate function, $U(x)$:

$$U = 2(1 - \alpha)U_0(1 - \frac{x}{L}) + 2\alpha U_0 \frac{x}{L}, \qquad (7)$$

where U_0 the mean uplift rate in the Whataroa catchment ($6 \, \text{km} \, \text{Myr}^{-1}$), L the distance between the Alpine Fault and the Main Divide and x the position varying between 0 and L. In this expression, α varies between 0 and 1 corresponding to

a maximum uplift rate near the Alpine Fault or the Main Divide, respectively (Fig. 7); $\alpha = 0.5$ corresponds to a spatially uniform uplift rate.

In the following inversion, four parameters were left free: α, K, m and n. The results show that the values of the parameter α that best minimizes the misfit function and thus constrains the landscape to remain at steady state all lie above 0.5. The PDF of α (Fig. 8) demonstrates that the geometry of the river profile of the Whataroa is best explained with α values 0.59 ± 0.004. This implies that the uplift rate should be increasing away from the Alpine Fault in accordance with the suggestion made by Braun et al. (2010). This result must, however, be considered with much caution as it relies on our assumption that spatial variations in K can be neglected. Although the rock type and rainfall distribution are relatively uniform within the Whataroa catchment, the level of fracturing is highly variable and strongly increases towards the Alpine Fault as a result of the very large strain that has been accumulated at depth in the ductile regime and near the surface in the brittle regime by the Alpine Fault and the adjacent 1–5 km wide region. It is therefore not unlikely that the asymmetry in uplift rate that we evidence through our inversion of the topographic data is not real but an artifact of our assumption that K is uniform. We do not feel confident, how-

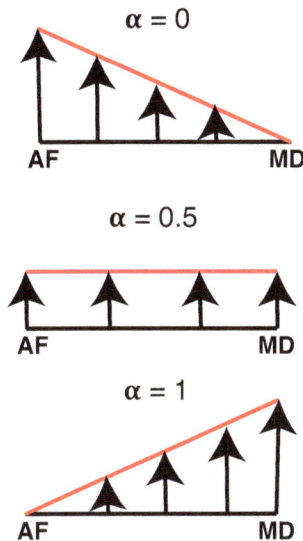

Figure 7. Three extreme cases controlled by the value of α. AF: Alpine Fault; MD: Main Divide.

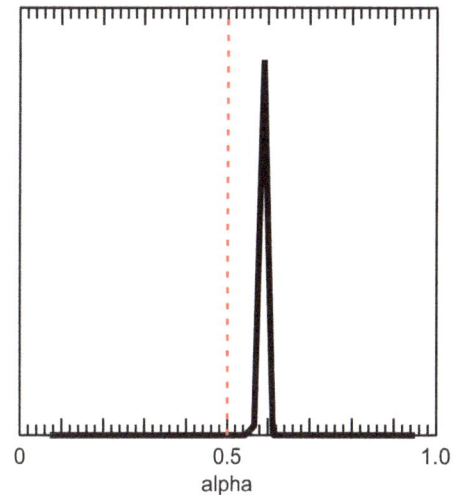

Figure 8. Results from inversion: PDF of the parameter α.

ever, about applying a similar treatment to K as we did for U, i.e., allowing it to vary linearly as a function to the Alpine Fault, as we have no hard constraint on its mean value, nor on how it may depend on fracturing. This implies that we would need to introduce two free parameters to represent a spatial variation in K, while we have shown that a single value cannot be constrained from topographic data only.

6 Discussion

6.1 Steady-state assumption

How to define whether a mountain range has reached steady state has been the object of numerous studies (Braun and Sambridge, 1997; Willett et al., 2001; Willett and Brandon, 2002), mostly based on the use of SPMs. True topographic steady state is reached when each point of the landscape remains at a constant elevation with respect to base level. Although such a situation might be achieved in a numerical model, whether it can be or has even been reached in a natural system remains difficult to imagine. One of the main reasons is that the horizontal advection of landforms prevents erosion and uplift rates from perfectly compensating each other on an orogenic scale (Willett et al., 2001). Better questions to ask might be on which spatial scale can we expect geomorphic steady state to develop and over which time frame.

The work we presented here is based on the assumption that geomorphic systems do reach steady state and that it has been reached in at least some parts of the Southern Alps in New Zealand. Under this assumption, the optimization scheme we have developed show that one can constrain the parameterization of channel incision from the geometry of the steady-state landscape. However, these constraints are

relatively weak and do not permit considering the SPL as a predictive tool, i.e., a law that could be transposed to a range of environments in a physically consistent manner.

To improve our method and extract more useful constraints on the SPL parameterization, we need to relax our hypothesis of steady state, which will imply longer, more computer-intensive simulations. We are in the process of doing so, but, concomitantly, we will need to find and utilize observational constraints on the time evolution of the system. Otherwise, simulating or predicting the rate of evolution of a landform from its shape only, and thus without a priori knowledge or independent constraints on uplift rate or the rate of landform evolution, is a futile exercise and therefore certain to fail.

6.2 Comparison to other studies

The use of optimization methods has been rather limited in geomorphology, mostly due to the long computational times required to run SPMs at the required resolution. Most previous attempts were limited to fitting 1-D longitudinal river profiles (van Der Beek and Bishop, 2003; Tomkin et al., 2003; Roberts and White, 2010) or to systematic search through low-dimension parameter search (van der Beek and Braun, 1998).

Many river incision models have been proposed over the past few decades. In their study, van Der Beek and Bishop (2003) used a unique data set on the evolution of river profiles in the Lachlan catchment of southeastern Australia to test which would better reproduce the natural behavior of the river. Their inversion procedure showed that the "detachment-limited" model presented here as well as the "undercapacity" model, which relates erosion rate to the river sediment carrying capacity (Beaumont et al., 1992) and takes into account channel width, best reproduced the data. Their approach did not lead to well-constrained values for the model parameters, which, they noted, are strongly controlled by lithology.

More recent studies have attempted to deduce uplift rate histories from longitudinal river profiles (Roberts and White, 2010; Roberts et al., 2012). In these studies, the SPL coefficients were derived from independent constraints (minimum residual misfit between theoretical observed and river profiles for m and n, and local incision estimate and known uplift histories for K) and used to both derive a simple relationship and reproduce knickpoint propagation. The two main locations where they have applied this method (Africa and the Colorado Plateau) have not been affected by recent tectonic events and their recent uplift history is assumed to be related to mantle processes (mantle plume impinging on the overlying lithosphere or dynamic topography caused by mantle circulation), implying that the inversion scheme must be performed over relatively long periods of time (i.e., 50 Myr). This approach neglects lithological control on knickpoint propagation and potential variations in climate (precipitation) and/or catchment geometry, which is difficult to justify over such long periods of time. Although Roberts and White (2010) and Roberts et al. (2012) do not explicitly aim at constraining the SPL coefficients, their studies demonstrate that present-day river profiles can be reproduced with great accuracy using arbitrarily defined model parameters and therefore do not contain all the information necessary to constrain the SPL parameterization.

The study we present here is the first that considers the problem in 2-D and lets the SPL coefficients vary over broad ranges in order to determine their best values in a quantitative manner. We also show that the method is able to retrieve information about the distribution of present-day uplift, even under the assumption of geomorphic steady state.

6.3 The stream power law

Although the SPL used in this study is able to reproduce many observations and natural processes (i.e., the concavity of river profiles to the migration of knickpoints), it must be regarded as a first-order parameterization of the integrated effects of river incision and, as such, cannot be expected to adequately represent the many and varied physical processes at play. For example, the SPL cannot include the effect of sediment being delivered to the river channel by hill-slope processes. In most of the Whataroa catchment, which is subjected to rapid tectonic uplift, valley sides are at the critical angle of repose, as indicated by their steep, V-shaped morphology (Herman and Braun, 2006) and supply vast quantity of sediments to the streams. Several formulations have been proposed to include the effect of sediment (both bed load and suspended load) on the erosional power of a stream, its transport capacity and the way sediment protects the bed from erosion (Beaumont et al., 1992; Sklar and Dietrich, 2001). Furthermore, tectonic uplift is accompanied by deformation and fracturing, which is clearly evidenced by the intense hy-

drothermal activity in the vicinity of the Alpine Fault (Allis and Shi, 1995). The resistance of rocks to erosion is very likely to be strongly influenced by fracturing (Molnar et al., 2007), but it cannot be included in the SPL in a physically meaningful manner; at best the coefficient K can be arbitrarily adjusted to represent the zeroth-order effect of fracturing on erodibility.

In most of our inversions, the minimum misfit value remains quite high, which could lead to one of two conclusions: that (a) the classical formulation of the SPL used here is not sufficiently complex to reproduce a steady-state landscape or (b) that geomorphic steady state does not exist or does not apply to the Whataroa catchment. Further investigation is required to determine whether introducing a better parameterization of channel width, the effect of sediment load on the incision power of a stream in the SPL or spatial variations in K related to lithology or precipitation patterns would lead to a substantial reduction in misfit.

Progressing in the testing and improving of stream incision laws also requires that we go beyond fitting topographic data and introduce in our inversions observational evidence on the temporal evolution of a stream profile. This includes a broad range of thermochronometric tools as well as exposure dating techniques. Sediment provenance data are another important tool that should provide independent information to constrain the SPL (or other potential parameterizations).

7 Conclusions and perspectives

We have presented here a novel approach to constrain the coefficients of the SPL parameters that combines a very efficient surface process model (FastScape) and an inversion method (the neighborhood algorithm). The inversion is constrained by a misfit function that compares a reference or observed topography with that predicted by the SPM under the assumption of geomorphic steady state. Using the method on synthetic landscapes, we have demonstrated its potential by an in-depth analysis of the resulting misfit function that is dependent on the number of degrees of freedom (or model parameters) in the inversion procedure. We proved that the method is accurate and efficient to retrieve the ratio between m and n, the two exponents in the SPL formulation, that strongly controls the concavity of river profiles.

We applied the method to a natural landscape from the South Island of New Zealand where geomorphic steady state is likely to be achieved due to the present-day, very high tectonically driven uplift rates. In this case, we show that the ratio m/n can be constrained but that the estimate we obtain is close to the upper bound of commonly "accepted" values, suggesting that the region may be subject to substantial spatial variations in uplift rate. We also show that the value of none of the SPL coefficients can be retrieved with confidence due to the presence of a spurious solution corresponding to $m = n = 0$ and $K = U$.

We also performed an inversion in which uplift rate was allowed to vary spatially around a fixed mean value that is relatively well constrained in New Zealand through the interpretation of many thermochronological data sets. The results show that the preferred solution (i.e., the one that minimizes the misfit between observed and predicted elevation) implies a decrease in uplift rate towards the Alpine Fault, which is consistent with a recent study demonstrating that the strong gradient in deformation east of the Alpine Fault indicates that uplift rate must increase with distance from the Alpine Fault (Braun et al., 2010).

We also conclude that although promising, the method we have developed needs to be improved to include transient effects, other observational constraints on the temporal evolution of landscapes, and the spatial distribution of erosion rate in and out of the channels, which will also require modification of the misfit function.

Acknowledgements. The authors wish to thank the Institut Universitaire de France for funding T. C. MSc stipend and Dimitri Lague for constructive comments on an earlier version of this manuscript. We are grateful to G. R. Roberts and A. Howard, whose comments helped to improve the manuscript. M. Sambridge is thanked for making the neighborhood algorithm code available.

Edited by: S. Castelltort

References

Adams, J.: Contemporary uplift and erosion of the Southern Alps, New Zealand, Geol. Soc. Am. Bull., 91, 1–114, 1980.

Allis, R. and Shi, Y.: New insights into temperature and pressure beneath the central Southern Alps, New Zealand, New Zealand, J. Geol. Geophys., 38, 585–592, 1995.

Batt, G. E., Braun, J., Kohn, B. P., and McDougall, I.: Thermochronological analysis of the dynamics of the Southern Alps, New Zealand, Geol. Soc. Am. Bull., 112, 250–266, 2000.

Beaumont, C., Fullsack, P., and Hamilton, J.: Erosional control of active compressional orogens, in: Thrust Tectonics, edited by McClay, K. R., pp. 1–18, Chapman and Hall, New York, 1992.

Braun, J. and Sambridge, M.: Modelling landscape evolution on geological time scales: a new method based on irregular spatial discretization, Basin Research, 9, 27–52, 1997.

Braun, J. and Willett, S. D.: A very efficient O (n), implicit and parallel method to solve the stream power equation governing fluvial incision and landscape evolution, Geomorphology, 180–181, 170–179, 2013.

Braun, J., Herman, F., and Batt, G.: Kinematic strain localization, Earth Planet. Sci. Lett., 300, 197–204, 2010.

Crave, A. and Davy, P.: A stochastic precipiton model for simulating erosion/sedimentation dynamics, Computers Geosci., 27, 815–827, 2001.

DeMets, C., Gordon, R. G., Argus, D. F., and Stein, S.: Current plate motions, Geophys. J. Internat., 101, 425–478, 1990.

Duvall, A., Kirby, E., and Burbank, D.: Tectonic and lithologic controls on bedrock channel profiles and processes in coastal California, J. Geophys. Res. (2003–2012), 109, doi: 10.1029/2003JF000086, 2004.

Griffiths, G. A. and McSaveney, M. J.: Distribution of mean annual precipitation across some steepland regions of New Zealand., N. Z. J. SCI., 26, 197–209, 1983.

Herman, F. and Braun, J.: Fluvial response to horizontal shortening and glaciations: A study in the Southern Alps of New Zealand, J. Geophys. Res. (2003–2012), 111, doi: 10.1029/2004JF000248, 2006.

Herman, F., Rhodes, E. J., Braun, J., and Heiniger, L.: Uniform erosion rates and relief amplitude during glacial cycles in the Southern Alps of New Zealand, as revealed from OSL-thermochronology, Earth Planet. Sci. Lett., 297, 183–189, 2010.

Howard, A. D.: A detachment-limited model of drainage basin evolution, Water resources research, 30, 2261–2285, 1994.

Kirby, E. and Whipple, K. X.: Quantifying differential rock-uplift rates via stream profile analysis, Geology, 29, 415–418, 2001.

Kirby, E. and Whipple, K. X.: Expression of active tectonics in erosional landscapes, J. Struct. Geol., 44, 54–75, 2012.

Kooi, H. and Beaumont, C.: Escarpment evolution on high-elevation rifted margins: Insights derived from a surface processes model that combines diffusion, advection, and reaction, J. Geophys. Res. (1978–2012), 99, 12191–12209, 1994.

Lague, D., Davy, P., and Crave, A.: Estimating uplift rate and erodibility from the area-slope relationship: Examples from Brittany (France) and numerical modelling, Phys. Chem. Earth, Part A, 25, 543–548, 2000.

Molnar, P., Anderson, R., and Anderson, S.: Tectonics, fracturing of rock, and erosion, J. Geophys. Res., 112, F03014, doi: 10.1029/2005JF000433, 2007.

Montgomery, D. R. and Brandon, M. T.: Topographic controls on erosion rates in tectonically active mountain ranges, Earth Planet. Sci. Lett., 201, 481–489, 2002.

Norris, R. J., Koons, P. O., and Cooper, A. F.: The obliquely-convergent plate boundary in the South Island of New Zealand: implications for ancient collision zones, J. Struct. Geol., 12, 715–725, 1990.

Roberts, G. G. and White, N.: Estimating uplift rate histories from river profiles using African examples, J. Geophys. Res., 115, B02406, doi: 10.1029/2009JB006692, 2010.

Roberts, G. G., White, N. J., Martin-Brandis, G. L., and Crosby, A. G.: An uplift history of the Colorado Plateau and its surroundings from inverse modeling of longitudinal river profiles, Tectonics, 31, TC4022, doi: 10.1029/2012TC003107, 2012.

Sambridge, M.: Geophysical Inversion with a neibourhood algorithm – I. Searching a parameter space, Geophys. J. Int., 138, 479–494, 1999a.

Sambridge, M.: Geophysical Inversion with a neibourhood algorithm – II. Appraising the ensemble, Geophys. J. Int., 138, 727–746, 1999b.

Sklar, L. and Dietrich, W.: Sediment and rock strength controls on river incision into bedrock, Geology, 29, 1087–1090, 2001.

Sklar, L. and Dietrich, W. E.: River longitudinal profiles and bedrock incision models: Stream power and the influence of sediment supply, GEOPHYSICAL MONOGRAPH-AMERICAN GEOPHYSICAL UNION, 107, 237–260, 1998.

Snyder, N. P., Whipple, K. X., Tucker, G. E., and Merritts, D. J.: Landscape response to tectonic forcing: Digital elevation model analysis of stream profiles in the Mendocino triple junction region, northern California, Geological Society of America Bulletin, 112, 1250–1263, 2000.

Stock, J. D. and Montgomery, D. R.: Geologic constraints on bedrock river incision using the stream power law, J. Geophys. Res. (1978–2012), 104, 4983–4993, 1999.

Tippett, J. M. and Kamp, P. J. J.: Fission track analysis of the late Cenozoic vertical kinematics of continental Pacific crust, South Island, New Zealand, J. Geophys. Res., 98, 16119–16148, 1993.

Tomkin, J. H., Brandon, M. T., Pazzaglia, F. J., Barbour, J. R., and Willett, S. D.: Quantitative testing of bedrock incision models for the Clearwater River, NW Washington State, J. Geophys. Res., 108, 2308, doi: 10.1029/2001JB000862, 2003.

Tucker, G., Lancaster, S., Gasparini, N., and Bras, R.: The channel-hillslope integrated landscape development model (CHILD), in: Landscape erosion and evolution modeling, Springer, 349–388, 2001.

Tucker, G. E. and Hancock, G. R.: Modelling landscape evolution, Earth Surf. Proc. Land., 35, 28–50, 2010.

Van Der Beek, P. and Bishop, P.: Cenozoic river profile development in the Upper Lachlan catchment (SE Australia) as a test of quantitative fluvial incision models, J. Geophys. Res. (1978–2012), 108, doi:10.1029/2002JB002125, 2003.

van der Beek, P. and Braun, J.: Numerical modelling of landscape evolution on geological time-scales: a parameter analysis and comparison with the south-eastern highlands of Australia, Basin Research, 10, 49–68, 1998.

Wellman, H. W.: An uplift map for the South Island of New Zealand, and a model for uplift of the Southern Alps, Bull. R. Soc. NZ, 18, 13–20, 1979.

Whipple, K. X.: Bedrock rivers and the geomorphology of active orogens, Annu. Rev. Earth Planet. Sci., 32, 151–185, 2004.

Whipple, K. X. and Tucker, G. E.: Dynamics of the stream-power river incision model: Implications for height limits of mountain ranges, landscape response timescales, and research needs, J. Geophys. Res., 104, 17661–17674, 1999.

Whittaker, A. C., Cowie, P. A., Attal, M., Tucker, G. E., and Roberts, G. P.: Bedrock channel adjustment to tectonic forcing: Implications for predicting river incision rates, Geology, 35, 103–106, 2007.

Willett, S. D. and Brandon, M. T.: On steady states in mountain belts, Geology, 30, 175–178, 2002.

Willett, S. D., Slingerland, R., and Hovius, N.: Uplift, shortening, and steady state topography in active mountain belts, Am. J. Sci., 301, 455–485, 2001.

13

Image analysis for measuring the size stratification in sand–gravel laboratory experiments

C. Orrú[1], **V. Chavarrías**[1,2], **W. S. J. Uijttewaal**[1], **and A. Blom**[1]

[1]Environmental Fluid Mechanics Section, Civil Engineering and Geosciences, Delft University of Technology, P.O. Box 5048, 2600 GA, Delft, the Netherlands
[2]Escola Tècnica Superior d'Enginyers de Camins, Canals i Ports de Barcelona, Universitat Politècnica de Catalunya, C/Jordi Girona, 31, 08034 Barcelona, Spain

Correspondence to: C. Orrú (c.orru@tudelft.nl)

Abstract. Measurements of spatial and temporal changes in the grain-size distribution of the bed surface and substrate are crucial to improving the modelling of sediment transport and associated grain-size selective processes. We present three complementary techniques to determine such variations in the grain-size distribution of the bed surface in sand–gravel laboratory experiments, as well as the resulting size stratification: (1) particle colouring, (2) removal of sediment layers, and (3) image analysis. The resulting stratification measurement method was evaluated in two sets of experiments. In both sets three grain-size fractions within the range of coarse sand to fine gravel were painted in different colours. Sediment layers are removed using a wet vacuum cleaner. Subsequently areal images are taken of the surface of each layer. The areal fraction content, that is, the relative presence of each size fraction over the bed surface, is determined using a colour segmentation algorithm which provides the areal fraction content of a specific colour (i.e. grain size) covering the bed surface. Particle colouring is not only beneficial to this type of image analysis but also to the observation and understanding of grain-size selective processes. The size stratification based on areal fractions is measured with sufficient accuracy. Other advantages of the proposed size stratification measurement method are (a) rapid collection and processing of a large amount of data, (b) a very high spatial density of information on the grain-size distribution, (c) the lack of disturbances to the bed surface, (d) only minor disturbances to the substrate due to the removal of sediment layers, and (e) the possibility to return a sediment layer to its original elevation and continue the flume experiment. The areal fractions are converted into volumetric fractions using an existing conversion model.

1 Introduction

Spatial and temporal variations in the grain-size distribution of the river bed surface are the outcome of grain-size selective processes and/or abrasion (e.g. Parker, 1991a, b). Having insight in such variations is crucial to understanding a river system and predicting the autonomous morphodynamic behaviour of a river system, as well as the morphodynamic effects of (future) human interventions. Spatial variations in the grain-size distribution are usually indicated with the term sorting. An example of streamwise sorting is downstream fining, which is characterized by a decrease in mean grain size in streamwise direction. Downstream fining is usually attributed to abrasion and grain-size selective transport (e.g. Parker, 1991a,b). An example of sorting in the vertical direction is armouring. Two types of armouring can be distinguished: static and mobile armour layers (e.g. Parker and Klingeman, 1982; Jain, 1990). A static armour forms at low shear stresses when the finer grains are mobile while the coarser ones are not, which under limited sediment supply conditions results in a coarsening of the bed surface. This coarsening prevents the underlying finer sediment from being entrained by the flow. Instead, a mobile armour develops at higher shear stresses at which coarser grains are also supplied from upstream. A coarse surface layer is then needed to increase the transport capacity of the coarse sediment which enables the sediment supplied from upstream to be transported downstream. Dune sorting is another example of vertical sorting, and originates through (a) avalanching of grains down the bedform lee face which results in an upward fining pattern (e.g. Bagnold, 1941; Allen, 1965) and (b) selective or partial transport, which can result in the formation of a coarse layer underneath the migrating dunes consisting of the less mobile sediment (e.g. Zanke, 1976; Blom et al., 2003). Sorting processes at local scale are also seen in bars, where coarser sediment tends to be found on bar heads and fines on bar tails (e.g. Leopold and Wolman, 1957; Ashworth and Ferguson, 1986; Lunt and Bridge, 2004). Lateral sorting can be observed in meander bends where the shallower inner bends are finer than the deeper outer bends (e.g. Parker and Andrews, 1985; Julien and Anthony, 2002). Monitoring techniques that allow for rapid and non-destructive measurement of sorting and the resulting size stratification are required to better understand and predict grain-size selective processes and the river bed evolution. In this study we focus on the development of such a method to measure the size stratification in sand–gravel laboratory experiments.

Before moving to techniques measuring the size stratification in the bed, we first discuss measuring the grain-size distribution of the bed surface. An overview of earlier methods to measure the grain-size distribution of the bed surface such as sieving and grid sampling were presented by Kellerhals and Bray (1971) and wax sampling by Diplas and Sutherland (1988). These procedures are commonly used to collect samples from a gravel bed surface. Wax sampling generally is limited to coarser mixtures because of the limited penetration of liquid wax in the pores of finer mixtures. In the case of grid sampling the grains lying below the grid points constitute the sample. Therefore application of the method is limited to collecting mostly coarse sediment. The major drawback of these techniques is their time-consuming procedure for determining the grain-size distribution due to the time required for taking the samples as well as for sieve analyses.

It is worth noticing that particularly in the case of determining the grain-size distribution of the bed surface the results of the areal sampling techniques (e.g. wax or clay sampling, image analysis) are not equivalent to the outcomes of sieving. These areal sampling techniques provides areal grain-size distributions that are not equivalent to the volumetric grain-size distributions originating from sieve analysis (Kellerhals and Bray, 1971). A conversion model (e.g. Kellerhals and Bray, 1971; Diplas and Sutherland, 1988; Fraccarollo and Marion, 1995) is then needed to convert areal fraction contents into volume fraction contents.

In the past years image analysis techniques have emerged as a reliable, practical and non-destructive procedure (McEwan et al., 2000; Sime and Ferguson, 2003; Graham et al., 2005a, b, 2010). The first image processing techniques still

required laborious work through manual measuring on the images, overlaying transparent grids on the image to count the grains (Adam, 1979) or manually digitizing large prints to delineate the particle edges (Ibbeken and Scheyler, 1986). The development of fully automated methods was obviously connected to the advancements in computer technology. The common automated procedures are based on detecting the edges of individual grains and determining the particle dimensions (McEwan et al., 2000; Sime and Ferguson, 2003; Graham et al., 2005a, b, 2010). Application of these techniques to a variety of grain shapes, lithotypes, texture, and sizes still needs to be assessed (Graham et al., 2005a). The limiting accuracy for the finer particles is controlled by the resolution of the image. Graham et al. (2010) remark that even when accounting for the improvement in the resolution of digital cameras, there will be constraints in the accurate identification of the edges of the smaller grains. A compromise between the minimum grain size that can be detected and the surface area covered by an image is required.

The restriction to the minimum grain size that can be detected within a certain surface area covered by the image was avoided by combining the image analysis with particle colouring (Wilcock and McArdell, 1993, 1997; Wu and Yang, 2004). In their laboratory experiments, the detection of the edges of the particles is replaced by the detection of colour. This has enabled application of image analysis to smaller grain sizes: sands and fine gravels (Wilcock and McArdell, 1993, 1997; Wu and Yang, 2004). Wilcock and McArdell (1993, 1997) introduced the so-called bed of many colours (BOMC) to evaluate the grain-size distribution of the bed surface. They placed a grid over the images and the colours of the grains positioned on the grid intersections were registered. Wu and Yang (2004) modified the procedure of Wilcock and McArdell (1993, 1997) and counted the number of grains present inside the grid cells in images taken at different times to also allow for the detection of grain mobility. An improvement in image analysis combined with particle colouring was suggested by Heays et al. (2010a, b). Their approach consists of recognizing sediment colours by setting RGB colour thresholds in order to detect a specific colour and the isolation of grain clusters based on colour. Detecting the fraction contents of the grain-size fractions as grain clusters rather than the size of individual grains allows to automatically and rapidly determine the areal fraction contents covered by each colour. Yet, the measurement of other parameters, such as particle velocity (Wu and Yang, 2004; Heays et al., 2010a, b) still requires the detection of individual or small groups of particles.

More recently, an innovative approach based on the characteristics of image pixels was developed for photographic techniques used in field campaigns. In ground-based photography (Rubin, 2004; Warrick et al., 2009) and airborne photography (Carbonneau et al., 2004, 2005; Dugdale et al., 2010), relationships were found between grain-size properties and the pixel properties. These properties, such as pixel brightness, are measured using approaches based on spatial statistics. Rubin (2004) presented a spatial autocorrelation method to measure the correlation between the pixel intensity of two regions of the image. The method was applied to measure subaqueously the grain-size distribution of the bed surface and vertical profiles. Images with an unknown grain-size distribution are analyzed comparing them with autocorrelation curves that are representative of a certain sediment grain size. This method used as well by Warrick et al. (2009) can be applied to a large range of grain sizes. The second approach based on spatial statistics consists in measuring the local semi-variance, which is a function of the difference in brightness and the distance between these pixels (Carbonneau et al., 2004). This second method was applied in airborne photography for the grain-size mapping of river catchment areas (Carbonneau et al., 2004, 2005; Dugdale et al., 2010). An application of the Carbonneau et al. (2005) method was used to study the evolution of braided river deposit in a flume experiment (Gardner and Ashmore, 2011).

A large variety of samplers and box corers is available to determine the size stratigraphy of a deposit formed out of cohesive sediment, whereas measuring the stratigraphy of a non-cohesive deposit has always been a challenging task as the sediment sample drops from the box corer unless the box is closed or the sample is fixed with some substance. The combination of using a box corer with sieve analysis is a common method to obtain information about the vertical variation of the grain-size distribution. Jonasson and Olausson (1966) proposed two devices with different closing mechanisms for sampling the uppermost layers of marine sediment. The method does not allow for undisturbed samples and the operation procedure together with the devices' large dimensions limit their use to field surveys. Blom et al. (2003) developed a box coring method to analyse the vertical sorting within a sand–gravel bed under laboratory conditions. The sampler is shaped as a rectangular box which is closed using a rectangular plate that is pushed obliquely into the sample. After it is closed, the box is removed from the bed. A drawback is the fact that some grains are pushed down when pushing the box into the bed and when pushing down the closing device.

In order to take samples out of non-cohesive sediment in a less disturbed way, many methods are based on pouring a liquid substance inside the pores and letting the sample solidify. In this manner the particles are attached to each other and to the sampling tool (i.e. box corer or probe pipe) creating a sample that can be lifted from the bed in one piece. Various fixing methods are used: wax coring (Sibanda et al., 2000); freezing samples by liquid nitrogen (Gomez, 1983; Pallara et al., 2006) or liquid carbon dioxide (Walkotten, 1976; Carling and Reader, 1981), and epoxy resin (Alexander et al., 2001). The resulting sample is then analysed melting the fixing substance (Walkotten, 1976; Gomez, 1983; Sibanda et al., 2000) or first cutting sections of the sample at various elevations (Carling and Reader, 1981) and sieving the sedi-

ment. The sampling method presented by Marion and Fraccarollo (1997) is one of the least invasive freezing methods since the sample containers are introduced inside the flume before starting the experiment. The samples are removed and later frozen. These frozen blocks are broken in four pieces and the inner undisturbed surfaces are photographed in order to measure the vertical variation of the grain-size distribution through image analysis. The parts are later recomposed and placed back into the flume and, after thawing, the experiment is restarted. Repeating this procedure several times allowed Marion and Fraccarollo (1997) to study the evolution of the size stratification of the bed.

A final option to define the size stratification in laboratory experiments consists of removing separate layers from the deposit and subsequently analyse the grain-size distribution through sieve analysis. This was realized by means of siphoning of sand layers by Ribberink (1987) or, for a coarser sand–gravel mixture, removing the sediment layers using a spatula by Kleinhans (2005).

The objective of this paper is to present a method for measuring the size stratification of a deposit in laboratory sand–gravel experiments. The method combines three of the above techniques: (1) particle colouring, (2) removal of thin sediment layers, and (3) image analysis. The image analysis is based on the approach by Heays et al. (2010a, b). The difference lies in the fact that (a) we extend the image analysis to analysing the grain-size distribution of the substrate rather than only the bed surface, (b) we use another algorithm for colour detection, and (c) we work with the wider subset of colours of the Lab colour space. In Heays et al. (2010a, b) colour detection is based on the isolation of a limited number of colours (equal to the number of grain sizes), with a consequent loss or non-detection of pixels (e.g. darker grains due to shadows). Instead our approach is based on colour segmentation and therefore assigns a colour to each pixel in the image. The areal fraction content of a certain grain-size fraction is measured rather than measuring the individual grain sizes. This is because the stratification measurement method was developed to compare future data from flume experiments with the results of models such as the surface-based transport model for mixed sediment (e.g. Wilcock and Crowe, 2003), and a sediment continuity model for mixed sediment (e.g. Hirano, 1971; Ribberink, 1987; Blom, 2008), which require information on the fraction content for a range of grain sizes at the bed surface and substrate.

2   The size stratification measurement method

In this section we present the three steps in the new method to measure the size stratification of a coloured sand–gravel deposit: (1) particle colouring, (2) removal of thin sediment layers, and (3) image analysis. The deposit used to assess the applicability of the size stratification measurement method

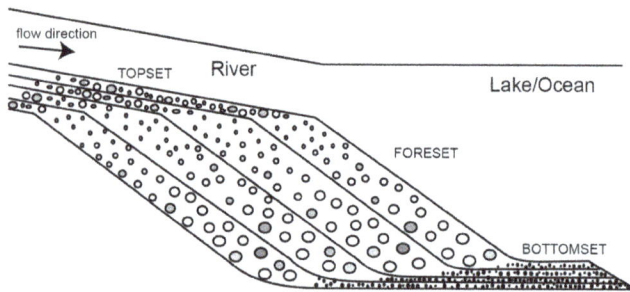

Figure 1. Schematic of a Gilbert-type delta (after Gilbert, 1890).

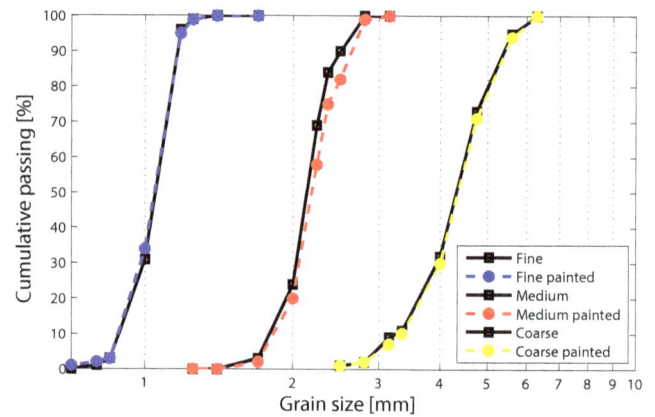

Figure 2. Grain-size distribution of the original sediment (black lines) and the coloured sediment (blue, red and yellow lines).

is a Gilbert or foreset-dominated delta formed in laboratory experiments (Chavarrías et al., 2013, 2014).

The stratigraphy of foreset-dominated deltas differs from the one of the well-known topset-dominated deltas (Edmonds et al., 2011). Foreset-dominated deltas are deltas that are governed by depositional processes over the delta front. The foreset deposit, comprises the larger part of the deposit. In topset-dominated deltas sediment deposition mainly takes place on the delta top or fluvial reach. This fluvial reach consists of distributary channels and levees, river mouth bars and inter-distributary bays (Edmonds et al., 2011).

A Gilbert delta (Fig. 1) is formed where (a) a river flows into a deep basin such as a lake or an ocean, and (b) sediment deposition is dominated by avalanches over its steep slip face. Its stratigraphy consists of three parts (Gilbert, 1885): (a) a topset that forms as a result of deposition over the fluvial reach, (b) a foreset that forms as a result of avalanches of grains (i.e. grain flows) over a steep lee face, and (c) a bottomset that forms out of fine sediment that deposits downstream from the foreset as a result of grain fall. The avalanches down the steep slope of the delta front typically result in an upward fining profile. Sediment transport over the fluvial reach results in the delivery of sediment to the delta brinkpoint. Due to the sudden expansion of the flow the sediment deposits over the brinkpoint forming a wedge. When the angle of repose is exceeded the sediment avalanches down the delta front. These avalanches makes the delta prograde. The delta progradation leads to the lengthening of the fluvial reach and the decrease of the river bed slope. This leads to aggradation over the fluvial reach and a gradual decrease in the delta progradation rate. The rate at which the delta prograndes is affected by sediment supply or sea level changes (Postma, 1995).

In a drained flume areal images of the Gilbert delta were taken at different elevations of its deposit to determine its size stratification. These elevations correspond to the top of layers of 1 cm thickness that were removed one by one using a wet vacuum cleaner. The grain-size distribution of the surface of each sediment layer is analysed through image analysis. We compare the results of the image analysis to the results of sieve analysis to assess the image analysis method. The required steps are explained in the next few sections.

2.1 Particle colouring

The sand–gravel deposit was composed of three well-sorted grain-size fractions within the range of coarse sand to fine gravel. We considered the fractions to have negligible overlap since sieve analysis indicated only a slight overlap of 1–2 % between medium fraction 2 and coarse 3 (Fig. 2). The grain-size fractions were non-overlapping to facilitate the sieve analysis, which is required to evaluate the image analysis technique. The original sediment mixture is composed of a fine size fraction ($D_{50,1} = 1.05$ mm), a medium size fraction ($D_{50,2} = 2.13$ mm) and a coarse size fraction ($D_{50,3} = 4.32$ mm).

The grains were coloured with a common wall paint using a concrete mixer. Each size fraction was coloured twice to obtain a resistant and bright colour and to fully coat the grains with paint. Sieve analysis was performed to determine the grain-size distribution of the painted sediment. Only little difference between the grain-size distribution of the original and the painted grain-size fractions is found (Fig. 2). The median grain sizes of the painted fractions are $D_{50,1P} = 1.04$ mm (fine fraction, blue), $D_{50,2P} = 2.19$ mm (medium fraction, red) and $D_{50,3P} = 4.36$ mm (coarse fraction, yellow).

The densities of the original and the painted grain-size fractions were measured using a hydrostatic balance. We noticed a slight decrease in mass density for the three grain-size fractions due to the painting, which was considered acceptable. The densities of the original sediment are 2614 kg m^{-3} (fine), 2614 kg m^{-3} (medium), 2600 kg m^{-3} (coarse). The densities of the painted fractions are 2590 kg m^{-3} (fine), 2564 kg m^{-3} (medium), 2549 kg m^{-3} (coarse).

2.2 Removal of sediment layers

The water was drained from the flume once the experiment was finished. The sampling procedure was started when the bed was dry. Two metal sheets with a thickness of 2 mm, a

Figure 3. Experimental set-up. Side view of the delta progradation experiment during the removal of sediment layers and sampling for sieving purposes.

Figure 4. Set-up for image analysis and sampling: top view of the sampling area with the sampling grid placed on the bed surface and the metal sheets inserted into the bed. The Plexiglas® grid and sampling per grid cell is only needed for the assessment phase of the image analysis method. Once the conversion method is clear, the sampling grid is no longer required and an entire thin sediment layer can be removed at once without using the grid.

width of 0.39 m and a height of 0.30 m were driven along the edges of the sampling area into the delta deposit to support the sediment faces while thin sediment layers in between the metal sheets were removed (Fig. 3).

A rigid Plexiglas® sampling grid was placed on the bed surface and was used to divide the reach into patches (Fig. 4). The reach was subdivided to enable a comparison of the local grain-size distribution determined through image analysis with the one determined through sieving. The Plexiglas® grid and the sieve analysis are only required when assessing the image analysis procedure. Please note that once the method is validated, the sampling grid is no longer required and an entire sediment layer can be removed at once without using the grid. A grid with a vertical grid size of 1 cm was indicated on the glass side walls of the flume as an aid to removing the 1 cm sediment layers. We took areal images of the deposit within each grid cell. After this a 1 cm sediment layer was removed from each grid cell and later sieved. Subsequently the camera position was lowered by 1 cm and the full procedure was repeated until reaching the parent material of the initial bed. We excluded the grid cells bordering the metal sheets from the analysis as the sediment in that area may have been disturbed by driving the metal sheets into the bed.

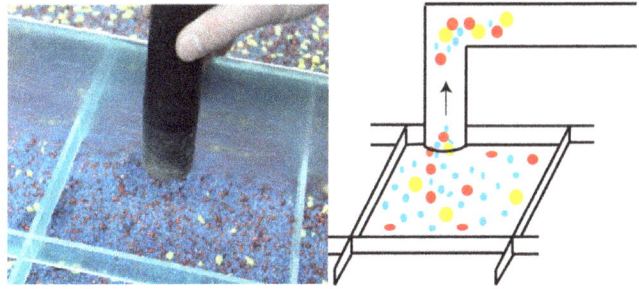

Figure 5. Nozzle of vacuum cleaner close to the bed surface removing the 1 cm sediment layers.

The 1 cm sediment layers were removed using a wet vacuum cleaner (Fig. 5). The vacuum nozzle was placed at a certain distance from the surface of the deposit to extract the sediment. The sediment extracted from each grid cell was collected from the vacuum cleaner cylinder, and stored for sieve analysis. The cylinder was created with a filter net having a mesh smaller than the finest fraction. This was done to reduce the time required to collect the sediment inside the vacuum cleaner. The sampling with the vacuum cleaner was not size-selective and did not show a preference for the fine fractions. The fact that the sediment was painted helped us to verify the removal of all size fractions. For instance, the coarser yellow particles were clearly distinguishable and it was visible when they were removed by the vacuum cleaner together with the finer fractions. The vacuum cleaner was used passing over the same area several times until the 1 cm thick layer was removed. This helped us to collect all fractions from that layer.

In case continuation of an experiment is required, the removed sediment patches can be placed back separately at the original elevations of the deposit after sampling. It is clear that this option is feasible only when sampling restricted areas of a deposit.

2.3 Image analysis

The areal images taken during the experiments are processed with an image analysis algorithm that is written using Matlab. The code is based on colour segmentation and provides the areal fraction content of a surface area covered by a certain colour, i.e. a certain grain size (Fig. 6). Colour segmentation indicates the division of all the pixels present in an image into a limited number of imposed colour groups (i.e. clusters). The camera lens used, the Canon EF-S 18–55 mm IS f/3.5–5.6 wide-angle to mild telephoto, has very small geometric distortions. The barrel distortion is 3.35 % at a focal length of 18 mm, which means that at the applied focal length of 55 mm the barrel distortion is lower than 3.35 %. The pincushion distortion at a focal length of 55 mm is 0.389 %. For these reasons we decided not to correct the images for geometric distortions.

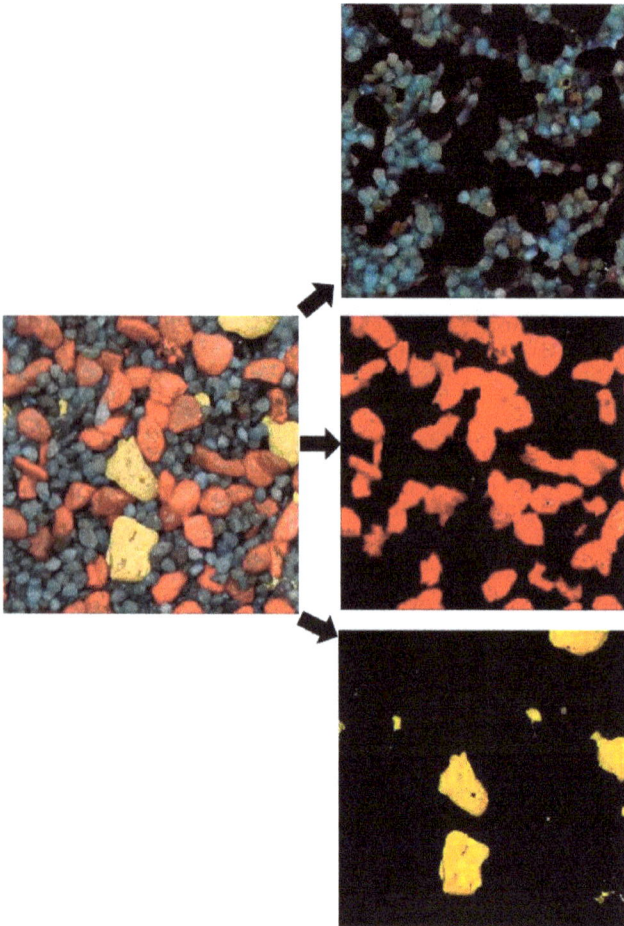

Figure 6. Colour segmentation for the three grain-size fractions.

average value for each colour (i.e. cluster centre). The colour segmentation is carried out in the Lab colour space. The Lab colour space is a colour model as the RGB, and is composed of three dimensions. In the case of RGB the dimensions are given by red, green and blue values. A colour results from the combination of the percentages of red, green, and blue values. The Lab colour space combines the luminosity dimension L and the colour (so chromatic) dimensions a and b. This means that the colour information is in the a and b dimensions. The parameter a indicates how the colour plots along the red–green axis, and b indicates how it plots along the blue–yellow axis.

The procedure to determine appropriate cluster centres based on a selection of images is summarized as follows:

1. image reading;

2. colour saturation (Zaman, 2011);

3. image conversion from RGB colour space to Lab colour space;

4. colour classification by k-means clustering to obtain the cluster centres;

5. taking the average of the cluster centres.

Below these steps are explained in more detail:

1. The RGB images chosen to determine appropriate cluster centres are loaded.

2. Colour saturation enhances the image colours and gives more uniformity to the different tones of the grains.

3. The RGB values of each pixel are converted into Lab colour space values. In the Lab colour space each pixel includes a combination of three indices consisting of a luminosity value L and two chromaticity values a and b that delineate the colours.

4. The segmentation is based on k-means clustering. The function checks the colours by means of the a and b values of the image and it separates them into the different clusters comparing their values. The cluster centre values consist of the mean a and b values for each cluster.

5. The cluster centre values of the selection of images are averaged using the k-means function to compute one mean centre value per cluster. Please note that in this case the k-means algorithm is applied to average the cluster centre values of the images instead of clustering the pixels of the images as done in point 4.

A preparatory step in the procedure concerns cropping of the images and the detection of cluster centre values. Each cluster centre corresponds to a colour (i.e. grain size) and indicates the mean values of its colour coordinates in colour space. In other words, if we have N grain-size fractions, we define N cluster centres. These cluster centres are determined analysing a limited number of images (in our case 5 % of all images). We consider images in which the presence of each colour is at least higher than 10 % to obtain appropriate cluster centres, thus allowing for a proper colour segmentation. The colours are detected automatically using k-means clustering while imposing the number of colours; k-means clustering (MacQueen, 1967) is a partitioning method aimed at dividing a set of objects into groups (i.e. clusters) based on their attributes. Each object has a location in space. The distance from this location to a representative point of the cluster (i.e. cluster centre) is used to define to which group the object belongs. Using an iterative algorithm the objects are moved between the clusters until the sum of distances from each object to its cluster centroid is minimized. Although the method can be applied to a wide range of cases, here we use it to divide a set of pixels into colour groups to determine an

Subsequently, in all images the colours are segmented imposing the cluster centre values determined under step 5. This a priori setting of the cluster values leads to a much faster computation during the image analysis and most importantly to a more robust colour segmentation. More specifically, the colour segmentation based on forced cluster centres shows good results even for cases in which when one of the colours has an areal fraction smaller than 10 %, whereas k-means clustering results in an incorrect clustering of the colours under these conditions. The procedure to determine the areal fraction content of each colour for all images based on the forced cluster centres is summarized as follows:

1. initialization of the cluster centre variables;

2. image reading;

3. colour saturation (Zaman, 2011);

4. image conversion from RGB colour space to Lab colour space;

5. calculation of the distances of a and b coordinates to the cluster centres;

6. colour segmentation defining the minimum distances;

7. calculation of areal fraction content covered by each colour.

Below these steps are explained in more detail:

1. The cluster centre values of the N colours, defined in steps 1-5, are set as input parameters.

2. The RGB images are loaded and read.

3. Colour saturation is performed to render the image colours uniform.

4. The RGB values of each pixel are converted into the Lab colour space values.

5. The chromaticity values a and b are used to classify the colours. The distances between the a and b coordinates and the cluster centres of the N colours (step 6) are calculated for each pixel in the image (i.e. for each pixel there are N values for the distance to each cluster centre).

6. Among the N distances calculated, a minimum distance is identified. The value of the minimum distance determines in which cluster (i.e. to which colour) every pixel is allocated (Fig. 7).

7. The pixels within each cluster are counted and the areal fraction covered by each colour in the image is computed.

The image analysis algorithm proposed here assigns a cluster to 100 % of the pixels of the image. In Heays et al. (2010a, b) colour detection is based on the isolation of each colour, with consequently a loss or non-detection of some pixels due to shadows in the image or particles with a darker colour. For these reasons their maximum surface detection reached for the images is 91 %. Moreover, rather than

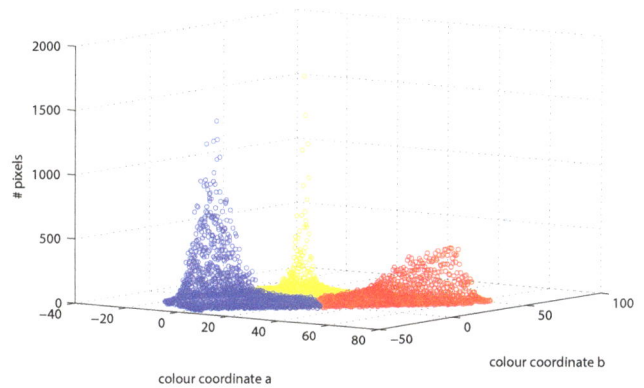

Figure 7. Three-dimensional view of the a and b domain. Example for one image. Clusters segmented in the three colours: blue, red and yellow. The position of each dot in the a and b plane is related to the colour and its vertical position indicates the number of all pixels with the same colour coordinates.

working in the RGB colour space like Heays et al. (2010a, b), we work in the wider subset of colours of the Lab colour space which includes all perceivable colours. Our algorithm considers the two-dimensional space composed of the two chromaticity values a and b. Using the Lab colour space instead of the three-dimensional RGB colour space results in a shorter computational time for the algorithm.

The image analysis procedure presented here was applied to measure the size stratification of a coloured sand–gravel deposit. Figure 8 shows a schematic of the method. Images of the deposit are taken at various elevations and each image is processed with the above described algorithm in order to obtain the areal fraction content covered by each colour (i.e. grain size). Before discussing the results of applying the method to the Gilbert delta (Sect. 4), the stratification measurement method will be assessed in the next section.

3 Assessment of the size stratification measurement method

The stratification measurement method was assessed in two sets of experiments conducted at the Environmental Fluid Mechanics Laboratory of Delft University of Technology. The image analysis technique was optimized and tested in an image analysis experiment (Fig. 9). Subsequently, the stratification measurement method was applied to a Gilbert delta progradation experiment. The latter analysis will be described in Sect. 4.

3.1 Image analysis experiment

The image analysis experiment was performed to determine the best settings to take the images and to develop, optimize, and test the image processing algorithm. Once the best settings (e.g. lights, camera settings) were chosen, images were

Figure 8. Application of the image analysis technique to stratigraphy measurements. Schematic of size stratification of a deposit evaluated with image analysis.

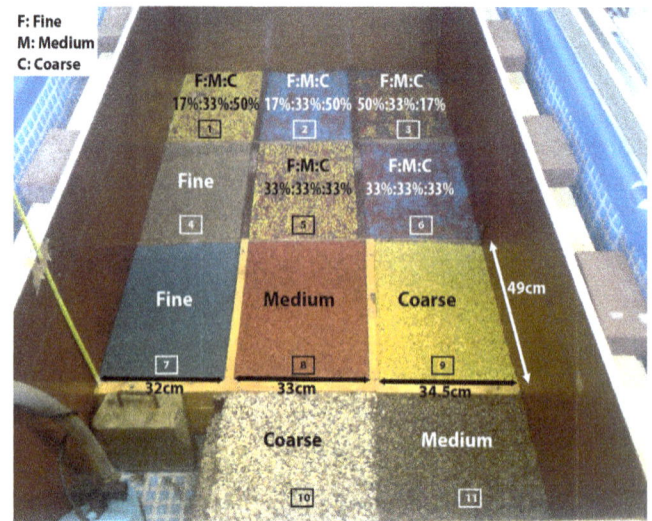

Figure 9. Settings of the image analysis experiment. Grain-size distribution within the compartments, compartment geometry and numbers.

taken of the same area of the bed for different water depths. The experiment was realized in a rectangular pool in which a wooden box of 2 m length, 1.04 m width, and 0.60 m height was placed. The box was divided into 12 compartments with vertical wooden panels and filled with sediment (Fig. 9). One compartment was left empty to place the water pump for varying the water depth in the pool. The experimental set-up was designed to test also other measurement techniques and patches 1–3 and 5–9 were used for testing the image analysis procedure.

We used the same three well-sorted non-overlapping grain-size fractions as described in Sect. 2.1. For some patches we used the finest fraction in its natural colour to evaluate if the image analysis can be applied to unpainted fines. This would avoid painting and potential stickiness of large quantities of relatively fine sediment.

The compartments used to test the image analysis were filled with different mixtures, that is, different volume fraction contents of the three grain-size fractions and by different colour combinations (Fig. 9). The colour combinations are (a) blue for the fine size fraction, red for the medium size fraction, and yellow for the coarse size fraction (patches 1, 3, 5), and (b) natural colour for the fine size fraction, red for the medium size fraction, and blue for the coarse size fraction (patches 2, 6). In the remaining compartments a single size fraction was present on the surface, that is, one colour (patches 7–9). It is important to note that in the compartments filled with sediment mixtures (patches 1–3, 5, 6), it was not possible to exactly reproduce the foreseen areal frac-

tion content of each size fraction on the bed surface. This is because, while pouring the sediment the finer grains, on average, deposit at lower elevations due to kinetic sieving.

The camera used in the experiment was a Canon EOS 550D, an 18.0 megapixel digital single-lens reflex camera with a CMOS sensor. It has a maximum resolution of 5184×3456 and 3 : 2 aspect ratio. Images were stored in colour JPEG format and taken in Live View mode. The camera was used with the Canon EF-S 18–55 mm IS f/3.5–5.6 wide-angle to mild telephoto zoom lens. Tests were carried out to assess the camera lens and the proper camera height. The distance between the camera and the bed surface was set to 50 cm. The focal length for the camera lens was 55 mm. All pictures were taken with the auto exposure program (PAE) and without flash and the lens was used with the image stabilizer function (IS), which automatically corrects for vibrations of the photo camera.

In order to apply this method to submerged beds, a Plexiglas® plate was placed on the water surface to reduce reflection and refraction caused by water surface fluctuations (Wu and Yang, 2004). We used four flood lights to illuminate the bed surface. The lights were switched on at least 20 min before taking the pictures to provide a steady illumination and to avoid reflections on the Plexiglas® plate from the different sources of light in the laboratory. In this way there was as little influence as possible from (naturally varying) natural light from outside.

In the image analysis experiment the area of the bed surface covered by the pictures was ca. 18 cm × 12 cm and the pixel resolution was 0.03 mm. The upper limit of the tested pixel resolution was 0.1 mm (pixel dimension). For this

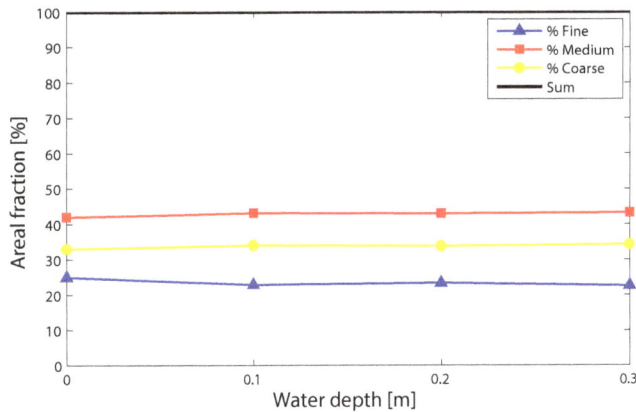

Figure 10. Example of the effect of water depth on the areal fraction in patch 5. The data represent the areal fraction of the three size fractions for different water depths obtained through processing of images covering the same surface area.

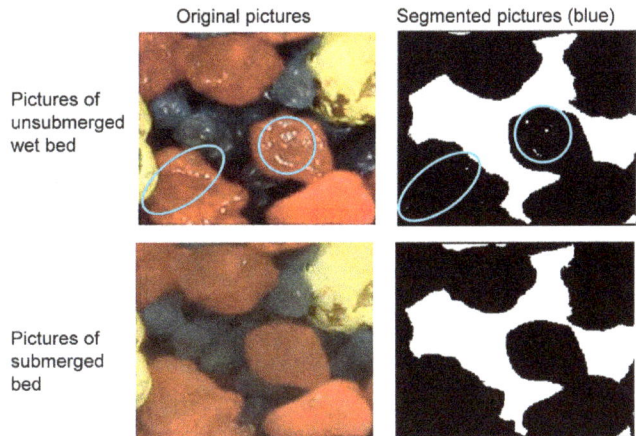

Figure 11. Origin of white pixels in images of an unsubmerged bed; left: original images; right: binary image of the segmented image for the colour blue; at the top: images for an unsubmerged bed; at the bottom: images for a submerged bed. The extra white pixels in the unsubmerged pictures, indicated with blue circles, are due to reflections of water drops.

resolution we still obtain a correct segmentation of colours in the image processing. With the camera settings for this maximum pixel resolution, we can cover a bed surface area of ca. $60\,cm \times 40\,cm$ in one image. A higher camera resolution would enable covering even bigger bed surface areas.

Mixtures composed of a blue fine fraction appeared to be segmented more accurately than the ones in which the fines were of a natural colour. The remaining part of this paper considers the colour combination blue (fine), red (medium), and yellow (coarse).

The camera was mounted on a carriage that was controlled by an operator and a computer to move the camera over the pool. The camera coordinates were registered to be able to reposition the camera at the same location to take pictures for different water depths. In this way we obtained images of equal surface areas for water depths equal to $0.0, 0.1, 0.2$, and $0.3\,m$. The water depth was increased by inserting a hose in the space between the wooden box wall and the swimming pool wall in order to have negligible flow velocities to not disturb the bed surface. The water depth appeared to not significantly affect the resulting grain-size distribution (Fig. 10). Yet, images for the submerged cases showed somewhat better results compared to the unsubmerged case, since in the presence of water the colours were more saturated. Certain images taken of an unsubmerged bed showed some white pixels in the segmented image for the colour blue, which are caused by reflections due to water drops (Fig. 11). For this reason it is advised to wait for the sediment to dry when taking pictures of an unsubmerged bed surface, which can require a few hours.

3.2 Accuracy of the size stratification measurements

The sources of inaccuracies in the size stratification measurement technique are due to a combination of inaccuracies in the (1) particle colouring, (2) removal of thin sediment layers, and (3) image analysis.

An imperfect or thin coating of the paint over the grains showing the original colour of the sediment may lead to inaccuracies in the colour segmentation. We have reduced this effect by colouring the grains twice. Another issue is due to the potential stickiness of grains which can be almost completely avoided by shaking and mixing the sediment several times while the paint is drying. Yet, the medium size fraction consisted of some flat grains which appeared to be difficult to separate during the shaking procedure (Fig. 2). An important aspect for the colour segmentation is the choice of the combination of the colours. For a proper colour segmentation it is essential to choose colours that are as different as possible. This means as distant as possible in the applied colour space. Also the original colour of the sediment plays a role and can cause the resulting colour of the painted grains to be slightly different from the paint colour.

The inaccuracies in the removal of the sediment layers are mostly related to using the wet vacuum cleaner. As discussed before the wet vacuum cleaner did not appear to be grain-size selective. A source of inaccuracy in the sediment removal is the precision in removing layers of 1 cm thickness. A vertical grid was created on the side walls of the flume to aid the estimation of the layer thickness. Based on our observations the inaccuracy in thickness of the sediment layers was between 1 to 2 mm. The imprecision is caused by an oblique view of the grid due to the position of the operator or the distortion due to the glass of the side walls. Another reason for

Results from image analysis

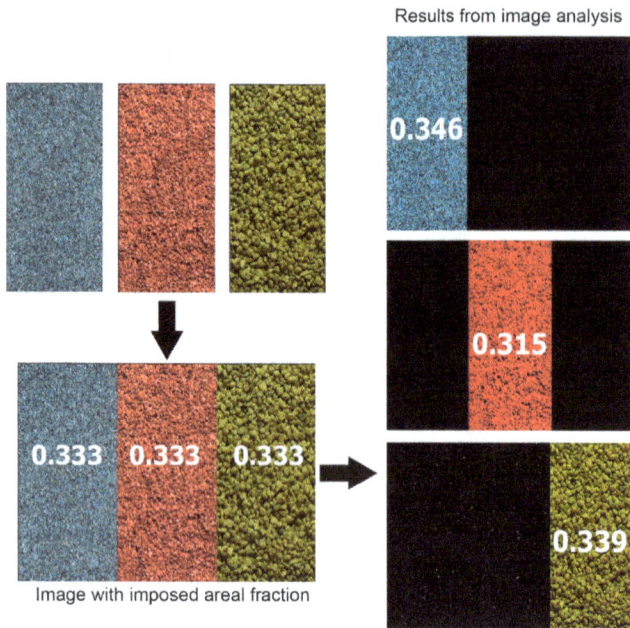

Image with imposed areal fraction

Figure 12. Accuracy test; left: Image with imposed areal fraction content created by combining three images of uniform sediment of the same pixel area; right: segmented images and mean areal fraction contents resulting from the image processing.

inaccuracy in the layer thickness is due to varying the elevation of the mouth of the wet vacuum cleaner above the bed surface. Please note that these inaccuracies in the thickness do not have a cumulative effect while removing more and more sediment layers.

The fact that there are instrinsic differences between areal and volumetric estimates of the grain-size distribution has been recognized since the earliest methods were presented (Kellerhals and Bray, 1971; Adams, 1979; Ibbeken and Schleyer, 1986) and later examined in depth in more recent studies (McEwan et al., 2000; Sime and Ferguson, 2003; Graham et al., 2005a,b; Graham et al., 2010). The main bias is related to the fact that the geometry of three-dimensional grains is estimated by a two-dimensional image (Sime and Ferguson, 2003). Graham et al. (2005b) referred to this error as the spatial distortion given by the lens projecting a three-dimensional object into a two-dimensional image. The errors described in the work of Graham et al. (2005b) are related to the image analysis technique and to the arrangement of the grains. In the case of image analysis based on particle edge detection the errors are related to the capability to detect and measure the sediment particles (Graham et al., 2005b). These inaccuracies are encountered for instance in (a) the detection of edges of a single grain, (b) the distinction between voids and parts of grains in the shadow, and (c) the estimation of the grain axes dimensions. With respect to the errors related to the arrangement of the exposed grains, Graham et al. (2010) distinguished the following three causes of

errors: partial burying of grains, overlapping of grains (imbrication), and grains appearing smaller when viewed from an angle (foreshortening). This generally results in the underestimation of the correct diameter of individual grains (Graham et al., 2010).

The image analysis method proposed here is accompanied by inaccuracies mostly related to colour segmentation. Verifying the proficiency of the image processing in the colour segmentation is not straightforward. First, the size of the finest grain-size fraction used in our experiment prevents the use of validation methods as "paint and pick" or the manual digitation used in other studies to verify the results of the image analysis (e.g. Graham et al., 2005b; Sime and Ferguson, 2003). Secondly, in the image analysis experiment it was difficult to reproduce the foreseen grain-size distribution on the bed surface as, due to kinematic sieving, in the pouring and installing process the fines tend to find lower elevations than the coarser ones.

As an alternative, we have created an image with an imposed grain-size distribution and analyse the results of the colour segmentation. The image combines the images of three areas of equal pixel size, each with a uniform grain size (colour). Each image was taken of a compartment filled with only one grain-size fraction. This means that the imposed areal fraction content equals 0.333 for each colour. The results after repeating the colour segmentation on several images show average areal fractions of 0.346 for the colour blue, 0.315 for the colour red, and 0.339 for the colour yellow (Fig. 12). The inaccuracy of 1–2 % originates from an incorrect colour segmentation for shadow-rich grains. These pixels are then incorrectly clustered with a darker colour. For example, when zooming in on the images we noticed that few red particles that suffer from shadows due to partial burial underneath other grains are recognized as blue in the colour segmentation (Fig. 13).

A final source of inaccuracies is observed when the grains show a bright colour tone due to light reflections or in some cases by a defect coating of the paint showing the original often lighter colour of the sediment. Despite these inaccuracies sources the maximum inaccuracy in the image analysis has appeared to be only 0.02 (2 %) of the measured segmented area. This can be considered to be a relatively high level of accuracy in determining the grain-size distribution. Overall, the stratification measurement method, despite the sum of inaccuracies in particle colouring, sediment removal, and image analysis, results in a significantly smaller inaccuracy in measuring the size stratification of a deposit than common box corers.

4   Size stratification measurements

In this section we describe the application of the stratification measurement method to two Gilbert delta experiments. The size stratification within the deposit measured through image

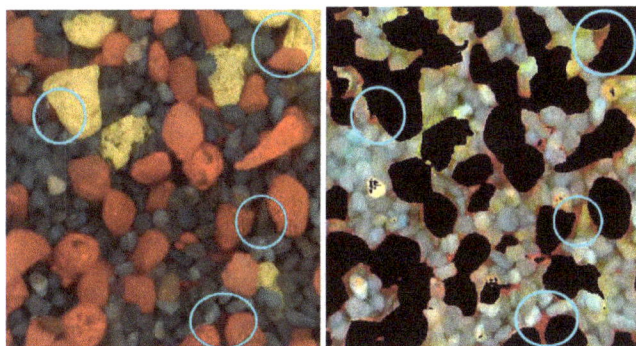

Figure 13. Example of partially buried and therefore shadow-rich red grains, indicated with blue circles, recognized as blue in the colour segmentation; left: original image of a submerged bed surface; right: corresponding segmented image for the blue colour. Please note that the blue colour in the segmented image was lightened to better illustrate the incorrect segmentation of some red grains.

analysis is compared to the one found through sieve analysis. We apply an existing conversion model (Parker, 1991a, b) that converts the areal fraction contents found through image analysis into volumetric fraction contents.

4.1 The Gilbert delta experiment

Laboratory experiments on the progradation of a Gilbert or foreset-dominated delta (Chavarrías et al., 2013, 2014) are used to assess the applicability of the stratification measurement method. In this paper we consider the data related to Experiments 1 and 2 (called E1 and E2) which are conducted with a constant base level, similar to Experiment I in Chavarrías et al. (2013, 2014). The experiments were conducted using the same initial and boundary conditions. The only difference among the two is the geometry of the resulting deltas. The length of the delta in Experiment E1 was about 1.4 m and in Experiment E2 was about 1.2 m. In the experiments the sediment transported over the brinkpoint was too coarse to form a bottomset and deposited over the foreset through avalanches (grain flows). The tests were conducted with the same three sediment fractions as used in the image analysis experiment (Fig. 2) with the colour combination blue (fine), red (medium), and yellow (coarse). The flume length was 14 m, the flume height 0.45 m, and the flume width 0.40 m. At the upstream end of the flume the water discharge ($13\,\mathrm{L\,s^{-1}}$) was maintained constant. The downstream water level was maintained constant and created a flow depth over the brinkpoint equal to 0.06 m, resulting in a flow velocity of $0.54\,\mathrm{m\,s^{-1}}$ over the brinkpoint. The sediment was fed at a constant rate of $280\,\mathrm{g\,min^{-1}}$ using a feeder, which was located at 2.5 m upstream from the initial delta brinkpoint. The volume fraction content in the sediment mixture fed to the flume as well as the one constituting the initial bed were 0.50 (fine), 0.35 (medium) and 0.15 (coarse). After the formation

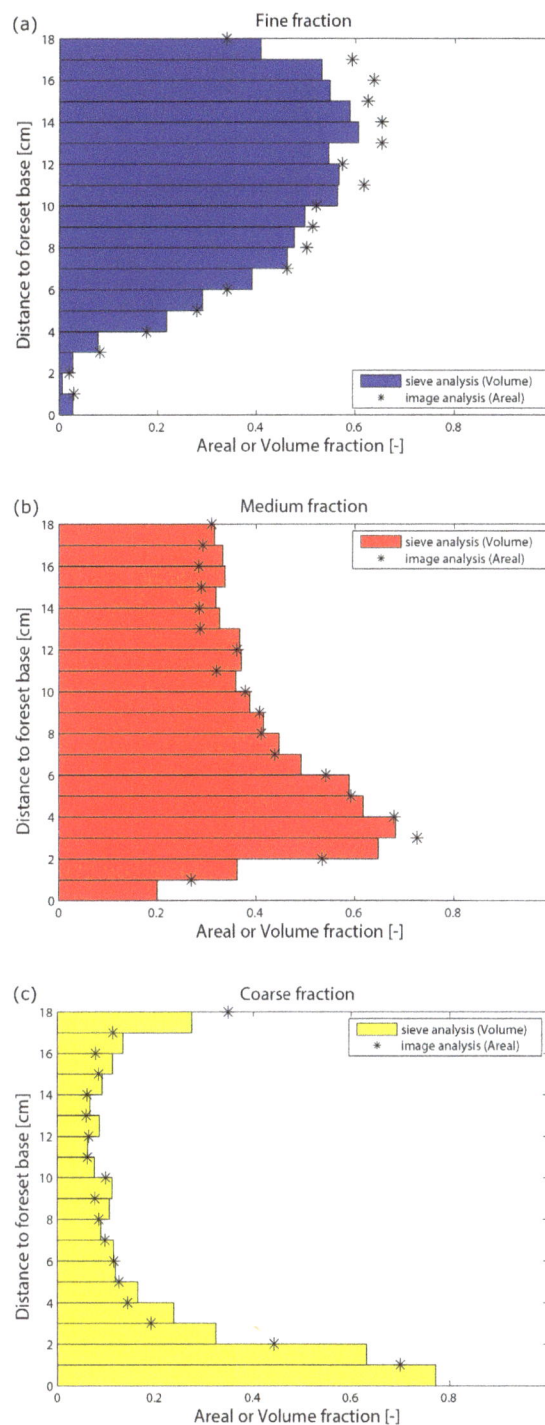

Figure 14. Vertical profiles of (1) the measured volumetric fraction contents measured through sieve analysis and (2) the areal fraction contents determined through image analysis for each grain-size fraction: (a) fine, (b) medium, and (c) coarse for Experiment E1, centre sample.

of a delta deposit, the experiment was stopped and the water was carefully drained from the flume.

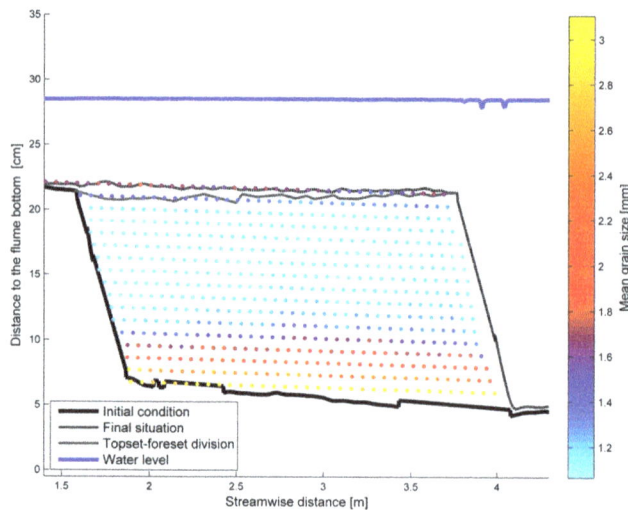

Figure 15. Size stratification within a Gilbert delta deposit measured using the stratigraphy measurement technique proposed in this paper. The colours indicate the local mean grain size based on areal fraction content.

4.2 The Gilbert delta size stratification

The deposits of the Gilbert deltas were analysed following the stratification measurement method described in Sect. 2.2, and using the Plexiglas® sampling grid to compare the results of the image analysis to the ones from sieve analysis. In this analysis two parts of the deposits are considered: one location at the centre of the flume and one location at the side, close to the flume walls. The two grid cells, and therefore the image, cover an area of about 24 cm by 18 cm (centre) and 24 cm by 11 cm (side). The height of the delta deposit (i.e. the distance from the top of the topset to the foreset base) was about 18 cm in Experiment E1 and therefore consisted of 18 sediment layers. The height of the delta deposit in Experiment E2 was about 16 cm, with consequently 16 layers. A coarser mixture was found at the side areas, which was due to a slightly parabolic front of the delta.

In Fig. 14 the results on the areal fraction contents based on the image analysis are indicated at the elevation corresponding to the top of a specific sediment layer. The volume fraction contents resulting from sieve analysis are represented by bars covering some vertical distance, as the data represent the volume fractions of each grain size within a certain layer. Before we consider the differences between the results of the two methods, we can see that the trends of the areal fraction contents determined through image analysis and the volumetric fractions due to sieve analysis are similar. A topset is present within the vertical range of 17 to 18 cm from the foreset base. The upper layer of the topset is characterized by a mobile armour layer (e.g. Fig. 14c). The topset appears to be coarser than the upper part of the foreset. Below the topset we find the foreset, which is the main part

of the delta deposit. It is characterized by a upward fining profile (Fig. 14a).

Let us now consider the differences between the image analysis and sieving methods. Please note how the mobile armour layer (i.e. the top layer of the topset) is more clearly identified in the image analysis. This is due to the fact that in the upper sediment layer analysed in the sieve analysis sediment from the mobile armour layer is mixed with the fine sediment just underneath. This is the reason for the top sample resulting from sieve analysis to be finer than the top image. Except for this top layer, we find that the areal fraction content of fines derived by image analysis is somewhat larger than the corresponding volume fraction content based on sieve analysis. For the medium and coarse fractions, we find the opposite trend. As mentioned before, there are some intrinsic differences between the two methods. The same factors encountered by previous authors using image analysis (Sect. 3.2) are valid for our method and related to the facts that (1) the image analysis outcomes represent a more discrete elevation are compared to the ones of sieve analysis which represents a sediment layer; (2) the particle arrangement, such as the orientation and/or partial burial of the grains of the medium and coarse fractions, influences the relative areal presence of the particles on the bed surface, which affects the number of pixels of a certain colour in an image. It is likely that the medium and coarse grains are covered by several fine particles, so reducing their areal fraction content when compared to the volume fraction content resulting from sieve analysis. To account for these intrinsic differences a conversion model is needed.

Figure 15 shows an example of the result of size stratification measurement method proposed in this manuscript (Chavarrías et al., 2013, 2014). The graph shows the size stratification of the delta in which the colours represent the mean grain size in an area within the deposit equal to $0.34\,\mathrm{m} \times 0.06\,\mathrm{m}$ (lateral vs. streamwise distance) at every 1 cm over the vertical. Figure 15 demonstrates the strength of the developed size stratification measurement method as it results in an unprecedented spatial density of information on the grain-size distribution.

4.3 Conversion from areal to volumetric fractions

The proposed method has appeared to be a powerful tool to determine areal fraction contents within a deposit. We use a conversion model to convert these areal fractions resulting from image analysis into volumetric fractions; furthermore, for modelling purposes we sometimes need information on volume rather than areal fraction contents.

In general, the results of the areal sampling techniques (e.g. wax or clay sampling, image analysis) are not equivalent to the outcomes of sieving. The only procedure that leads to a grain-size distribution that is directly comparable to the results of sieve analysis is the grid by number technique (Kellerhals and Bray, 1971). The grid by number technique

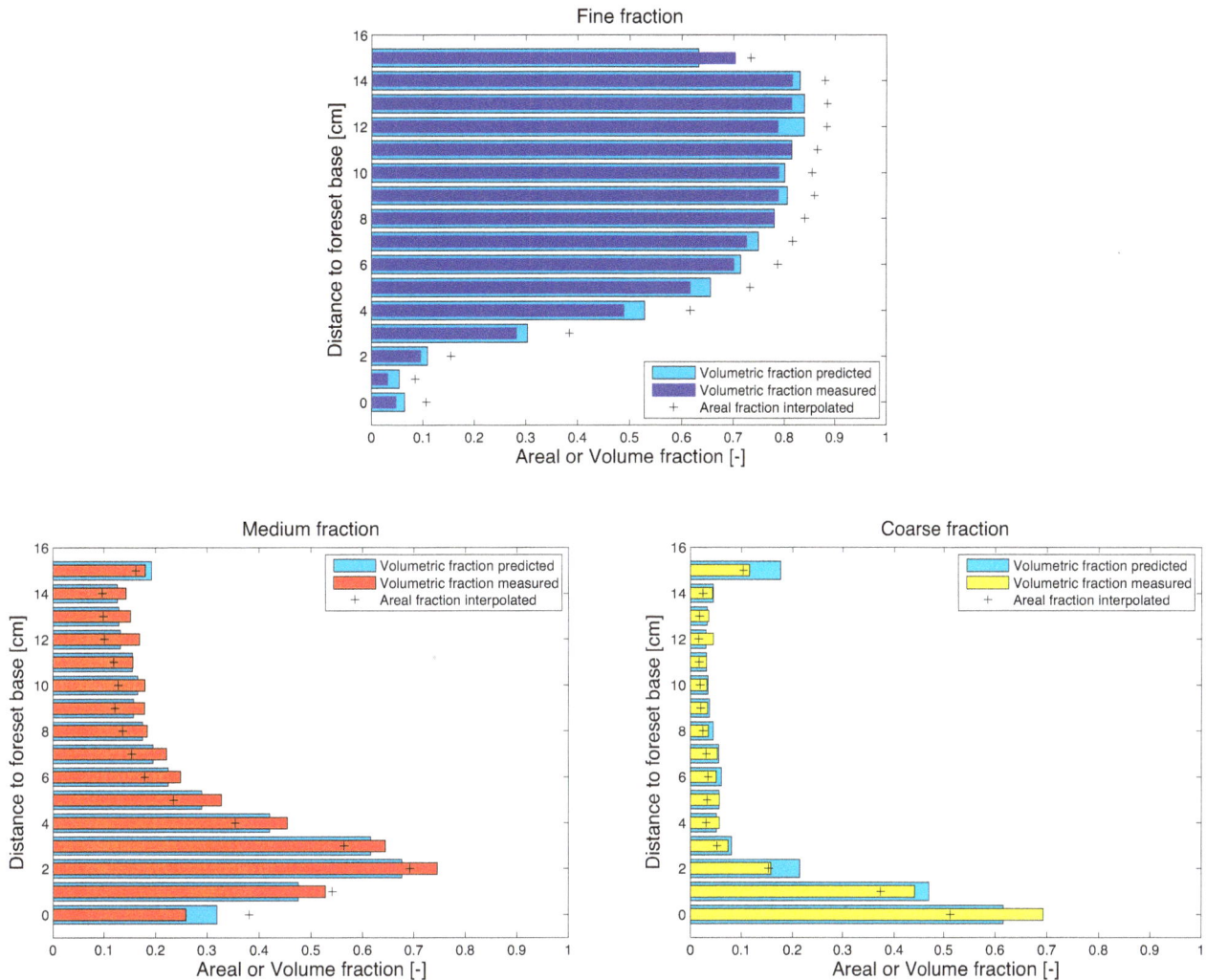

Figure 16. Vertical profiles of (1) the predicted volume fraction contents using Eq. (2), (2) the measured volume fraction contents measured through sieve analysis, (3) the areal fraction contents through image analysis for each grain-size fraction: (a) fine, (b) medium, and (c) coarse for Experiment 2, side sample. Note the areal fractions have been linearly interpolated. The crosses are now located at the elevation corresponding to the middle of the layer.

consists of collecting the particles below the nodes of a grid and counting the number of grains in the various size classes. The frequency of each size class is expressed as the percentage by number compared to the total number of particles. The method used to sample the particles in the grid by number is therefore equivalent to the resulting grain-size distribution (i.e. frequency by number). For the other areal sampling techniques (e.g. wax or clay sampling), the sampling methods are not equivalent to sieve analysis. Each sample has its own bias, which often not only depends on the method used for sampling and for the analysis, but also on the grain-size distribution of the deposit. Kellerhals and Bray (1971) were the first to provide a conversion model valid for a wide series of methods. They suggested an idealized cube to model the arrangement of mixed size particles without pores. Diplas and Sutherland (1988) modified the model to account for

porosity. Fraccarollo and Marion (1995) present a conversion model that takes into account the irregular shape and random arrangement of the bed material. Therefore, unlike the cube model of Kellerhals and Bray (1971) and the modified version of Diplas and Sutherland (1988), the application of the model by Fraccarollo and Marion (1995) is not restricted to a fixed amount of grain-size fractions, certain particle sizes and shapes.

The conversion model used here is the one proposed by Parker (1991a, b). The model originates from the ones developed by Proffitt (1980) and Diplas and Sutherland (1988) and provides an estimate of the areal fraction content based on a given volume fraction content. The model by Parker (1991a, b) rewritten in the form as presented by Parker

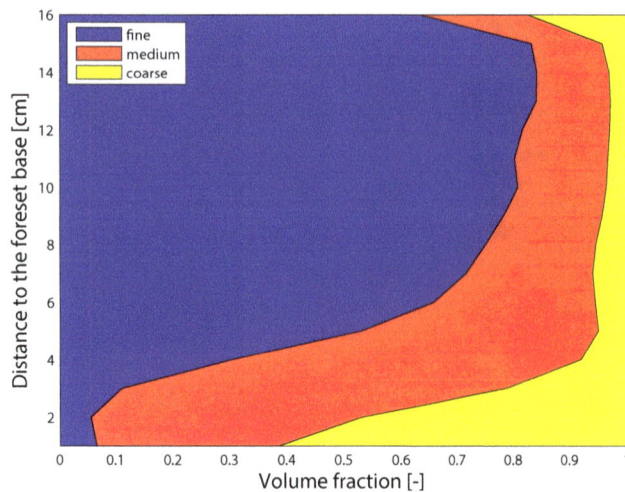

Figure 17. Vertical profile of the volume fraction contents for Experiment 2 (side sample) estimated through conversion of measured areal fraction contents to volume fraction contents using the Parker (1991a, b) conversion model.

and Cui (1998b) and Cui and Parker (2005) is

$$F_{Ai} = \frac{F_{Vi} / \sqrt{D_i}}{\sum\limits_{k=1}^{N} \left(F_{Vk} / \sqrt{D_k}\right)}, \tag{1}$$

where F_{Vi} denotes the volume fraction content of size fraction i in a sediment layer $[-]$, F_{Ai} is the areal fraction content of size fraction i of the same sediment layer $[-]$, and D_i is the grain size of size fraction i [m]. Like the subscript i, the subscript k denotes the size fraction.

We now write Eq. (1) in the reverse form to determine the volume fraction content of each grain-size fraction, F_{Vi}, from the areal fraction content resulting from image analysis, F_{Ai},:

$$F_{Vi} = \frac{F_{Ai} \sqrt{D_i}}{\sum_{k=1}^{N} \left(F_{Ak} \sqrt{D_k}\right)}. \tag{2}$$

Equation (2) was then applied to the data set of the Gilbert delta Experiments E1 and E2. The areal fraction contents resulting from image analysis were linearly interpolated between vertical elevations to provide data on the areal fractions at the same elevations as the values of the volumetric fraction contents. Figure 16 shows the results of the conversion model for the near-wall sample of Experiment E2. We now compare (1) the predicted volume fraction contents using Eq. (2), (2) the volume fraction contents measured through sieve analysis, and (3) the areal fraction contents determined through image analysis. Observing these three profiles, we notice that the conversion model reduces the overestimation of the fine fraction and the underestimation of the medium and coarse fractions. After converting the results of the image analysis into the corresponding volume fraction

contents, we plot the variation of these volume fractions over the vertical (Fig. 17). We recognize a topset deposit over the elevation range between 15 and 16 cm from the foreset base. The topset is somewhat coarser than the sediment just below that belongs to the foreset deposit. The resulting foreset deposit is characterized by upward fining as a result of the avalanching process.

5   Conclusions

In this paper we present a new method to measure the entire size stratification of a sand–gravel deposit in a laboratory experiment. The method is based on a combination of particle colouring, removal of sediment layers, and image analysis. Colour segmentation of areal images of a bed surface covered by coloured particles is combined with a nondestructive and flexible method to remove sediment layers using a wet vacuum cleaner.

A clear advantage of the method is its high spatial resolution of the resulting data on the grain-size distribution, which is unequalled in other methods. Other methods generally provide information on the grain-size distribution in a limited number of locations.

We observed small inaccuracies in the sediment removal technique which are mainly attributable to imprecisions in the exact thickness of a sediment layer. Disturbances caused by driving a box corer into the bed, such as compressing sediment layers and/or displacing grains to lower elevations, are avoided here.

Using the proposed method a large amount of data is quickly collected, time consuming sieve analyses are avoided, and the data is quickly processed using an automatic and fast script. The image analysis processing is based on colour segmentation within the Lab colour space which results in colour detection of all pixels in an image. The accuracy in determining the areal fraction contents using image analysis is of the order of 1–2 %. A conversion model has been applied to convert areal fraction contents derived by image analysis into volume fraction contents.

The image analysis method was successfully used to determine the grain-size distribution of a submerged bed surface, while creating no disturbances to the bed surface. Measurement of the grain-size distribution of the bed surface during a running laboratory experiment, at flows with transported sediment, is currently being tested by the author.

Particle colouring by itself appeared to be of great help in the interpretation of grain-size selective processes based on observations of the flume experiments.

Overall, the size stratification measurement method, despite the sum of inaccuracies in particle colouring, sediment removal, and image analysis, results in a significantly smaller inaccuracy in measuring the size stratification of a deposit than a box corer.

Acknowledgements. This research is funded by scholarship 10.015 of the Dr ir Cornelis Lely Stichting and Aspasia scholarship 015.007.051 of the Netherlands Organisation for Scientific Research (NWO). The authors especially thank the technicians of the Environmental Fluid Mechanics Laboratory of Delft University of Technology for their assistance during the experiments.

Edited by: F. Metivier

References

Adams, J.: Gravel size analysis from photographs, J. Hydraulics Division, ASCE, 105, 1247–1255, 1979.

Alexander, J., Bridge, J. S., Cheel, R. J., and Leclair, S. F.: Bedforms and associated sedimentary structures formed under supercritical water flows over aggrading sand beds, Sedimentology, 48, 133–152, 2001.

Allen, J. R. L.: Sedimentation to the lee of small underwater sand waves: An experimental study, J. Geol., 73, 95–116, 1965.

Ashworth, P. J. and Ferguson, R. I.: Interrelationships of channel processes, changes and sediments in a proglacial braided river, Geografiska Annaler, Series A, Phys. Geogr., 68, 361–371, 1986.

Bagnold, R. A.: The physics of blown sand and desert dunes, Methuen, New York, 1941.

Blom, A.: Different approaches to handling vertical and streamwise sorting in modeling river morphodynamics, Water Resour. Res., 44, W03415, doi:10.1029/2006WR005474, 2008.

Blom, A., Ribberink, J. S., and de Vriend, H. J.: Vertical sorting in bed forms: Flume experiments with a natural and a trimodal sediment mixture, Water Resour. Res., 39, 1025, doi:10.1029/2001WR001088, 2003.

Carbonneau, P. E., Lane, S. N., and Bergeron, N. E.: Catchment-scale mapping of surface grain size in gravel bed rivers using airborne digital imagery, Water Resour. Res., 40, W07202, doi:10.1029/2003WR002759, 2004.

Carbonneau, P. E., Bergeron, N. E., and Lane, S. N.: Automated grain size measurements from airborne remote sensing for long profile measurements of fluvial grain sizes, Water Resour. Res., 41, W11426, doi:10.1029/2005WR003994, 2005.

Carling, P. A. and Reader, N. A.: A freeze sampling technique suitable for coarse river bed material, Sediment. Geol., 29, 223–239, 1981.

Chavarrías, V., Blom, A., Orrú, C., and Viparelli, E.: Laboratory experiment of a mixed-sediment Gilbert delta under varying base level, RCEM Symposium, 9–13 June, Santander, Spain, Book of abstracts, 114 pp., 2013.

Chavarrías, V., Orrú, C., Viparelli, E., Martín-Vide, J. P., and Blom, A.: Size stratification of a laboratory Gilbert delta due to base level changes, J. Geophys. Res., submitted, 2014.

Cui, Y. and Parker, G.: Numerical Model of Sediment Pulses and Sediment-Supply Disturbances in Mountain Rivers, J. Hydraul. Eng., 131, 646–656, 2005.

Diplas, P. and Sutherland, A.: Sampling Techniques for Gravel Sized Sediments, J. Hydraul. Eng., 114, 484–501, 1988.

Dugdale, S. J., Carbonneau, P. E., and Campbell, D.: Aerial photosieving of exposed gravel bars for the rapid calibration of airborne grain size maps, Earth Surf. Process. Landforms, 35, 627–639, 2010.

Edmonds, D. A., Shaw, J. B., and Mohrig, D.: Topset-dominated deltas: A new model for river delta stratigraphy, Geology, 39, 1175–1178, 2011.

Fraccarollo, L. and Marion, A.: Statistical Approach to Bed-Material Surface Sampling, J. Hydraul. Eng., 121, 540–545, 1995.

Gardner, J. T. and Ashmore, P.: Geometry and grain size characteristics of the basal surface of a braided river deposit, Geology, 39, 247–250, 2011.

Gilbert, G. K.: The topographic features of lake shores, US Geol. Surv., Annu. Rep., 5, 75–123, 1885.

Gilbert, G. K.: Lake Bonneville, US Geol. Surv. Monogr., 1, 438 pp., 1890.

Gomez, B.: Representative sampling of sandy fluvial gravels, Sediment. Geol., 34, 301–306, 1983.

Graham, D. J., Reid, I., and Rice, S. P.: Automated sizing of coarse grained sediments: Image-processing procedures, Math. Geol., 37, 1–28, 2005a.

Graham, D. J., Rice, S. P., and Reid, I.: A transferable method for the automated grain sizing of river gravels, Water Resour. Res., 41, W07020, doi:10.1029/2004WR003868, 2005b.

Graham, D. J., Rollet, A.J., Piégay, H., and Rice, S. P.: Maximizing the accuracy of image-based surface sediment sampling techniques, Water Resour. Res., 46, W02508, doi:10.1029/2008WR006940, 2010.

Heays, K., Friedrich, H., and Melville, B. W.: Re-Evaluation of image analysis for sedimentary processes research. In: International Association of Hydraulic Engineering-Asia Pacific Division, Auckland, 21–24 February, 2010a.

Heays, K., Friedrich, H., and Melville, B. W.: Advance particle tracking for sediment movement on river beds: A Laboratory study, 17th Australasian Fluid Mechanics Conference Auckland, New Zealand 5–9 December, 2010b.

Hirano, M.: River bed degradation with armouring, Trans. Jpn. Soc. Civ. Eng., 3, 194–195, 1971.

Ibekken, H. and Schleyer, R.: Photo-sieving: A method for grain-size analysis of coarse-grained, unconsolidated bedding surfaces, Earth Surface Proc. Landforms, 11, 59–77, 1986.

Jain, S.: Armor or Pavement, J. Hydraul. Eng., 116, 436–440, 1990.

Jonasson, A. and Olausson, E.: New devices for sediment sampling, Mar. Geol., 4, 365–371, 1966.

Julien, P. Y. and Anthony, D. J.: Bed load motion and grain sorting in a meandering stream, J. Hydr. Res., 40, 125–133, 2002.

Kellerhals, R. and Bray, D. I.: Sampling Procedures for coarse fluvial sediments, J. Hydr. Div., ASCE, 97, 1165–1180, 1971.

Kleinhans, M. G.: Grain-size sorting in grainflows at the lee side of deltas, Sedimentology, 52, 291–311, 2005.

Leopold, L. B. and Wolman, M. G.: River channel patterns: Braided, meandering, and straight: US Geological Survey Professional Paper 282-B, 39–85, 1957.

Lunt, I. A. and Bridge, J. S.: Evolution and deposits of a gravelly braid bar, Sagavanirktok River, Alaska, Sedimentology, 51, 415–432, 2004.

MacQueen, J.: Some methods for classification and analysis of multivariate observations, Proceedings of the Fifth Berkeley Symposium on Mathematical Statistics and Probability, Statistics, University of California Press, Berkeley, California, 1, 281-297, 1967.

Marion, A. and Fraccarollo, L.: Experimental investigation of mobile armouring development, Water Resour. Res., 33, 1447–1453, 1997.

McEwan, I. K., Sheen, T. M., Cunningham, G. T., and Allen, A. R.: Estimating the size composition of sediment surfaces through image analysis, Proc. Inst. Civ. Eng. Water Mar. Energy, 142, 189–195, 2000.

Pallara, O., Froio, F., Rinolfi, A., and Lo Presti, D.: Assessment of strength and deformation of coarse grained soils by means of penetration tests and laboratory tests on undisturbed samples, Solid Mechanics and Its Applications, 2007, Geotechnical Symposium in Rome, 146, 201–213, 2006.

Parker, G.: Selective sorting and abrasion of river gravel, I: theory, J. Hydraul. Eng., 117, 131–149, 1991a.

Parker, G.: Selective sorting and abrasion of river gravel, II: applications, J. Hydraul. Eng., 117, 150–171, 1991b.

Parker, G. and Andrews, E. D.: Sorting of bed load sediment by flow in meander bends, Water Resour. Res., 21, 1361–1373, 1985.

Parker, G. and Cui, Y.: The arrested gravel front: stable gravel-sand transitions in rivers. Part 2: General numerical solution, J. Hydr. Res., 36, 159–182, 1998b.

Parker, G. and Klingeman, P.: On why gravelbed streams are paved, Water Resour. Res., 18, 1409–1423, 1982.

Postma, G.: Sea-level-related architectural trends in coarse-grained delta complexes, Sediment. Geol., 98, 3–12, 1995.

Proffitt, G. T.: Selective transport and armoring of non-uniform alluvial sediments, Report 80/22, University of Canterbury, Christchurch, New Zealand, 1980.

Ribberink, J. S.: Mathematical modelling of one-dimensional morphological changes in rivers with non-uniform sediment, Ph.D. thesis, Delft University, Delft, Netherlands, 1987.

Rubin, D. M.: A simple autocorrelation algorithm for determining grain size from digital images of sediment, J. Sediment. Res., 74, 160–165, 2004.

Sibanda, E., McEwan, I., and Marion, A.: Measuring the structure of mixed grain size sediment beds, J. Hydr. Engin., 126, 347–353, 2000.

Sime, L. C. and Ferguson, R. I.: Information on grains sizes in gravel-bed rivers by automated image analysis, J. Sediment. Res., 73, 630–636, 2003.

Walkotten, W. J.: An improved technique for freeze sampling streambed sediments, USDA For. Serv. Res. Note, PNW-281, 11 pp., 1976.

Warrick, J. A., Rubin, D. M., Ruggiero, P., Harney, J. N., Draut, A. E., and Buscombe, D.: Cobble cam: grain-size measurements of sand to boulder from digital photographs and autocorrelation analyses, Earth Surf. Process. Landforms, 34, 1811–1821, 2009.

Wilcock, P. R. and Crowe, J. C.: Surface-based transport model for mixed-size sediment, J. Hydraul. Eng., 129, 120–128, 2003.

Wilcock P. R. and McArdell B. W.: Surface-based fractional transport rates: Mobilization thresholds and partial transport of a sand-gravel sediment, Water Resour. Res., 29, 1297–1312, 1993.

Wilcock, P. R. and McArdell, B. W.: Partial transport of a sand/gravel mixture, Water Resour. Res., 33, 235–245, 1997.

Wu, F. C. and Yang, K. H.: Entrainment probabilities of mixed-size sediment incorporating near-bed coherent flow structures, J. Hydraul. Eng., 130, 1187–1197, 2004.

Zaman, T.: [Saturation] Saturate or Desaturate an image in Matlab (http://www.timzaman.com/?p=545), licensed under a Creative Commons Attribution-Non Commercial-Share Alike 3.0 Unported License, 2011.

Zanke, U.: Über den Einfluss von Kornmaterial, Strömungen und Wasserstanden auf die Korngrossen von Transportörpern in offenen Gerinnen, Tech. Rep. 44, Mitt. Franzius Inst., Hannover, Germany, 1976.

14

Comparison between experimental and numerical stratigraphy emplaced by a prograding delta

E. Viparelli[1], A. Blom[2], C. Ferrer-Boix[3], and R. Kuprenas[1]

[1]Department of Civil and Environmental Engineering, University of South Carolina,
Columbia, South Carolina, USA
[2]Faculty of Civil Engineering & Geosciences, Delft University of Technology, Delft, the Netherlands
[3]Department of Geography, University of British Columbia, Vancouver, Canada

Correspondence to: E. Viparelli (viparelli@cec.sc.edu)

Abstract. A one-dimensional model that is able to store the stratigraphy emplaced by a prograding delta is validated against experimental results. The laboratory experiment describes the migration of a Gilbert delta on a sloping basement into standing water, i.e., a condition in which the stratigraphy emplaced by the delta front is entirely stored in the deposit. The migration of the delta front and the deposition on the delta top are modeled with total and grain-size-based mass conservation models. The vertical sorting on the delta front is modeled with a lee-face-sorting model as a function of the grain size distribution of the sediment deposited at the brinkpoint, i.e., at the downstream end of the delta top. Notwithstanding the errors associated with the grain-size-specific bedload transport formulation, the comparison between numerical and experimental results shows that the model is able to reasonably describe the progradation of the delta front, the frictional resistances on the delta top, and the overall grain size distribution of the delta top and delta front deposits. Further validation of the model in the case of variable base level is currently in progress to allow for future studies, at field and laboratory scale, on how the delta stratigraphy is affected by different changes of relative base level.

1 Introduction

A fluvio-deltaic deposit can be considered composed of two parts: the delta top, where sediment is transported, eroded, and deposited by fluvial-type processes, and the delta front, where sedimentary processes are characterized by avalanches, deposition of sediment from suspension, and particle entrainment and deposition by submarine currents (Swenson et al., 2000). Depending on the dominant submarine processes, delta front deposits can either be characterized by upward-fining, upward-coarsening, or more complex vertical depositional patterns (e.g., Rohais et al., 2008, for the Gilbert deltas in the Gulf of Corinth, Greece). This paper focuses on relatively coarse grained Gilbert deltas, with steep fronts and a thick delta front deposit compared to the delta top deposit (Edmonds et al., 2011). Angles of the delta front of about 20–35° are observed in the Serra Ciciniello section of the Potenza Basin, Italy (Longhitano, 2008).

In this study we present the validation against experimental observations of a one-dimensional delta migration model that is able to predict and record the spatial, i.e., vertical and streamwise, variation of the grain size distribution within the deltaic deposit, under the assumption that grain flows are the dominant delta front sedimentation process. In particular, the model describes (i) grain flow deposition on the delta front and the emplacement of upward-fining units, as well as (ii) fluvial deposition on the delta top. It can thus be of aid in the interpretation of ancient and modern deltaic deposits in which grain flows are the dominant delta front deposition process.

The work of Kleinhans (2005), Blom et al. (2003, 2013), and Blom and Kleinhans (2006) on grain flow deposits emplaced on the lee face of dunes, bars, and on Gilbert delta fronts demonstrates that the vertical sorting pattern of the grain flow deposits does not significantly change from migrating bedforms to prograding Gilbert deltas. It can indeed be modeled with the same mathematical relations as a function of the height of the lee face and of the sediment transport on the topmost part of the lee face, the brinkpoint. Thus, the procedure for the storage of stratigraphy, validated herein for the case of a single downstream-migrating lee face (or delta front), may be extended in the future to multiple migrating lee faces, such as those observed in the cases of stacked deltaic complexes and downstream-migrating dunes or bars.

The upward-fining pattern of grain flow deposits is the result of several processes that can be summarized as follows (Fig. 1). Bedload sediment is deposited in a wedge on the topmost part of the slip face until the static angle of repose of the material is reached. This mechanism is termed grain fall, and it is characterized by preferential deposition of coarser grains in the upstream part of the grain fall deposit. When the static angle of repose is exceeded, the wedge collapses, a grain flow is initiated, and the remobilized sediment avalanches down the lee face. During the grain flow, sediment sorting takes place, and coarse sediment is deposited over the lower part of the lee face, while fine sediment remains trapped in its upper portion. The formation and emplacement of grain-fall–grain-flow deposits is schematically represented in Fig. 1a and b, respectively. In Fig. 1 the delta is assumed to migrate over a horizontal substrate for illustration purposes only.

A numerical model that needs to reproduce the stratigraphy emplaced by a prograding delta is composed of three sub-models (types 1–3) that respectively describe (1) the total sediment mass conservation in the system (e.g., Wright and Parker, 2005a, b), (2) the mass conservation of sediment in each grain size range (e.g., Hirano, 1971), and (3) the sorting process on the lee face (e.g., Blom and Parker, 2004; Blom et al., 2006). Each sub-model has its purpose. In particular, total sediment mass conservation models, type 1, predict the rates of channel bed aggradation and delta front migration. Type 2 models account for the mobility of sediment particles of different sizes on the delta top. Finally, lee-face-sorting models, type 3, synthetically describe the grain-fall–grain-flow mechanism that occurs on the lee face. Bookkeeping procedures to store stratigraphy are implemented to record the spatial variation of the characteristics of the deposited material.

The comparison between experimental measurements and one-dimensional numerical predictions of stratigraphy, defined as the vertical and streamwise variation of grain size distribution within the deposited sediment (Viparelli et al., 2010a), is performed for the case of an experimental Gilbert delta prograding into standing water (Ferrer-Boix et al., 2013). The experiment is characterized by the following conditions: (i) the system is always net depositional; (ii) the stratigraphy emplaced by the delta front is entirely stored in the deeper portion of the deposit; (iii) the stratigraphy emplaced on the delta top is stored in the upper portion of the

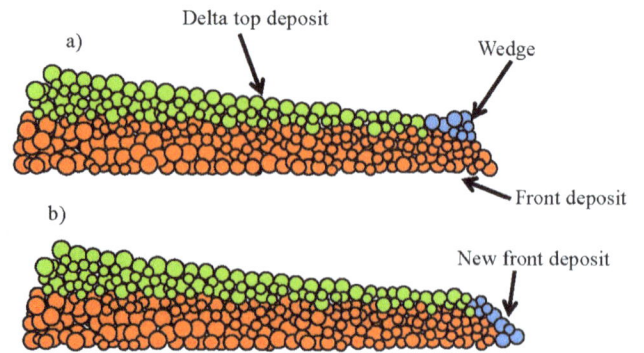

Figure 1. Schematic representation of the Gilbert delta stratigraphy. (**a**) Grain flow deposit, (**b**) grain fall deposit.

deposit; (iv) the mode of sediment transport on the delta top is lower plane bed bedload regime, i.e., migrating dunes and bars that would partially rework the upper portion of the delta top are not present; and (v) grain flows are the dominant deposition process on the delta front.

The mass conservation sub-model (type 1) is validated by comparing measured and predicted longitudinal profiles of delta elevation and delta front migration rates. Due to the different depositional processes on the delta front and on the delta top (Fig. 1), these experimental results allow for the validation of both the grain-size-specific mass conservation sub-model (type 2) and the lee-face-sorting sub-model (type 3). In particular, the type 2 sub-model is validated with the comparison between measured and predicted grain size distributions of the topmost layer of the delta top deposit at the final state of the experiment, while the lee-face-sorting sub-model (type 3) is validated by comparing the grain size distribution of the front deposits, i.e., the deposit below the delta top.

The paper is organized as follows: the relevant characteristics of the laboratory experiment and the numerical model are presented in Sects. 2 and 3, respectively. The comparison between measured and numerical results is discussed in Sect. 4. The results of the study and the plans for future work are summarized in the last section of the manuscript. A brief discussion of the model sensitivity to the grain-size-specific bedload model is presented in the Appendix.

2   The laboratory experiment

The laboratory experiment was performed in the 12 m long and 0.60 m wide tilting flume at the Hydrosystems Laboratory, University of Illinois, Urbana-Champaign (Ferrer-Boix et al., 2013). The parent material was a mixture of sand and pea gravel with a geometric mean diameter, D_g, of 3.43 mm, and geometric standard deviation of 1.75, shown in Fig. 2, where the blue line represents the cumulative grain size distribution. The yellow diamonds denote the fractions of sediment finer than the bound diameters, D_{bi}, used in the numerical runs presented in Sect. 4. The red line connects

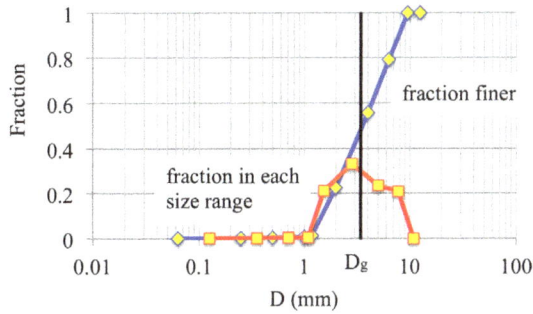

Figure 2. Grain size distribution of the parent material. D denotes the grain diameter in millimeters. The blue line is the cumulative distribution, and the yellow diamonds denote the bound diameters used in the numerical calculations. The yellow squares indicate the fractions of parent material contained in each characteristic grain size range, i.e., between two bound diameters. The vertical black line is the geometric mean diameter, D_g, of the parent material.

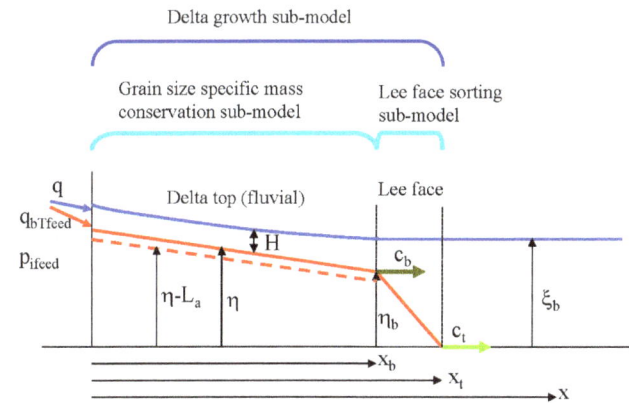

Figure 3. The numerical sub-models and the relevant model parameters. x is a streamwise coordinate, x_b and x_t are the coordinates of the brinkpoint and of the delta toe. H is the water depth on the delta top. q is the input water discharge per unit channel width, q_{bTfeed} is the total (i.e., summed over all the grain sizes) volumetric sediment feed rate per unit channel width, p_{ifeed} characterizes the grain size distribution of the fed material, and ξ_b is the elevation of the downstream standing water above the datum. η represents the elevation of the delta top above the datum, η_b is the elevation of the brinkpoint, and L_a denotes the active layer thickness. c_b denotes the migration rate of the brinkpoint, and c_t is the migration rate of the delta toe.

yellow squares denoting the fractions of sediment contained in each characteristic grain size range, i.e., the grain size range bounded by two consecutive bound diameters, D_{bi} and D_{bi+1}. The sediment in each characteristic grain size range is modeled as uniform and with characteristic grain size D_i equal to the geometric mean of the bound diameters (e.g., Parker, 2004).

The laboratory flume operated in sediment feed mode; that is, water and sediment were fed at a constant rate at the upstream end of the flume. In particular, the flow rate of $47\,\mathrm{L\,s^{-1}}$ was controlled with an electromagnetic flow meter, and the feed rate of parent material was set at $800\,\mathrm{g\,min^{-1}}$ with a screw-type feeder.

The flume was tilted with a bottom slope of 2 %. An initial 10 cm thick layer of parent material was placed on the bottom of the flume for the entire length of the experimental section. The downstream water elevation was set at 26 cm above the initial deposit by means of a transverse wall located 9 m downstream of the flume entrance. The experiment started when the flow and the sediment feeder were simultaneously turned on. A downstream-migrating Gilbert delta formed and migrated downstream. The experiment terminated when the Gilbert delta reached the transverse wall, i.e., after 10.42 h.

Longitudinal profiles of delta elevation were periodically recorded (i.e., after 0.15, 0.67, 1.05, 2.30, 3.97, 5.98 ,and 8.50 h from the beginning of the experiment) with four ultrasonic transducer probes (Wong et al., 2007). At the end of the experimental run, a last longitudinal profile was measured and core samples were collected in six locations of the deposit (i.e., 3.5, 4.5, 5.5, 6.5, 7.5, and 8.5 m downstream of the flume entrance) with the metallic box described by Blom et al. (2003). Each core sample was then sliced into 2 cm thick layers. Each layer was oven dried and its grain size distribution was measured to characterize the emplaced

stratigraphy. Further details on the experiment are reported in Ferrer-Boix et al. (2013).

3   The numerical model

The numerical model to predict the stratigraphy emplaced by a downstream-migrating Gilbert delta is built by means of coupling (i) a delta progradation model, type 1 (Wright and Parker, 2005a, b); (ii) an active layer model for mass conservation of nonuniform sediment on the delta top, type 2 (Hirano, 1971; Parker, 1991a, b); (iii) a lee-face-sorting model for the delta front, type 3 (Blom et al., 2013); and (iv) a procedure for the storage of stratigraphy (Viparelli et al., 2010a). The role of each model is schematically represented in Fig. 3 with the definition of the relevant model parameters and boundary conditions.

As discussed by Blom (2008) different modeling approaches can be used to couple grain-size-specific mass conservation models (type 2) and lee-face-sorting models (type 3). The active layer approximation (Hirano, 1971, as modified by Parker, 1991a) is used herein because (i) it can be implemented with reasonably large spatial and temporal steps, allowing for future laboratory- and field-scale applications (Blom, 2008), and (ii) it reasonably reproduces the stratigraphy emplaced under lower plane bed bedload transport conditions on the delta top (Viparelli et al., 2010a, b).

To simplify the schematic representation of the system, in Fig. 3 the delta prograde on a horizontal basement. The model, however, is designed to handle an arbitrarily sloping basement with constant slope S_b.

The model parameters represented in Fig. 3 are the streamwise coordinate x and the streamwise locations of the brinkpoint and the delta toe, x_b and x_t, respectively. As further discussed in the remainder of this section, the model boundary conditions are expressed in terms of volumetric feed rate of water, q, and sediment, q_{bT}, and elevation of the downstream standing water above the datum, ξ_b. The grain size distribution of the sediment feed is characterized in terms of fraction of sediment in the generic grain size range i, p_{ifeed}. The water depth on the delta top is denoted by H, the elevation of the deltaic deposit above the datum with η, and the thickness of the active layer with L_a. The elevation of the brinkpoint above the datum is η_b, and its migration rate is denoted by c_b. Similarly, the migration rate of the delta toe is denoted by c_t.

The numerical model is a one-dimensional (laterally averaged) model of delta growth based on the standard shallow water equations of open channel flow and on the equation of sediment conservation (e.g., Parker, 2004). Before outlining the governing equations and the numerical scheme for the storage of grain size stratigraphy, the simplifying assumptions are listed below. Some of these assumptions are introduced to apply the model at laboratory scale and can be relatively easily relaxed for field scale applications:

1. the volume bedload transport rate is orders of magnitude smaller than the flow discharge, so that the quasi-steady approximation (De Vries, 1965) holds and the bed elevation profile can be considered as unchanging in the hydraulic calculations;

2. the channel cross section is rectangular, with constant width B and vertical smooth sidewalls;

3. the flow is Froude-subcritical and the shallow water equations are reduced to the equation for a backwater curve, so that the equations can be integrated upstream starting from the brinkpoint, i.e., the downstream end of the delta top;

4. the laboratory flume is long enough that entrance effects can be reasonably neglected;

5. grain flows are the dominant depositional process on the delta front.

3.1   Calculation of the flow

As in Viparelli et al. (2010a), the shallow water momentum equation is modified to account for the different shear stresses acting on the smooth flume sidewalls and on the rough delta top. This correction is necessary to properly model bedload transport in a laboratory flume (Vanoni,

1975). The water mass and momentum conservation equations for open channel flow take the form

$$\frac{\partial H}{\partial t} + \frac{\partial UH}{\partial x} = 0, \tag{1}$$

$$\frac{\partial UH}{\partial t} + \frac{\partial U^2 H}{\partial x} = -gH\frac{\partial H}{\partial x} + gHS - \frac{1}{\rho}\tau_e, \tag{2}$$

respectively, where H denotes the water depth, U is the cross-sectionally averaged flow velocity, g is the acceleration of gravity, S denotes the slope of the delta top, ρ is the water density, τ_e is the effective shear stress associated with the resistances of the smooth sidewalls and of the rough bed, and t and x are a temporal and a streamwise coordinate, respectively.

The simplified version of the Vanoni–Brooks decomposition (e.g., Francalanci et al., 2008) of the shear stress into a sidewall and a bed component to estimate the effective shear stress, τ_e, implemented in the previous versions of the model is substituted with the complete Vanoni and Brooks (Vanoni, 1975) decomposition, as modified by Chiew and Parker (1994). These formulations are based on the assumption that the cross section can be decomposed into two non-interacting regions, the bed region and the sidewall region, where the mean flow velocity and the energy gradient are respectively equal to the mean flow velocity, U, and the energy gradient, S_f, of the entire cross section. The continuity equation thus takes the form

$$A = A_w + A_b, \tag{3}$$

where A is the cross-sectional area, and A_w and A_b respectively denote the area of the wall and of the bed region.

The main difference between the Chiew–Parker and the simplified Vanoni–Brooks decomposition (e.g., Francalanci et al., 2008) is related to the partition of the cross section between the smooth sidewall region and the rough bed region. The underlying assumption of the simplified Vanoni and Brooks formulation is that the boundary between the sidewall region and the bed region is a 45° straight line. In the complete formulation (Vanoni, 1975; Chiew and Parker, 1994) the areas of the sidewall and of the bed regions are computed from the flow characteristics, with a better estimate of the shear stress on the rough boundary (Chiew and Parker, 1994).

The Chiew–Parker decomposition is based on the following form of the momentum balance for the cross section:

$$\tau_e = \frac{\tau_b P_b + \tau_w P_w}{P}, \tag{4}$$

where τ_b and τ_w respectively denote the shear stresses on the rough bed and on the smooth sidewalls, and P, P_b, and P_w are the wetted perimeters of the entire cross section $(B+2H)$, the bed (B), and the sidewall region $(2H)$.

In the case of turbulent open channel flow, the shear stresses can be expressed as the product of the water density, the mean velocity squared, and a nondimensional friction coefficient C_f, $\tau = \rho C_f U^2$. Since the mean flow velocity is assumed to be the same in the bed region, in the sidewall region, and in the entire cross section, Eq. (4) can be rewritten as

$$C_{fe} = \frac{C_{fb} P_b + C_{fw} P_w}{P},\tag{5}$$

where C_{fe} is an effective nondimensional friction coefficient associated with the resistances on the sidewalls and on the bed, and C_{fb} and C_{fw} denote the nondimensional friction coefficients for the bed and the sidewall region, respectively.

Under the assumption that the Darcy–Weisbach relation can be applied to the entire cross section, to the bed, and to the sidewall region, the energy gradient is given as

$$S_f = \frac{C_{fe} U^2}{gr} = \frac{C_{fb} U^2}{gr_b} = \frac{C_{fw} U^2}{gr_w},\tag{6}$$

where r, r_b, and r_w denote the hydraulic radii (i.e., the ratios between the cross-sectional areas and the wetter perimeters) for the entire cross section, for the bed, and for the sidewall region, respectively. Recalling that the Reynolds number of the cross section is defined as $Re = rU/\nu$, with ν denoting the kinematic viscosity of the fluid, Eq. (6) can be rewritten as

$$\frac{C_{fe}}{Re} = \frac{C_{fb}}{Re_b} = \frac{C_{fw}}{Re_w},\tag{7}$$

where Re_b and Re_w are the Reynolds numbers of the bed and the sidewall region, respectively.

The unknowns in Eqs. (3), (5), and (7) are the friction coefficients, C_{fe}, C_{fb}, and C_{fw}; the area of the bed region, A_b; and the area of the wall region, A_w. Two closure relations are thus needed to solve the problem.

The first closure relation expresses C_{fb} as a function of the hydraulic radius of the bed region, r_b, and of the roughness height of the delta top, k_s, as

$$C_{fb}^{-1/2} = 8.1 \left(\frac{r_b}{k_s}\right)^{1/6}.\tag{8}$$

Figure 3 in Viparelli et al. (2010a) shows that this bed resistance model is appropriate to describe flow resistances in the bed region with Ferrer-Boix et al. (2013) sediment mixture and flow conditions if (i) the roughness height is assumed to be equal to $1.5 D_{s90}$ and (ii) the active layer thickness is assumed to be equal to D_{s90}. Here D_{s90} denotes the diameter of the sediment stored in the active layer such that 90 % of the active layer sediment is finer.

The second closure is the relation for hydraulically smooth walls given by Vanoni (1975) to compute the Darcy–Weisbach sidewall friction coefficient $f_w = 8 C_{fw}$ as a function of the Reynolds number of the wall region

$$\frac{1}{\sqrt{f_w}} = 0.86 \ln\left(Re_w \sqrt{f_w}\right) - 0.8.\tag{9}$$

Equations (1) and (2) are reduced to the classical backwater form using (i) the quasi-steady approximation (De Vries, 1965) to drop the time dependence as well as (ii) the definition of effective shear stress as a product of water density, mean flow velocity square, and effective friction coefficient, $\tau_e = \rho C_{fe} U^2$. The backwater equation thus takes the form

$$\frac{\partial H}{\partial x} = \frac{S - C_{fe} Fr^2}{1 - Fr^2},\tag{10}$$

where Fr denotes the Froude number defined as $U/(gH)^{0.5}$. In the numerical run described below, Eq. (10) is integrated in the upstream direction with the downstream boundary condition $\xi = \xi_b = 0.26$ m, with ξ denoting the water surface elevation above the datum and the subscript b indicating the downstream end of the delta top, i.e., the brinkpoint (see Fig. 3).

3.2 Calculation of sediment transport and deposition on the delta top

Bedload sediment transport on the delta top is modeled with the version of the Ashida–Michiue bedload relation of Viparelli et al. (2010b). This grain-size-specific bedload relation is derived for mobile bed equilibrium conditions obtained in the same laboratory flume and with the same sediment mixture of Ferrer-Boix et al. (2013).

During the Viparelli et al. (2010b) experiment the flume was operated in sediment-recirculating mode, meaning that the sediment collected in the sediment trap was recirculated to the upstream end of the flume. Thus, during the condition of nonequilibrium, the sediment input rate and its grain size distribution were not constant in time (Viparelli et al., 2010a, b).

In addition, in a sediment-recirculating flume, the total volume of sediment in the system does not change in time, and so the grain size stratigraphy of the bed deposit and the equilibrium conditions are dependent on the initial experimental conditions (Parker and Wilcock, 1993). Ferrer-Boix et al. (2013) operated the laboratory flume in sediment feed mode, i.e., with constant grain-size-specific sediment input rate, and with a volume of sediment in the flume that increased in time.

In a sediment feed flume the conditions of mobile bed equilibrium are independent of the initial condition of the experiment and are dictated by the upstream input of water and sediment only (Parker and Wilcock, 1993). It is thus reasonable to expect that disequilibrium bedload transport conditions in a sediment feed flume, such as those of the Ferrer-Boix et al. (2013) experiment, are somewhat different from those observed in a sediment-recirculating flume (Viparelli et al., 2010a).

As shown in Fig. 2, the parent material is divided in M ($M = 9$ for the numerical run presented herein) grain size ranges with characteristic diameters D_i. The volumetric bedload transport rate per unit channel width, q_{bT}, is equal to the sum of the volumetric bedload transport rates per unit width in the M grain size ranges, q_{bi},

$$q_{bT} = \sum_{i=1}^{M} q_{bi}. \tag{11}$$

Grain-size-specific nondimensional volumetric bedload transport rates per unit width, q_{bi}^*, (Einstein parameters) are defined as (Parker, 2008)

$$q_{bi}^* = \frac{q_{bi}}{F_i \sqrt{R g D_i} D_i}, \tag{12}$$

where F_i represents the fraction of active layer sediment in the generic (ith) grain size range, and R denotes the submerged specific gravity of the parent material, i.e., $(\rho_s - \rho)/\rho$, with ρ_s denoting the density of the sediment. $R = 1.58$ for the experiment discussed herein.

The grain-size-specific Einstein parameters are computed as

$$q_{bi}^* = 17\beta \left(\tau_{bi}^* - \tau_{ci}^*\right) \left(\sqrt{\tau_{bi}^*} - \sqrt{\tau_{ci}^*}\right), \tag{13}$$

where β is an adjustment coefficient equal to 0.27 for the considered experimental conditions, τ_{bi}^* is the grain-size-specific nondimensional shear stress on the bed region (Shields number), and τ_{ci}^* is its reference value for significant bedload transport of sediment in the generic (ith) grain size range.

The grain-size-specific Shields number is defined as (Parker, 2008)

$$\tau_{bi}^* = \frac{\tau_b}{\rho R g D_i}; \tag{14}$$

its reference value for significant bedload transport is estimated with the hiding/exposure function derived by Viparelli et al. (2010b) that is valid for the Ferrer-Boix et al. (2013) experimental conditions

$$\frac{\tau_{ci}^*}{\tau_{scg}^*} = \begin{cases} \left(\dfrac{D_i}{D_{sg}}\right)^{-0.98} & \text{for} \quad \dfrac{D_i}{D_{sg}} \leq 1 \\ \left(\dfrac{D_i}{D_{sg}}\right)^{-0.68} & \text{for} \quad \dfrac{D_i}{D_{sg}} > 1, \end{cases} \tag{15}$$

where D_{sg} is the geometric mean diameter of the active layer sediment, and τ_{scg}^* represents the reference Shields number for significant motion in the case of uniform sediment. τ_{scg}^* is equal to 0.043.

The Exner equation of conservation of total (i.e., summed over all the grain sizes) sediment mass conservation takes the form (Parker, 2004)

$$\left(1 - \lambda_p\right) \frac{\partial \eta}{\partial t} = -\frac{\partial q_{bT}}{\partial x}, \tag{16}$$

where η denotes the elevation of the delta top above the datum (Fig. 3) and λ_p is the bulk bed porosity, equal to 0.35 in

the numerical run discussed below (Viparelli et al., 2010a). Equation (16) is solved to compute the aggradation rate of the delta top and thus update the longitudinal profile of the Gilbert delta top at each time step.

The grain-size-specific equation of conservation of sediment in the generic (ith) grain size range takes the form (e.g., Hirano, 1971; Parker, 2004)

$$\left(1 - \lambda_p\right) \left[L_a \frac{\partial F_i}{\partial t} + (F_i - f_{Ii}) \frac{\partial L_a}{\partial t} + f_{Ii} \frac{\partial \eta}{\partial t}\right] = -\frac{\partial q_{bi}}{\partial x}, \tag{17}$$

where L_a denotes the thickness of the active layer, F_i is the fraction of sediment in the ith grain size range in the active layer, and f_{Ii} represents the fraction of sediment in the generic grain size range exchanged between the active layer and the deposit during channel bed aggradation or degradation.

In the case of delta top erosion, the grain size distribution of the sediment exchanged between the emplaced deposit and the active layer, f_{Ii}, is equal to the grain size distribution of the deposit. However in the case of an aggrading delta top, the grain size distribution of the sediment transferred to the deposit is assumed to be a weighted average between the grain size distribution of the active layer and of the bedload (Hoey and Ferguson, 1994),

$$f_{Ii} = \alpha F_i + (1 - \alpha) p_i, \tag{18}$$

where p_i denotes the fraction of sediment in the generic grain size range in the bedload, i.e., $p_i = q_{bi}/q_{bT}$. Toro-Escobar et al. (1996) discuss the reason why the parameter α should be greater than 0 and smaller than 1. When $\alpha = 0$ the grain size distribution of the sediment transferred to the substrate during channel bed aggradation is equal to the grain size distribution of the bedload, and the downstream fining observed in gravel-bed rivers cannot be modeled. However, if $\alpha = 1$ the surface material is directly transferred to the substrate during channel bed aggradation, and the formation of the coarse pavement observed in gravel-bed rivers that regulates the mobility of particles differing in size (Parker et al., 1982; Parker and Klingeman, 1982) cannot be modeled.

To model the Ferrer-Boix et al. (2013) Gilbert delta experiment with Eqs. (8), (13), and (15), the parameter α is equal to 0.2 (Viparelli et al., 2010a).

Equation (17) is solved to compute the time rate of change of F_i and thus update the grain size distribution of the active layer at each time step.

3.3 Calculation of sediment transport and deposition on the delta front

The bedload transport rate that reaches the brinkpoint is deposited as grain fall deposit on the upper part of the delta front. Thus, the overall grain size distribution of the grain fall deposit is equal to the grain size distribution of the bedload at the brinkpoint at the specific time. When the static angle of repose of the sediment is exceeded, a grain flow is initiated, and sediment is distributed over the delta front. In particular, coarse sediment is deposited more abundantly in the lowermost part of the front and finer sediment is trapped more abundantly in the upper portion of the lee face.

Vertical sorting of sediment on the lee face of the delta front is modeled with the lee face model of Blom et al. (2013). In particular, it is described in terms of a sorting function, ω_i, defined as

$$\omega_i = \frac{p_{i,\mathrm{b}}}{f_{si}}, \tag{19}$$

where f_{si} represents the volume fraction content of sediment in the ith grain size range on the slip face at elevation z above the datum, and $p_{i,\mathrm{b}}$ represents the volume fraction content of sediment in the generic grain size range in the bedload at the brinkpoint. In the Blom et al. (2013) the sorting function is assumed to linearly vary with the nondimensional elevation $z^* = (z - \eta_{\mathrm{ba}})/\Delta$, where η_{ba} is the average elevation of the slip face and Δ is the height of the slip face deposit relative to a vertical coordinate,

$$\omega_i = 1 + \delta_i z^*. \tag{20}$$

δ_i is the lee sorting parameter, defined as

$$\delta_i = 2 p_{i,\mathrm{b}}^{0.5} \frac{\phi'_{\mathrm{reli}}}{\sigma_{\mathrm{qbb}}^{0.7}} \left(\tau^*_{\mathrm{bbsg}} \right)^{-0.3}, \tag{21}$$

with σ_{qbb} denoting the standard deviation on the sedimentological φ scale of the bedload at the brinkpoint, and τ^*_{bbsg} representing the Shields parameter at the brinkpoint evaluated with the geometric mean diameter of the active layer, D_{sg}. φ'_{reli} is the adjusted relative arithmetic grain size defined as

$$\phi'_{\mathrm{reli}} = \phi_i - \phi'_{\mathrm{mtop}}, \tag{22}$$

where φ_i is the characteristic grain size D_i on the φ scale, $\varphi_i = -\log_2 D_i$, and φ'_{mtop} is the adjusted arithmetic mean grain size of the lee face deposit

$$\phi'_{\mathrm{mtop}} = \frac{\sum_{i=1}^{M} \phi_i p_{i,\mathrm{b}}^{1.5}}{\sum_{i=1}^{M} p_{i,\mathrm{b}}^{1.5}}. \tag{23}$$

3.4   Grids for the storage of the stratigraphy

The delta growth problem is characterized by a moving boundary at the downstream end of the delta top, the brinkpoint. Thus, Eqs. (10), (16), and (17) could be integrated in a moving-boundary coordinate system, in which the streamwise coordinate, x, is made nondimensional with the coordinate of the brinkpoint, x_{b} (Swenson et al., 2000). In this

moving-boundary system the distance between the computational nodes does not change in time but, due to the movement of the brinkpoint, it varies in the dimensioned coordinate system x (e.g., Wright and Parker, 2005a).

In an active layer model, the moving-boundary transformation requires cumbersome interpolations of the size distributions associated with each computational node. The grain size distributions associated with each node change for (i) fluxes of sediment in the streamwise direction due to the changing dimensioned spatial distance between the computational nodes Δx, as well as for (ii) vertical fluxes of sediment due to aggradation and degradation of the bed deposit. The streamwise fluxes of sediment in the active layer and in the bed deposit are estimated by interpolating the grain size distributions associated with the computational nodes, with a consequent loss of stratigraphic information.

Since the ultimate scope of the numerical model is to store and access the stratigraphy emplaced by the prograding delta, the governing equations are not solved in the moving-boundary coordinate system. A grid with a fixed distance between the computational nodes, Δx, is used to model sediment transport and deposition upstream of the brinkpoint (Eke et al., 2011; Viparelli et al., 2011a). The distance between the brinkpoint and the last grid node is denoted as $\Delta x_{\mathrm{brink}}$. As the brinkpoint moves downstream, $\Delta x_{\mathrm{brink}}$ increases. When $\Delta x_{\mathrm{brink}} > \Delta x$, a new grid node is added to the fixed grid, as shown in Fig. 4. As in the previous figures, the delta for Fig. 4 prograde on a horizontal basement; however the formulation holds for an arbitrary sloping basement with slope S_{b}.

The migration rate of the brinkpoint, c_{b}, is computed under the assumptions that (i) all the sediment is trapped on the delta front and (ii) the lee face has a constant slope S_{l} (e.g., Wright and Parker, 2005a),

$$c_{\mathrm{b}} = \frac{1}{S_{\mathrm{l}} - S|_{x = x_{\mathrm{b}}}} \left[\frac{q_{\mathrm{bbT}}}{(1 - \lambda_{\mathrm{p}})(x_{\mathrm{t}} - x_{\mathrm{b}})} - \frac{\partial \eta_{\mathrm{b}}}{\partial t} \right], \tag{24}$$

where q_{bbT} denotes the total bedload transport rate at the brinkpoint, η_{b} is the elevation of the brinkpoint above the datum, and x_{b} and x_{t} respectively denote the streamwise coordinates of the brinkpoint and of the delta toe.

Equation (24) is derived by integrating the Exner equation, Eq. (16), on the delta front. As sediment is deposited on the delta front, the delta toe migrates downstream with velocity c_{t}. The migration rate of the delta toe is computed by imposing the continuity of bed elevation, i.e., the elevation of the lowermost point of the delta front must be equal to the elevation of the basement. The continuity condition for the movement of the delta toe takes the form (e.g., Wright and Parker, 2005a)

$$c_{\mathrm{t}} = \frac{1}{S_{\mathrm{l}} - S_{\mathrm{b}}} \left[(S_{\mathrm{l}} - S|_{x - x_{\mathrm{b}}}) c_{\mathrm{b}} + \frac{\partial \eta_{\mathrm{b}}}{\partial t} \right], \tag{25}$$

where S_{b} denotes the basement slope.

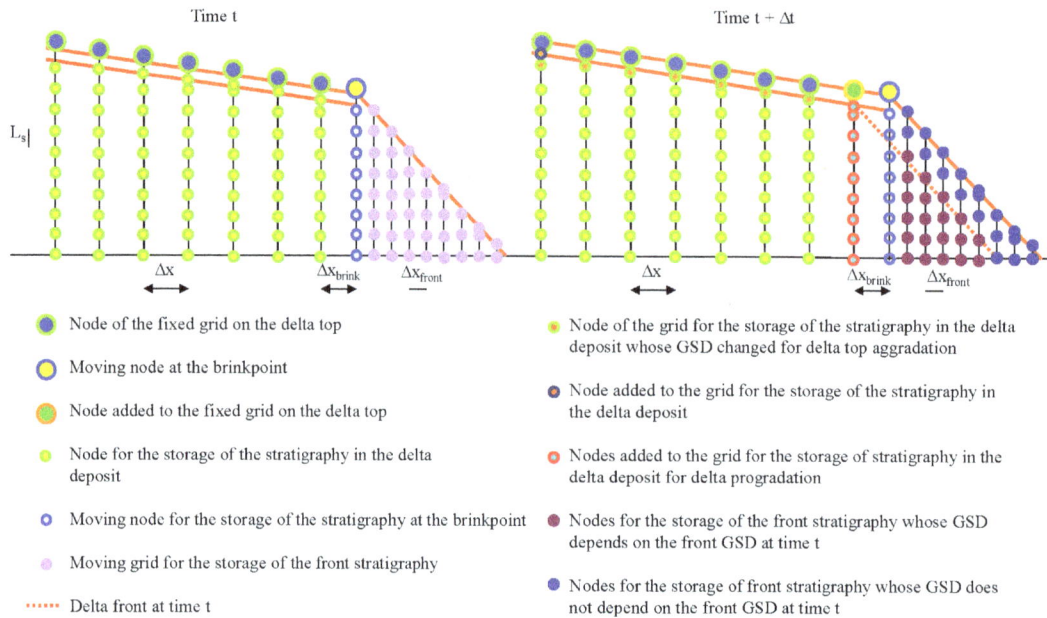

Figure 4. Model grids. GSD stands for grain size distribution.

Due to the assumption of the constant slope of the delta front, Eqs. (24) and (25) are solved to update the streamwise coordinates of the brinkpoint and of the delta toe and thus the longitudinal profile of the delta front.

The bookkeeping procedure of stratigraphy in the delta deposit, i.e., upstream of the brinkpoint point, is that of Viparelli et al. (2010a). The bed is divided in two parts – the relatively thin and well-mixed (i.e., no vertical variation of the grain size distribution) active layer and the substrate, whose grain size distribution can vary in the vertical direction. The grain size distribution of the substrate is stored in the grid represented in Fig. 4 at time t. The substrate deposit is divided into horizontal well-mixed layers. The lowermost grid node, node 1, is located on the datum; the uppermost grid node, node N, is at the active-layer–substrate interface, i.e., at elevation $\eta - L_a$ above the datum. The grain size distribution associated with the grid node j is representative of the layer bounded by the grid nodes j and $j - 1$. The vertical distance between the consecutive grid nodes from node 1 to node $N - 1$ is L_s, equal to 2 cm in the numerical run presented herein.

The vertical distance between node $N - 1$ and node N is $\Delta z < L_s$. As the delta top aggrades, sediment is stored in the topmost part of the substrate and Δz increases. When Δz becomes greater than L_s, a new grid node is added to the grid (see Fig. 4 at time $t + \Delta t$). The distance between node N and node $N - 1$ is equal to L_s and the new node $N + 1$ is added to the grid at the active-layer–substrate interface. The grain size distribution of the material stored in each layer is a weighted average over the thicknesses of the topmost layer of the grid and of the sediment deposited at each time step. The grain

size distribution of the sediment transferred to the substrate during channel bed aggradation is f_{Ii}, given by Eq. (18).

The stratigraphy of the delta front is stored in a moving grid (Fig. 4). The streamwise distance between the N_f grid nodes is equal to $\Delta x_{\text{front}} = (x_t - x_b)/(N_f - 1)$, which may vary in time due to the different migration rates of the brinkpoint and of the delta toe. Horizontal fluxes of sediment from the front to the fixed grid one time step later and between the nodes of the moving gird are estimated by interpolating the grain size distributions of the sediment stored at the same elevation above the datum. Vertical fluxes of sediment are computed to transfer the newly deposited sediment to the existing front substrate with the same averaging procedure implemented for the delta deposit, as shown in Fig. 4 (Viparelli et al., 2011a).

4   Results and discussion

The numerical simulation of the Ferrer-Boix et al. (2013) experiment is performed with a fixed distance between the computational nodes on the delta top, Δx, of 0.1 m; a temporal interval, Δt, of 10 s; and 40 moving grid nodes on the delta front.

It is important to note here that the parameters of the total and grain-size-based sediment conservation models, i.e., α, k_s, β, and L_a, are those of the Viparelli et al. (2010b) experiments. The parameters of the lee-face-sorting model, i.e., exponents and coefficient of Eq. (21), are those of the Blom et al. (2013) analysis of laboratory and field data on vertical sorting on the lee face of dunes and Gilbert deltas. Thus, it

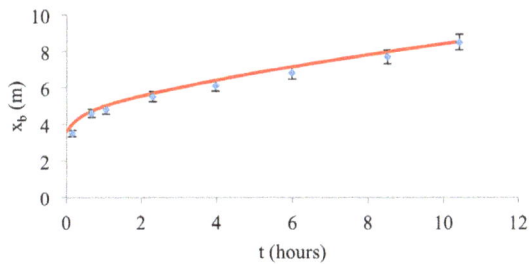

Figure 5. Comparison between measured (diamonds) and predicted (red line) brinkpoint position. Error bars denote ±5 % of the brinkpoint position.

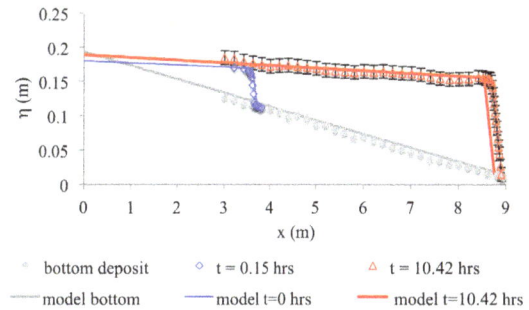

Figure 6. Comparison between measured (diamonds and triangles) and numerical (lines) longitudinal profiles. The profile of the bottom deposit (grey) is a model boundary condition. The delta profile at $t = 0.15$ h (blue) is the model initial condition. The delta profile at $t = 10.42$ h (red) is a model result. Error bars of the measured profile at $t = 10.42$ h denote ±0.01 m.

was not necessary to tune or calibrate any model parameter to validate the numerical model against experimental results.

The validation of the delta growth sub-model (type 1) is performed by comparing (i) the brinkpoint location, x_b, in time and (ii) the longitudinal profile of delta elevation at the end of the experiment. Measured and numerical temporal variations of brinkpoint location are represented in Fig. 5, where the vertical bars denote a ±5 % error. The comparison between longitudinal profiles of bed elevation is reported in Fig. 6, where the elevation of the initial deposit is represented in grey, the initial condition for the numerical run is in blue, and the final longitudinal profile is in red. Error bars in Fig. 6 show that the numerical delta profile at the end of the numerical run approximates the experimental data within a ±1 cm interval. Figures 5 and 6 show that the total (i.e., summed over all the grain sizes) sediment mass conservation model is able to reasonably capture the temporal evolution of the longitudinal profile and thus the total bedload transport rates on the delta top. In addition, Fig. 6 shows that the bed resistance model reasonably reproduces the experimental conditions, since the slopes of the numerical and the experimental delta top are reasonably similar.

The validation of the grain-size-specific mass conservation model for the delta top (type 2) is presented in Fig. 7, where the error bars denote a ±5 % interval around the measured points. Due to the lack of experimental data on the grain size distribution of the active layer, the comparison is done in terms of measured grain size distributions of the topmost 2 cm of the experimental delta top (diamonds in Fig. 7) and the average grain size distribution of the topmost portion of the numerical deposit, i.e., the active layer and the two uppermost layers of the grid for the storage of grain size stratigraphy (red line in Fig. 7). The numerical results are averaged over a volume thicker than the experimental samples to have a relative robust estimate of the grain size distribution in a well-mixed layer characterizing the grain size distribution of the active layer and of the topmost portion of the substrate.

The comparison in Fig. 7 shows an overall reasonable agreement between measurements and numerical predictions, which is rarely obtained with a 1-D active layer model. In the sampling sections at 3.5, 6.5 ,and 7.5 m from the

entrance of the flume the model underestimates the fraction of sediment in the 1.53 mm size range. This is balanced in the 3.5 m section by a slight overestimation of the sediment in the 5.02 and 7.74 mm ranges, and by a more severe overestimation of the sediment in the 2.83 mm range in the sections 6.5 and 7.5 m downstream of the flume entrance. We suspect that the differences between the numerical results and the experimental data are related to the grain-size-specific sediment transport model, i.e., Eqs. (13) and (15).

As mentioned above, the Viparelli et al. (2010b) model is based on sediment-recirculating flume experiments. In this experimental setting, mobile bed equilibrium is reached through a rotation of the longitudinal profile around the center of the flume. In other words, only the topmost part of the deposit is reworked and the grain size distribution of the transported sediment is constrained by the grain size distribution of the mobilized sediment (Viparelli et al., 2010a). The flume in the Ferrer-Boix et al. (2013) experiment is operated in sediment feed mode, i.e. with a constant input rate of parent material. It is thus reasonable to expect that the sediment mobility in the Ferrer-Boix et al. (2013) experiment is slightly different than in the Viparelli et al. (2010b) experiments. Unfortunately no experimental data are available to further validate the bedload transport model.

The numerical stratigraphy of the bed deposit is represented in Fig. 8 (with $\Delta x = 5$ cm and $L_s = 1$ cm for illustration purposes only), where the dots represent the grid nodes for the storage of stratigraphy and the color scale represents the geometric mean diameter of the substrate layer. The blue oval indicates the portion of the delta deposit whose stratigraphy is affected by the model initial condition, i.e., a well-mixed deposit of parent material with the longitudinal profile represented in Fig. 6. The stratigraphy of this initial profile does not significantly change in time during the numerical runs because the delta front migrates downstream and only a thin layer of sediment is deposited on top of the initial deposit. The black line in Fig. 8 represents the elevation of the

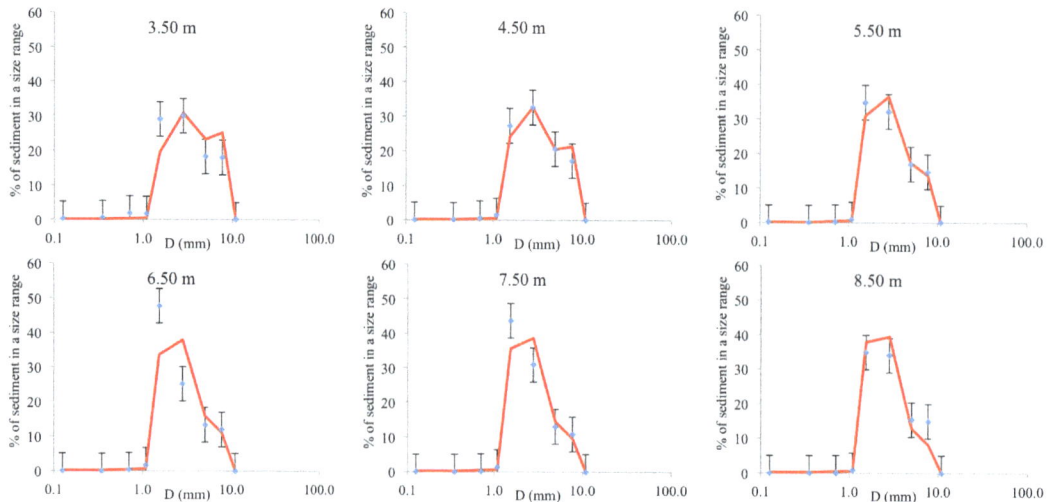

Figure 7. Comparison between predicted and measured grain size distributions of the topmost part of the delta top. The diamonds represent the measurements and the red line is the numerical result. Error bars denote a ±5 % error.

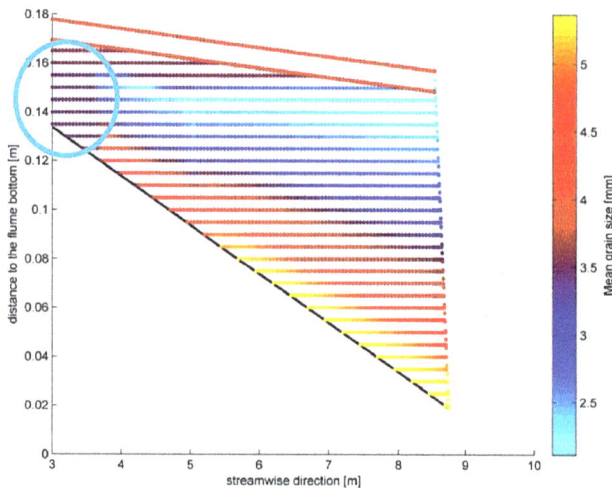

Figure 8. Numerical stratigraphy of the deposit. The dots represent the grid nodes, and the color scale is associated with the geometric mean diameter in millimeters of the substrate layers. The black line represents the top of the initial layer of parent material. The blue oval indicates the stratigraphy affected by the initial conditions.

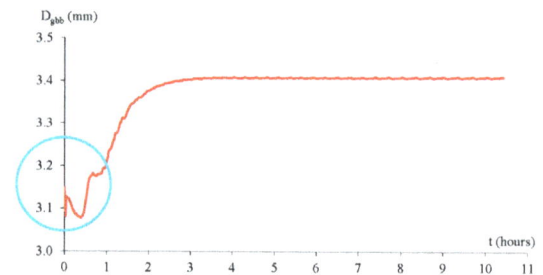

Figure 9. Temporal variation of the geometric mean diameter of the bedload at the brinkpoint. The blue oval indicates the initial numerical adjustment of the model, mostly related to the development of a coarse mobile active layer.

initial layer of parent material placed on the bottom of the flume. The two lines of red dots in the upper part of the delta top denote the active layer thickness.

The color scheme of Fig. 8 shows that the model is able to reproduce the upward-fining profile emplaced by the downstream-migrating lee face. A closer look at the figure reveals that the delta top deposit has a geometric mean diameter similar to the parent material and finer than the active layer, as observed in gravel bed rivers (e.g., Viparelli et al., 2011b). Further, as observed by Ferrer-Boix et al. (2013) during the experimental work, the sediment stored in the

lowermost part of the front deposit appears to become coarser in the downstream direction.

Figure 9 shows the temporal variation of the geometric mean diameter of the bedload at the brinkpoint, D_{gbb}. After an initial adjustment due to the development of the mobile armor on the initial deposit, D_{gbb} and thus the grain size distribution of the bedload at the brinkpoint remain reasonably constant in time. Thus, the observed downstream coarsening cannot be the result of an increasingly coarser bedload transport rate at the brinkpoint in time. Instead, we interpret the apparent downstream coarsening as the result of the increasing delta front elevation. As the Gilbert delta progrades on the steep basement, the front becomes higher, and there is more space to sort the bedload material that reaches the brinkpoint, as modeled with Eq. (20). Numerical experiments are currently in progress to investigate whether a similar downstream coarsening can be driven by relative base-level rise.

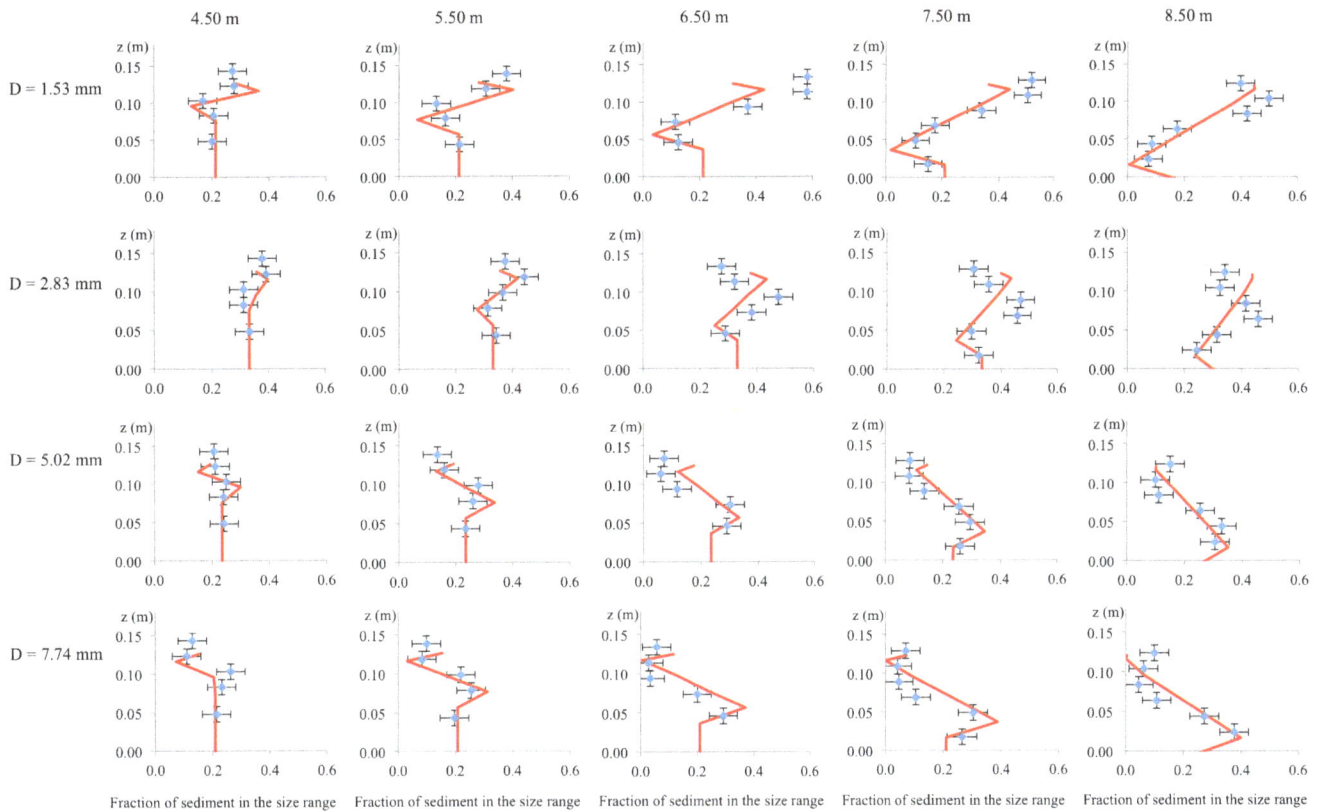

Figure 10. Comparison between measured and numerical grain size distribution of the front deposit. The diamonds are the experimental data, and the lines are the numerical predictions. Horizontal error bars denote a ±5 % error. The vertical elevation above the datum, z, of the diamonds corresponds to the elevation of the center of the sample. The vertical error bars at ±1 cm denote the thickness of the sampled layer.

The blue ovals in Figs. 8 and 9 identify the area in which the bedload transport rate at the brinkpoint is affected by the initial model condition, i.e., the development of a coarse active layer on the unarmored initial delta of parent material.

The lee-face-sorting model (type 3) is validated by comparing experimental and numerical grain size distributions of the delta front deposit in the cross sections at 4.5, 5.5, 6.5, 7.5 ,and 8.5 m from the entrance of the flume. Data collected in the measuring section at 3.5 m are not used in the comparison because, as shown in Figs. 8 and 9, the numerical results are affected by the initial condition.

The comparison between experimental and numerical data is represented in Fig. 10 by vertical profiles of sediment fractions in the characteristics grain size ranges 1.53, 2.83, 5.02, and 7.74 mm. The diamonds represent the experimental data, the red lines are the model results, and the horizontal error bars denote a ±5 % error. The vertical elevation of the diamonds, z, corresponds to the elevation of the center of each sampling layer above the datum, and the vertical error bars identify the thickness of the sampling layer, i.e., ±1 cm.

The comparison in Fig. 10 shows that – notwithstanding the uncertainties related to the grain-size-specific bedload transport relation, and thus the grain size distribution of the bedload passing the brinkpoint – the model is able to

reasonably capture the overall grain size distribution of the delta front deposit. Significant differences between the fractions of fine sediment, i.e., 1.52 and 2.83 mm, stored in the deposit in the 6.5 and 7.5 m positions confirm what we have previously observed for the grain size distribution of the delta top deposit, i.e., that the bedload transport model may not always be able to properly reproduce the transport of the finer components of the sediment mixture.

The bedload transport model, i.e., the predicted grain size distribution of the sediment at the brinkpoint, is certainly one of the major sources of error in the prediction of the grain size distribution of the delta front deposit. An additional source of error can be hidden in the lee-face-sorting model. As the delta front migrates on a 2 % sloped basement, the delta front height increases (see Fig. 6). The Blom et al. (2013) lee-face-sorting model is a linear model, Eq. (20), in the nondimensional elevation z^*,; thus nonlinear effects due to an increasing front height are not explicitly accounted for. The study of vertical sorting on an increasingly high lee face goes well beyond the scope of this paper, but we suspect that it may partially explain the differences between the numerical and the experimental stratigraphy of the considered Gilbert delta.

5 Conclusion and future work

The comparison between numerical and experimental delta stratigraphy was conducted for the case of a Gilbert delta prograding on a sloping basement into standing water. These experimental conditions are appropriate for the validation of a model of delta morphodynamics because the stratigraphy emplaced by the migrating delta front (i.e., the lee face) is entirely stored within the deposit. In other words, a train of migrating bedforms, such as bars or dunes, does not rework the lee face deposit.

The comparison was done in three steps. First, the flow and the total (i.e., summed over all the grain sizes) sediment conservation models were validated against profiles of channel bed elevation and migration rates of the brinkpoint. This comparison shows the following results:

1. Numerical predictions of the streamwise coordinate of the brinkpoint were in reasonable agreement with the experimental measurements (Fig. 5). Since the migration rate of the brinkpoint was computed with an integral shock condition of the equation of total sediment conservation, the model was able to reasonably predict the total bedload transport rates at the brinkpoint.

2. Measured and predicted slopes of the delta top were reasonably similar (Fig. 6); thus frictional resistances on the channel bed were properly captured by Eqs. (3), (5), and (7)–(9).

The results of the grain-size-specific sediment conservation model on the delta top were validated against the grain size distributions of the topmost part of the delta top deposit. The results of Fig. 7 show that the model was able to reasonably capture the overall grain size distribution of the delta top, but it tended to underestimate the fractions of fine material deposited in the topmost part of the experimental Gilbert delta. This was probably due to a failure of the bedload transport model, based on sediment-recirculating flume experiments and applied to simulate a sediment feed flume experiment. Laboratory measurements on the grain size distribution of the active layer were unfortunately not available to further validate the bedload transport model.

Finally, the numerical stratigraphy emplaced by the delta front was compared with the laboratory data in Fig. 10. The model reasonably reproduced the upward fining observed in the laboratory, but due to the errors in the predictions of the grain size distributions at the brinkpoint, the differences between the numerical predictions and the experimental measurements were sometimes larger than 5 %.

The results presented herein represent the first step in the validation of the numerical model. Further validation against laboratory experiments is currently in progress to study the stratigraphy of a Gilbert delta under different scenarios of base level change. In the near future we plan to modify the code to not only store but also to access the stratigraphy emplaced by prograding deltas and model (1) Gilbert delta progradation at the field scale and (2) the formation of stacked Gilbert delta complexes at the laboratory and field scale.

Appendix A: Model sensitivity to the bedload transport formulation

Due to the uncertainties related to the grain-size-specific bedload transport formulation discussed in the previous section, in this section we explore the model sensitivity to the bedload transport relation itself. As illustrated in Sect. 3, a grain-size-specific bedload transport relation, in general, consists of two elements: the relation to compute the volumetric bedload transport rate in each grain size range, e.g., Eq. (13), and the hiding/exposure function to estimate the threshold for significant bedload transport of sediment particles in each grain size range, e.g., Eq. (15).

Noting that the Viparelli et al. (2010b) bedload transport relation is based on the surface-based version of the Ashida and Michiue model (as in Parker, 2004), we repeat the numerical simulation of the Ferrer-Boix et al. (2013) experiment by replacing the Viparelli et al. (2010b) formulation with the surface-based Ashida and Michiue model, which is not appropriate to simulate the Ferrer-Boix et al. (2013) experiment, as further discussed in this section.

To ensure the numerical stability of the model, the Ashida and Michiue simulation is performed with a distance between the computational delta top nodes of 0.1 m, a temporal increment of 0.05 s, and an active layer thickness of $1.5 D_{s90}$. All the other model parameters were equal to those used in the simulation of the Ferrer-Boix et al. (2013) experiment presented in the paper.

The Viparelli et al. (2010b) relation to compute the grain-size-based volumetric bedload transport rate has the same form of the Ashida and Michiue relation as in Parker (2004). In the original Ashida and Michiue formulation the parameter β in Eq. (13) is equal to 1, while in the Viparelli et al. (2010b) formulation it is set equal to 0.27. The hiding/exposure function of Viparelli et al. (2010b) is given in Eq. (15), while the Ashida and Michiue hiding/exposure function is a modified version of the Egiazaroff (1965) relation of the form

$$\frac{\tau_{ci}^*}{\tau_{scg}^*} = \begin{cases} 0.843 \left(\frac{D_i}{D_{sg}} \right)^{-1} & \text{for} \quad \frac{D_i}{D_{sg}} \leq 0.4 \\ \left[\dfrac{\log(19)}{\log\left(19 \frac{D_i}{D_{sg}} \right)} \right]^2 & \text{for} \quad \frac{D_i}{D_{sg}} > 0.4, \end{cases} \tag{A1}$$

where τ_{scg}^* denotes the threshold for significant bedload transport of uniform sediment equal to 0.05. In the Viparelli et al. (2010b) formulation τ_{scg}^* is equal to 0.043.

The temporal evolution of the longitudinal profile of the deltaic deposit is represented in Fig. A1, where lines with different colors represent the delta at different times, as indicated in the legend on the bottom. Due to the unarmored condition of the initial bed deposit, significant erosion occurs on the delta top at the beginning of the numerical run while the mobile armor develops. In this initial erosional phase, the large divergence in bedload transport rate on the delta top results in the formation of two small downstream-prograding

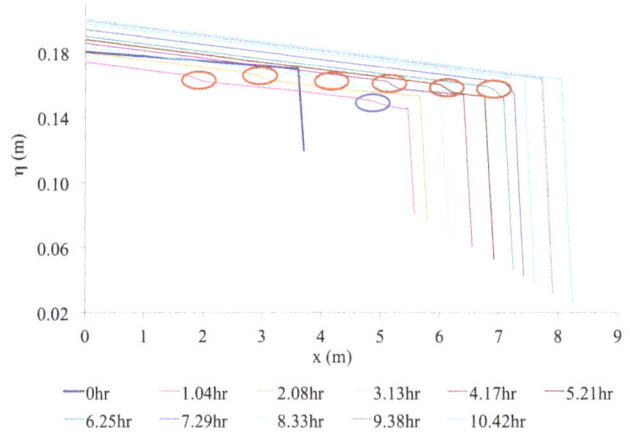

Figure A1. Temporal evolution of the longitudinal profile of delta elevation. The blue and red blue ovals respectively denote the location of the first and second small delta fronts. Time = 0 h denotes the model initial condition. The results are obtained using the Ashida and Michiue grain-size-based bedload relation, which is not appropriate for modeling the Ferrer-Boix et al. (2013) experiment.

deltas on the delta top (blue and red ovals in Fig. A1), which were not observed in the Ferrer-Boix et al. (2013) laboratory experiment. The first front (blue oval in Fig. A1) reaches the brinkpoint in the first simulated 2 h. After 7–8 h, the second small front (red oval in Fig. A1) reaches the brinkpoint. When a small front reaches the brinkpoint, the two fronts are merged in a thicker Gilbert delta.

The spatial variation of grain size distribution in the deposit is presented in Fig. A2 in terms of geometric mean diameter of the deposit. Although the model does not capture vertical sorting on the small delta migrating on the initial deposit, the figure clearly shows that most of the coarse sediment is trapped in the small delta deposit. The finer sediment reaches the initial delta top, is transported downstream, and is responsible for the progradation of the thicker and lowermost delta.

When a small delta front reaches the brinkpoint of the underlying Gilbert delta, after a phase of rapid aggradation of the delta top, in which lenses of very fine sediment characterize the front stratigraphy, as indicated by the orange ovals in Fig. A2, the coarser material reaches the shoreline. The arrival of coarser material at the shoreline results in a coarser lowermost part of the delta front deposit compared to the stacked deltas case.

The comparison between Figs. 8 and A2 shows that the morphodynamic evolution of the deltaic deposit obtained with two different grain-size-based bedload formulations is significantly different. When the Viparelli et al. (2010b) formulation is implemented a single Gilbert delta migrates downstream in the laboratory flume. In the Ashida and Michiue simulation, two small delta fronts form on the initial deposit, prograde, and reach the brinkpoint of the underlying delta deposit. The formation of stacked deltas results

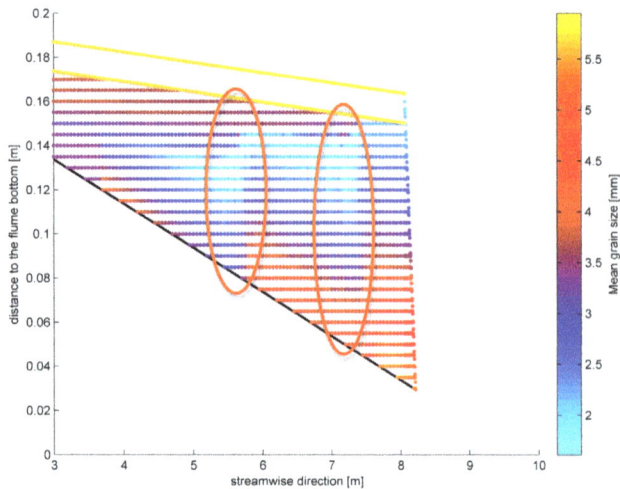

Figure A2. Numerical stratigraphy of the deposit for the Ashida and Michiue simulation. The dots represent the grid nodes, and the color scale associated with the geometric mean diameter in millimeters of the substrate layers is the same as in Fig. 8. The black line represents the top of the initial layer of parent material. The orange ovals indicate the stratigraphic record emplaced when the stacked small deltas reach the brinkpoint of the initial deltaic deposit. The geometric mean diameter of the parent material is 3.43 mm. The results are obtained using the Ashida and Michiue grain-size-based bedload relation, which is not appropriate for modeling the Ferrer-Boix et al. (2013) experiment.

in the emplacement of finer delta front deposits in the upstream part of the system compared to the case of a single prograding Gilbert delta. More than a sensitivity analysis on the bedload relation, this exercise becomes the comparison between the stratigraphy emplaced by a single Gilbert delta and by stacked deltas, which is an interesting problem that we hope to study in the relatively near future.

Acknowledgements. The authors sincerely thank Gary Parker for his help during the experimental and modeling work; Victor Chavarrias for the discussion and the help with the figures; and Maarten Kleinhans, Raleigh Martin, and Francois Metivier for their reviews and comments. Rachel Kuprenas was supported through NSF grant EAR-1250641.

Edited by: F. Metivier

References

Blom, A.: Different approaches to handling vertical and streamwise sorting in modeling river morphodynamics, Water Resour. Res., 44, W03415, doi:10.1029/2006WR005474, 2008.

Blom, A. and Kleinhans, M.: Modeling sorting over the lee face of individual bed forms, in: Proceedings River Flow 2006, edited by: Ferreira, R. M. L., Alves, E. C. T. L., Leal, J. G. A. B., and Cardoso, A. H., Taylor & Francis Group, London, 807–816, 2006.

Blom, A. and Parker, G.: Vertical sorting and the morphodynamis of bed form-dominated rivers: A modeling framework, J. Geophys. Res., 109, F02007, doi:10.1029/2003JF000069, 2004.

Blom, A., Ribberink, J. S., and de Vriend, H. J.: Vertical sorting in bed forms: Flume experiments with a natural and a trimodal sediment mixture, Water Resour. Res., 39, 1025, doi:10.1029/2001WR001088, 2003.

Blom, A., Parker, G., Ribberink, J. S., and de Vriend, H. J.: Vertical sorting and the morphodynamics of bed-form-dominated rivers: An equilibrium sorting model, J. Geophys. Res., 111, F01006, doi:10.1029/2004JF000175, 2006.

Blom, A., Kleinhans, M., and Viparelli, E.: Modelling sorting over the lee face of a single dune, bar of Gilbert delta, Water Resour. Res., in review, 2013.

Chiew, Y. and Parker, G.: Incipient sediment motion on non-horizontal bed slopes, J. Hydraul. Res., 32, 649–660, 1994.

De Vries, M.: Considerations about non-steady bed-load transport in open channels, Proceedings 11th IAHR Congress, Leningrad, 381–388, 1965.

Edmonds, D. A., Shaw, J. B., and Mohrig, D.: Topset-dominated deltas: A new model for river delta stratigraphy, Geology, 39, 1175–1178, 2011.

Egiazaroff, I. V.: Calculation of nonuniform sediment concentrations, J. Hydraul. Eng., 91, 225–247, 1965.

Eke, E., Viparelli, E., and Parker, G.: Field-scale numerical modeling of breaching s a mechanism for generating turbidity currents, Geosphere, 7, 1063–1076, 2011.

Ferrer-Boix, C., Martin-Vide, J. P., and Parker, G.: Bed material sorting in a prograding delta composed of sand-gravel mixture, Sedimentology, in review, 2013.

Francalanci, S., Solari, L., and Parker, G.: Effect of Seepage-Induced Nonhydrostatic Pressure Distribution on Bed-Load Transport and Bed Morphodynamics, J. Hydraul. Eng., 134, 378–389, 2008.

Hirano, M.: On riverbed variation with armoring, Proc. Japan Soc. Civ. Eng., 195, 55–65, 1971.

Hoey, T. B. and Ferguson, R. I.: Numerical simulation of downstream fining by selective transport in gravel bed rivers: Model development and illustration, Water Resour. Res., 30, 2251–2260, 1994.

Kleinhans, M. G.: Grain-size sorting in grainflows at the lee side of deltas, Sedimentology, 52, 291–311, 2005.

Longhitano, S. G.: Sedimentary facies and sequence stratigraphy of coarse grained Gilbert-type deltas within the Pliocene thrust-top Potenza Basin (Southern Apennines, Italy), Sediment. Geol., 219, 87–110, 2008.

Parker, G.: Selective Sorting and Abrasion of River Gravel, I: Theory, J. Hydraul. Eng., 117, 131–149, 1991a.

Parker, G.: Selective Sorting and Abrasion of River Gravel, II: Applications, J. Hydraul. Eng., 117, 150–171, 1991b.

Parker, G.: 1D Morphodynamics of Rivers and Turbidity Currents, available at: http://hydrolab.illinois.edu/people/parkerg//?q=_people/parkerg/ (last access: 26 May 2014), 2004.

Parker, G.: Transport of gravel and sediment mixtures, in: Sedimentation Engineering processes: Measurements, modeling and practice, 3, edited by: Garcia, M. H., ASCE, Reston, VA, 165–251, 2008.

Parker, G. and Klingeman, P.: On why gravel bed streams are paved, Water Resour. Res., 18, 1409–1423, 1982.

Parker, G. and Wilcock, P. R.: Sediment Feed and Recirculating Flumes: Fundamental Difference, J. Hydraul. Eng., 119, 1192–1204, 1993.

Parker, G., Dhamotharan, S., and Stefan, S.: Model experiments on mobile paved gravel bed streams, Water Resour. Res., 18, 1395–1408, 1982.

Rohais, S., Eschard, R., and Guillocheau, F.: Depositional model and stratigraphic architecture of rift climax Gilbert-type fan deltas (Gulf of Corinth, Greece), Sediment. Geol., 210, 132–145, 2008.

Swenson, J. B., Voller, V. R., Paola, C., Parker, G., and Marr, J. G.: Fluvio-deltaic sedimentation: A generalized Stefan problem, Eur. J. Appl. Math., 11, 433–452, 2000.

Toro-Escobar, C. M., Parker, G., and Paola, C.: Transfer function for the deposition of poorly sorted gravel in response to streambed aggradation, J. Hydraul. Res., 34, 35–53, 1996.

Vanoni, V. (Ed.): Sedimentation engineering, ASCE Manuals and Reports on Engineering Practice 54, ASCE, New York, 1975.

Viparelli, E., Sequeiros, O. E., Cantelli, A., Wilcock, P. R., and Parker, G.: River morphodynamics with creation/consumption of grain size stratigraphy, 2: Numerical model, J. Hydraul. Res., 48, 727–741, 2010a.

Viparelli, E., Haydel, R., Salvaro, M., Wilcock, P. R., and Parker, G.: River morphodynamics with creation/consumption of grain size stratigraphy, 1: laboratory experiments, J. Hydraul. Res., 48, 715–726, 2010b.

Viparelli, E., Blom, A., and Parker, G.: Numerical prediction of the stratigraphy of bedload-dominated deltas: preliminary results, Proceedings River, Coastal and Estuarine Morphodynamics RCEM 2011, Tsinghua University, Beijing, China, 2011a.

Viparelli, E., Gaeuman, D., Wilcock, P., and Parker, G.: A model to predict the evolution of a gravel bed river under an imposed cyclic hydrograph and its application to the Trinity River, Water Resour. Res., 47, W02533, doi:10.1029/2010WR009164, 2011b.

Wong, M., Parker, G., DeVries, P., Brown, T., and Burges, S.: Experiments on dispersion of tracer stones under lower-regime plane-bed equilibrium bed load transport, Water Resour. Res., 43, 1–23, 2007.

Wright, S. and Parker, G.: Modeling downstream fining in sand-bed rivers, I: formulation, J. Hydraul. Res., 43, 612–619, 2005a.

Wright, S. and Parker, G.: Modeling downstream fining in sand-bed rivers, II: application, J. Hydraul. Res., 43, 620–630, 2005b.

15

Controls on the magnitude-frequency scaling of an inventory of secular landslides

M. D. Hurst, M. A. Ellis, K. R. Royse, K. A. Lee, and K. Freeborough

British Geological Survey, Keyworth, Nottingham, NG12 5GG, UK

Correspondence to: M. D. Hurst (mhurst@bgs.ac.uk)

Abstract. Linking landslide size and frequency is important at both human and geological timescales for quantifying both landslide hazards and the effectiveness of landslides in the removal of sediment from evolving landscapes. The statistical behaviour of the magnitude-frequency of landslide inventories is usually compiled following a particular triggering event such as an earthquake or storm, and their statistical behaviour is often characterised by a power-law relationship with a small landslide rollover. The occurrence of landslides is expected to be influenced by the material properties of rock and/or regolith in which failure occurs. Here we explore the statistical behaviour and the controls of a secular landslide inventory (SLI) (i.e. events occurring over an indefinite geological time period) consisting of mapped landslide deposits and their underlying lithology (bedrock or superficial) across the United Kingdom. The magnitude-frequency distribution of this secular inventory exhibits an inflected power-law relationship, well approximated by either an inverse gamma or double Pareto model. The scaling exponent for the power-law scaling of medium to large landslides is $\alpha = -1.71 \pm 0.02$. The small-event rollover occurs at a significantly higher magnitude (1.0–$7.0 \times 10^{-3}\,\mathrm{km}^2$) than observed in single-event landslide records ($\sim 4 \times 10^{-3}\,\mathrm{km}^2$). We interpret this as evidence of landscape annealing, from which we infer that the SLI underestimates the frequency of small landslides. This is supported by a subset of data where a complete landslide inventory was recently mapped. Large landslides also appear to be under-represented relative to model predictions. There are several possible reasons for this, including an incomplete data set, an incomplete landscape (i.e. relatively steep slopes are under-represented), and/or temporal transience in landslide activity during emergence from the last glacial maximum toward a generally more stable late-Holocene state. The proposed process of landscape annealing and the possibility of a transient hillslope response have the consequence that it is not possible to use the statistical properties of the current SLI database to rigorously constrain probabilities of future landslides in the UK.

1 Introduction

This paper describes the generation and analysis of a secular landslide inventory (SLI) derived from the UK National Landslide Database (NLD) (Foster et al., 2012). We tackle two basic questions. First, does this secular landslide inventory reflect similar or different statistical properties as generally better constrained, single-event driven inventories and local-scale historical inventories (cf. Van Den Eeckhaut et al., 2007)? Second, what role is played by the underlying lithology and type of landslide in controlling the statistical properties of the inventory? The drivers for the current analysis include the need to quantify landslide hazards and to better understand erosional processes in long-term landscape evolution.

Landslides pose a significant hazard to human life and infrastructure. In the US, Japan, Italy and India landslides have been estimated to result in economic losses for each in excess of (1990 USD) USD 1.0 billion per annum (Schuster, 1996). Between 2004 and 2010 there were at least 2600 fatal landslides globally, with 32 000 associated fatalities (Petley, 2012). Whilst loss of life due to landsliding in the UK is relatively rare, landslides pose a risk to infrastructure and are relevant in land use planning (Gibson et al., 2013). Landslides also have the potential to disrupt transport links (Winter et al., 2010), and land use change has been acknowledged to influence the occurrence of landslides throughout the world (Glade, 2003). Given that landslide behaviour is in part dictated by levels and frequency characteristics of precipitation, there is concern that the patterns and severity of landsliding may be affected by future climate change (Crozier, 2010; Keiler et al., 2010; Korup et al., 2012).

Landslides are important geomorphic processes which generate and transport significant volumes of rock, regolith and soil (e.g. Korup et al., 2010; Larsen et al., 2010). Landslides occur in a variety of styles, dictated by a web of interrelated factors, including material properties (e.g. soil type and thickness, bedrock type, the orientation and spacing of discontinuities), landscape morphology (e.g. slope, topographic convergence, aspect) and climate (e.g. freeze–thaw and shrink–swell cyclicity, pore-water pressures). Whilst large landslides are often perceived to be the most hazardous, small landslides occur most frequently; therefore quantifying the size-frequency distribution for landslide events is important to the assessment of landslide hazard and to land use planning (Malamud et al., 2004). Landslides may also be a significant component of the sediment budget in a landscape and hence understanding their size-frequency characteristics is important to studies of long-term landscape evolution (Stark and Guzzetti, 2009). In a recent review, Guzzetti et al. (2012) recognised that detailed inventories of landslides are lacking, advocating them as a vital tool in assessing susceptibility and risk at a variety of time and length scales. Inventories may focus at a variety of temporal and spatial scales, from a single drainage basin (Guzzetti et al., 2008) to national scale (Trigila et al., 2010; Van Den Eeckhaut and Hervás, 2012); from single event-triggered landslide clusters (Parker et al., 2011) to multi-temporal historical records (Galli et al., 2008) with unconstrained landslide ages.

Several studies have proposed that the non-cumulative size-frequency distribution for landslides (i.e. the number of slides of a give size occurring over a given length of time or within a given area) follows a negative power-law relationship for medium to large landslides (e.g. Guzzetti et al., 2002; Hovius et al., 1997; Pelletier et al., 1997; Stark and Hovius, 2001; Turcotte et al., 2002). Estimates of the exponent α for power-law scaling of large events vary from $\alpha = 1.4$ up to $\alpha = 3.3$ (Van Den Eeckhaut et al., 2007). Van Den Eeckhaut et al. (2007) report an average value of $\alpha = 2.3 \pm 0.6$ based on a compilation of inventories, and Malamud et al. (2004) suggested $\alpha = 2.4$ might be universally applicable to event-triggered inventories based on consensus between three contrasting event-driven data sets. Larsen et al. (2010) caution that estimates of volume of material transported by landslides may be very sensitive to this scaling exponent, resulting in prediction errors of over an order of magnitude. The scaling exponents may vary with underlying geology (e.g. Frattini and Crosta, 2013; Guzzetti et al., 2008), and the type of fail-

ure event (e.g. Brunetti et al., 2009; by analysis of landslide volume rather than area statistics).

A negative power-law model only holds for landslides larger than a particular size, and this minimum size will vary between different inventories. Landslide size-frequency distributions from around the world consistently exhibit a rollover to a positive relationship for smaller landslides, with the size of landslide at which the rollover occurs varying between inventories (e.g. Brardinoni and Church, 2004; Guzzetti et al., 2008; Malamud et al., 2004; Van Den Eeckhaut et al., 2007). For complete landslide inventories the rollover has been interpreted as resulting from the interplay of cohesion and friction, whereby these forces offer resistance to landsliding for small and large landslides respectively (Guzzetti et al., 2002; Malamud et al., 2004; Pelletier et al., 1997; Stark and Guzzetti, 2009). Alternatively (or perhaps additionally), the rollover has been attributed to the under-sampling of small landslides when compiling the inventory. Under-sampling might occur due to evidence of small landslides being rapidly removed through erosion, the reworking of deposits and recolonisation by vegetation (Brardinoni and Church, 2004), difficulties in identification of smaller landslides, or resolution issues with remotely sensed data sets (Malamud et al., 2004; Stark and Hovius, 2001). The rollover occurs at larger landslide sizes in historical inventories where evidence of smaller landslides has been lost from the geomorphic record (Malamud et al., 2004; Van Den Eeckhaut et al., 2007). Additionally Van Den Eeckhaut et al. (2007) demonstrated that a historical inventory of the Flemish Ardennes in Belgium was well characterised by the superposition of two power laws, one characterising the size-frequency of large, old landslides driven by natural processes and another for recent landslide activity influenced by human impacts on the landscape. Two statistical distributions have been proposed to model the rollover in size-frequency distributions of terrestrial landslides. Stark and Hovius (2001) found landslide inventories from New Zealand and Taiwan could be fit by a double Pareto distribution. Malamud et al. (2004) favour fitting an inverse gamma function, which can also account for the rollover. An inverse gamma function provided a good approximation of the size frequency distribution of data sets from Italy, Guatemala and the USA, with different trigger conditions (snowmelt, storm and earthquake triggers respectively) (Malamud et al., 2004). The three inventories were considered to be complete (i.e. the rollover is real and not a result of under-sampling of small landslides) thereby leading Malamud et al. (2004) to suggest the model as a general fit for any complete event-driven landslide inventory.

The universality of such a general model for landslide distributions has not been verified. Malamud et al. (2004) suggest it has applicability to historic, multi-trigger inventories since the model can be fitted to the large landslide tail of a historical inventory which is more likely to be a substantially complete record, since evidence of larger landslides will persist for longer time periods in a landscape. As a result, by comparison to the proposed general distribution, the total number of landslides missing from an inventory can be predicted even for an incomplete landslide inventory (Malamud et al., 2004). Historic inventories (e.g. Guzzetti et al., 2008; Trigila et al., 2010) show similar power-law scaling with $\alpha \approx 2.4$ but with the location of the rollover offset towards larger landslides. Guzzetti et al. (2008) interpret the offset as due to difficulty in documenting smaller landslides from aerial photos and their tendency to amalgamate (i.e. incompleteness of the record). The difference might also relate to the loss of smaller landslides from the record due to landscape annealing by erosion, reworking of deposits and recolonisation by vegetation (Brardinoni and Church, 2004).

The concept of a general model for a landslide size-frequency relationship may seem at odds with the range of factors expected to influence landslide occurrence, such as climate, vegetation, material properties of bedrock/regolith and the type/style of failure. Clarke and Burbank (2010) compared the size-frequency distribution of two landslide inventories in Fiordland and the Southern Alps in New Zealand, which are dominated by igneous and high-grade metamorphic lithologies, and low-grade metamorphic lithologies respectively. Whilst power-law scaling exponents were similar between the two sites ($\alpha \approx 1.07$ and 1.16, respectively), the sizes of the largest landslides were roughly an order of magnitude larger, and the position of the rollover in frequency was also shifted toward larger landslides in the Southern Alps compared to Fiordland. Frattini and Crosta (2013) constructed synthetic size-frequency distributions using slope stability analysis to suggest that less resistant materials tend to promote more shallow landslides whilst more resistant lithologies tend toward deeper landslides with limited numbers of smaller landslides. This has important implications for the volume of materials transported by landslides in different materials. Larsen et al. (2010) compiled a global data set of landslide geometries and observed that scaling of volume with area for shallow, soil landslides has a lower exponent than for deep-seated bedrock landslides. A general model for the distribution of landslides (Malamud et al., 2004) may not take into account lithologic variability and differences in the type of mass movement processes (which are likely linked themselves).

In this study, we attempt to quantify and explain the statistical properties of a national-scale secular landslide inventory, test the geomorphic completeness (i.e. degree of landscape annealing) of such an inventory and estimate the number of landslides that might be missing from the geomorphic record. We link the frequency distribution of landslide sizes to the lithology or deposits in which they occur to assess whether particular lithologies may have been more prone to landsliding. Finally, we select landslides in which the type of failure was known in order to assess whether scaling relationships are a function of landslide type. We achieve this by generating a national-scale SLI from the NLD in the United Kingdom, which we combine with mapped landslide deposits, and maps of underlying geology.

2 Data and methods

2.1 Landslide data

The NLD is an extensive inventory of ancient and recent landslides in the UK (Fig. 1) (Foster et al., 2012). The database is managed by the British Geological Survey (BGS), having inherited and expanded a database initially compiled from a desk study carried out by Geomorphological Services Limited in the late 1980s to document the occurrence of landslides in the UK on behalf of the UK Government's Department of the Environment (Jones and Lee, 1994). The database consisted of records compiled from journal articles and reports, and maps and reports held by the BGS (Foster et al., 2012). The NLD has expanded from this origin and is maintained and managed by the BGS. The database now comprises a series of points ($n = 16\,808$; November 2012) recording the location of known landslides, the precise timing of which is often unknown. Many of these points have been through a quality assurance (QA) procedure ($n = 13\,108$; November 2012) verifying their location by reference to previous studies, maps or field surveys.

The NLD includes a detailed record of many attributes of a particular landslide event including landslide type, slide material, presence of vegetation, hillslope gradient and estimated age (see Foster et al., 2012, for further details), but the availability of these data depend on when and by whom the individual landslide was recorded. During QA the points are related to mapped landslide deposits recorded by geological mapping at 1 : 10 000 and 1 : 50 000 scales over the last century by the BGS (DiGMap; British Geological Survey, 2009, 2010). These mapped deposits have been recorded by a multitude of geologists whose primary concern may not have been the precise recording of landslide deposits, thus the quality of the data set is likely to be highly variable. The NLD is not considered a complete inventory of landslide occurrence in the UK. The data were collated from a multitude of sources and therefore approaches to mapping are unlikely to have been consistent. Additionally the spatial coverage of landslide mapping is unlikely to have been consistent. As part of the continuing collection, updating and verification of landslide information by the BGS, existing landslide data are subjected to a standardised QA procedure designed to improve the consistency and reliability of the these data sets, and areas with poor coverage are identified for detailed resurveying (Foster et al., 2012). These activities are vital to planning and development within the UK (Foster et al., 2012), and is a fundamental component in the nationwide assessment of landslide susceptibility (Walsby, 2008).

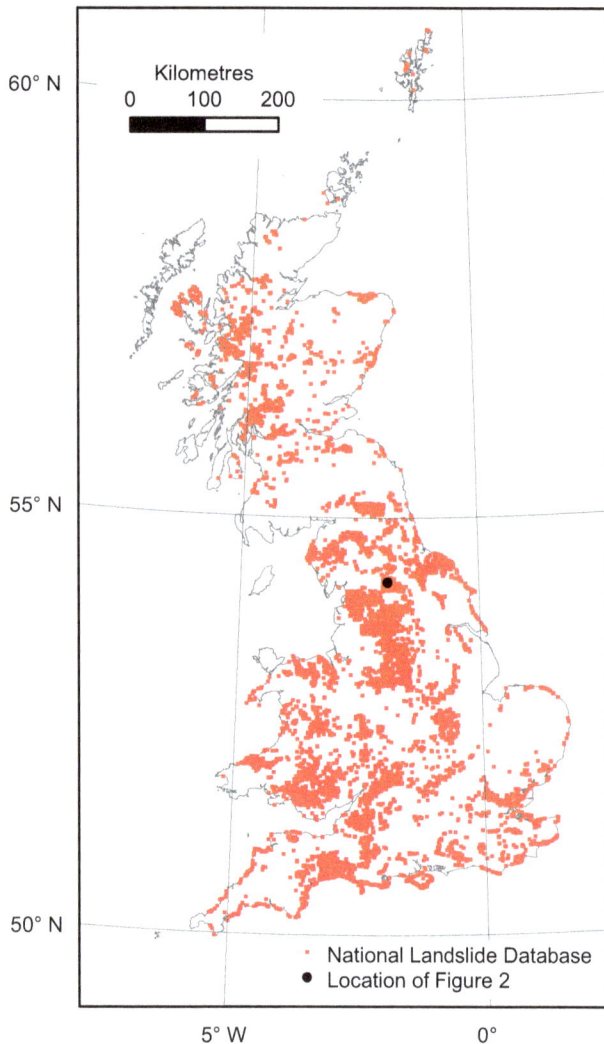

Figure 1. UK Map showing the distribution of landslide points in the NLD which have undergone quality assurance control at the British Geological Survey. Black circle shows the location of Fig. 2.

2.2   Sampling methods

The magnitude-frequency relationship for landslides in the UK was quantified based on linking quality assured landslides in the NLD (Fig. 1) to their associated mapped landslide deposits (for either 1 : 10 000 and 1 : 50 000 scale maps), where available, using GIS software. We emphasise that in this contribution we are analysing the *deposit area* rather than the total failure plus run-out area as is more commonly reported (Malamud et al., 2004), because only landslide deposit areas have been recorded as part of BGS's geological mapping.

During the quality assurance procedure, if a landslide reported in the NLD can be allied to a mapped landslide deposit, the coordinates of the point record in the NLD are moved to the location of highest elevation at the edge of the mapped landslide polygon. In some cases where the head

scarp of the landslide is visible in aerial photographs or topographic data, the point coordinates in the NLD will alternatively be moved to the highest point on the observed scarp. In order to link points in the NLD to mapped landslide polygons we used ArcGIS software to measure the shortest distance between records in the NLD and their nearest deposit area polygon. Where this distance was less than 50 m we considered that the point and polygon were related and hence attributes of the mapped landslide deposit polygons were linked to the NLD (e.g. Fig. 2; box 2) to generate the sample used in the current analysis.

There were a number of caveats to this linking procedure requiring consideration. Firstly, mapped deposits may consist of the amalgamation of several proximal landslide runouts, or be the result of landslide reactivation and therefore have multiple associated events in the NLD (e.g. Fig. 2; box 1). In order to isolate individual event deposits, we omitted records where multiple points from the NLD were associated with a single landslide deposit polygon ($n = 1944$). Similarly, we omitted occurrences of a landslide deposit polygon that had no associated nearby (< 50 m) records in the NLD ($n = 1177$). We also omitted records in the NLD with no associated mapped landslide deposit (Fig. 2; red points, $n = 6026$). Finally, we also omitted coastal landslides (via a 500 m buffer from the UK coastline) in order to restrict our analysis to strictly terrestrial landslides ($n = 386$). The resulting sampled data set consists of 8452 single landslide event-deposit area pairs. We subsequently refer to this filtered landslide data set to as the SLI.

To quantify landslide size, we used ArcGIS to measure the aerial extent of each mapped landslide deposit polygon retained in the SLI. We used the centroid points of mapped deposit polygons to sample the underlying lithology and the presence/absence of superficial material from digital geological maps (DiGMap; British Geological Survey, 2009, 2010), following the BGS's standardised rock classification scheme (RCS) (Styles et al., 2006). The geology of the United Kingdom is quite diverse, with over 180 separate RCS codes identified during sampling. In order to look for lithologic control on landslides, we split these into seven broad lithologic groups: superficial deposits, mudstones, interbedded sedimentary units, coarse clastic sedimentary units (sandstones and coarser), carbonates, metamorphics and igneous (Table 1). Superficial deposits refer to young (Quaternary age) geological deposits (glacial or alluvial) which rest on bedrock (we are not able to distinguish soil from bedrock landslides). Interbedded units refer to sequences where fine grained clastics (i.e. mudstones) are interbedded with coarser, usually more resistant layers.

Table 1. Parameter estimates of power-law model for the entire SLI data set and subsets grouped by lithology. Power laws were fitted following Clauset et al. (2009) with error ranges reported as standard errors.

Data set	No. Landslides	α
SLI	8453	1.71 ± 0.01
Superficial	2497	1.82 ± 0.03
Mudstone	2339	1.69 ± 0.02
Interbedded	1986	1.71 ± 0.02
Clastic	1188	1.67 ± 0.03
Carbonate	268	1.58 ± 0.04
Metamorphic	111	1.53 ± 0.05
Igneous	64	1.82 ± 0.12

2.3 Statistical analysis

2.3.1 Quantifying landslide size distributions

The non-cumulative frequency density (F_D) of a landslide inventory is given by the number of landslides dN over the range of areas dA. Probability density (P_D) can be estimated for a landslide data set as F_D normalised to the total number of landslides in the inventory N_T according to

$$P_D = \frac{1}{N_T} F_D = \frac{1}{N_T} \frac{dN}{dA}, \tag{1}$$

in which P_D is the probability of a landslide with area A [L^2] (Malamud et al., 2004). We calculated F_D and P_D for the SLI data set by sorting the data into bins spaced evenly in A in logarithmic space.

2.3.2 Models for landslide size distribution

As previously noted, the scaling of probability density P_D with landslide size for medium-large landslides can be described by a power law:

$$P_D = \frac{\alpha - 1}{A_{min}} \left(\frac{A}{A_{min}} \right)^{-\alpha}, \tag{2}$$

where A_{min} is the landslide size cutoff for power-law scaling and α is a dimensionless scaling exponent. Stark and Hovius (2001) proposed using a double Pareto model to describe the size distribution of observed landslides, which accounts for under-sampling of smaller landslides. In the model, P_D is a function of two scaling exponents, α_p and β, which describe the rate of decay for large and small landslides respectively, either side of a peak landslide area A_{peak}:

$$P_D = \frac{\beta}{A_{peak}} \times \frac{\left[1 + \left(\frac{A_{max}}{A_{peak}} \right)^{-\alpha_p} \right]^{\frac{\beta}{\alpha_p}}}{\left[1 + \left(\frac{A_{max}}{A_{peak}} \right)^{-\alpha_p} \right]^{1+\frac{\beta}{\alpha_p}}} \times \left(\frac{A}{A_{peak}} \right)^{-(\alpha_p+1)}, \tag{3}$$

where A is landslide area, A_{max} is the largest landslide in the data set, A_{peak} is the area at which the rollover occurs,

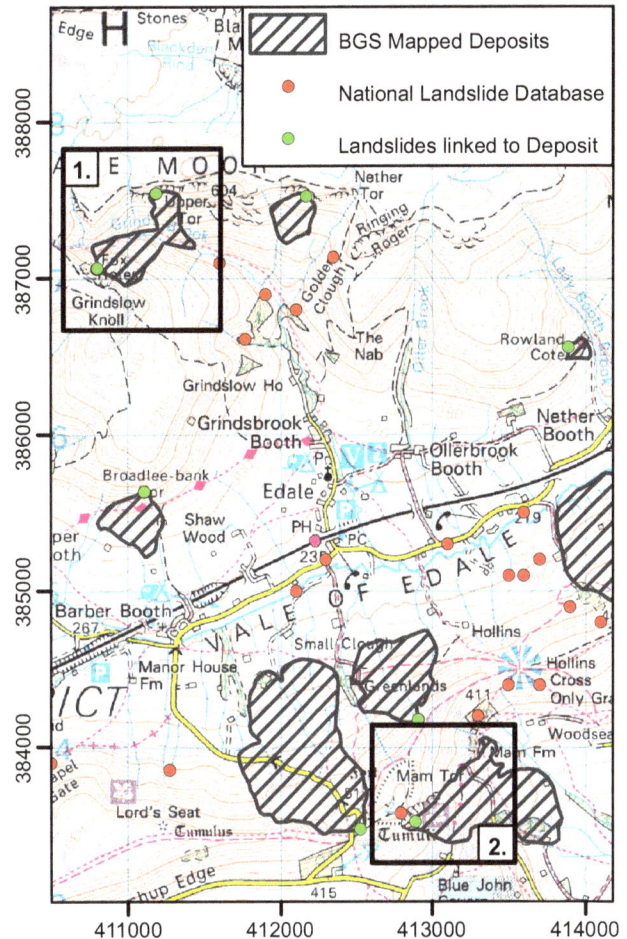

Figure 2. Example map from the Vale of Edale in Derbyshire, showing the locations of points in the NLD (red), and mapped landslide deposits (hatched). Here, there were significantly more events in the database than there were polygons of mapped deposits, probably due to the scale at which mapping took place (1 : 50 000). Landslide events (red) with no associated, mapped deposit were removed from subsequent analysis. Box 1 highlights a scenario where a single mapped deposit polygon is a composite of two separate landslide events; hence these data are not included in later analysis. Box 2 shows an occasion where a landslide event (red), placed on the back scarp during QA, has been associated with a nearby polygon and linked (green). The spatial reference system is British National Grid; the units are metres. OS topography © Crown Copyright. All rights reserved. 100017897/2010.

α_p is the exponent controlling negative power-law scaling for $A_{peak} < A < A_{max}$, and β is the exponent controlling positive power-law scaling when $0 < A < A_{peak}$ (Stark and Hovius, 2001). Note that the negative power law scaling α in Eq. (2) is equivalent to $\alpha_p + 1$. Similarly, Malamud et al. (2004) modelled the probability density of a landslide inventory with a three-parameter inverse gamma function, which acts as an inverse power law for medium-large landslides:

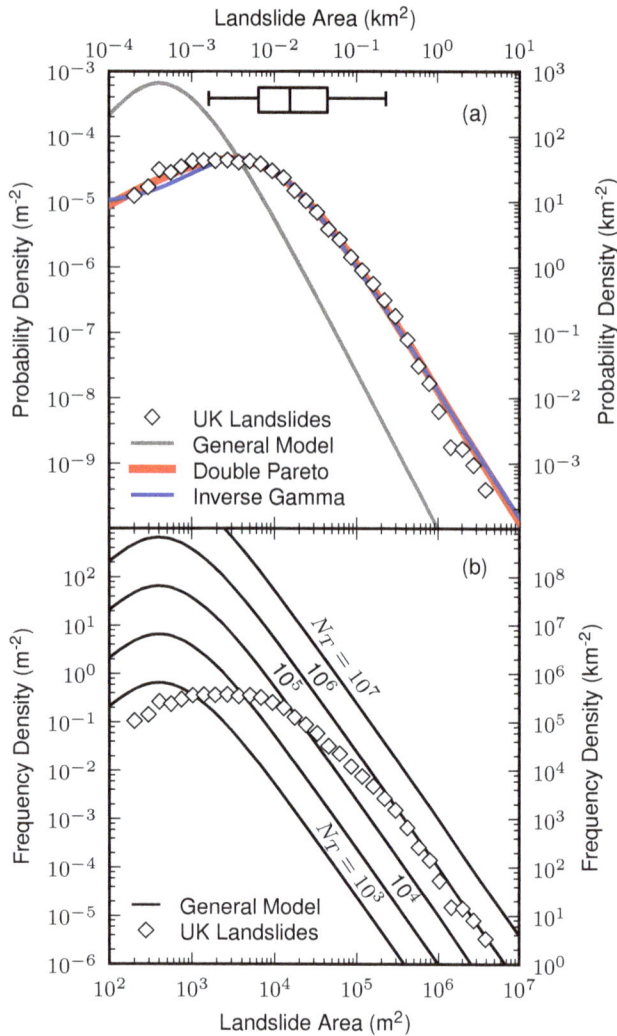

Figure 3. (a) Probability distribution of landslide deposit area for $n = 8453$ landslides in the UK organised into bins spaced evenly in logarithmic space (open diamonds). Solid red and blue lines show MLE of a double Pareto function ($\alpha_p = 1.01 \pm 0.01$; $\beta = 1.71 \pm 0.07$; $A_{peak} = 8.09 \pm 0.6 \times 10^{-3}$ km^2) and inverse gamma function ($\alpha_g = 0.95 \pm 0.02$; $r = 10.9 \pm 0.4 \times 10^{-3}$ km^2; $s = -1.91 \pm 0.08 \times 10^{-3}$ km^2), respectively (error ranges based on one standard deviation of bootstrapped MLE parameters). The grey line is a proposed general distribution for landslides put forward by Malamud et al. (2004). Box plot shows the median (central line), upper/lower quartiles (extent of rectangle) and 5th and 95th percentiles (whiskers) of area data with a median value of 1.53×10^{-2} km^2. (b) Frequency density distribution for landslides in the UK. Solid lines represent the general distribution proposed by Malamud et al. (2004) for varying total number of landslides N_T.

$$P_D = \frac{1}{r\Gamma(\alpha_g)} \left(\frac{r}{A-s}\right)^{\alpha_g+1} \exp\left(\frac{r}{A-s}\right), \qquad (4)$$

where α_g is the exponent setting the inverse power-law scaling for large landslides (again note that α is equivalent to

α_g+1), r [L^2] is a parameter controlling the location of the peak in the probability distribution and s [L^2] controls the rate of decay for small landslide areas.

2.3.3 Statistical analysis

To visualise the size distribution of the SLI we calculated F_D and P_D following Sect. 2.3.1 for the data set as a whole and for subsets grouped by lithology and landslide type (see Figs. 3, and 4). We use a maximum likelihood estimation (MLE) approach to fit statistical models in Eqs. (2–4) to the data and various subsets but apply this to the raw data rather than the binned frequencies and probabilities shown in Figs. 3 and 4. To do so we calculate the log-likelihood L according to

$$L = \ln P_D(A|\theta) = \sum_{i=1}^{n} \ln P_D(A_i|\theta), \qquad (5)$$

where P_D is a probability density model (e.g. Eqs. 2–4), and θ are the parameters to optimise. The log-likelihood of a particular set of parameters θ is therefore the sum of probabilities for all landslides in the data set or subset (of size n). Finding the combination of parameters that optimise L gives the MLE of parameters for a given model.

To constrain the uncertainty on MLE parameters we perform bootstrap analyses in which we repeat the MLE method on 10 000 data sets sampled by replacement from the SLI (and subsets). We therefore generate 10 000 estimates of the most likely parameter combinations for the respective models and use the mean and standard deviation to report our most likely parameter combinations. For fitting of power-law distributions we use the MLE solutions for α provided by Clauset et al. (2009). Testing their solutions against our bootstrapping approach for fitting power-laws yields identical parameter estimates but with larger standard errors. We use these analytical error estimates for power law MLE, and report standard deviations about bootstrapped parameter means when performing MLE for double Pareto and inverse gamma functions (Eqs. 3 and 4, respectively).

3 Results

3.1 Statistical distribution of landslide size in the UK

The size-frequency distribution of the SLI is shown in Fig. 3. The probability distribution of landslides increases with landslide area, peaking at 1.0–7.0×10^{-3} km^2, before appearing to diminish in a power-law fashion (Fig. 3a). Previously documented event-driven landslide inventories show similar humped probability distributions (Brardinoni and Church, 2004; Guzzetti et al., 2008; Malamud et al., 2004; Pelletier et al., 1997; Stark and Hovius, 2001). A double Pareto distribution (Stark and Hovius, 2001) and a truncated inverse gamma function (Malamud et al., 2004) have also been plotted in Fig. 3a using MLE to find the best fit parameters. These

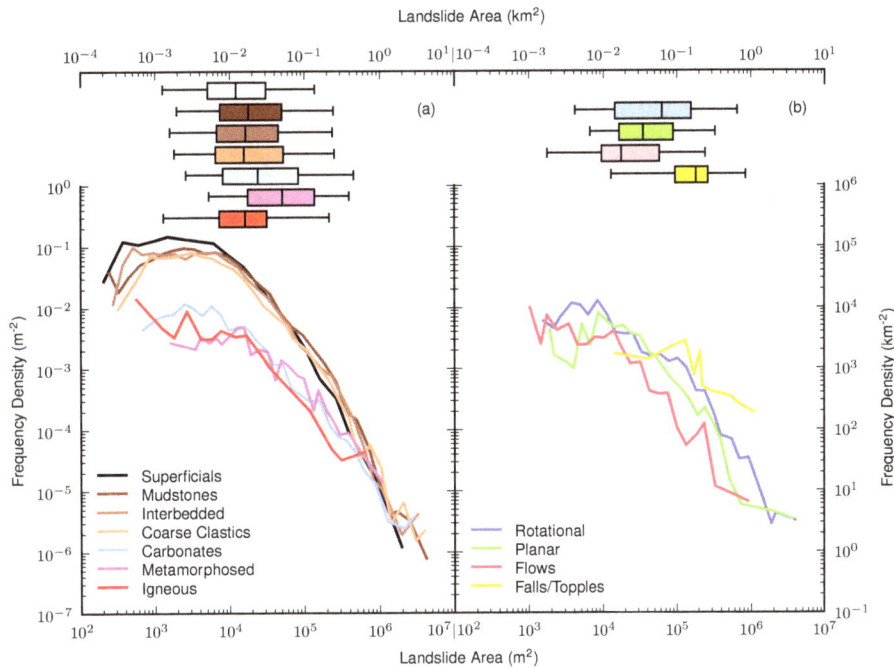

Figure 4. (a) Frequency distributions classified into broad lithologic groups for bins spaced evenly in logarithmic space. Table 1 provides details of the number of landslides in each group and the MLE parameter estimate for α for an assumed power-law distribution of medium-large landslides. Box plots show the median and lower/upper quartile statistics of area data for each lithologic group (whiskers are 5th and 95th percentiles). **(b)** Frequency distributions classified by type of mass movement process; rotational slides, planar slides, flows and falls/topples. Box plots show the median and lower/upper quartile statistics of area data for each lithologic group (whiskers are 5th and 95th percentiles).

functions coincide well with the observed probability distribution for UK landslides at areas $< 10^0$ km^2, although there is discrepancy between the data and model distributions for the largest mapped deposits. The median landslide size is 1.53×10^{-2} km^2 and the most frequent landslides are of the order 10^{-3}–10^{-2} km^2. Figure 3a also shows the general distribution model postulated by Malamud et al. (2004) and attributed to complete landslide inventories associated with a trigger event (e.g. earthquake or storm). The peak probability in landslide size in the SLI is offset by roughly an order of magnitude compared to the general model, indicating fewer small landslides ($< 10^{-2}$ km^2) in the SLI compared to complete event-driven landslide inventories. The general model of Malamud et al. (2004) is able to produce a reasonable fit for the frequency distribution of the largest landslides in the data set ($> 10^0$ km^2) with $N_T = 10^6$ (Fig. 3b). The implications of this alternative fit are discussed in Sect. 4.

3.2 Statistical distribution grouped by lithology and landslide type

We subdivided the SLI into broad lithologic groups (Fig. 4a) sampled from BGS 1 : 50 000 scale geological maps (British Geological Survey, 2010). The majority of landslides occur in superficial material, or clastic sedimentary rocks, particularly fine grained clays and muds and fines interbedded

with coarser units. Based on the abundance of landslides (Table 1) and the distribution in Fig. 4a we refer to these as less-resistant lithologies. Landslides in carbonates, metamorphic and igneous units make up a relatively small part of the data set (see Table 1) and we refer to these as more resistant lithologies. Small landslides ($< 10^{-2}$ km^2) are most abundant in superficial deposits, but medium-large landslides are more common in clastic sedimentary bedrock.

For more resistant lithologic groups there are similar numbers of large landslides ($\sim 10^0$ km^2) as there are in clastic sedimentary rocks, but smaller landslides are relatively infrequent (Fig. 4a). We quantified these observations by fitting a power-law relationship of the form of Eq. (2) by lithology fixing $A_{\min} = 10^{-2}$ km^2. Table 1 shows the MLE α determined following Clauset et al. (2009). Whilst a power law might not be the optimal distribution for every data set, we assume a power law in order to quantify an equivalent α for comparison. Our data indicate that landslides in more resistant lithologies may have lower exponents and therefore a proportionately greater number of large landslide events.

Finally, a subset of the SLI was plotted where information about the type of mass movement process was available ($n = 854$). Figure 4b shows the probability distribution for the four most common types of landslide; rotational slides ($n = 373$), planar slides ($n = 303$), flows ($n = 131$), and falls/topples

($n = 47$). Despite a much smaller sample size, these categories still broadly display rollover-power-law scaling for the landslide size-frequency relationship. The median event size decreases from rotational slides ($A = 0.058 \, \text{km}^2$) to planar slides ($A = 0.033 \, \text{km}^2$) and down to flow events ($A = 0.021 \, \text{km}^2$). This is not unexpected as rotational landslides tend to be large, deep-seated events involving significant amounts of bedrock, whilst flows tend to be hydraulically driven and mobilise material at the near-surface.

4   Discussion

Whilst we do not present a complete data set of all landslide occurrences in the UK, several features emerge from the results that speak to difference and similarities between event-driven and secular landslide inventories, and to the important part that geology plays, each of which have implications for landslide hazard management.

4.1   Landscape annealing and the small-landslide rollover

Comparison of the SLI magnitude-frequency relationship (Fig. 3) to the proposed, general distribution for event-triggered landslides (Malamud et al., 2004; Fig. 3a, grey line) reveals an order of magnitude offset between the peak areas. We interpret this to indicate the relative incompleteness of the SLI due to under-representation of small landslides. The causes behind this could include differences in the methods and coverage of landslide mapping resulting in the under-representation of small landslides, or difficulties in recognising small events in the field due to recolonisation by vegetation or subsequent erosion and redistribution of the deposit. We note that the NLD is not a complete landslide inventory and is constantly growing with the addition of newly observed historic and new landslides (Foster et al., 2012). Mapping of landslide deposits as part of the geological mapping program at the BGS is a continuing process and it is not expected that a complete coverage of landslide deposits in the UK has yet been achieved. This is demonstrated by a number of events in the NLD that did not link to an associated mapped deposit (Fig. 2) and hence were not incorporated into the SLI.

To test the extent to which small landslides are under-represented in the SLI, we analysed separately a subset of the landslide data recently mapped in the North Yorkshire Moors, which are considered to be a substantially complete historic inventory (a comprehensive record of landslide deposits mapped through analysis of high resolution topography, aerial photography and field mapping). The area at which peak probability is observed is only slightly offset between the North Yorkshire data set ($2660 \, \text{m}^2$) and the SLI ($3100 \, \text{m}^2$), compared to the order of magnitude offset observed in relation to the general distribution proposed by Malamud et al. (2004) (Fig. 5). We suggest this slight dif-

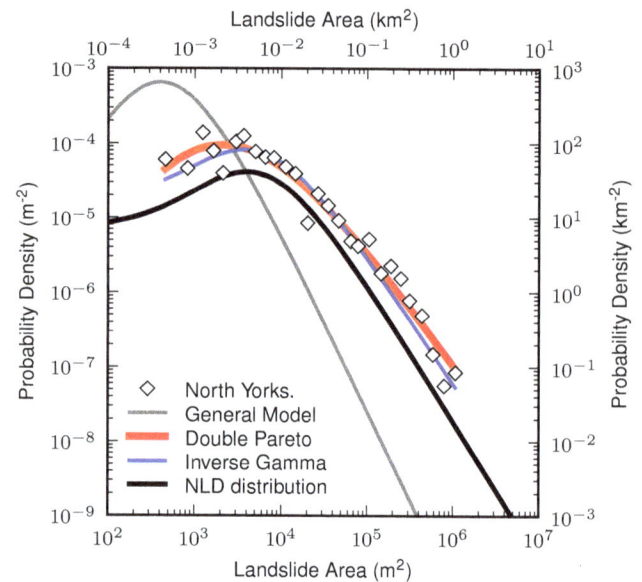

Figure 5. Probability distribution of landslide deposit area for landslides in North Yorkshire organised into logarithmically spaced bins (open diamonds). Solid black and grey lines show maximum likelihood estimates of a double Pareto function ($\alpha_p = 0.61$; $\beta = 3.7$; $A_{\text{peak}} = 927 \, \text{m}^2$) and inverse gamma function ($\alpha_g = 0.71$; $r = 9.57 \times 10^{-3} \, \text{km}^2$; $s = -1.83 \times 10^{-3} \, \text{km}^2$) respectively, dashed grey and black lines show the general distribution proposed by Malamud et al. (2004) and the UK distribution from Fig. 3a, respectively. Note the similarity in shape between the North Yorkshire data set and the UK NLD fit (the vertical offset in probability density, is due to a difference in the range of landslide sizes considered and does not indicate relative probability).

ference may relate to the completeness of these data sets, with the likelihood that there are some smaller events in the NLD that were omitted when compiling the SLI. We stress, however, that there remains a large offset between the SLI and the general event based model proposed by Malamud et al. (2004), suggesting that this offset is real and likely the result of landscape annealing due to the loss of evidence of small events from the landscape.

4.2   Role of lithology

The frequency distributions for soft lithologies appear slightly curved in log-log space (Fig. 4a) suggesting that a power-law distribution may not be the most appropriate fit to the data. However, assuming a power-law model allows us to make comparisons between most likely parameters for different lithologic groups. Landslides in superficial deposits and soft lithologies dominate the SLI, whilst harder lithologic groups exhibit distinct magnitude-frequency scaling characterised by lower values of α setting a lower scaling gradient in log-log space (Fig. 4a; Table 1). This result has important implications for landslide size and associated hazard. Whilst there is significantly lower probability of small landslides in more resistant lithologies, the difference is minimal for larger landslides ($\sim 10^6$ m^2). Perhaps unsurprisingly the largest proportion of landslides and in particular smaller landslides ($< 10^3$ m^2) occurs in poorly consolidated superficial deposits and hence characterisation of superficial materials will be important to site-based investigation of landslide susceptibility. Frattini and Crosta (2013) predict that size distributions should differ with material properties, such that weaker materials should result in more small, shallow landslides whilst stronger materials may promote relatively more large, deep seated landslides. Our findings also suggest that geology will play an important role in setting the size distribution of landslides, suggesting that the influence of lithology should be further explored.

Magnitude-frequency scaling of landslides classified by the type of mass movement have power-law scaling exponents ($\alpha = -1.3$ to -1.7) lower than the data set as a whole ($\alpha = -1.71$). Lower exponents suggests the subset of data may be biased towards larger events, and indeed it seems likely (and reasonable) that detailed field studies to determine the style of failure are preferentially carried out for larger failure events. Unfortunately there are few observations of landslide type below areas of 10^3 m^2 (Fig. 4b) yet there are a large number of landslides in the NLD of this magnitude (Fig. 3b). However, whilst the sample sizes are small, there is a suggestion that the gradient of the most likely power law decreases with landslide type. This would be broadly consistent with the results of Brunetti et al. (2009) who performed analysis of landslide *volume* distributions rather than area as in the present study. Brunetti et al. (2009) performed magnitude-frequency analysis of 19 landslide inventories and found that the scaling exponents for landslide *volume* were lower for rock falls and rock slides than for slides and soil slides. More data are needed to provide an empirical test that more resistant lithologies preferentially yield deep-seated landslides whilst weaker materials preferentially yield shallow landslides, as found theoretically by Frattini and Crosta (2013).

4.3 The large-landslide deficit

A national landslide inventory for Italy comprising $\sim 377\,000$ landslides (Trigila et al., 2010) exhibits power-law scaling above 10^{-2} km^2 similar to the SLI (Fig. 3a). Interestingly both data sets show deviation from fitted scaling relationships for the largest landslides ($> 10^0$ km^2 for the UK; $> 10^1$ km^2 in Italy) suggesting that either we are under-sampling with respect to the largest landslides or large events are less frequent than power-law scaling would predict. The difference in cutoff areas between the two data sets may be the result of only reporting the areas of mapped deposits in the UK whilst in Trigila et al. (2010) *area* refers to the combined source and sink outline.

There are a number of possible explanations for this apparent deficit of large landslides. Firstly, the data set is expected to be incomplete and there may be some large landslides that have not been recorded. It seems unlikely, however, that the deficit represents observational bias, since large landslides should be the most prominent in the landscape. The deficit may, in fact, be larger, as inspection shows that some of the largest mapped deposit areas consist of amalgamated deposits of numerous smaller events. Therefore, we suggest the deficit is real. Possible explanations for the deficit aside from an incomplete database, which we describe below, include an incomplete landscape, and/or a temporal transience in the occurrence of landslides since the last glacial maximum (LGM).

For a large landslide to occur requires a large slope. The availability of the highest relief is spatially limited in the UK to central and northern Wales, the Lake District and the Scottish Highlands. As previously observed, data coverage in these regions is not as extensive as in other parts of the British Isles. The relative paucity of large slopes (and associated large landslides) elsewhere in the country may result in the large-landslide deficit. We refer to this as an incomplete landscape. Figure 6 shows the spatial density of mapped landslides comprising the SLI. Whilst we would anticipate that areas with the greatest relief and steepest hillslope gradients might contain the highest density of landslides, this seems not to be the case in the SLI. Coverage of landslide mapping in the Scottish Highlands and parts of northern Wales are sparser than other low relief areas of the UK. This is of particular relevance since these areas will have large slopes which may yield large landslides, and these areas tend to be underlain by more resistant lithologies.

It is likely that the bulk of landslides range in age from the LGM (~ 27 ka; Clark et al., 2012) to the present day. During this time, climate will have varied as the British Ice Sheet receded, and mass movement processes are likely to have been initially more active as soils and regolith both warmed and lost structural support from ice cover and permafrost. We speculate, therefore, that many landslides and certainly most of the larger landslides would occur early in this LGM-to-present time span, during the paraglacial transition (Ballantyne, 2002; Dadson and Church, 2005). Unlike active mountain belts, steep slopes will not be regenerated by continued rock uplift and erosion, and therefore the drivers for those landslides are gradually reduced over time as the emerging landscape passes through a period of readjustment to new and more stable conditions. Instability likely continued through the variable climate immediately prior to the Holocene, and returned again during the latter part of the Holocene (Neolithic times, in particular) as extensive anthropogenic forest clearance and land-use changes occurred. These latter processes, all else being equal, would lead to an increase in the rate of landslide activity, consistent with rapid Neolithic valley sedimentation observed in many parts of the UK (Brown, 2009). We suggest, therefore, that the population of landslides in the SLI may be dominated by the relatively rapid denudation of early post-LGM and early anthropogenic times.

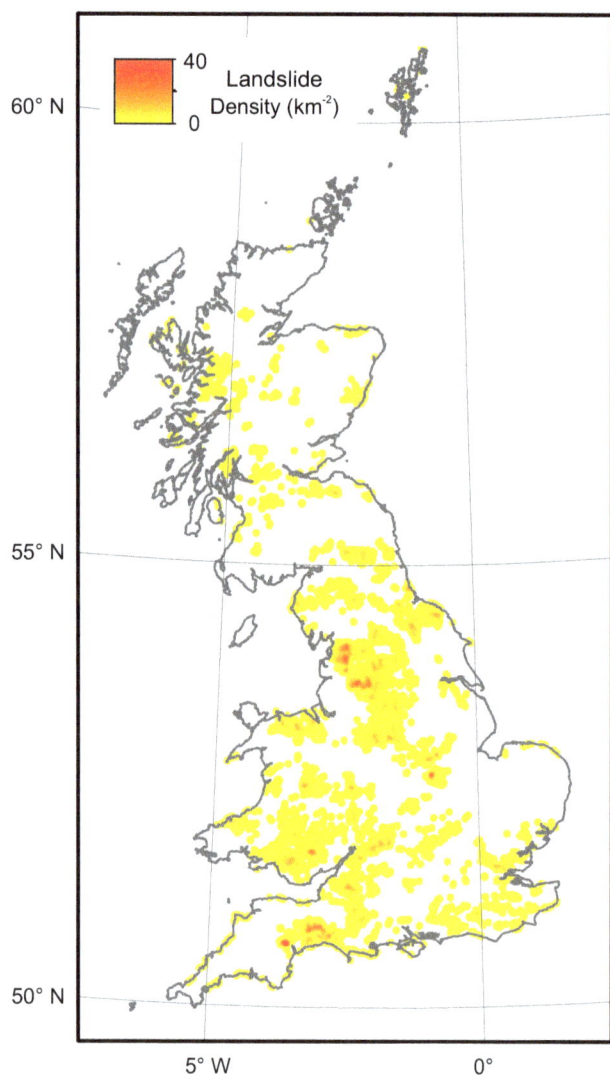

Figure 6. Map of landslide density (number of landslides per km²) across the UK derived from mapped landslide deposits. Data is gridded to 1 km and calculated using a 5 km search radius. Low landslide densities in Scotland and NE Wales may indicate that coverage is particularly poor.

As a result, relatively large landslides show a deficit with respect to a model fit that is derived principally from the relatively greater number of smaller- to moderate-sized landslides.

An alternative perspective is provided by the area-frequency analysis (Fig. 3b), which would suggest that large-landslide deficit is only apparent, and that it is smaller- and moderate-sized landslides that are in deficit. To reach this conclusion would require the assumption that the general model proposed by Malamud et al. (2004) was appropriate to represent the probability density for all landslides in the UK since the LGM. Moreover, it would suggest that the landscape annealing processes by which small events `are lost from the geomorphic record not only act to offset the posi-

tion of the "hump" in historic landslide inventories, but also reduce the exponent α through time. Data for historic inventories presented by Malamud et al. (2004) (after Guzzetti et al., 2003; Ohmori and Sugai, 1995) suggest that this is not the case because α appears to be conserved in those historic inventories (see also Guzzetti et al., 2008; Trigila et al., 2010). Thus it remains unclear whether Malamud's general model is appropriate for a secular landslide inventory spanning several thousand years and a highly variable external forcing, during which time there is reason to suspect variation in the frequency (and possibly size) of landslides.

4.4 Implications for landslide hazards

The combined results here have implications for the assessment of landslide hazards and ultimately for landslide risk management. At face value, for example, the model-fits (double Pareto or inverse gamma; see Fig. 3) presented here yield a low frequency of small landslides relative to complete, event-triggered landslide databases. This size category includes anything from 3 to 30 m in equivalent radius, which can be hazardous to a wide variety of infrastructure. It is more likely, however, that the SLI significantly under-represents landslides of this size, as we argue above. In other words, the national-scale, small-landslide hazard is greater than the SLI would suggest (Fig. 3a; Table 1). Our results also tell us that the occurrence of landslides of any particular size is largely independent of type (i.e. scaling between size and frequency still follows a power law with rollover), but that type and magnitude are linked, with deep-seated rotational landslides tending to be larger than planar slides and flows. The role of lithology emerges as control by two broad classes of bedrock, (resistant: carbonates, metamorphic and igneous rocks vs. less resistant: superficial, mudstones, interbedded, and coarse clastics) each characterised by a distinct power-law distribution and each (with the exception of igneous rocks) showing a rollover at relatively small landslides. We also suggest that the discrepancy between model and observations for relatively large landslides may be a function of a transient landslide response as the UK emerged from glacial conditions and into an initially variable (e.g. Allerød warming and Younger Dryas events) then relatively stable Holocene climate. In other words, the national-scale large-landslide hazard is lower than predicted by the model fit to the SLI (Fig. 3a and Table 1). It is important to emphasise that the SLI is a sample of the NLD (itself an incomplete record of all past landslides in the UK), and that it does not include coastal landslides. Importantly, the proposed process of annealing and the potentially transient response of hillslopes have the consequence that it is not possible to use the statistical properties of the current SLI database to rigorously constrain probabilities of future landslides in the UK.

5 Conclusions

Analysis of a national (UK) secular landslide inventory reveals a statistical distribution that can be well characterised by an inverse gamma or double Pareto distribution with a well-defined rollover at a landslide area between 10 and 30 m^2. The power-law component for medium-large landslides has a scaling exponent $\alpha = -1.71 \pm 0.02$. This general form of the distribution is similar to that found for many single-event driven landslides, although there are two important specific differences. First: the magnitude of the small-landslide rollover occurs at a significantly larger size than in single-event samples. We interpret this as a reflection of landscape annealing processes (e.g. recolonisation by vegetation, reworking of landslide deposits), with the corollary that the model fit would underestimate the frequency of relatively small landslides. Second: we observe a deficit, relative to the model fit, in the largest landslides. Possible explanations for this deficit include (i) poor data coverage in areas where large landslides might be expected; (ii) spatial limitation in the occurrence of large landslides due to lack or limited occurrence of large hillslopes; and (iii) temporal transience or non-linear response of the UK landscape as it emerged from the LGM; and, during the Neolithic, accelerated landscape change due to human activity. In such a scenario, most of the landslides, certainly the larger ones, are likely to have formed early in the post-LGM time span as the soil–regolith–bedrock column lost support of both ice and/or permafrost.

Acknowledgements. We thank Claire Dashwood, Catherine Pennington, Vanessa Banks, Helen Reeves and Bruce Malamud for discussions that greatly improved the sampling techniques and interpretations presented here. Thanks also to Ken Lawrie for assistance with the NLD, to Murray Lark and Ben Marchant for guidance on statistical analyses, and the Associate Editor, Dimitri Lague, and two anonymous reviewers for the many suggestions that improved the final paper. This paper is published with the permission of the Executive Director of the British Geological Survey and was supported in part by the Climate and Landscape Change research programme at the BGS.

Edited by: D. Lague

References

Ballantyne, C. K.: Paraglacial geomorphology, Quaternary Sci. Rev., 21, 1935–2017, 2002.

Brardinoni, F. and Church, M.: Representing the landslide magnitude–frequency relation: Capilano River basin, British Columbia, Earth Surf. Proc. Land., 29, 115–124, 2004.

British Geological Survey: Digital Geological Map of Great Britain 1:10 000 scale (DiGMapGB-10) data. Version 2.18. Keyworth, Nottingham, British Geological Survey, Release date: 15 January 2009, 2009.

British Geological Survey: Digital Geological Map of Great Britain 1:50 000 scale (DiGMapGB-50) data. Version 6.20. Keyworth,

Nottingham, British Geological Survey, Release date: 14 October 2010, 2010.

Brown, A. G.: Colluvial and alluvial response to land use change in Midland England: An integrated geoarchaeological approach, Geomorphology, 108, 92–106, 2009.

Brunetti, M. T., Guzzetti, F., and Rossi, M.: Probability distributions of landslide volumes, Nonlin. Processes Geophys., 16, 179–188, doi:10.5194/npg-16-179-2009, 2009.

Clark, C. D., Hughes, A. L. C., Greenwood, S. L., Jordan, C., and Sejrup, H. P.: Pattern and timing of retreat of the last British-Irish Ice Sheet, Quaternary Sci. Rev., 44, 112–146, 2012.

Clarke, B. A. and Burbank, D. W.: Bedrock fracturing, threshold hillslopes, and limits to the magnitude of bedrock landslides, Earth Planet. Sc. Lett., 297, 577–586, 2010.

Clauset, A., Shalizi, C. R., and Newman, M. E.: Power-law distributions in empirical data, SIAM review, 51, 661–703, 2009.

Crozier, M.: Deciphering the effect of climate change on landslide activity: A review, Geomorphology, 124, 260–267, 2010.

Dadson, S. J. and Church, M.: Postglacial topographic evolution of glaciated valleys: a stochastic landscape evolution model, Earth Surf. Proc. Land., 30, 1387–1403, 2005.

Foster, C., Pennington, C. V. L., Culshaw, M., and Lawrie, K.: The national landslide database of Great Britain: development, evolution and applications, Environ. Earth Sci., 66, 941–953, doi:10.1007/s12665-011-1304-5, 2012.

Frattini, P. and Crosta, G. B.: The role of material properties and landscape morphology on landslide size distributions, Earth Planet. Sc. Lett., 361, 310–319, 2013.

Galli, M., Ardizzone, F., Cardinali, M., Guzzetti, F., and Reichenbach, P.: Comparing landslide inventory maps, Geomorphology, 94, 268–289, 2008.

Gibson, A. D., Culshaw, M. G., Dashwood, C., and Pennington, C. V. L.: Landslide management in the UK—the problem of managing hazards in a "low-risk" environment, Landslides, 10, 599–610, 2013.

Glade, T.: Landslide occurrence as a response to land use change: a review of evidence from New Zealand, CATENA, 51, 297–314, 2003.

Guzzetti, F., Malamud, B. D., Turcotte, D. L., and Reichenbach, P.: Power-law correlations of landslide areas in central Italy, Earth Planet. Sc. Lett., 195, 169–183, 2002.

Guzzetti, F., Reichenbach, P., Cardinali, M., Ardizzone, F., and Galli, M.: The impact of landslides in the Umbria region, central Italy, Nat. Hazards Earth Syst. Sci., 3, 469–486, doi:10.5194/nhess-3-469-2003, 2003.

Guzzetti, F., Ardizzone, F., Cardinali, M., Galli, M., Reichenbach, P., and Rossi, M.: Distribution of landslides in the Upper Tiber River basin, central Italy, Geomorphology, 96, 105–122, 2008.

Guzzetti, F., Mondini, A. C., Cardinali, M., Fiorucci, F., Santangelo, M., and Chang, K. T.: Landslide inventory maps: New tools for an old problem, Earth-Sci. Rev., 112, 42–66, 2012.

Hovius, N., Stark, C. P., and Allen, P. A.: Sediment flux from a mountain belt derived by landslide mapping, Geology, 25, 231–234, 1997.

Jones, D. K. C. and Lee, E. M.: Landsliding in Great Britain, Department of the Environment, London, 1994.

Keiler, M., Knight, J., and Harrison, S.: Climate change and geomorphological hazards in the eastern European Alps, Philos. T. R. Soc. A, 368, 2461–2479, 2010.

Korup, O., Densmore, A. L., and Schlunegger, F.: The role of land-slides in mountain range evolution, Geomorphology, 120, 77–90, 2010.

Korup, O., Görüm, T., and Hayakawa, Y.: Without power? Land-slide inventories in the face of climate change, Earth Surf. Proc. Land., 37, 92–99, 2012.

Larsen, I. J., Montgomery, D. R., and Korup, O.: Landslide erosion controlled by hillslope material, Nat. Geosci., 3, 247–251, 2010.

Malamud, B. D., Turcotte, D. L., Guzzetti, F., and Reichenbach, P.: Landslide inventories and their statistical properties, Earth Surf. Proc. Land., 29, 687–711, 2004.

Ohmori, H. and Sugai, T.: Toward geomorphometric models for estimating landslide dynamics and forecasting landslide occurrence in Japanese mountains, Z. Geomorphol. Supp., 101, 149–164, 1995.

Parker, R. N., Densmore, A. L., Rosser, N. J., De Michele, M., Li, Y., Huang, R., Whadcoat, S., and Petley, D. N.: Mass wasting triggered by the 2008 Wenchuan earthquake is greater than oro-genic growth, Nat. Geosci., 4, 449–452, 2011.

Pelletier, J. D., Malamud, B. D., Blodgett, T., and Turcotte, D. L.: Scale-invariance of soil moisture variability and its implications for the frequency-size distribution of landslides, Eng. Geol., 48, 255–268, 1997.

Petley, D.: Global patterns of loss of life from landslides, Geology, 40, 927–930, 2012.

Schuster, R. L.: Socioeconomic Significance of Landslides, Land-slides: Investigation and Mitigation, National Academies Press, Transportation Research Board Special Report, 12–35, 1996.

Stark, C. P. and Guzzetti, F.: Landslide rupture and the probability distribution of mobilized debris volumes, J. Geophys. Res.-Earth, 114, F00A02, doi:10.1029/2008JF001008, 2009.

Stark, C. P. and Hovius, N.: The characterization of landslide size distributions, Geophys. Res. Lett., 28, 1091–1094, 2001.

Styles, M. T., Gillespie, M. R., Bauer, W., and Lott, G. K.: Current status and future development of the BGS Rock Classification Scheme, British Geological Survey, Nottingham, UK, 2006.

Trigila, A., Iadanza, C., and Spizzichino, D.: Quality assessment of the Italian Landslide Inventory using GIS processing, Land-slides, 7, 455–470, 2010.

Turcotte, D. L., Malamud, B. D., Guzzetti, F., and Reichenbach, P.: Self-organization, the cascade model, and natural hazards, P. Natl. Acad. Sci. USA, 99(Suppl. 1), 2530–2537, 2002.

Van Den Eeckhaut, M. and Hervás, J.: State of the art of national landslide databases in Europe and their potential for assessing landslide susceptibility, hazard and risk, Geomorphology, 139–140, 545–558, 2012.

Van Den Eeckhaut, M., Poesen, J., Govers, G., Verstraeten, G., and Demoulin, A.: Characteristics of the size distribution of recent and historical landslides in a populated hilly region, Earth Planet. Sc. Lett., 256, 588–603, 2007.

Walsby, J. C.: GeoSure; a bridge between geology and decision-makers, Geological Society, London, Special Publications, 305, 81–87, 2008.

Winter, M., Dent, J., Macgregor, F., Dempsey, P., Motion, A., and Shackman, L.: Debris flow, rainfall and climate change in Scot-land, Q. J. Eng. Geol. Hydroge., 43, 429–446, 2010.

Permissions

List of Contributors

W. Schwanghart
Geohazards, University of Potsdam, Institute of Earth and Environmental Science, Karl-Liebknecht-Str. 24–25, 14476 Potsdam-Golm, Germany
Department of Agroecology, Aarhus University, Blichers Allé 20, 8830 Tjele, Denmark

D. Scherler
Geological and Planetary Sciences, California Institute of Technology, 1200 E California Blvd, Mailcode 100-23, Pasadena, CA 91125, USA

M. Nones
Research Center for Constructions – Fluid Dynamics Unit, University of Bologna, Italy

M. Guerrero
Hydraulic Laboratory, University of Bologna, Italy

P. Ronco
University of Padova, Department of Civil, Environmental and Architectural Engineering ICEA, Padova, Italy

S. Dutta
Dept. of Civil and Environmental Engineering, University of Illinois at Urbana-Champaign, Urbana, USA

M. I. Cantero
Centro Atómico Bariloche and Instituto Balseiro, Consejo Nacional de Investigaciones Científicas y Técnicas (CONICET) and Comisión Nacional de Energía Atómica (CNEA), San Carlos de Bariloche, Río Negro, Argentina

M. H. Garcia
Dept. of Civil and Environmental Engineering, University of Illinois at Urbana-Champaign, Urbana, USA
Dept. of Geology, University of Illinois at Urbana-Champaign, Urbana, USA

P. J. Morris
Soil Research Centre, Department of Geography and Environmental Science, University of Reading, RG6 6DW, UK

A. J. Baird
School of Geography, University of Leeds, Leeds, LS2 9JT, UK

L. R. Belyea
School of Geography, Queen Mary University of London, 327 Mile End Road, London, E1 4NS, UK

H. Patton
Institute of Geography and Earth Sciences, Aberystwyth University, Aberystwyth, SY23 3DB, UK

A. Hubbar
Institute of Geography and Earth Sciences, Aberystwyth University, Aberystwyth, SY23 3DB, UK

T. Bradwell
British Geological Survey, Murchison House, West Mains Road, Edinburgh, EH9 3LA, UK

N. F. Glasser
Institute of Geography and Earth Sciences, Aberystwyth University, Aberystwyth, SY23 3DB, UK

M. J. Hambrey
Institute of Geography and Earth Sciences, Aberystwyth University, Aberystwyth, SY23 3DB, UK

C. D. Clark
Department of Geography, University of Sheffield, Sheffield, S10 2TN, UK

M. Liang
Department of Civil, Environmental, and Geo-Engineering, National Center for Earth Surface Dynamics, Saint Anthony Falls Laboratory, University of Minnesota, Twin Cities, Minneapolis, Minnesota, USA

V. R. Voller
Department of Civil, Environmental, and Geo-Engineering, National Center for Earth Surface Dynamics, Saint Anthony Falls Laboratory, University of Minnesota, Twin Cities, Minneapolis, Minnesota, USA

C. Paola
Department of Geology and Geophysics, National Center for Earth Surface Dynamics, Saint Anthony Falls Laboratory, University of Minnesota, Twin Cities, Minneapolis, Minnesota, USA

M. Liang
Department of Civil, Architectural and Environmental Engineering and Center for Research in Water Resources, The University of Texas at Austin, Austin, Texas, USA
Department of Geological Sciences, The University of Texas at Austin, Austin, Texas, USA

N. Geleynse
Department of Geological Sciences, The University of Texas at Austin, Austin, Texas, USA

D. A. Edmonds
Department of Geological Sciences, Indiana University, Bloomington, Indiana, USA

P. Passalacqua
Department of Civil, Architectural and Environmental Engineering and Center for Research in Water Resources, The University of Texas at Austin, Austin, Texas, USA

E. B. Goldstein
Division of Earth and Ocean Sciences, Nicholas School of the Environment, Center for Nonlinear and Complex Systems, Duke University, P.O. Box 90227, Durham, NC 27708, USA

G. Coco
Environmental Hydraulics Institute, "IH Cantabria", c/ Isabel Torres no. 15, Universidad de Cantabria, 39011 Santander, Spain

A. B. Murray
Division of Earth and Ocean Sciences, Nicholas School of the Environment, Center for Nonlinear and Complex Systems, Duke University, P.O. Box 90227, Durham, NC 27708, USA

M. O. Green
National Institute of Water and Atmospheric Research (NIWA), P.O. Box 11-115, Hamilton, New Zealand

M. W. Schmeeckle
School of Geographical Sciences and Urban Planning, Arizona State University, Tempe, Arizona, USA

Z. Zhou
Environmental Hydraulics Institute, "IH Cantabria", University of Cantabria, Santander, Spain

L. Stefanon
Department of Civil, Environmental and Architectural Engineering, University of Padova, Padova, Italy

M. Olabarrieta
Department of Civil and Coastal Engineering, University of Florida, Florida, USA

A. D'Alpaos
Department of Geosciences, University of Padova, Padova, Italy

L. Carniello
Department of Civil, Environmental and Architectural Engineering, University of Padova, Padova, Italy

G. Coco
Environmental Hydraulics Institute, "IH Cantabria", University of Cantabria, Santander, Spain

M. J. Westoby
Geography, Engineering and Environment, Northumbria University, Newcastle upon Tyne, UK

J. Brasington
School of Geography, Queen Mary, University of London, London, UK

N. F. Glasser
Department of Geography and Earth Sciences, Aberystwyth University, Aberystwyth, Wales, UK

M. J. Hambrey
Department of Geography and Earth Sciences, Aberystwyth University, Aberystwyth, Wales, UK

J. M. Reynolds
Reynolds International Ltd, Suite 2, Broncoed House, Broncoed Business Park, Mold, UK

M. A. A. M. Hassan
HR Wallingford Ltd, Howberry Park, Wallingford, Oxfordshire, UK

A. Lowe
CH2M HILL, 304 Bridgewater Place, Warrington, Cheshire, UK

T. Croissant
ISTerre, Université Grenoble-Alpes and CNRS, BP 53, 38041 Grenoble Cedex 9, France

J. Braun
ISTerre, Université Grenoble-Alpes and CNRS, BP 53, 38041 Grenoble Cedex 9, France

C. Orrú
Environmental Fluid Mechanics Section, Civil Engineering and Geosciences, Delft University of Technology, P.O. Box 5048, 2600 GA, Delft, the Netherlands

V. Chavarrías
Environmental Fluid Mechanics Section, Civil Engineering and Geosciences, Delft University of Technology, P.O. Box 5048, 2600 GA, Delft, the Netherlands
Escola Tècnica Superior d'Enginyers de Camins, Canals i Ports de Barcelona, Universitat Politècnica de Catalunya, C/Jordi Girona, 31, 08034 Barcelona, Spain

W. S. J. Uijttewaal
Environmental Fluid Mechanics Section, Civil Engineering and Geosciences, Delft University of Technology, P.O. Box 5048, 2600 GA, Delft, the Netherlands

A. Blom
Environmental Fluid Mechanics Section, Civil Engineering and Geosciences, Delft University of Technology, P.O. Box 5048, 2600 GA, Delft, the Netherlands

E. Viparelli
Department of Civil and Environmental Engineering,
University of South Carolina,
Columbia, South Carolina, USA

A. Blom
Faculty of Civil Engineering&Geosciences, Delft
University of Technology, Delft, the Netherlands

C. Ferrer-Boix
Department of Geography, University of British Columbia,
Vancouver, Canada

R. Kuprena
Department of Civil and Environmental Engineering,
University of South Carolina,
Columbia, South Carolina, USA

M. D. Hurst
British Geological Survey, Keyworth, Nottingham, NG12
5GG, UK

M. A. Ellis
British Geological Survey, Keyworth, Nottingham, NG12
5GG, UK

K. R. Royse
British Geological Survey, Keyworth, Nottingham, NG12
5GG, UK

K. A. Lee
British Geological Survey, Keyworth, Nottingham, NG12
5GG, UK

K. Freeborough
British Geological Survey, Keyworth, Nottingham, NG12
5GG, UK